2nd International Seminar on Metallurgy and Materials (ISMM 2018)

IOP Conference Series: Materials Science and Engineering
Volume 541

Tangerang Selatan, Indonesia
25 – 26 September 2018

ISBN: 978-1-5108-9086-2
ISSN: 1757-8981

Printed from e-media with permission by:

Curran Associates, Inc.
57 Morehouse Lane
Red Hook, NY 12571

Some format issues inherent in the e-media version may also appear in this print version.

Printed by Curran Associates, Inc. (2019)

For permission requests, please contact the Institute of Physics at the address below.

Institute of Physics
Dirac House, Temple Back
Bristol BS1 6BE UK

Phone: 44 1 17 929 7481
Fax: 44 1 17 920 0979

techtracking@iop.org

Additional copies of this publication are available from:

Curran Associates, Inc.
57 Morehouse Lane
Red Hook, NY 12571 USA
Phone: 845-758-0400
Fax: 845-758-2633
Email: curran@proceedings.com
Web: www.proceedings.com

Table of contents

Volume 541

International Seminar on Metallurgy and Materials

25–26 September 2018, Tangerang Selatan, Indonesia

Accepted papers received: 10 April 2019
Published online: 3 July 2019

Preface

Papers

IOP Conf. Series: Materials Science and Engineering **541** (2019) 011001 doi:10.1088/1757-899X/541/1/011001

Preface

International Seminar on Metallurgy and Materials 2018 (ISMM 2018)

We are honored to bring you the collection of articles from the 2nd International Seminar on Metallurgy and Materials (ISMM 2018) which was held in Puspiptek Area, Tangerang Selatan, Indonesia from 25-26 September 2018.

The primary focus of the conference was to bring together scientists, researchers, academicians and industries to share their knowledge in various areas of metallurgy and materials. The ISMM 2018 as a good platform for the scientific community to meet, share and exchange their ideas.

During the two days of the seminars, the researchers presented the most recent achievements and discoveries in metallurgy and materials as well as established networking for possible joint researches and collaborations among the scientists.

We are very pleased to our outstanding keynote speakers Prof. Dr.-Ing Ching An Huang, Prof. Dr. Azizah Hanom Ahmad, Assoc. Prof. Dr. Akbar Rhamdani, Ass. Prof. Dr. Grzegor Lisak, Dr. Efendi Mabruri and invited speakers for their valuable sharing and knowledge in metallurgy and material fields.

The editor would like to thanks the participants who have contributed to the articles. We also express our gratitude to all the organizing committee for their hard work during the seminar and also to the reviewers for their valuable revision of the papers.

Murni Handayani, M.Sc, Ph.D
Chairman of International Seminar on Metallurgy and Materials 2018
(ISMM 2018)

IOP Conf. Series: Materials Science and Engineering **541** (2019) 011002 doi:10.1088/1757-899X/541/1/011002

Peer review statement

All papers published in this volume of *IOP Conference Series: Materials Science and Engineering* have been peer reviewed through processes administered by the proceedings Editors. Reviews were conducted by expert referees to the professional and scientific standards expected of a proceedings journal published by IOP Publishing.

IOP Conf. Series: Materials Science and Engineering **541** (2019) 012001 doi:10.1088/1757-899X/541/1/012001

Investigation of corrosion rate on the modified 410 martensitic stainless steel in tempered condition

S Prifiharni[1], M T Sugandi[2], R R Pasaribu[2], Sunardi[2], E Mabruri[1]
[1] Research Center of Metallurgy and Material, Indonesia Institute of Science
Building 470, Puspiptek Area, Tangerang Selatan, Banten, 15314, Indonesia
[2] Departement of Mechanical Engineering, Engineering Faculty University of Sultan Ageng Tirtayasa
Jl. Jend. Sudirman Km.3, Cilegon – Banten, Indonesia, 42435

E-mail : siska.prifiharni@lipi.go.id

Abstract. Martensitic stainless steels were usually used for turbine blade. Their properties can be improved in various ways, such as by heat treatment. This paper reports the influences of tempering temperature and holding time in various concentration and temperature of NaCl solution on corrosion rate of the modified 13Cr martensitic stainless steels. Samples were austenized at 1050°C for 1 hour followed by oil quenching. The austenized sample were tempered at 600°C, 650°C, and 700°C for 1, 3, and 6 hours. All tempered specimens were immersed in 20% NaCl concentration for 120 hours and tempered specimen at 700°C for 6h was prepared for another corrosion test. Samples were immersed in different concentration of NaCl solution 3.5%; 10%; and 20% at room temperature, 60°C, and 80°C. Increasing temperature and concentration of solution will lead to corrosion rate increased. However, corrosion rate slightly decreases at higher temperature due to dissolved oxygen was decreased.

1. Introduction

Low pressure steam turbines (LPSTs) are used as the terminal energy extraction devices in steam cycle systems, such as those that exist in thermal power plants (nuclear and fossil fuelled). The steels used in fabricating the disks/rotors and the turbine blades are commonly AISI Type 403 SS (Fe13Cr martensitic stainless steel), with the latter being chosen primarily because of its advantageous vibrational mechanical damping characteristics [1,2]. However, it is often found failure due to mechanical combination with corrosion due to turbine blades working at high rotation and corrosive environment of the main mechanism of pitting corrosion, corrosion fatigue (CF) and corrosion cracking (stress corrosion cracking / SCC) [3,4].

Martensitic stainless steels are generally heat treated to provide moderate corrosion resistance and a good combination of mechanical properties. This often involves an austenitizing heat treatment followed by air or fan quenching. It is well known that tempering of the as-quenched martensitic steel can bring about secondary hardening when the softening effect due to annealing is offset by the precipitation of alloy carbides in the material [5]. The precipitation and coarsening of secondary phases during heat treatment including austenitizing, quenching and tempering, are then a key point to obtain the desirable microstructure and, hence, to achieve high temperature properties under such serious severe condition [6]. Depending on the composition and processing history, the microstructure of martensitic stainless steel consists of martensite, undissolved carbides as well as precipitated carbides, retained austenite and δ-ferrite. It is well-known that properties obtained in these steels are strongly influenced by such treatments [7].

In recent years, new types of martensitic stainless steels can increase toughness and corrosion resistance for certain applications with the addition of minor alloys. The most widely used alloying elements consist of nitrogen (N) [8,9], nickel (Ni) [10], and molybdenum (Mo) [11,12]. Mo is particularly effective, but only in the presence of Cr. It was suggested that Mo acts by adsorbing on the surface as molybdate or by blocking active sites during active dissolution [13]. Previous research has been explained about the effect of addition Mo and Ni elements on mechanical properties and corrosion resistance [14–16]. The effects of austenization and tempering temperatures may also affect the mechanical properties and corrosion resistance of this type of stainless steel [17–19]. However, there has been no research on the corrosion rate in different temperature of the modified 13Cr3Mo3Ni martensitic stainless steels. Therefore, the aim of this study is to determine the corrosion rate after tempering that occurs in various concentrations and temperature of the solution.

2. Materials and method

2.1. Preparation and heat treatment
Material used in this study is stainless steel 13Cr3Mo3Ni ingot. The chemical composition is shown in Table 1. The material is then hot forged at a temperature range of 900 – 1100°C. Then, the sample was austenized at 1050°C for 1 hour, followed by oil quenched. The austenized samples were then tempered at temperatures of 600°C, 650°C, and 700°C with holding time for 1, 3, and 6 hours.

Table 1. Chemical composition (% wt) the modified 410 martensitic stainless steel.

C	S	P	Mn	Si	Cr	Mo	Ni
0.10	0.001	0.01	0.69	0.24	12.50	2.72	2.89

2.2. Corrosion Test
All tempered specimens were immersed in 20% NaCl for 120 hours. And tempered specimen at 700°C with a holding time of 6 hours, were selected for corrosion test at various NaCl concentration and immersion time. The test specimen was immersed for 120 hours at a concentration variation of 3.5% NaCl solution; 10%; and 20% with temperature variations 60°C, 80°C, and room temperature.

Corrosion rate was calculated by weight loss method according to ASTM G1, with the following formula

$$\text{CR (mmpy)} = \frac{\Delta W \; x \; K}{D \; x \; A \; x \; T} \tag{1}$$

where ΔW is the weight loss in grams, K is the constant ($8.76 x 10^4$), D is the metal density (g/cm^3), A is the area of the specimen (cm^2), and t is the exposure time (days).

3. Results and discussion

3.1. Corrosion Rate
Figure 1 shows the corrosion rate of the modified 410 martensitic stainless at various temperature and holding time. Figure 1 shows that corrosion rate decreases with increasing of tempering temperature and holding time.

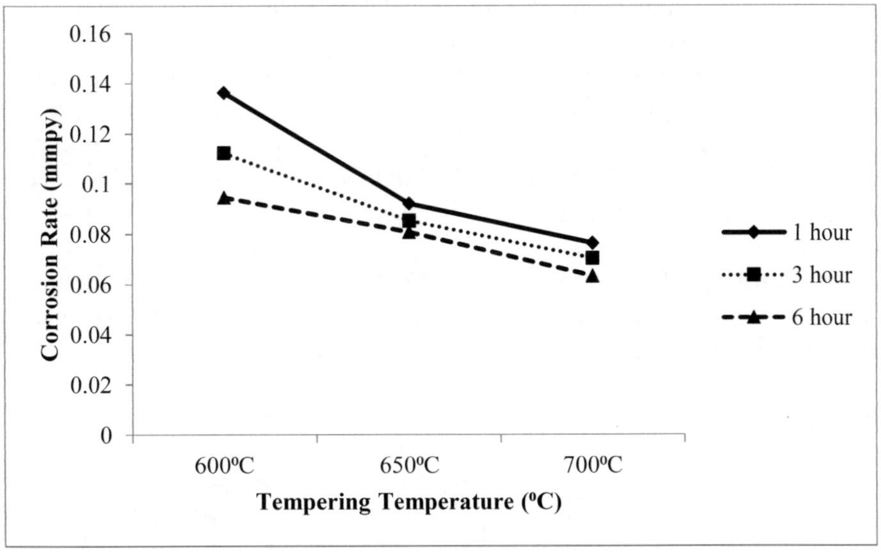

Figure 1. Corrosion rate of the modified 410 martensitic stainless steel at various tempering temperatures and holding time.

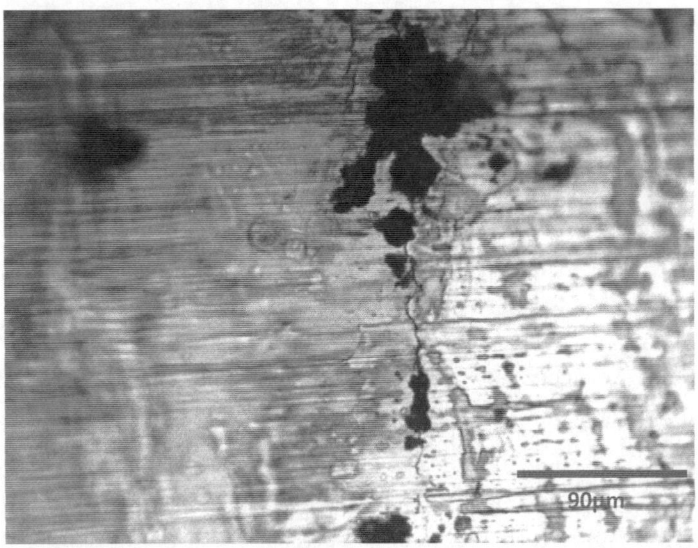

Figure 2. Optical micrograph of pitting corrosion on the modified 410 martensitic stainless steel tempered at 600ºC for 1 hour after immersion in 20% NaCl for 120 hours (magnification 500x).

The highest corrosion rate was obtained at tempering 600ºC for 1 hour and the lowest corrosion rate was obtained at tempering 700ºC for 6 hours. The decreasing corrosion rate could be attributed to chromium carbide precipitation. Dissolution of carbides in the matrix causing in more homogenous microstructure is an important factor in reduction of corrosion rate [20]. The dissolution of chromium carbide precipitation increases the contents of Cr and Mo, therefore the corrosion resistance enhanced [21]. The current density in both active and passive regions will decreases significantly with addition of Mo [22]. Figure 2 shows the microstructure on the modified 410 martensitic stainless steel tempered at

600°C for 1 hour after immersion in 20% NaCl for 120 hours. Visible pitting corrosion on the surface appears in dark contrast or contrast due to high oxygen content and there is a tendency to inter-granular corrosion. It is suggested that the chromium level in the affected area is reduced to below the critical value which is necessary for passivation, then the steel becomes sensitized to inter-granular corrosion [23].

Corrosion rate of the modified 410 martensitic stainless steel in various NaCl concentration and temperature is shown in Figure 3. The modified of tempered 410 martensitic stainless steel tempered at 700°C for 6 hours was chosen for corrosion test. As seen in Figure 3, the curve showed that corrosion rate increases as increasing NaCl concentration and temperature operation. The minimum value of corrosion rate is 0.198 mmpy that was obtained in 3.5% NaCl concentration at room temperature and the maximum is 1.335 mmpy that was obtained in 20% NaCl concentration at 80°C.

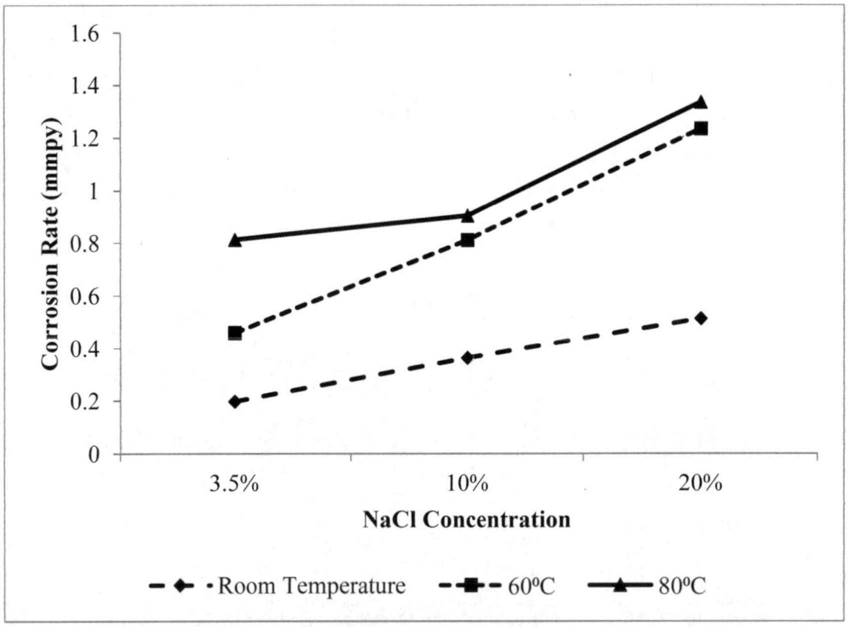

Figure 3. Corrosion rate of the modified 410 martensitic stainless steel tempered sample at 700°C for 6 hours in various NaCl concentration and temperature.

Temperature of the process plays an important role to accelerate corrosion rate of metals. Increasing temperature will lead to higher corrosion rate. Dissolved oxygen in the solution will increase as the temperature in the solution is rising. The rate of oxygen diffusion increases with increasing temperature [24]. When corrosion is controlled by diffusion of oxygen, the corrosion rate at a given oxygen concentration approximately doubles for every 30°C rise in temperature [25]. In an open vessel, allowing dissolved oxygen to escape, the corrosion rate increases with temperature about 80°C and then falls to a very low value at the boiling point. The lower corrosion rate above 80°C is related to a marked decrease of oxygen solubility in water as the temperature is increase, and this effect eventually overshadows the accelerating effect of temperature alone [26].

Increasing chloride concentration also increases the corrosion rate (Figure 3). The addition of chloride ions has a strongly effect on the corrosion behavior of stainless steel in acidic solutions, since in the presence of this ion, a passivity breakdown process takes place above a certain potential that decreases with the chloride concentration. Chloride ions slowly penetrate into the surface, at which the protective film is destroyed and the steel begins to corrode. Chloride corrosion tends to evenly pit the entire surface area of the specimens with shallow, flat-bottomed, irregular shaped pits. Chloride penetration in metal is due to the presence of different mechanisms, mainly diffusion and capillary absorption [27].

4. Conclusion

Tempering temperature and holding time affect on corrosion resistance of the modified 410 martensitic stainless steel. Corrosion rate of the modified 410 martensitic stainless steel after tempering at 700°C for 6 hours was increase as increasing NaCl concentration and temperature operation. The minimum value of corrosion rate is 0.198 mmpy that was obtained in 3.5% NaCl concentration at room temperature. Rise in temperature lead to higher corrosion rate due to the rate of oxygen diffusion increases with increasing temperature. And also, the chloride ion has a strongly effect on the corrosion behavior of the modified 410 martensitic stainless steel. Hence, the highest corrosion rate was obtained in 20% NaCl concentration at 80°C.

References

[1] Zhang Y, Urquidi-MacDonald M, Engelhardt G R and MacDonald D D 2012 *Electrochim. Acta* **69** 1–11
[2] Taji I, Moayed M H and Mirjalili M 2015 *Corros. Sci.* **92** 301–8
[3] Schönbauer B M, Stanzl-Tschegg S E, Perlega A, Salzman R N, Rieger N F, Zhou S, Turnbull A and Gandy D 2014 *Int. J. Fatigue* **65** 33–43
[4] Rodríguez J A and Clemente C M 2016 *IOSR J. Eng.* **06** 19–24
[5] Lim L C, Lai M O, Ma J, Northwood D O and Miao B 1993 *Mater. Sci. Eng. A* **171** 13–9
[6] Tao X, Gu J and Han L 2014 *J. Nucl. Mater.* **452** 557–64
[7] Rajasekhar A, Madhusudhan Reddy G, Mohandas T and Murti V S R 2009 *Mater. Des.* **30** 1612–24
[8] Leda H 1995 Nitrogen In Martensitic Stainless Steels *J. Mater. Process. Technol.* **53** 263–72
[9] Kaluba W J, Kaluba T and Taillard R 1999 *Scr. Mater.* **41** 1289–93
[10] Thursdiyanto V, Bae E and Baek E 2008 *J. Mater. Sci. Technol.* **24** 343–6
[11] Singh A K, Reddy G M and Rao K S 2015 *Def. Technol.* **11** 299–307
[12] Jargelius-Petterson R F A and Pound B G 1998 *J. Electrochem. Sci.* **145** 1462–9
[13] Frankel G S 1998 *J. Electrochem. Soc.* **145** 2186–98
[14] Mabruri E, Anwar M S, Prifiharni S, Romijarso T B and Adjiantoro B 2015 *Maj. Metal.* **30** 133–40
[15] Mabruri E, Anwar M S, Prifiharni S, Romijarso T B and Adjiantoro B 2016 *AIP Conference Proceedings* **1725**
[16] Prifiharni S, Perdana H, Romijarso T B, Adjiantoro B, Juniarsih A and Mabruri E 2017 *Int. J. Eng. Technol.* **8** 2672–5
[17] Prifiharni S, Anwar M S and Mabruri E 2016 *Widyariset* **2** 9–16
[18] Prifiharni S, Ahmad D, Juniarsih A and Mabruri E 2017 *Maj. Metal.* **32** 83–90
[19] Anwar M S, Prifiharni S and Mabruri E 2016 *AIP Conference Proceedings* **1725**
[20] Isfahany A N, Saghafian H and Borhani G 2011 *J. Alloys Compd.* **509** 3931–6
[21] Park J Y and Park Y S 2007 *Mater. Sci. Eng. A* 448–451 1131–4
[22] Hashimoto K, Asami K and Kawashima A 2007 *Corros. Sci.* **49** 42–52
[23] Abbasi-Khazaei B and Mollaahmadi A 2017 *J. Mater. Eng. Perform.* **26** 1626–33
[24] Gerasimov V V and Rozenfeld I L 1957 *Bull. Acad. Sci. USSR* **6** 1192–7
[25] Skaperdas G T and Uhlig H H 1942 *Ind. Eng. Chem. Res.* **34** 748–54
[26] Revie R W and Uhlig H H 2008
[27] Prawoto Y, Ibrahim K and Nik W B W 2009 Arab. *J. Sci. Eng.* **34** 115–27

Acknowledgment
This paper is part of development of high temperature material for power plant research. The authors would like to be obliged to Research Center for Metallurgy and Material of Indonesian Institute of Sciences for providing laboratory facility and funding this research.

IOP Conf. Series: Materials Science and Engineering **541** (2019) 012002 doi:10.1088/1757-899X/541/1/012002

Effect of reduction temperature and sodium-based additives on nickel upgrading process of laterites ores

A Shofi[1], A Rahmahwati[2], F Nurjaman[1], B Suharno[2]
[1]Research Unit for Mineral Technology, Indonesian Institute of Sciences, Jl. Ir. Sutami Km. 15 Tanjung Bintang, South Lampung, Lampung, Indonesia
[2]Department of Metallurgy and Materials Engineering, Faculty of Engineering, Universitas Indonesia, Depok, Indonesia

E-mail: fajar.nurjaman@lipi.go.id

Abstract. In this present work, the selective reduction of nickel laterite was conducted by using 10 wt. % of sodium carbonate, sodium chloride and sodium sulphate as additives. The 5 wt. % of palm shell charcoal was used as the reductant. The selective reduction process for nickel laterite was carried out from 950°C to 1150°C for 60 minutes followed by wet magnetic separation using 500 Gausses permanent magnet. The characterization of reduced ore was performed by X-Ray Diffraction, optical microscope and Scanning Electron Microscope with Energy Dispersive X-Ray Spectroscopy, while the chemical composition of concentrate or ferronickel was identified by X-Ray Fluorescence. The results showed that the nickel grade was increased with the increasing of reduction temperature for nickel ore with sodium sulfate additive. Nevertheless, it was relatively constant for sodium carbonate and sodium chloride additive. Nickel recoveries found increased with the increasing of reduction temperatures of nickel ore with sodium-based additives. The ferronickel particle was agglomerated and growth with the increasing of reduction temperature, thus it provided favourable conditions for the separation of ferronickel from impurities in the magnetic separation process.

1. Introduction

Nickel laterite is closely associated with iron oxide and silicate minerals as an isomorphous substitution for iron and magnesium in the lattice. Laterite ores are not able to process by physical beneficiation methods because of their complex mineralogy. The inherent difficulty of the beneficiation of laterite ores leads to their higher relative processing costs [1-2]. Pyrometallurgical or hydrometallurgical methods are usually applied in nickel laterite processing. Hydrometallurgical processes, including high-pressure acid leaching (HPAL), atmospheric acid leaching (AL) and Caron process [3]. The pyrometallurgical method has been applied commercially to extract nickel from the nickel the laterite ore such as blast furnace (BF) smelting, rotary kiln-electric furnace (RKEF) process, Krupp-Renn (Oheyama) process [4].

The BF and RKEF process is considered to be energy intensive for producing ferronickel from laterite ores as it involves several high-temperature steps, especially when laterite ores with low nickel content are treated [5]. As for the Oheyama process, it is less energy consumable compared to the RKEF process but problem encountered by the Oheyama process is the ringing of the reactor wall caused by the hot-sticky agglomerates [4,6].

A new method for producing ferronickel from nickel laterite ores by reducing metal oxides at moderate temperature and followed by magnetic separation process was called selective reduction process. By controlling the reduction conditions, it should be possible to obtain reasonable enrichment and high nickel recoveries while minimizing the reduction of the iron oxides.

Several investigations on the reduction roasting of nickel laterite ore using coal-based reductant followed by magnetic separation to produce ferronickel concentrate had been performed. A variety of reducing conditions and chemical additives in selective reduction process had been used to increase the nickel grade and recovery in the ferronickel alloy. According to Li et al., a concentrate of ferronickel containing 9.48% Ni with a nickel recovery of 83.01% was obtained from the nickel laterite ore containing 1.91% Ni. The reduction was carried out at 1100°C for 60 minutes with the addition 20 wt. % of sodium sulfate [7]. Jiang et al. (2013) reported that the addition of 10 wt. % of sodium sulfate and 2 wt.% of coal in nickel laterite ore (1.49% Ni and 34.69% Fe) processing by reduction process at 1200°C for 50 minutes which reported a nickel grade of 10.86% and recovery of 88.56% [8].

However, there is still less information in the selective reduction of nickel laterite processing using other sodium-based additives. In this present work, the extracting ferronickel concentrate from nickel laterite ore was carried out with added some sodium-based additives such as sodium carbonate, sodium chloride and sodium sulfate in the reduction process continued with the magnetic separation processes. The effects of various additives and reduction temperature on this selective reduction process were investigated clearly.

2. Materials and Methods

2.1. Materials

The nickel laterite ore was from South East Sulawesi, Indonesia. The X-ray fluorescence (XRF) was performed to determine the chemical compositions of the nickel ore, as showed in table 1.

Table 1. Chemical analysis of laterite ores.

Element	Fe	Ni	Al	Si	Ca	Mg	Cr	Mn	Co
wt. %	50.5	1.4	4.86	6.5	0.18	1.81	2.68	0.85	0.07

A 10 wt. % of palm shell charcoal was used as the reductant. The composition of the reductant is listed in table 2. The sodium-based additives which were used in this reduction process were sodium carbonate, sodium chloride, and sodium sulfate.

Table 2. Chemical analysis of palm kernell shell charcoal.

Chemical analysis of ash content (wt. %)						Proximate analysis (wt. %)				S	Calories (cal/g)
SiO_2	CaO	MgO	Al_2O_3	Na_2O	Fe_2O_3	Moisture	Volatile content	Ash content	Fixed carbon content		
54.09	20.71	3.37	3.12	1.6	6.72	0.43	22.57	21	77	0.01	6.244

2.2. Reduction and Magnetic Separation Process

The nickel laterite ore and palm shell charcoal were crushed and ground to less than 149 μm. The fine nickel laterite and 5 wt. % palm shell charcoal was thoroughly mixed with different additives, i.e. sodium carbonate, sodium chloride and sodium sulfate with 5 to 15 wt. %. The mixture was pelletized into 10-15 mm diameter and dried in an oven at 100°C for 4 hours. The 60 gram of pellets were placed in a graphite crucible and the reduction process was carried out in the laboratory using muffle furnace. It was heated using various reduction temperatures, i.e. 950°C, 1050°C and 1150°C for 60 minutes. After the reduction process, the reduced pellets were quenched into water and dried at 100°C for 4 hours followed by grinding into less than 74 µm. Then, the wet magnetic separation process was performed to the fine-reduced ore using 500 Gauss permanent magnet. The magnetic particle of reduced ore or concentrate was rich in ferronickel, while non-magnetic or tailing was rich in impurities/slag.

2.3. Characterization

The characterization of concentrate was identified by X-ray Fluorescence (XRF). Polished sections were prepared to observe the morphological changes and analyzed the element content of the reduced ore using Scanning Electron Microscopy-Energy Dispersive Spectroscopy (SEM-EDS). X-ray Diffraction (XRD) analysis was used to determine the different mineralogical phases present in the reduced ore.

3. Results and Discussion

3.1. Effect of reduction temperature

The effect of reduction temperatures of nickel laterite ore on iron-nickel grade and recovery using various additives is given on figure 1(a-d). The 10 wt. % of additives and 5 wt. % of palm shell charcoal reductants were used in this experiment. The increasing of reduction temperature increased the nickel grade in nickel ore concentrate, as shown in figure 1(a). The optimum temperature reduction was 1150°C produced 4.6 wt. % of nickel grade using sodium sulfate additive. The additives addition in selective reduction process promoted the recovery of nickel, as shown in figure 1b. Nickel recovery also increased along with the increased temperature from 950°C to 1150°C. The optimum nickel recovery, i.e. 73.2%, was resulted from the reduction of nickel ore at temperature 1150°C using sodium sulfate additive.

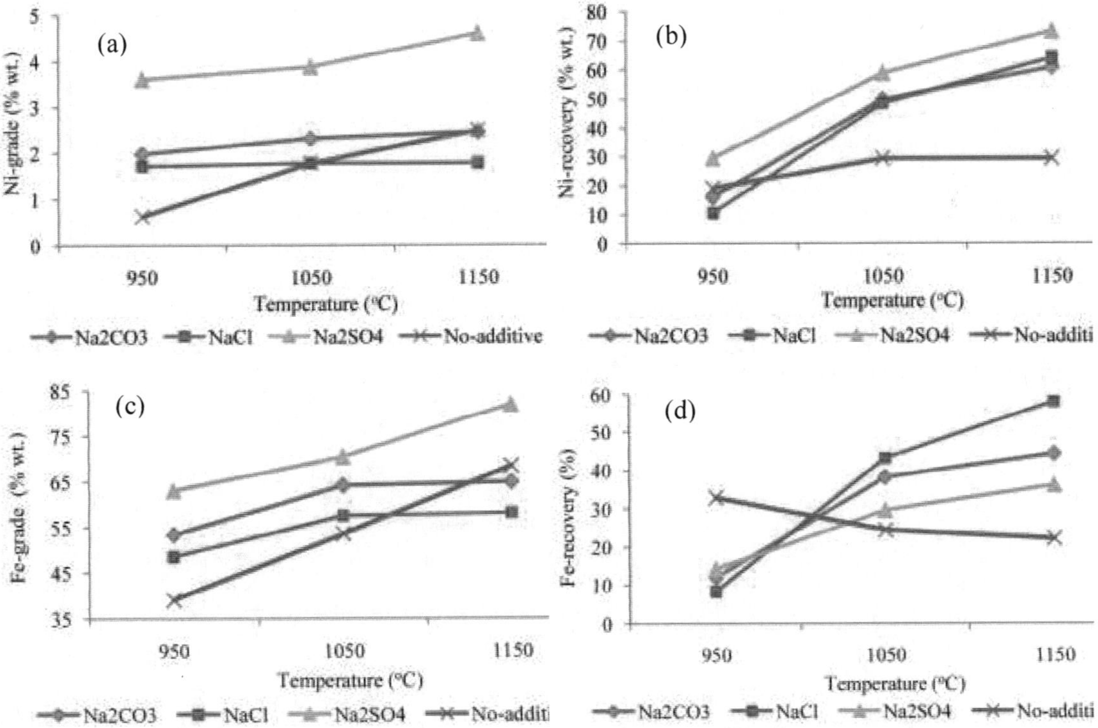

Figure 1. Effect of reduction temperature on iron-nickel grade and recovery using 10 wt. % additives.

The increase of iron grade was found in with and without additives during the increasing of reduction temperature, as shown in figure 1(c). The optimum iron grade in this selective reduction process, i.e.

81.9 wt. %, was resulted from the reduction process at temperature 1150°C using sodium sulfate additive. In the addition of sodium chloride and sodium carbonate, the iron grade in concentrate was relatively constant after the reduction process at temperature 1050°C. Nevertheless, at temperature reduction 1150°C, the iron grade was lower than no-additive addition.

The iron recovery was decreased for nickel ore reduction without additives by the increasing of reduction temperature, as shown in figure 1(d). By the addition of the additives, the iron recovery was increased with the increasing reduction temperature. The addition of sodium chloride additive in nickel ore promoted a significant increase of iron recovery due to the pyro-hydrolysis reaction of sodium chloride which was rising along with the increase of temperature [9]. The increased temperature up to 1150°C should increase the nickel-iron grade and recovery with additives. The increase of temperature should improve the degree of metallization on nickel and iron [10].

From figures 1(b) and (d), the iron and nickel recovery of reduced ore containing sodium carbonate was increased with the increasing of reduction temperature. From XRD analysis, as shown in figure 2, the peak of ferronickel diffraction increased along with the increasing of reduction temperature up to 1150°C. Nevertheless, from figure 1(a), the optimum nickel grade of reduced ore with sodium carbonate additive was only 2.5%. The low nickel grade in this reduced ore due to the high of iron recovery resulted from the reduction process. The decomposition of sodium carbonate produced carbon dioxide (CO_2) gas, as expressed in equation (1) [11], promotes the formation of wustite (FeO) due to the Bouduard reaction by producing carbon monoxide (CO) gas to enhance the reduction process. Furthermore, the wustite will reduce into iron. As shown in figure 2, The wustite was decreased with the increase of temperature reduction due to the more carbon monoxide resulted from decomposition of sodium carbonate. The peak of quartz (SiO_2) was decreased along with the increase of temperature and it was disappeared at the temperature of 1150°C. The quartz was decreased with the increasing temperature reduction also indicated that the decomposition rate of sodium carbonate enhanced by the increasing of reduction temperature.

$$Na_2CO_3 \rightarrow Na_2O + CO_2 \quad (1)$$

$$Na_2CO_3 + MgSiO_3 \rightarrow Na_2MgSiO_4 + CO_2 \quad (2)$$

From SEM and EDAX analysis, as shown in figure 3, the sodium silicate (Na_2SiO_3) which resulted from the decomposition of sodium carbonate was found as dark area (point C) with the appearance element of Na, Si, and O, while the ferronickel was found as light-white area which was indicated by the element of Fe and Ni. Sodium silicate has a low melting temperature, i.e. 1089°C. Thus, it gives a beneficial condition for migrations and aggregations of metallic iron and nickel in liquidus state. As shown in figure 3, a less amount of ferronickel particle was found in reduction process at temperature 950°C and it was found the more and larger ferronickel particle at temperature 1150°C.

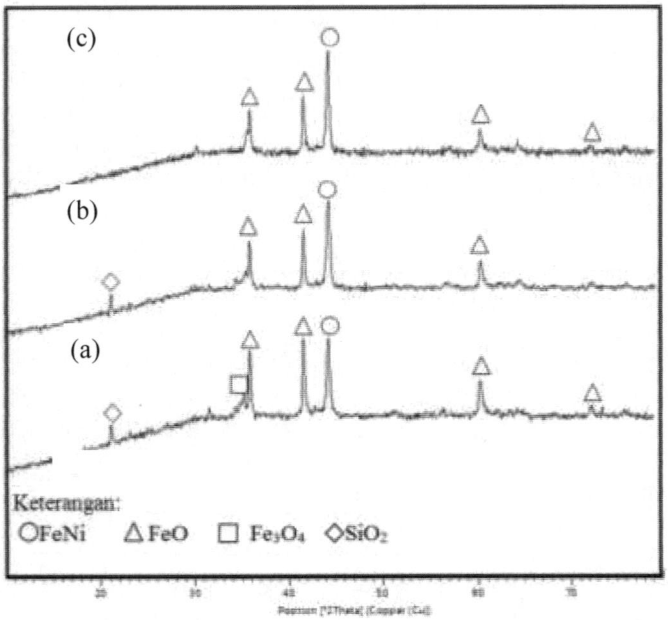

Figure 2. XRD pattern of the reduced nickel laterite with 10 wt. % additive sodium carbonate at various temperature (a) 950°C, (b) 1050°C (c) 1150°C.

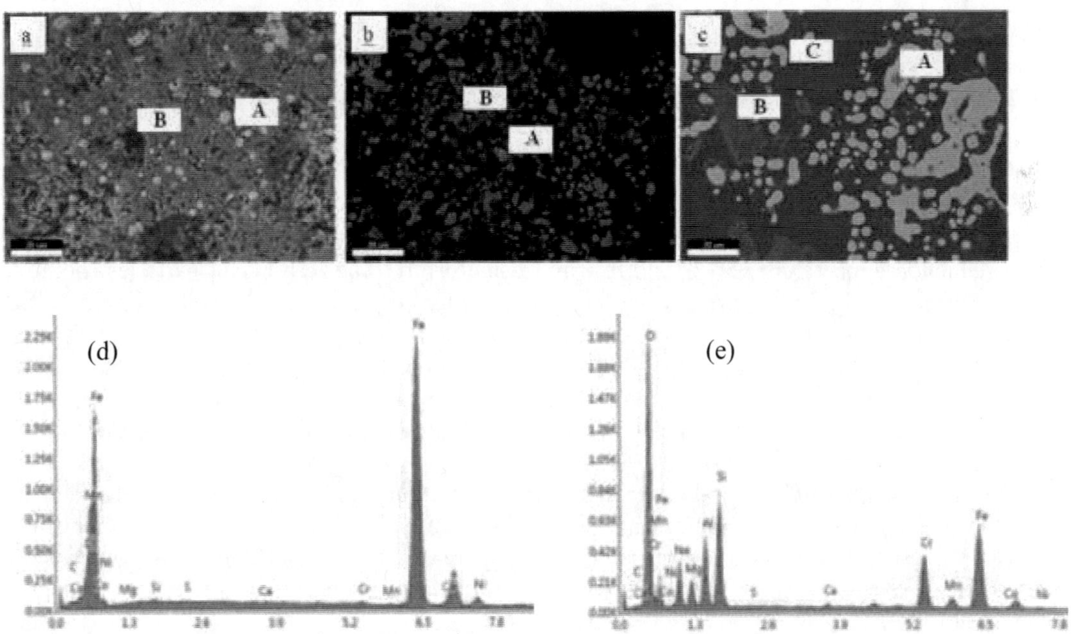

Figure 3. SEM analysis of reduced laterite in presence of 10 wt. % sodium carbonate with reduction temperature: (a) 950°C, (b) 1050°C and (c) 1150°C; EDAX analysis of point: (d) B, (e) A

Figure 4(a-c) presented the XRD analysis of reduced nickel laterite at various reduction temperatures with sodium chloride additive. Along with the increase of reduction temperature, the peak of fayalite ($(Fe, Mg)_2SiO_4$) decreased due to the reaction of fayalite with carbon monoxide forming magnetite.

There was no-wustite, such as in sodium carbonate, due to the limited of CO gas sources. As shown in figure 4, magnetite was formed at 1050°C followed by reducing the peak of fayalite. At 1150°C, the peak of magnetite was reduced due to the transformation of metallic iron from Fe_3O_4 by the reduction process, thus the peak of ferronickel increased. Hydrochloride gas was formed by the pyro-hydrolysis of sodium chloride. Nevertheless, the hydrochloride gas tended to be difficult to directly chlorinate iron oxides during the process of chlorination in this reduction process, thus iron chloride was not found in reduced nickel ore, as shown in XRD pattern (figure 4). It was also agreed with Zhou et al., that the chlorination of iron oxide was not occurred even at 1150°C of reduction temperature [9]. In this reduction process, the metallic iron was resulted from the reduction mechanism of fayalite which transform into magnetite, then reduced into metallic iron.

The increasing of reduction temperature of nickel ore with sodium chloride additive promoted the growth of ferronickel particle in reduced nickel ore, as shown in figure 5. From SEM and EDAX observation, it was found that fayalite as the main gangue in the reduced ore was found as a grey area (point B) which indicated by the element Fe, Mg, and Si, as shown in figure 5(d), while ferronickel shown as white area (point A) which indicated by element Fe and Ni. Sodium chloride addition promoted the aggregation of fayalite during the process of reduction, shown in figure 5. Therefore, it had an effect on the increase of iron grade and recovery in the concentrates through the reduction of fayalite into metallic iron.

The result of XRD testing at various temperatures with additive sodium sulfate is shown in figure 6(a-c). The peak of wustite (FeO) increased along with the increase of temperatures due to the reduction of hematite to wustite. However, most of the wustite was not transform into metallic iron during the increasing of temperature in this reduction process. As expressed in equation (3), the sodium sulfate was reduced into Na_2S. It reacts with wustite to form non-magnetic troilite (FeS), as expressed in equation (4), thus it inhibits the reduction of wustite into metallic iron. Therefore, it suppressed the iron recovery and resulted in low-grade iron in concentrate. The peak of ferronickel on XRD patterns decreased at the increasing temperature up to 1150°C. Nevertheless, the nickel grade was increased as given on figure 1(a-b). This indicates that there is more nickel than iron content in the concentrates.

$$Na_2SO_4 + 4CO \leftrightarrow Na_2S + 4CO_2 \tag{3}$$

$$Na_2S + FeO + 2SiO_2 \leftrightarrow Na_2Si_2O_5 + FeS \tag{4}$$

The growth of ferronickel size in reduced nickel laterite at temperatures 950°C to 1150°C using sodium sulfate additive was observed and presented on figure 7 (a-c). Fine ferronickel particles were found at reduction temperature 950°C and distributed uniformly. The ferronickel particles (point A) were agglomerated and grown larger during the increasing temperature. The diameter of particle size was 30.6 μm surrounded by troilite (point B) at reduction temperature 1150°C, as shown in figure 7. From XRD analysis, the troilite was also found in the reduction of nickel laterite with sodium sulfate additive, as shown in figure 6. At temperature 1050°C, the metallization occurred in liquid condition and it promoted the reduction rate of nickel and iron oxide into metallic phase due to the low eutectic temperature of troilite, which was 985°C [12].

Therefore, the nickel and iron grade of reduced ore with sodium sulfate additive was higher than the one with no-additive, even with sodium carbonate and sodium chloride additives. The decomposition rate of $(Fe,Mg)_2SiO_4$ into wustite and quartz was also increased in liquidus state. Thus, wustite was increased along the increasing of reduction temperature, while quartz was found at reduction temperature of 1150°C.

The investigation of ferronickel size in nickel ore concentrate toward reduction temperature was reported in this study. The growth of ferronickel size was increased by the increasing of reduction temperature from 950°C to 1150°C, as shown in figure 8. The particles size was 30.6, 12.8, and 0.8 μm at temperature 1150°C for reduction nickel ore with the addition 10 wt. % of sodium sulfate, sodium carbonate, and sodium chloride additives, respectively. The addition of sodium sulfate additives produced ferronickel size three times larger than the one with sodium carbonate and sodium chloride.

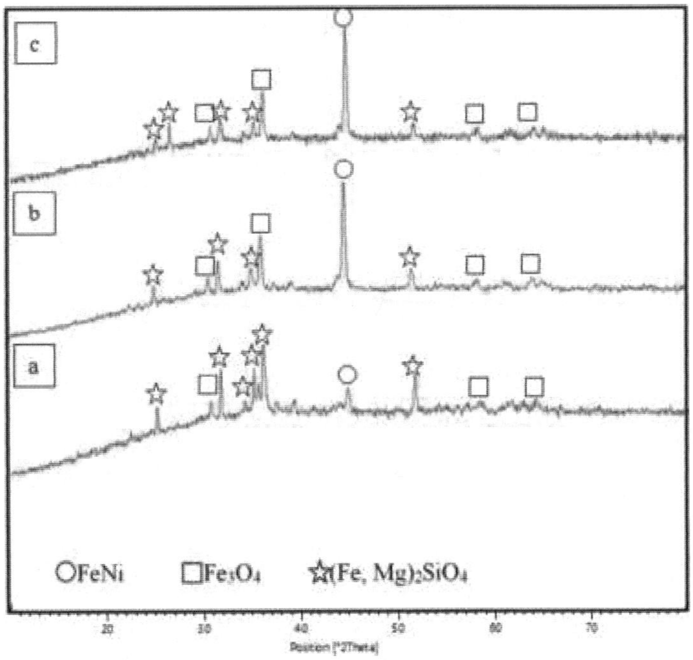

Figure 4. XRD pattern of the reduced laterite with 10% wt. additive sodium chloride and various temperature: (a) 950°C, (b) 1050°C (c) 1150°C.

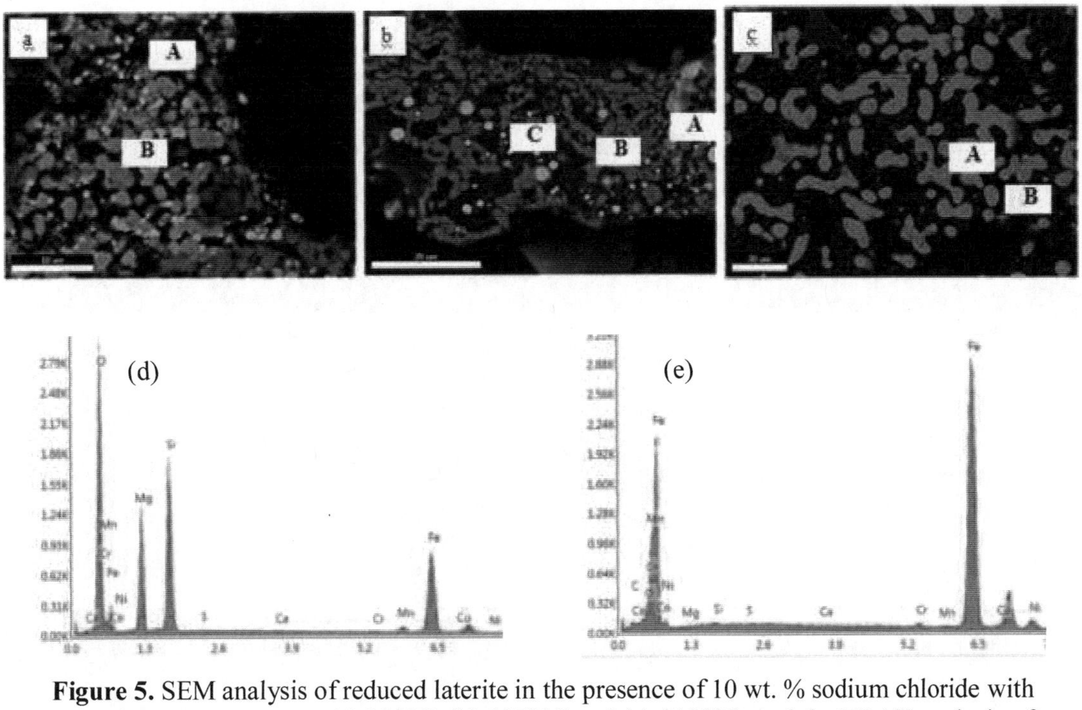

Figure 5. SEM analysis of reduced laterite in the presence of 10 wt. % sodium chloride with reduction temperature at: (a) 950°C, (b) 1050°C and (c) 1150°C; and the EDAX analysis of point: (d) B, (e) A.

IOP Conf. Series: Materials Science and Engineering **541** (2019) 012002 doi:10.1088/1757-899X/541/1/012002

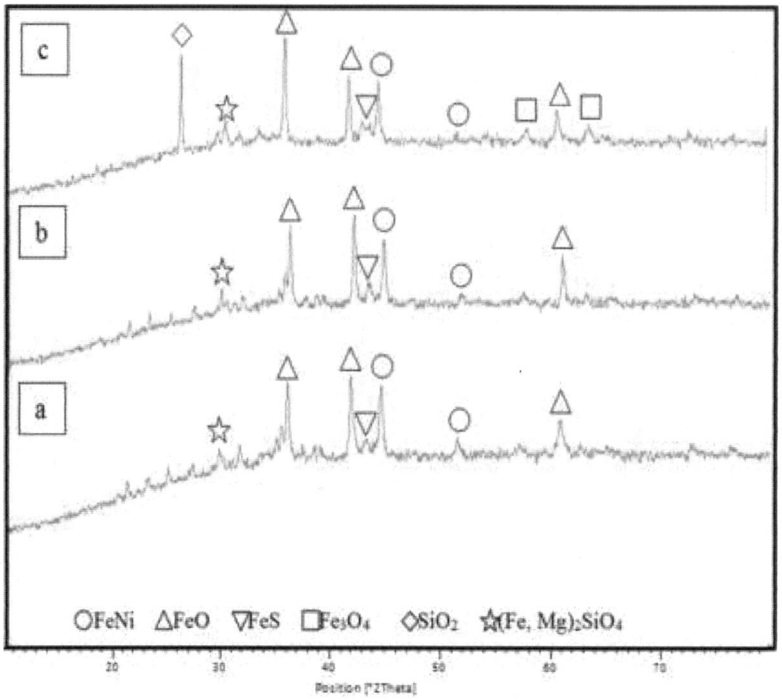

Figure 6. XRD pattern of the reduced laterite with the additive sodium sulphate and various temperature (a)950ºC, (b)1050ºC, (c)1150ºC.

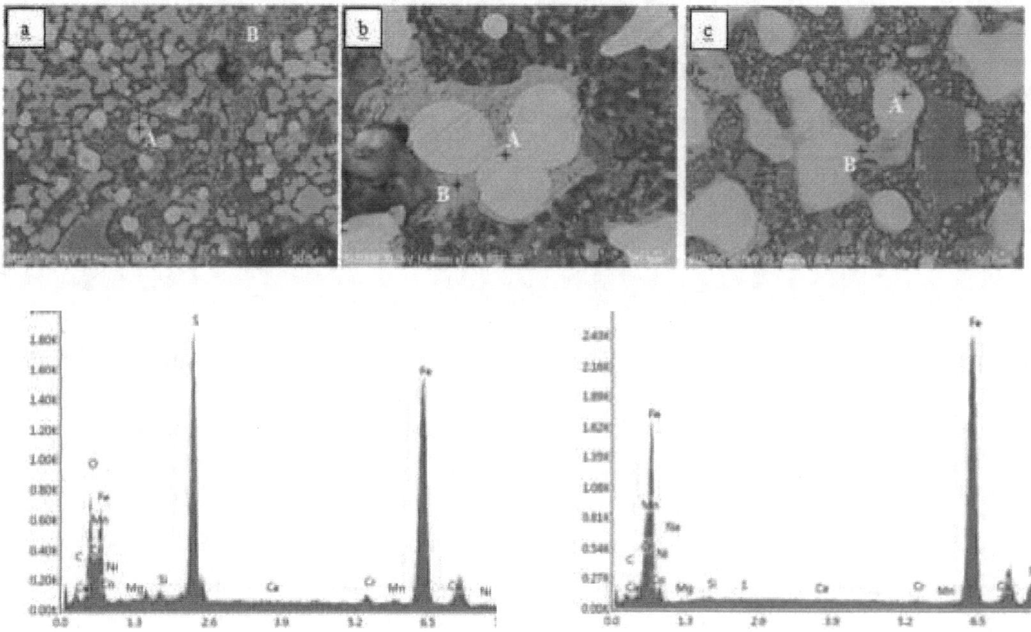

Figure 7. SEM analysis of reduced laterite in the presence 10 wt. % of sodium sulphate with reduction temperature at: (a) 950°C, (b) 1050°C and (c) 1150°C; and EDAX analysis of point: (d) B, (e) A.

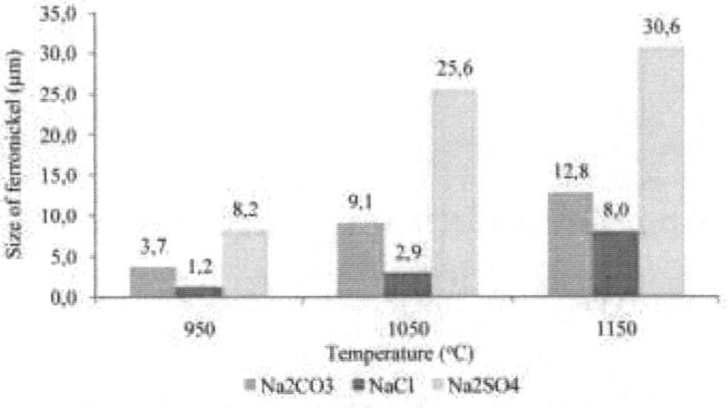

Figure 8. Ferronickel particles size of reduced nickel ore at various temperatures for 60 min with 10 wt. % additives.

4. Conclusion

The addition of sodium sulfate additive in the reduction of nickel ore containing 1.4%Ni-50.5% Fe resulted in higher nickel grade and recovery than sodium carbonate and sodium chloride additives, due to the troilite (FeS) formation for inhibiting the metallization of iron oxide, thus lowering the iron recovery resulted in high nickel grade in concentrate. The nickel grade was increased with the increasing reduction temperature for nickel ore with sodium sulfate additive. Nevertheless, it was relatively constant for sodium carbonate and chloride additives, due to the high iron recovery was resulted. The growth of ferronickel particles in reduced ore was increased with the increasing of reduction temperature, thus it gave the beneficial condition for nickel and iron separation through the magnetic separation process.

References

[1] Dalvi A, Bacon W G and Osborne R C 2004 *PDAC International Convention* 1
[2] Warner A., Diaz C, Dalvi A, Mackey P and Tarasov A 2006 *JOM* **58** 11
[3] Mudd G 2009 Proc. Int. Conf. of Metallurgist **48** 1
[4] Rao M, Li G, Jiang T and Luo J 2013 *JOM* **65** 1573
[5] Nayak J 1985 *Trans Indian Inst Met* **38** 241
[6] Watanabe T, Ono S and Matsumori T 1987 *Int J Miner Process* **19** 173
[7] Li G, Shi T, Rao M, Jiang T and Zhang Y 2012 *Miner. Eng.* **32** 19
[8] Jiang M, Sun T, Liu Z, Kou J, Liu N, and Zhang S 2013 *Int J Miner Process* **123** 32
[9] Zhou S, Wei Y, Li B, Wang H, Ma B and Wang C 2015 *Metall Mater Trans B* **47** 145
[10] Zhu D Q, Cui Y, Vining K, Hapugoda S, Douglas J, Pan J and Zheng G L 2012 *Int J Miner Process* **106-109** 1
[11] Zhou S, Dong J, Lu C, Li B, Li F, Zhang B, Wang H and Wei Y 2017 *Mater. Trans* **58** 790
[12] Rao M, Li G, Zhang X, Luo J, Peng Z and Jiang T 2016 *Sep. Sci. Technol.* **51** 1727

Acknowledgments

The authors would like thanks to the Ministry of Research, Technology and Higer Education for funding this research through Indonesia's National Innovation System (INSINas) Programme with Grant Number: 06/INS-1/PPK/E4/2018 and The Research Unit for Natural Product Technology for SEM-EDAX utilization in this experiment.

The effect of nitrogen dopant on hydrogenated graphene for hydrogen storage application

M Ganta[1], D N A Susilo[1], G K Sunnardianto[1,2], M Handayani[3]

[1]Faculty of Engineering and Technology, Sampoerna University, Jl. Raya Pasar Minggu, Kav 16, Jakarta, Indonesia
[2]Research Center for Chemistry, Indonesian Institute of Science (LIPI), Puspiptek Serpong, Tangerang Selatan, 15314, Indonesia
[3]Research Center for Metallurgy and Material, Indonesian Institute of Science (LIPI), Puspiptek Serpong, Tangerang Selatan, 15314, Indonesia

E-mail: muhammadganta@gmail.com

Abstract. We investigated the effect of nitrogen dopant on the charge transfer rate (CTR) and reaction pathways of hydrogen molecule dissociation on graphene and nitrogen-doped graphene based upon first principle calculation. We found that the nitrogen atom form covalent bonds with the carbon atoms with bond length of C-N of 1.41 Å. The geometry optimization results in nitrogen doped pristine graphene revealed that nitrogen pulled downward the nearest neighbor of carbon atom, thus the structure is a bit downward. We found the charge transfer rates from hydrogen to carbon atom is decrease with respect to pristine graphene and nitrogen doping also can decrease the activation barrier of hydrogen molecule dissociation on graphene surfaces.

1. Introduction

As promising material, graphene provides unique properties among other potential materials. Harder than some material, but flexible and give very thin area as graphene is 2D symmetry material. Refinement of graphene properties and performance becoming promising scheme in order to face world's globalization which still remain challenging duty. Controlling properties by electron addition to the pristine graphene could be a solution.

By modifying electron bands of graphene, tailoring the properties become realizable idea. Doping is an efficient way to significantly change graphene's electronic properties and chemical reactivity with different atoms like Nitrogen, Boron, Silicon and etc [1]. Those material substitutes carbon in pristine graphene and could significantly shift graphene properties. N-doping classified into the most common doping configuration and gives interesting prospect for various application. N-doped could induce additional charge and n-type carriers to the carbon system, it is crucial for semiconductor devices, energy conversion and storage system [1].

It was revealed that N-doped graphene exhibits superior electrocatalytic than most other electrocatalyst [2]. N-doping not only contributes to the additional charge and shifting the electronic properties but also carbon system was enhanced by N-doping. It would increase the performance of molecular sensing application. Nitrogen doped graphene could efficiently identify organic molecules and provide highly sensitive surface for controlled modifications. Nitrogen atom as doping would help stabilize structural defects like vacancies. Finally, synthesizing this Nitrogen doped graphene is via a simple and efficient way AP-CVD method (Atmospheric-pressure chemical vapor deposition) [1]. Density functional theory was used to inquire stability, adsorption and desorption process in nitrogen-doped graphene as hydrogen storage system. Nitrogen position related to properties of graphene, effect of nitrogen substitution observed by density function theory. In addition, Gao et al. revealed a simple model of stability for hybrid carbon-conjugated material and computational was verified [3].

Importantly, various N-doped graphene including single vacancy in graphene have been previously synthesized [4]. For further investigation of Nitrogen-doped graphene, STM/STS experiment combined with ARPES and XPS was conducted and proves the effect of charge distribution and electron from nitrogen substitution into nitrogen-doped graphene [5]. Those experiment added with theoretical argument allows us to observed the effect of nitrogen doping. Electron and charge distribution classified into two major state based on the concentration which are localized state (high nitrogen oxidized) and delocalized state.

Many researchers have modified the pristine graphene in order to meet the need of hydrogen storage applications by creation of single vacancy on the graphene surfaces[6-8]. Here, we would like to modify the pristine graphene by inserting nitrogen doping into graphene surface to reduce the activation barrier of hydrogen molecule dissociation and to know the adsorption-desorption characteristic under nitrogen doping. As a part of the adsorption process, we also investigated the charge transfer rate (CTR) from hydrogen to graphene with and without nitrogen doping.

2. Method

We performed the optimization using density functional theory [9]. The simulations have been done using the quantum espresso (QE) code [10]. We used the projector-augmented- wave method and the ultra-soft pseudo potential [11]. The local density approximation by the Perdew-Zunger parameterization [12] was adopted. The parameters for hydrogenated graphene with and without nitrogen doping in a super cell were the energy cut off of 40 Ry for the plane wave expansion of the wave function and 400 Ry for the expansion of the augmented charge. The convergence criterion for the structural optimization was that the total absolute value of the inter-atomic force vector became less than 10^{-4} Ry/a.u.

The distance between graphene planes was separated by 10 Å in orders to avoid interaction between layers. Distribution of k-points on a mesh of 12x12x1 was selected via Monkhrost-Pack scheme [13]. The structure models used for simulation are graphene sheets consisting of 48 carbon atoms in rectangular super cells. We used lowdin charge analysis to calculate the charge transfer rates from hydrogen to graphene and n-doped graphene [14]. In order to predict the reaction pathways and calculate the energy barrier, we used NEB method to investigate the reaction pathways of adsorption-desorption process of hydrogenated graphene and nitrogen-doped graphene [15].

3. Results

3.1 Charge transfer rates

When hydrogen interacts with N-doped graphene, there is a charge transfer from hydrogen to graphene. Thus, it is important to calculate the charge transfer rate from hydrogen to graphene. We investigated the charge transfer from hydrogen to N-doped graphene and compared the result of CTR from hydrogen to graphene surface without doping. The optimized structure is showed in Figure 1.

Table 1 shows the CTR from hydrogen to pristine graphene is about 0.20e, while the CTR from hydrogen to N-doped graphene is decrease to be 0.14e. The results indicated that N doped decreasing the CTR from hydrogen to graphene due to the presence of N doping. It can be analyzed from the optimized structure of hydrogenated graphene and hydrogenated N-doped graphene.

IOP Conf. Series: Materials Science and Engineering **541** (2019) 012003 doi:10.1088/1757-899X/541/1/012003

Table 1. Value of CTR from hydrogen to graphene with and without N doping.

Pristine graphene	CTR [e]	N-doped graphene	CTR [e]
48+3H	0.2035	47+N+3H	0.1428
72+3H	0.2041	71+N+3H	0.1430
96+3H	0.2041	95+N+3H	0.1433

3.2 Deformed structure

It is necessary to understand the reason why the CTR is decreasing in hydrogenated N-doped graphene. Figure 1 shows the optimum structure of hydrogenated graphene with and without nitrogen doping. We found that the location of nitrogen atom is a bit shifted downward around -0.13 Å and hydrogenated carbon atom around nitrogen a bit shifted upward around 0.37 Å

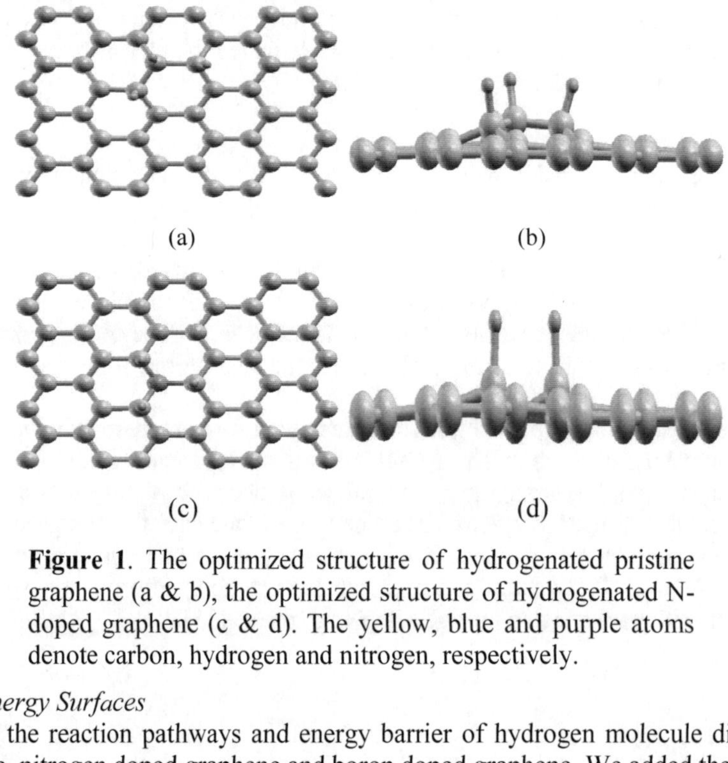

(a) (b) (c) (d)

Figure 1. The optimized structure of hydrogenated pristine graphene (a & b), the optimized structure of hydrogenated N-doped graphene (c & d). The yellow, blue and purple atoms denote carbon, hydrogen and nitrogen, respectively.

3.3. Potential Energy Surfaces

We investigated the reaction pathways and energy barrier of hydrogen molecule dissociation on the pristine graphene, nitrogen doped graphene and boron doped graphene. We added the result of reaction pathways of H_2 dissociation on boron doped graphene to compare the activation barrier of the most stable of doped graphene (boron and nitrogen doping).

Firstly, we have the same of the initial state of hydrogen molecule on pristine graphene and doped graphene, the hydrogen molecule is located on the vacuum. Along the reaction, we can see the energy of the system is increase followed by dissociation of hydrogen molecule into hydrogen atoms with an energy barrier of 1.3 eV (pristine graphene). Here, we reported that due to the presence of nitrogen doping, the activation barrier of hydrogen dissociation is decrease, which is a new finding that we could decrease the activation barrier only by nitrogen doping.

We also added another doping (boron doping) for the comparison since either nitrogen doping or boron doping of graphene have been observed experimentally. Thus, single nitrogen doping into graphene surface can be regarded as a "doped-catalytic" reaction center, which has a role to reduce the energy barrier. We observed that the reaction of hydrogen molecule dissociation to form two hydrogen atoms and adsorbed to the graphene surface with or without doping is exothermic reaction with difference energy between initial state and final state is < 1.5 eV.

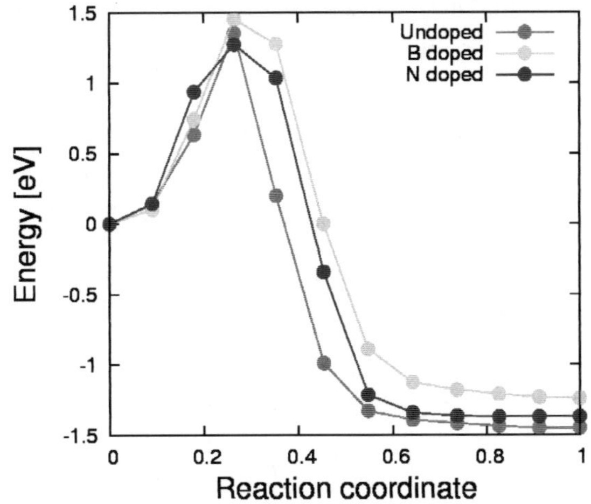

Figure 2. Minimum energy pathway of dissociation of hydrogen molecule on pristine graphene, nitrogen-doped graphene and boron-doped graphene.

4. Conclusion

The effect of nitrogen dopant on the interaction between hydrogen molecules on the graphene surface was investigated by using density functional theory calculation. We found a new finding on how to reduce the activation barrier of hydrogen molecule dissociation on graphene surface by single nitrogen doping.

Reference

[1] Lv R, Li Q, Botello-Méndez A R, Hayashi T, Wang B, Berkdemir A, Hao Q, Eléas A L, Cruz-Silva R, Gutiérrez H R, Kim Y A, Muramatsu H, Zhu J, Endo M, Terrones H, Charlier J C, Pan M and Terrones M 2012 *Sci. Rep.* **2** 1–8
[2] Xu H, Ma L and Jin Z 2018 *J. Energy Chem.* **27** 146–60
[3] Gao X, Zhang S B, Zhao Y and Nagase S 2010 *Chemie - Int. Ed.* **49** 6764–7
[4] Yu Y-X 2013 *Phys. Chem. Chem. Phys.* **15** 16819
[5] Joucken F, Tison Y, Le Fèvre P, Tejeda A, Taleb-Ibrahimi A, Conrad E, Repain V, Chacon C, Bellec A, Girard Y, Rousset S, Ghijsen J, Sporken R, Amara H, Ducastelle F and Lagoute J 2015 *Sci. Rep.* **5** 1–10
[6] Sunnardianto G K, Maruyama I and Kusakabe K 2016 *J. Comput. Theor. Nanosci.* **13** 4883–7
[7] Sunnardianto G K, Maruyama I and Kusakabe K 2017 *Int. J. Hydrogen Energy* **42** 23691–7
[8] Sunnardianto G K, Maruyama I and Kusakabe K 2016 *Adv. Sci. Eng. Med.* **8** 421–6
[9] Martin R M 2004 Electronic Structure, Basic Theory, and Practical Method (Cambridge University Press)
[10] Kresse G and Hafner J 1993 *Phys. Rev. B* **47** 558-61
[11] Blochel P E 1994 *Phys. Rev. B* **50** 17953-79
[12] Perdew J P and Zunger A 1981 *Phys. Rev. B* **23** 5048
[13] Pack J D and Monkhorst H J 1977 *Phys. Rev. B* **16** 1748–9
[14] Sanchez-Portal D, Artacho E and Soler J M 1995 *Solid State Commun.* **95** 685–90
[15] Henkelman G and Jónsson H 2000 *J. Chem. Phys.* **113** 9978–85

IOP Conf. Series: Materials Science and Engineering **541** (2019) 012003 doi:10.1088/1757-899X/541/1/012003

Acknowledgements

The calculations were carried out using the local computer at Sampoerna University and supported by supercomputer facilities of Graduate school of Engineering Science, Osaka University. Author (GKS) gratefully acknowledges to Prof. Koichi Kusakabe from Graduate School of Engineering Science, Osaka University, Japan for his kind continuous support.

IOP Conf. Series: Materials Science and Engineering **541** (2019) 012004 doi:10.1088/1757-899X/541/1/012004

The effect of boron dopant on hydrogenated graphene for hydrogen storage application

D N A Susilo[1], M Ganta[1], G K Sunnardianto[1,2], M Handayani[3]

[1]Faculty of Engineering and Technology, Sampoerna University, Jl. Raya Pasar Minggu, Kav 16, Jakarta, Indonesia
[2]Research Center for Chemistry, Indonesian Institute of Science (LIPI), Kawasan Puspiptek Serpong, Tangerang Selatan, 15314, Indonesia
[3]Research Center for Metallurgy and Material, Indonesian Institute of Science (LIPI), Puspiptek Serpong, Tangerang Selatan, 15314, Indonesia

E-mail: dwinaufullarifsusilo@gmail.com

Abstract. We investigate the effect of boron dopant on the charge transfer and reaction pathways of hydrogenated graphene based upon density functional theory calculation. We focused on the particularly the charge transfer rate of trimer hydrogen adsorption and its reaction pathways. Firstly, we investigated the effect of B dopant on the pristine graphene which is revealed that B-C bond length prior to hydrogenation is around 1.49 Å resulting the deformed structure of graphene since the size of boron is a bit larger that carbon atom. We have also calculated the charge transfer (CTR) from hydrogen to carbon atom after boron subtitution, we found that there is an increasing the CTR with respect to the pristine graphene indicated that the strong bond length between hydrogen and carbon.

1. Introduction

Graphene or well-known as a single layer carbon atom in two-dimensional (2D) honeycomb lattice [1] has been increasingly attached attention since it has fascinating physical properties as well as the unique electronics, thermal [2] and mechanical properties [3,4]. Those unique properties are gain a big promising for future application in such a many technology such that nanocomposite, sensors, biomedical and hydrogen storage application [5]. It is well known that hydrogen-based energy has been taking part in the scientist life. Hydrogen storage was one of the key technologies in the application as the future energy carrier. The reason is when the hydrogenated graphene in the form of graphene was demonstrated by Elias *et al.* [6] and shows that the process of hydrogenation is reversible, making the hydrogenated graphene a potential candidate for hydrogen storage material. The advantage of graphene hydrogenation as a means for storing hydrogen. However, it is fact that many of H storage has a problem with the low storage capacity.

In recent years, carbon-based materials especially graphene have been developed among the scientist in order to increase the capacity of H storage. One of plausible way to modify the graphene surface is by creating a single vacancy, which is good for hydrogen storage application [7,8]. Other facile way to modify the graphene surface to meet the need of diverse applications is chemical doping.

There are two popular chemical doping that commonly used such that B- or N- substitution in graphene materials [9] using Density Functional Theory (DFT) in order to calculate the stability, adsorption and desorption energy. Zhou *et al.* [9] shows that B- doping forms an electron-deficient six-membered ring structure and decreases the H atomic adsorption energy. Meanwhile, the N-doping has electron-rich six-membered ring structure and increase the H atomic adsorption energy. In addition, the size of boron and nitrogen are almost the same with the carbon atoms, so that, it is might be possible to do the B and N-doping in the graphene. However, the study of the B and N-doping on H adsorption is controversial since the study in this field is still rarely.

One important parameter to identify the interaction between hydrogen and graphene is charge transfer rate (CTR) since the CTR mediated the local bonding between carbon and hydrogen (C-H bonding). Gagus *et al.* have been observe the CTR rule from hydrogen to graphene surface [10]. However, based on our best knowledge, the CTR from hydrogen on doped graphene especially boron-doped graphene still not explore yet. Thus, in this study, we would like to study the CTR from hydrogen to graphene with and without boron doping.

2. Method

All the calculation including the structural and electronic structure optimization were performed using density functional theory (DFT) calculation[11], as implemented using the quantum espresso (QE) code [12]. We used the ultra-soft pseudopotential and projector-augmented- wave method [13]. Then we adopt the Perdew-Zunger parameterization [14] using local density approximation. The parameters for hydrogenated graphene with and without boron doping in a super cell were the energy cut off of 40 Ry for the plane wave expansion of the wave function and 400 Ry for the expansion of the augmented charge. The convergence criterion for the structural optimization was that the total absolute value of the inter-atomic force vector became less than 10-4 Ry/au.

The distance between graphene planes was separated by 10 Å in orders to avoid interaction between layers. Distribution of k-points on a mesh of 12x12x1 was selected via Monkhrost-Pack scheme [15]. The structure models used for simulation are graphene sheets consisting of 48 carbon atoms in rectangular super cells. We used lowdin charge analysis to calculate the charge transfer rates from hydrogen to graphene and boron-doped graphene [16] and NEB method to calculate the reaction pathways of the hydrogen adsorption on graphene with and without boron doping [8].

3. Result and Discussion

3.1 Charge Transfer rate

We investigated the trimer hydrogen adsorption on graphene with and without boron doping and calculated its charge transfer rates (CTR).

Table 1. Value of CTR from hydrogen to pristine graphene and B-doped graphene.

Pristine Graphene	CTR [e]	B-doped Graphene	CTR [e]
48C+3H	0.2035	47C+B+3H	0.2366
72C+3H	0.2041	71C+B+3H	0.2386
96C+3H	0.2041	95C+B+3H	0.2388

Table 1 shows that B doped increasing the CTR from hydrogen to graphene from 0.20e to 0.24e. CTR enhances by boron doping and suggests higher reactivity in a sense. In addition, enhanced CTR also might contribute stronger bonding between a hydrogen atom and a carbon atom. Since the size of boron is a little bigger than the size of carbon atom, thus there is a little bit distortion in the optimized of hydrogenated graphene with boron doping. The distortion effect could affect the CTR from hydrogen to graphene.

3.2 Deformed Structure

We found the reason why the CTR is increasing in hydrogenated B-doped graphene. Figure 1 shows the structure of hydrogenated graphene with and without boron doping.

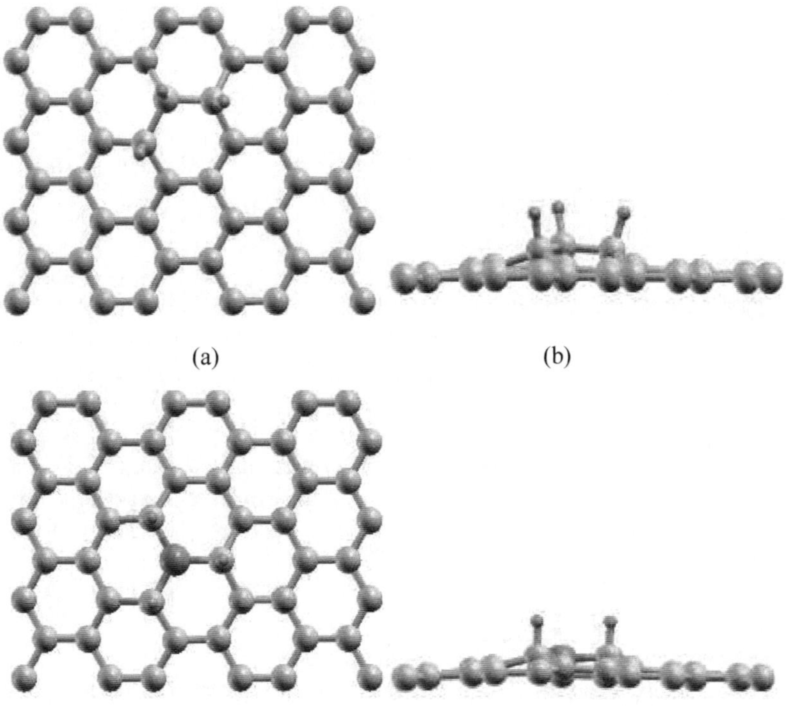

(a) (b)

(c) (d)

Figure 1. The optimized structure of hydrogenated pristine graphene (a & b), the optimized structure of hydrogenated B- doped graphene (c & d). The yellow, blue and purple atoms denote carbon, hydrogen and boron, respectively.

We found that the location of boron and hydrogenated carbon around boron atom is a bit shifted upward around 0.42 Å and 0.52 Å respectively. This is the reason why the CTR is increase on graphene with boron doping.

3.3 Potential Energy Surface

We investigated the influence of boron doping on the hydrogenation of graphene sheet by investigated the reaction pathways of trimer hydrogen adsorption on pristine graphene and boron-doped graphene.

As seen in Figure 2 (a) and (b) show the reaction of trimer hydrogen adsorption on pristine graphene and boron-doped graphene are exothermic, which indicated that the reaction is plausible in the nature.

The potential energy of the final state of trimer hydrogen adsorption on pristine graphene is around -4.5 eV, while the potential energy of the final state of trimer hydrogen adsorption on boron-doped graphene is around -2.4 eV. It means that in the boron-doped graphene the structure a little bit unstable if we compare with the pristine one. Since this is a bit unstable, thus the structure a bit distorts due to the presence of boron doped graphene. Thus, the results of deformed structure combining with the reaction pathway profile show clearly the reason of increasing the CTR from hydrogen to boron-doped graphene.

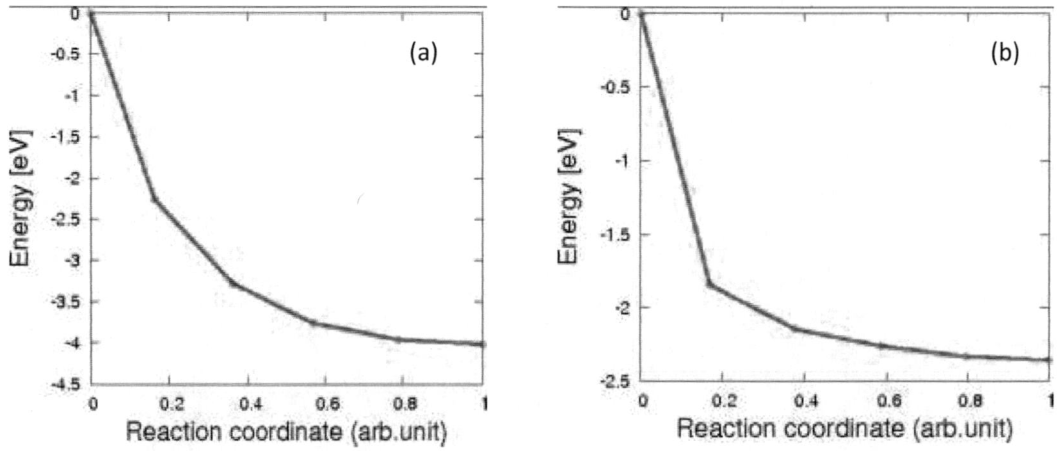

Figure 2. Minimum energy pathway of trimer hydrogen adsorption on pristine graphene (a) and trimer hydrogen adsorption on boron-doped graphene (b).

4. Conclusion

We have investigated the effect of boron dopant on the charge transfer rate (CTR) of trimer hydrogen adsorption on graphene using density functional theory calculation. We found the deformed structure of boron-doped graphene increasing the CTR. It was indicated that the CTR mostly affected by deformed structure due to the presence of boron doping.

References

[1] A K Geim, K S Novoselov 2007 *Nat. Mater.* **6** 183–91
[2] Balandin A A, Ghosh S, Bao W, Calizo I, Teweldebrhan D, Miao F and Lau C N 2008 *Nano Lett.* **8** 902–7
[3] Dikin D A, Stankovich S, Zimney E J, Piner R D, Dommett G H B, Evmenenko G, Nguyen S T and Ruoff R S 2007 *Nature* **448** 457–60
[4] Stankovich S, Dikin D A, Dommett G H B, Kohlhaas K M, Zimney E J, Stach E A, Piner R D, Nguyen S B T and Ruoff R S 2006 *Nature* **442** 282–6
[5] Oh W-C and Zhang F-J 2011 *Asian J. Chem.* **23** 875–9
[6] Elias D, Nair R R, Mohiuddin T M G, Morozov S V, Blake P, Halsall M P, Ferarri A C, Boukhvalov D W, Katsnelson M I, Geim A K and Novoselov K S 2009 *Sci.* **323** 610–3
[7] Sunnardianto G K, Maruyama I and Kusakabe K 2017 *Int. J. Hydrogen Energy* **42** 23691–7
[8] Sunnardianto G K, Maruyama I and Kusakabe K 2016 *Adv. Sci. Eng. Med.* **8** 6–11
[9] Zhou Y G, Zu X T, Gao F, Nie J L and Xiao H Y 2009 *J. Appl. Phys.* **105** 0–4
[10] Sunnardianto G K, Maruyama I and Kusakabe K 2016 J. Comput. Theor. Nanosci. **13** 4883–7
[11] Martin R M 2008 Electronic Structure: Basic Theory and Practical Methods (Cambridge University Press
[12] Kresse G and Hafner J 1993 *Phys. Rev. B* **47** 558–61
[13] Blöchl P E 1994 *Phys. Rev. B* **50** 17953–79
[14] Perdew J P and Zunger A 1981 *Phys. Rev. B* **23** 5048–79
[15] Pack J D and Monkhorst H J 1977 *Phys. Rev. B* **16** 1748–9
[16] Sanchez-Portal D, Artacho E and Soler J M 1995 Projection of plane-wave calculations into atomic orbitals Solid State Commun. **95** 685–90

Acknowledgment

The calculations were carried out using the local computer at Sampoerna University and supported by supercomputer facilities of Graduate school of Engineering Science, Osaka University. GKS gratefully acknowledges to Prof. Koichi Kusakabe at Osaka University, Japan for his kind continuous support.

IOP Conf. Series: Materials Science and Engineering **541** (2019) 012005 doi:10.1088/1757-899X/541/1/012005

The effect of single and double vacancy on hydrogenated graphene for hydrogen storage application

M S Yahya[1], G K Sunnardianto[1,2], M Handayani[3]

[1]Faculty of Engineering and Technology, Sampoerna University, Jl. Raya Pasar Minggu, Kav 16, Jakarta, Indonesia
[2]Research Center for Chemistry, Indonesian Institute of Science (LIPI), Kawasan Puspiptek Serpong, Tangerang Selatan, 15314, Indonesia
[3]Research Center for Metallurgy and Material, Indonesian Institute of Science (LIPI), Puspiptek Serpong, Tangerang Selatan, 15314, Indonesia

Email: sulthon_yahya@yahoo.com

Abstract. We investigated the effect of single and double vacancy on the charge transfer rate of dissociation of hydrogen molecule adsorb on graphene based upon density functional theory calculation for hydrogen storage application. Here, we would like to compare the effect of the single and double vacancy on the charge transfer rates of hydrogen molecule dissociation adsorbed on graphene. Before investigated the charge transfer rates, we optimized the structure of graphene, graphene with single and double vacancy and observed the interaction with hydrogen molecule. Preliminary result revealed that once hydrogen molecule is dissociated to be hydrogen atoms and adsorbed to the dangling bond, the charge transfer rates is decrease compared with the charge transfer from hydrogen to pristine graphene indicated that hydrogen atom absorbed on dangdling bond changing the orientation of hydrogen adsorbed. As for the C-H bond length is decrease around to be 1.11 Å compare to the pristine one 1.12 Å. We need to explore more on the reaction pathways to get the global picuture on the effect of single and double vacancy on the dissociation of hydrogen molecule on graphene surface.

1. Introduction

Nowadays, hydrogen-based technology is one of the best alternative energy to decrease the using of fossil energy. Graphene is molecules composed of pure carbon atoms. These atoms are related to each other, forming a two-dimensional hexagonal pattern resembling a honeycomb. In short, it can be said that graphene is a structural unit of graphite. The functionality of hydrogenated graphene was predicted by Roman et al. [1] and Sofo et al. [2] by using density functional theory calculation. Hydrogen molecules is interacting weakly with graphene. In molecular hydrogen adsorption, it is physisorption that occurs primarily, and the dissociation rate of hydrogen molecules on pristine graphene is very low. This is because the energy barriers against dissociating a H_2 molecule on pristine graphene are high [3]. One of facile way to increase the interaction between of hydrogen molecule on graphene surfaces. It has been observed the reaction of hydrogen molecule dissociation on hydrogenated graphene with single vacancy [4,5]. However, to best of our knowledge, the effect of single and double vacancy on the charge transfer rate still not observed yet in detail.

Gagus et al. reported the rule that determined the CTR from hydrogen to pristine graphene over a wide range of coverage [6]. Thus, in this study, we would like to investigate the effect of single and double vacancy from hydrogen to graphene since the interaction between hydrogen and graphene is mediated by CTR.

2. Method

We used density functional theory to performed the structural and electronic structure [7] by using density functional theory calculation framework and Quantum Espresso (QE) code [8]. We used the ultra-soft pseudo potential and the projector-augmented-wave method [9] and adopt the Perdew-Zunger parameterization [10] for local density approximation. The parameter for hydrogenated graphene with a vacancy in super cell were the energy cut off 400 Ry for the expansion of the augmented charge and 40 Ry for the plane wave expansion of the wave function. The total absolute value of the inter-atomic force vector is less than 10^{-4} Ry/a.u. for the convergence criterion.

The distance between graphene planes was separated by 10 Å in order to avoid the interaction between layers. Distribution of k-points on a mesh of 12x12x1 was chosen via Monkhrost-Pack scheme [11] and the structure of the model for the simulation are graphene that used 48 carbon atoms in rectangular super cells for pristine graphene, 47 carbon atoms for graphene single vacancy and 46 carbon atoms for graphene with double vacancy. The charge transfer from hydrogen to graphene is calculated by using lowdin charge analysis [12].

3. Result and Discussion

3.1 Charge transfer rate (CTR)

Hydrogenation should happen both at each vacancy and on the flat graphene surface. For the comparison, we will discuss hydrogenated graphene at first and then hydrogenated graphene vacancy either with single or double vacancy.

We observed the effect of double vacancy on graphene to the charge transfer rates from hydrogen to graphene by comparing the CTR from hydrogen to pristine graphene and graphene with single vacancy.

Table 1. Value of Ave-CTR on Graphene, Graphene with Single and Double Vacancy

Materials	Nc	Ave-CTR [e]
Graphene	48C+3H	0.2035
Graphene single vacancy	47C+3H	0.1864
Graphene double vacancy	46C+3H	0.1869

The results indicated that single vacancy decreasing the CTR from hydrogen to the graphene. As for the di-vacancy in graphene, a similar trend to single vacancy is found for the charge transfer rate. Thus, there is no meaningful changes are seen in the graphene di-vacancy. We could analyze the result of decreasing the CTR from the optimized structure of hydrogenated graphene with single vacancy and hydrogenated graphene with double vacancy.

3.2 Deformed structure

Once the structure of hydrogenated graphene with and without vacancy are optimized, we could predict why the vacancy either single or double decreasing the CTR. The optimized structures can be seen in Figure 1.

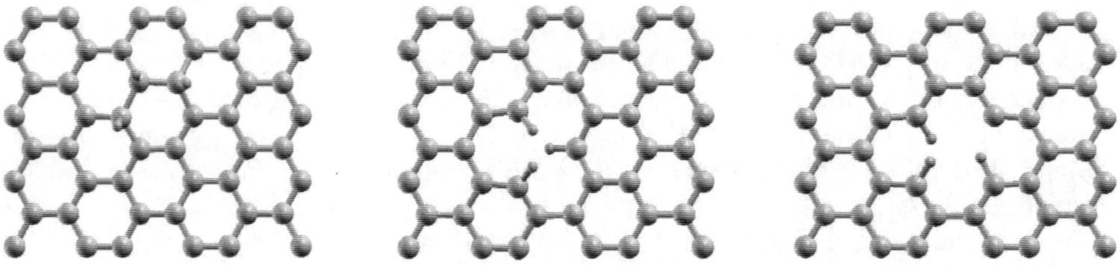

(a) (b) (c)

Figure 1. Structures of hydrogenated pristine graphene (a), hydrogenated graphene with single vacancy (b), and hydrogenated graphene with double vacancy (c). The yellow and blue atoms denote carbon and hydrogen, respectively.

We analyzed from the optimized structure, the selection of structured based on the most stable of hydrogenated graphene with single vacancy, where each carbon atom is mono-hydrogenated at the edge of vacancy (V_{111}), there are 3 hydrogen atoms adsorbed at vacancy site, thus we consider trimer hydrogen adsorption of pristine graphene and graphene with single and double vacancies.

We found the orientation of hydrogen atom adsorb on graphene surface determine the amount of the charge transferred. If the hydrogen orientation in the "up" position (figure 1 (a)), the charge transfer rate is bigger than if the hydrogen atom adsorbed to the vacancy in "side" orientation.

3.3 The formation of C-H bonding

Table 2. The optimized distance *d*C-H and buckling in Graphene, Graphene with Single and Double Vacancy

Materials	***d*C-H (Å)**	**Buckling (Å)**
Graphene	1.1189	0.63
Graphene single vacancy	1.0806	0.23
Graphene double vacancy	1.1073	0.17

The optimized distance *d*C-H of graphene with single vacancy is 1.0806 Å, which is shorter than the *d*C-H of pristine graphene (1.1189 Å). As for the puckered of hydrogenated carbon atom of graphene with single vacancy is 0.23 Å, which is lower than the puckered of hydrogenated carbon atom of pristine graphene (0.63 Å) indicate that the reduced of charge transfer rates in the graphene with single vacancy.

4. Conclusion

We have investigated the effect of single and double vacancy on the charge transfer rate from hydrogen to graphene by using Density Functional Theory. We found that the charge transfer rates and C-H bond length are decrease compared with the pristine one, but we need to explore more about the reaction pathways of double vacancy and charge to system for hydrogen storage application.

IOP Conf. Series: Materials Science and Engineering **541** (2019) 012005 doi:10.1088/1757-899X/541/1/012005

References

[1] Roman T, Dino W A, Nakanishi H, Kasai H, Sugimoto T and Tange K l **1765** 1–4

[2] Sofo J O, Chaudhari A S and Barber G D 2007 1–4

[3] Elias D C, Nair R R, Mohiuddin T M G, Morozov S V., Blake P, Halsall M P, Ferrari A C, Boukhvalov D W, Katsnelson M I, Geim A K and Novoselov K S 2009 **323**

[4] Sunnardianto G K, Maruyama I and Kusakabe K 2017 *Int. J. Hydrogen Energy* **42** 23691–7

[5] Sunnardianto G K, Maruyama I and Kusakabe K 2016 *Adv. Sci. Eng.Med.* **8** 421–6

[6] Sunnardianto G K, Maruyama I and Kusakabe K 2016 *J. Comput. Theor. Nanosci.* **13** 4883–7

[7] Martin R M 2008 *Electronic Structure: Basic Theory and Practical Methods* (England: Cambridge University Press)

[8] Kresse G and Hafner J 1993 *Phys. Rev. B Condens. Matter.* **47 558-561.**

[9] Blochl P E 1994 *Bull. Mater. Sci.* **50** 33–41

[10] Perdew J . 1981 *Phys. Rev. B* **23** 5048

[11] Pack J D and Monkhorst H J 1977 *Phys.Rev. B* **16** 1748–9

[12] Sanchez-Portal D, Artacho E and Soler J M 1995 *Solid State Commun.* **95** 685–90

Acknowledgements

The calculation was carried out using the local computer at Sampoerna University especially for visualization by using XCrySDen and supported by supercomputer facilities of Graduate School of Engineering Science, Osaka University. GKS gratefully acknowledges to Prof. Koichi Kusakabe at Osaka University, Japan for his kind continuous support.

Corrosion Behaviour of Aluminum Alloy AA7075-T651

R Q Dzakyprasetyo and A Anawati*
Department of Physics, Faculty of Mathematics and Natural Sciences, Universitas Indonesia, Depok 16424, Indonesia

*e-mail : anawati@sci.ui.ac.id

Abstract. The present work investigates the corrosion behavior of a high strength aluminum alloy AA7075-T651 which is widely used in the aerospace industry. The alloy composed of Zn, Mg, and Cu as the main alloying elements. The T651 referred to a peak strength temper. X-ray diffraction (XRD) analysis detected $MgZn_2$ as the second phase next to the Al matrix. The corrosion behavior of AA7075 was studied by conducting an immersion test in 30 g/l NaCl and 10 ml/l HCl solution at 35°C based on ASTM B597 for 8, 16, 24 h. The specimen was then characterized by an optical microscope and scanning electron microscope (SEM). The results showed that the specimen experienced a severe corrosion attack as a result of immersion test, as indicated by the change in the surface appearance, the development of surface roughness, as well as significant weight loss. The surface became remarkably rough and grain embrittlement following the metal grain boundaries occurred down to 150 μm depth. The near surface layer became disintegrate and easily detached from the surface after immersion 16 and 24 h. The results indicated that the alloy was sensitive to intergranular corrosion.

1. Introduction

The high strength aluminum alloy AA7075 has been extensively used in aircraft industry owing to its excellent mechanical properties and high specific density [1-3]. The main alloying elements in AA075 are zinc (5.1-6.1 wt.%), magnesium (2.1-2.9 wt.%), and copper (1.2-2.0 wt.%). The microstructure of the alloy contained intermetallic precipitates such as Al_7Cu_2Fe, $(Al, Cu)_6(Fe, Cu)$ phases which were the dominant precipitates, and Mg_2Si which present in small number [3]. The strengthening precipitate was $MgZn_2$ which existed as nanoparticles in the alloy. The high strength property was, however, counterbalance with the susceptibility of the alloy to local corrosion like pitting, intergranular, and exfoliation corrosion [4, 5].

The strengthening precipitates which segregated at the metallic grain boundaries of the alloy affected the resistance to intergranular corrosion. The intermetallic precipitates may change in morphology, composition, and type with thermo-mechanical treatment [6, 7]. The solution treatment and aging modified the distribution and size of intermetallic precipitates. The alteration in the microstructure can be characterized by measuring the electrical conductivity and mechanical hardness tests [3]. Meanwhile, the corrosion behavior was studied by performing an aqueous immersion test in chloride containing solution [8]. At peak strength condition of T6 and at overage T76 tempers, the precipitates Al_7Cu_2Fe, $(Al, Cu)_6(Fe, Cu)$ phases induced pitting corrosion in the alloy [3]. The pit initiated due to the galvanic coupling of the intermetallic phases with the surrounding matrix. The potential difference between the intermetallic and matrix promote microgalvanic corrosion where the phase with lower potential corroded selectively. The Cu-rich phase was detrimental to most of Al alloys as the phase tended to segregate along grain boundaries and induced corrosion of the surrounding matrix which depleted in Cu [6, 7]. The Cu-rich intermetallic is unavoidably formed in aluminum alloy although present in low concentration [9, 10].

The current work investigates the corrosion behavior of AA7075 alloy in the T651 temper condition. The T651 indicated the basic temper T6 that the alloy has been solution treated, quenched, and artificially aged and in addition, the alloy has been stress relieved by stretching ½ to 2%. The main purpose was to clarify the dominant type of corrosion occurred in the alloy. The corrosion behavior was studied by performing an immersion test in chloride solution.

2. Experimental procedure

The material used was a rolled plate AA7075-T651 with a thickness of 2 mm. The plate was cut to give an area of 2.6 cm^2. The specimens were ground with #600 to #1500 grit paper followed by polishing with 1 μm diamond paste. Afterward, the specimens were degreased in acetone and ethanol for 3 min in an ultrasonic bath.

For microstructure investigation, the specimen was etched in 10 wt% NaOH solution for 2 min. The metallographic microstructure was studied by using an optical microscope Olympus CX31. For a higher magnification image, observation on the specimen morphology was performed by using FE-SEM (FEI INSPECT F-50). The elemental composition of the alloy was analyzed by EDX (EDAX EDS Analyzer).

The corrosion behavior of the specimen was examined by weight loss method in a mixture solution of 30 g/l NaCl and 10 ml/l HCl solution at 35°C for 8, 16, 24 h, based on ASTM B597 [11]. After immersion, the specimen was cleaned in running water followed by DI water and then dried in an air stream.

3. Results and Discussion

The microstructure of the as-polished specimen was studied by an optical microscope and the result is presented in figure 1. The microstructure consisted of numerous fine intermetallic particles as indicated by the dark spots with the size in the range 5-10 μm distributed randomly as shown in figure 1. Some of them were agglomerated to form bigger spots. The spots are voids where previously occupied by the intermetallic particles. The particles were possibly detached from the surface during etching and desmutting. The remaining intermetallic particles were observed as white particles, pointed by arrows in Fig. 1. The small spherical particles were in the size of few μm while the bigger particles might reach several 10th μm. The metallic grain boundaries were quite small between 10-20 μm. The grain boundaries which appeared dark were also occupied by continuous segregated intermetallic particles. The typical grain boundaries precipitate in the AA7075 alloy is the strengthening $MgZn_2$ phase [3].

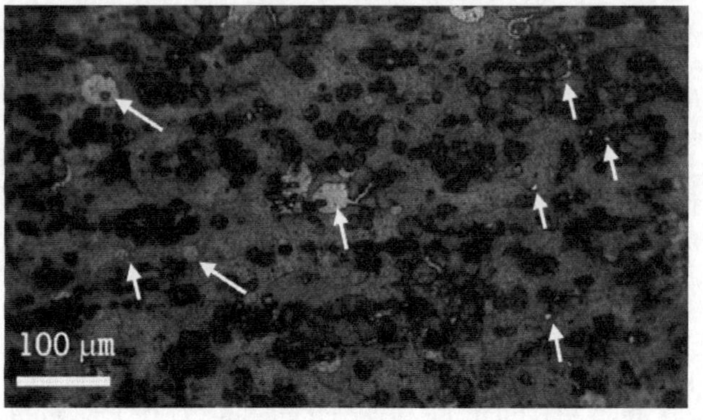

Figure 1. The optical microscope image of the as-polished AA7075 surface.

IOP Conf. Series: Materials Science and Engineering **541** (2019) 012006 doi:10.1088/1757-899X/541/1/012006

The elemental composition of the alloy is listed in Table 1 which is obtained from the EDX analysis. The main alloying elements detected in the alloy are Mg, Cu, and Zn and some impurities of Fe and Mn. The signal from Zn was rather low probably existed in solid solution matrix or as small precipitates. The result indicated that the intermetallic precipitates shown in Fig. 1 was mainly composed of Cu-rich and Mg-rich phases. The phases existed in the AA7075 alloy are $MgZn_2$, Al_2CuMg, and Al_7Cu_2Fe [12].

Table 1. The composition of AA7075 analyzed by EDX.

Element	Wt%	At%
Mg	2.64	2.82
Si	0	0
Mn	0.52	0.24
Fe	0.25	0.12
Cu	2.42	0.99
Zn	0.29	0.12
Al	Bal.	Bal.

The corrosion mechanism of the AA7075 alloy was studied by performing an immersion test in NaCl-HCl solution at a temperature of 35°C for 8, 16, and 24 h. The result of weight loss measurements is plotted as a function of time as shown in Fig. 2 while the corresponding FE-SEM cross-sectional corrosion morphologies are displayed in Fig. 3. The weight loss increased linearly with exposure time with the slope of 27 $mg.cm^{-2}.h^{-1}$. The alloy showed weight losses of 0.01, 0.04, and 0.07 $g.cm^{-2}$ after an immersion time of 8, 16, and 24 h, respectively.

The images of cross-section corrosion morphology as displayed in Fig. 3 confirmed the weight loss results in Fig. 2. A significant corrosion attack along grain boundaries was already observed after 8 h immersion. The depth of attack was about 150 μm from the specimen surface. The grain boundaries were corroded as indicated by thicker and darker boundaries relative to the uncorroded parts. Prior to the corrosion test, the grain boundaries of the as-polished alloy were occupied by either cathodic or anodic phases. The type of intermetallic precipitates will be investigated in our future work. The used of electrochemical microcell system was proved to be the useful method to investigate local corrosion phenomena induced by small precipitates [13].

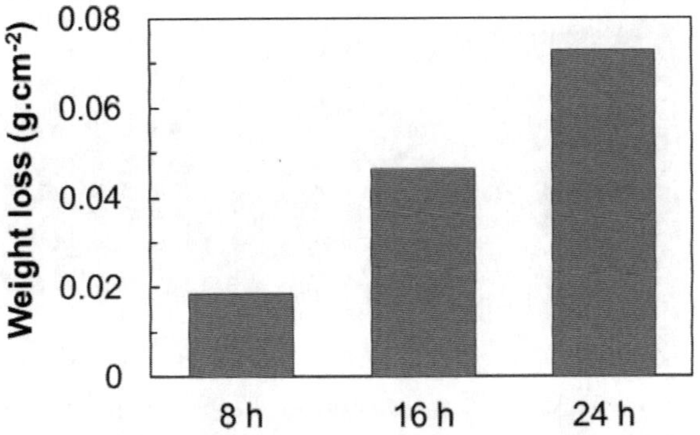

Figure 2. The weight loss of AA7075 after immersion in 3.5% NaCl solution for 8, 16, and 24 h.

The precipitates existed in aluminum alloys typically have different potential, higher or lower than the solid solution matrix. The potential gradient causes galvanic coupling which promotes galvanic corrosion. The matrix next to the cathodic precipitates was susceptible to corrosion while the anodic precipitates tend to corrode than the matrix. The corrosion morphology of AA7075 alloy shown in Fig. 3 indicated that the precipitates at grain boundaries were anodic type corresponded to the $MgZn_2$ phase since the grain boundaries were selectively attacked by corrosion leaving corrosion products. At an extended immersion time of 16 and 24 h, the depth of attack was relatively the same as that of 8 h but the degree of corrosion attack increased. The outer layer which was severely corroded was detached from the specimen surface resulting in a rough surface layer covered by the corrosion product. The corrosion product appeared as darker areas relative to the surrounding in Fig. 3. The corrosion product was presumably composed of aluminum oxide and chloride complexes similar to the reported work in ref [7]. The metallic grains near the surface were disintegrated as the areas along grain boundaries suffered from severe corrosion. The grain boundaries embrittlement occurred down to ~100 μm depth from the surface after 16 h immersion while deeper ~200 μm after 24 h exposure time. The corrosion test results indicated that the AA7075-T651 alloy was susceptible to intergranular corrosion.

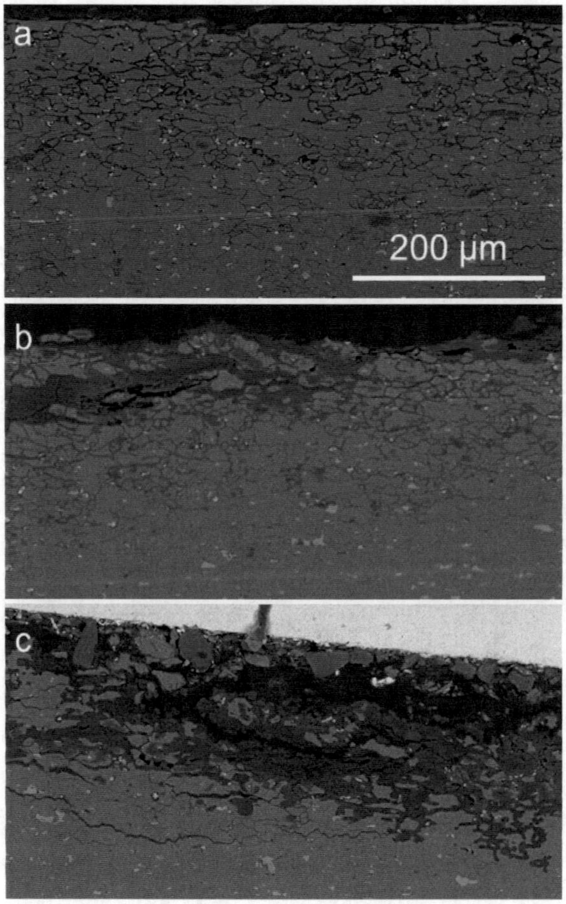

Figure 3. The cross-section FE-SEM images of AA7075 specimen after immersion in 3.5% NaCl for a) 8 h, b) 16 h, and c) 24 h.

4. Conclusion

The corrosion study on AA7075-T651 alloy revealed that the metallic grain boundaries were sensitized by the segregation of intermetallic phases along the grain boundaries. The microstructure study indicated that the phase existed along the grain boundaries was mainly the strengthening phase $MgZn_2$ which is more anodic than the surrounding Al matrix while other intermetallic precipitates such as Cu-rich and Fe-rich phases distributed randomly in the alloy as spherical particles in the order of 10 μm size. The grain boundaries were susceptible to intergranular corrosion. The $MgZn_2$ phase which segregated along the grain boundaries was selectively corroded during exposure in aqueous corrosive solution. At a short exposure time of 8 h, corrosion led to intergranular attack down to 150 μm depth from the surface. The grain boundaries were disintegrated and further detached from the surface after prolong exposure time of 16 and 24 h. A corrosion prevention method is necessary to be developed for the AA7075 alloy to protect the alloy from intergranular corrosion.

References

[1] Luiggi N J, Valera M V 2017 J. *Therm. Anal. Calorim.* **130** 1885-1902
[2] Tiringer U, Kovac J, Milosev I 2017 *Corros. Sci.* **119** 46-59
[3] Andreatta F, Terryn H, Wit J H W 2003 *Corros. Sci.* **45** 1733-1746
[4] Ngai S, Ngai T, Vogel F, Story W, Thompson G B, Brewer L N 2018 *Corros. Sci.* **130** 231-240
[5] Thomson M S, Frankel G S 2017 *Electrochem. Soc.* **164(12)** C653-C663
[6] Svenningsen G, Larsen M H, Walmsley J C, Nordlien J H, Nisancioglu K 2006 *Corros. Sci.* **48** 1528-1543
[7] Svenningsen G, Larsen M H, Nordlien J H, Nisancioglu K 2006 *Corros. Sci.* 258-272
[8] Liu L L, Pan Q L, Wang X D, Xiong S W 2018 *J. Alloys Compd.* **735** 261-276
[9] Anawati A, Diplas S, Nisancioglu K 2011 *Electrochem. Soc.* **158(5)** C158-C163
[10] Anawati A, Diplas S, Holme B, Solberg J K, Mathiesen R H, Walmsley J C, Nisancioglu K 2011 *Electrochem. Soc.* **158 (6)** C178-184
[11] ASTM B597-90 (1990), Standard Practice for Heat Treatment of Aluminum Alloys, ASTM, Philadelphia, PA.
[12] Binesh B, Aghaie-Khafri M 2016 *Metals* **6** 42
[13] Andreatta F, L. Fedrizzi 2016 *Electrochim. Acta* **203** 337-349

Acknowledgement

This work was funded by Hibah Publikasi Internasional Terindeks untuk Tugas Akhir Mahasiswa (PITTA) Universitas Indonesia with contract no. 2227/UN2.R3.1/HKP.05.00/2018.

Effect of Heat Treatment on Microstructure and mechanical hardness of aluminum alloy AA7075

A N Aliyah, A Anawati*
Department of Physics, Faculty of Mathematics and Natural Sciences, Universitas Indonesia, Depok 16424, Indonesia.

*E-mail: anawati@sci.ui.ac.id

Abstract. The mechanical properties of high strength aluminum alloy AA7075 were greatly affected by heat treatment. This paper was intended to clarify the effect of heat treatment on the microstructure and the mechanical hardness of commercial alloy AA7075-T651. Heat treatment was performed at a variation temperature of 300, 400, 500, and 600°C for 1 h in a vacuum furnace followed by a water quench. The surface observation was done by using optical and electron microscopy and the mechanical hardness was measured by the Vickers Hardness test. The XRD result on as-received alloy revealed the precipitates in the alloy was $MgZn_2$, Al_7Cu_2Fe, and Al_2CuMg. The average hardness of the as-received alloy was 136 HV. The hardness became significantly lower (78.5 HV) after heat treatment at 400°C while heat treatment at others temperature gave hardness in the range 116-122 HV. Heat treatment induced grain growth and dissolution of metastable precipitates. Reduction in the number of $MgZn_2$ and Fe-rich phase decreased the hardness significantly. Heat treatment above 400°C promoted segregation of fresh precipitates which enhanced the metal hardness again. The metallic grain size increased with heat treatment temperature. At 600°C, most of the precipitates were segregated along the equiaxed grain boundaries.

1. Introduction

The high strength aluminum alloy AA7075 has been widely used as a structural material in the aerospace industry. The mechanical characteristic of high strength to weight ratio was obtained by the precipitation hardening mechanism. The alloy composed of alloying element Zn, Mg, and Cu. The intermetallic and strengthening precipitates were formed in the alloy as a result of interaction between the alloying elements and impurities in the alloy [1]. The main precipitates existed in the alloy are the Cu- and Fe-rich phases which include Al_7Cu_2Fe and $(Al, Cu)_6(Fe, Cu)$ with the size in the range of 1-20 μm [1-3]. The strengthening precipitate in the alloy was the $MgZn_2$ phase which segregated as small particles in the order of a few nm at the metallic nanograin boundaries [3]. The other precipitates such as Al_2CuMg and Mg_2Si were present in small quantities.

The microstructure of AA7075 alloy was sensitive to heat treatment. The heat treatment and aging were reported to change the size, composition, and distribution of the intermetallic precipitates in the alloy [1, 2]. The microstructure change can be monitored by measuring the electrical conductivities and tensile strength. The Cu- and Fe-rich phases exhibited higher potential than that of the Al matrix while the Mg-rich phases are more anodic to the matrix. The presence of nobler elements in the Al tends to enhance the potential of the alloy [4]. Solution heat treatment tended to enhance the potential difference between the intermetallic and the surrounding matrix of AA7075-T6 as proved by the detail studied using electrochemical micro-cell as explained in ref [5] and by a more advanced technique of Scanning

Kelvin Probe (SKP) describe in ref [6]. The change in microstructure in relation to the distribution of the cathodic and anodic intermetallic precipitates can be characterized by measuring the electrical conductivity of the alloy. Solution heat treatment also induced dissolution of the strengthening phase of $MgZn_2$ in the Al matrix which led to the supersaturated solid solution [1]. The supersaturated solid solution of AA7075-T651 was characterized by a higher conductivity relative to the T6 temper due to the high concentration of Mg, Zn, Cu dissolved in the solid solution [3]. The potential gradient between the intermetallic and Al matrix enhanced the susceptibility to galvanic corrosion and pitting [5-7]. The Cu-rich phase, in particular, is detrimental to corrosion attack in Al alloys although present in a small concentration [8].

Other than conductivity tests, the tensile strength measurement was also utilized to monitor the microstructure change in the Al alloy as a result of heat treatment and aging. Heat treatment followed by water quench resulted in a lower mechanical strength of the supersaturated solid solution AA7xxx series than that of the T6 temper [9]. This work is intended to clarify if the alteration in the microstructure as a result of heat treatment can be characterized by measuring the microhardness of the alloy. Heat treatment was conducted on AA7075-T651 at various temperature of 300°, 400°, 500°, and 600°C.

2. Experimental Methods

The material used is a commercial rolled plate AA7075-T651 having a thickness of 2 mm. The plate was cut to give an area of 2 cm^2. The specimens were ground from #600 to #1200 grit papers. Finally, the specimens were polished by using an alumina paste to get a mirror finished. Prior to heat treatment, the specimens were cleaned in acetone followed by ethanol in an ultrasonic bath for 3 min each step.

Heat treatment was performed at 300, 400, 500, and 600°C for 1 h in a furnace. After heat treatment, the specimens were subsequently quenched in a cold water. To minimize aging, the heat treated specimens were stored in a freezer before further used.

For microstructure observation, the specimens were electropolished in 20% $HClO_4$ solution using 0.25 A for 4 min at 10°C and then degreased in 30% HNO_3 solution for 20 s. The microstructure was studied by an optical microscope (Zeiss) and FE-SEM-EDX (FEI INSPECT F-50 & EDAX EDS Analyzer). The crystalline phases existed in the heat treated specimens were analyzed by XRD (Panalytical X'Pert Pro MPD). For measuring the mechanical hardness, Vickers hardness indenter (Struers DuraScan) was utilized by using a 3 kgf load. Five indentations were conducted on each specimen to obtain the average value.

3. Results and Discussion

Figure 1 shows the microstructure development occurred on the AA7075 specimens as a result of heat treatment at 300, 400, 500, and 600°C for 1 h. All of the specimens were priory electropolished to reveal the grain boundaries and the intermetallic precipitates. The as-received specimen and the specimens heat treated below 600°C did not show a clear grain boundary (Figs. 1a-1d). The grain boundaries became clearly observed after heat treatment at 600°C as they were occupied by the segregation of intermetallic precipitates as shown in Fig. 1e.

The as-received alloy was hardened by intermetallic precipitates which included $MgZn_2$, Al_2CuMg, and Al_7Cu_2Fe, pointed by arrows in Fig. 1a. The intermetallic composition was confirmed by EDX analysis. The precipitate Al_7Cu_2Fe showed an irregular shape while the Al_2CuMg and $MgZn_2$ phases tended to have a spherical shape (Fig. 1). In agreement with the reported work in ref [10] that the Al_7Cu_2Fe, $Al_6(Cu, Fe)$ exhibited irregular shape while the Al_2CuMg was observed as either irregular or spherically shaped particles. Heat treatment at 300°C induced the formation of fine precipitations as indicated by the numerous black spots in the surrounding larger precipitates (Fig. 1b). The voids which appeared as black spots were formerly occupied by intermetallic which probably detached during electropolishing. The Mg-rich phase is more anodic to Al [3] and therefore easier to dissolve into the solution during electropolishing. Heat treatment at a higher temperature of 400°C enlarged the intermetallic precipitates size and the surrounding matrix became clearer from small precipitates (Fig. 1c). The big precipitates coarsened in the expense of the smaller ones. Increasing heat treatment temperature to 500°C promoted the formation of fresh intermetallic precipitates as indicated by the numerous fine black spots in the perimeter of the bigger precipitates (Fig. 1d). The microstructure of the specimens resulted from heat treatment at 500°C was somewhat similar to that of obtained after heat treatment at 300°C (Fig. 1b). Heat treatment at 600°C led to a significant metallic grain growth with an equiaxed shape as can be seen in Fig. 1e. The recrystallization of metallic grain to result in an equiaxed grain type was only obtained in the AA7075 alloy as a result of heat treatment at a high temperature of 600°C [10]. The grain size was in the range of 100-200 μm. The Cu-rich intermetallic precipitates were segregated along the grain boundaries forming a nearly continuous network. The grain bodies were relatively free from precipitates as shown in Fig. 1e.

The coarsening of the microstructure as a result of heat treatment in the AA7075 specimens followed the Ostwald ripening mechanism where the larger precipitates grew at the expense of the small one [11]. The driving force for such a phenomenon is the gradient of solubility of the solute atom (Cu and Zn) between the big and small precipitates. At an increasing temperature, the solute atoms move from the small to the big precipitates by a diffusion mechanism.

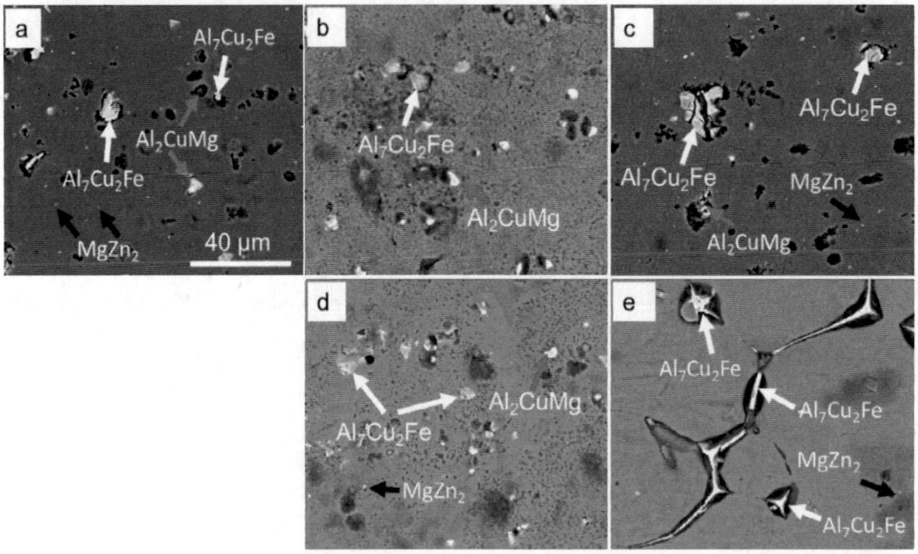

Figure 1. FE-SEM images of the AA7075 specimens showing the microstructure of a) as-received, and after heat treatment at b) 300°C, c) 400°C, d) 500°C, and e) 600°C.

Confirming the EDX analysis results, Figure 2 presents the XRD pattern of AA7075 after heat treatment at 300, 400, 500, and 600°C showing the phases existed in the alloys in comparison to the as-received one. All of the curves showed a similar serial peaks position for the Al matrix phase at the angles of 38.4°, 44.6°, 64.9°, 78.1°, and 82.4°. In addition to the Al phase, the as-received specimen composed of hardening precipitates $MgZn_2$, Al_2CuMg, and Al_7Cu_2Fe as indicated by the small peaks in between the angle 15°-85° in the XRD pattern. The peaks for Al_7Cu_2Fe phase at the angles of 39.0, 65.0. 78.2° and 82.6° were slightly overlapped with the peaks for Al. The $MgZn_2$, Al_2CuMg, and Al_7Cu_2Fe phases remained in the alloy after the heat treatment of 300, 400, and 500°C. However, the specimen heat-treated at 400°C showed more peaks for Al_2CuMg especially at low angles between 15° and 30°. The corresponding peaks did not exist in the specimen heat-treated at 300° and 500°. In agreement to the microstructure observation, heat treatment at 300 and 500°C induced the growth of $MgZn_2$ and Al_7Cu_2Fe precipitate while oppositely heat treatment at 400°C enlarged the Al_2CuMg precipitate. The peaks for $MgZn_2$ and Al_7Cu_2Fe phases partly disappeared for the specimen heat-treated at 400°C. Heat treatment at the highest temperature of 600°C left mainly $MgZn_2$ and Al_7Cu_2Fe precipitates as shown in the Fig. 2. The result which confirmed the FE-SEM and EDX results shown in Fig. 1. The Al_2CuMg phase was no longer detected at the specimen heated at 600°C.

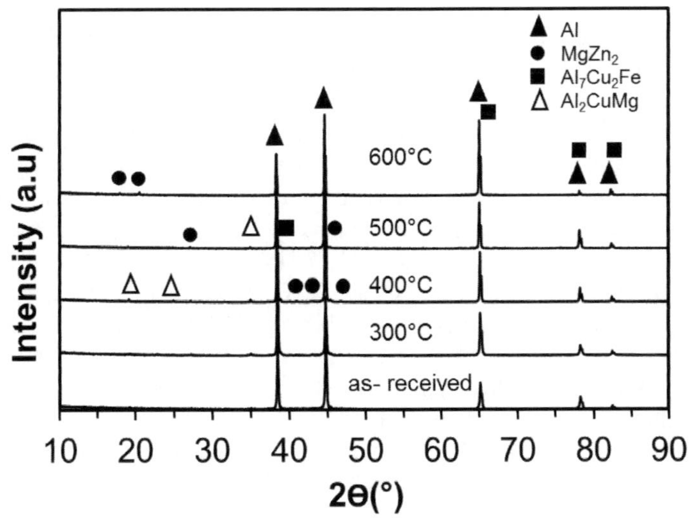

Figure 2. The XRD pattern of AA7075-T651 after heat treatment at 300, 400, 500, and 600°C for 1 h.

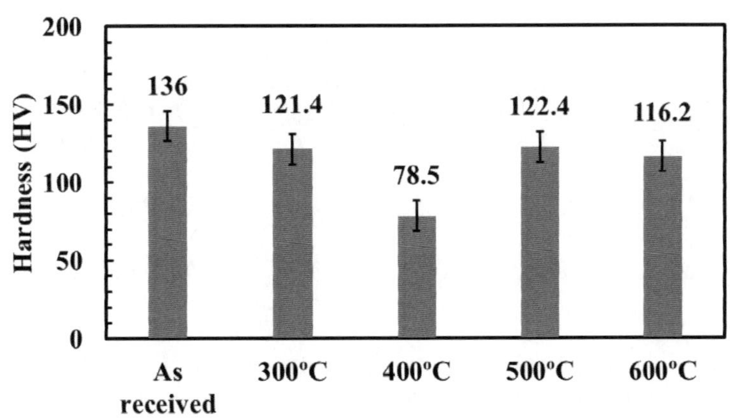

Figure 3. Mechanical hardness of AA7075-T651 after heat treatment at 300, 400, 500, and 600°C for 1 h.

Figure 3 shows the effect of heat treatment on the mechanical hardness of AA7075 specimens. The as-received alloy exhibited an average hardness of 136 HV. In general, the hardness tended to be lower as a result of heat treatment. Similar to the reported [1, 2, 9] investigation on the effect of solution heat treatment, the microhardness decreased as a result of heat treatment. The specimens which heat-treated at 300 and 500°C had a nearly similar hardness of 121.4 and 122.4 HV, respectively, as expected since they exhibited similar microstructure (Figs. 1 and 2). Heat treatment at 400°C resulted in the lowest hardness of 78.5 HV which was nearly half of the as-received specimens. Significant reduction in the number of $MgZn_2$ and Fe-rich precipitate was accused to be the reason for the corresponding hardness depression. The specimen which was heat treated at 600°C exhibited a hardness of 116.2 HV which was slightly lower than that of heat-treated at 300 and 500°C. The hardening precipitates were not distributed in the metallic grain as a result of heat treatment at 600°C but aggregated as a continuous precipitate along the grain boundaries.

4. Conclusion

The effect of heat treatment on the microstructure of AA7075 alloy has been investigated. The results showed that heat treatment at temperature 300-600°C gave a significant impact on the microstructure transformation and the resulting mechanical hardness of the alloy. A slight reduction (~15 HV) in the hardness was obtained as a result of heat treatment at 300° and 500°C. At both temperatures, the Fe-rich precipitates grew larger while the Mg-rich precipitates became smaller following the Ostwald ripening mechanism. Oppositely at 400°C, the dominant precipitate observed in the alloy was the Mg-rich phase which formed a big particle of about 10 μm while the matrix was relatively clean from small precipitates. Such microstructure resulted in a low hardness of 78.5 HV. A significant different microstructure was attained as a result of heat treatment at 600°C where the Cu-rich precipitates segregated along the grain boundaries forming a nearly continuous layer while the grain interior was relatively free from precipitates. The segregation of precipitates along grain boundaries did not give a significant contribution on the hardness of the bulk specimen and therefore the hardness was slightly lower (116.2 HV) than that of resulted from 300° and 500°C heat treatment (121-122 HV). The results indicated that the transformation in the microstructure of AA7075 alloy can be characterized by measuring the microhardness of the alloy.

References

[1] Andreatta F, Terryn H Wit J H W *Corros. Sci.* **45** 2003 1733-1746.
[2] Luiggi N J, Valera M V J. Therm. Anal. Calorim. **130** 2017 1885-1902.
[3] Tiringer U, Kovac J, Milosev I *Corros. Sci.* **119** 2017 46-59.
[4] Anawati A, Diplas S, Nisancioglu K *Electrochem. Soc.* **158(5)** 2011 C158-C163.
[5] Andreatta F, L. Fedrizzi *Electrochim. Acta* **203** 2016 337-349.
[6] Thomson M S, Frankel G S *Electrochem. Soc.* **164(12)** 2017 C653-C663.
[7] Ngai S, Ngai T, Vogel F, Story W, Thompson G B, Brewer L N *Corros. Sci.* **130** 2018 231-240.
[8] Svenningsen G, Larsen M H, Walmsley J C, Nordlien J H, Nisancioglu K *Corros. Sci.* **48** 2006 1528-1543.
[9] Liu L L, Pan Q L, Wang X D, Xiong S W *J. Alloys Compd.* **735** 2018 261-276.
[10] Binesh B, Aghaie-Khafri M *Metals* 6 **42** 2016.
[11] Hosford W F 2010 *Physical metallurgy 2nd Ed.* (CRC press; Taylor & Francis Group).

Acknowledgement

This work was fully supported by Hibah Publikasi Internasional Terindeks untuk Tugas Akhir Mahasiswa (PITTA) Universitas Indonesia with contract no. 2227/UN2.R3.1/HKP.05.00/2018.

IOP Conf. Series: Materials Science and Engineering **541** (2019) 012008 doi:10.1088/1757-899X/541/1/012008

Effect of solution treatment on microstructure and corrosion behaviour of Mg-9Al-1Zn alloys

R Naldi and **A Anawati***
Department of Physics, Faculty of Mathematics and Natural Sciences, Universitas Indonesia, Depok 16424, Indonesia

*E-mail: anawati@sci.ui.ac.id

Abstract. The effect of solution treatment on the microstructure and corrosion behavior of cast Mg-9Al-1Zn (AZ91) were studied by using electro-optical microscopy, X-ray diffraction (XRD) technique, and weight loss method. The solution treatment was done at 415°C for 2 h followed by a water quench. The as-cast AZ91 alloy composed of α-Mg phase and β-phase ($Mg_{17}Al_{12}$). Most of the β-phase was segregated along the grain boundaries. A significant reduction in the volume fraction of β-phase was obtained as a result of solution treatment. The β-phase precipitate was distributed randomly both at grain boundaries and in the grain interior as discrete particles of a few μm sizes. The solution treatment induced grain growth resulting in a significantly larger equiaxed grain in the range 100-500 μm. A higher corrosion rate was obtained for the solution treated alloy than the untreated one. The β-phase precipitate played a dual role in the corrosion of AZ91 alloy depending on its volume fraction in the alloy. A nearly continuous distribution of β-phase along grain boundaries in the as-cast alloy served as a barrier for corrosion across grain while the distribution of discrete particles in the solution-treated alloy accelerated corrosion in the surrounding matrix.

1. Introduction

Magnesium (Mg) has attracted considerable attention in engineering application particularly in the automotive and aerospace industry where the high strength to weight ratio is demanding [1]. Mg is the lightest metal on earth with a low density of 1.7 g/cm^3 having a high strength, great damping capability, and excellent fluidity in casting [2]. Mg is usually used in the form of an alloy which exhibited an improved in the mechanical and corrosion properties. The conventional Mg alloys typically contain alloying elements Al, Zn, Mn, Fe, Cu, and Si [2]. The recent development of Mg alloy uses rare earth elements to boost the mechanical and heat properties [3, 4]. The AZ91 alloy which composed of 9% Al and 1% Zn is widely investigated considering the potential application for low-temperature application [5, 6].

The microstructure affects the mechanical properties and corrosion behavior of Mg alloys. The conventional AZ series alloys typically consist of an α-Mg matrix which is decorated by intermetallic precipitates β-phase ($Mg_{17}Al_{12}$). The strengthening and hardening of Mg alloys are performed by solution treatment [6-8]. The strengthening and hardening mechanism was controlled by the number, size, and distribution of the intermetallic particles in the alloy [8]. From the corrosion viewpoint, the intermetallic precipitates existed in Mg alloys play role in either accelerating the corrosion rate or reducing the corrosion propagation depending on the distribution and the volume fraction of the intermetallic precipitates [7-9]. The precipitates which mostly are more cathodic than the Mg matrix induced potential gradient which accelerated local corrosion in the surrounding matrix. The micro galvanic corrosion is the common type of corrosion occurred in the Mg alloy [7, 10].

This work is aimed to throw lights on the effect of solution treatment on the development of microstructure in the alloy and its role in determining the mechanical hardness and corrosion behavior of as-cast Mg-9Al-1Zn. The corrosion behavior was studied by measuring weight loss as a result of immersion in NaCl solution.

2. Experimental procedure

The material used was as-cast Mg-9Al-1Zn prepared by gravity casting producing a cylindrical ingot of 15 cm long and 2 cm in diameter. The as-cast alloy was cut into a disk with 1 cm thickness. Some of the specimens were heat treated. The solution treatment was conducted in a furnace at a temperature of 415°C for 2 h followed by water quenching. Prior further used, the specimens were ground with #600, #800, and #1000 grit papers followed by polishing with alumina paste to give a mirror reflection. The polished specimens were then degreased in acetone and subsequently in ethanol in an ultrasonic bath for 3 min for each step.

For observing the metallurgical microstructure, the specimen was etched in 4% HNO_3 solution for 20 s to reveal the metallic grain. The specimen was then rinsed in DI water and then dried in an air stream. The microstructure was studied by an optical microscope of CX31 from Olympus. The alloy composition was analyzed by X-ray diffraction spectroscopy (XRD, Panalytical X'Pert Pro MPD). The mechanical hardness was determined by Vickers Microhardness tester (Struers DuraScan) by using 2 kgf load. The average hardness was obtained from 5 measurement points on the specimen.

The corrosion behavior of both as-cast and solution-treated specimens was studied by conducting an immersion test in 3.5% NaCl solution at 30°C. The specimen was immersed individually in 10 ml solution in a glass bottle inside a water bath controller. The average corrosion rate was determined based on the weight loss measured after 24, 48, and 72 h immersion.

3. Results and Discussion

The microstructure of the as-cast and solution-treated Mg-9Al-1Zn alloys was investigated by an optical microscope. The results are presented in Fig. 1. The as-cast alloy consisted of α-Mg, α-eutectic, and intermetallic precipitate β-phase ($Mg_{17}Al_{12}$). The as-cast alloy showed a relatively small metallic grain boundary below 100 μm size. The grain boundaries were occupied by a nearly continuous β-phase ($Mg_{17}Al_{12}$) which surrounded by the α-eutectic phase, while the matrix composed of α-Mg phase. The grain size became significantly larger in the range 200-600 μm after solution treatment as shown in Fig. 1b. Enlargement of grain size as a result of heat treatment was reported [6]. Most of the β-phase dissolved in the solid solution matrix as a result of solution treatment. The remaining β-phase intermetallic segregated at the grain boundaries and decorated the grain interior as small spherical particles. The solution treatment resulted in a remarkably lower volume fraction of β-phase relative to the as-cast.

Confirming the microstructure observation, the XRD patterns of the as-cast and solution-treated Mg-9Al-1Zn alloys revealed that the main composition of the alloys was α-Mg and β-$Mg_{17}Al_{12}$ phases as shown in figure 2. The existence of α-phase was recognized by the serial distinct peaks at the angles 32.6°, 34.7°, 37.0°, 48.4°, 58.1°, 63.8°, 68.3°, 69.4°, 70.9°, and 79.0° in the XRD curves for both as-cast and solution-treated alloys. The peaks intensity of the α-phase was higher for the solution-treated alloy indicating larger crystalline phase than that for the as-cast one which confirmed the microstructure observation in figure 1. The peaks corresponded to β-phase was rather small although unambiguously resolved at the angles of 34.4°, 36.1°, 40.4°, 44.8°, and 65.4° for the as-cast alloy. Some peaks for β-phase did not arise in the curve of solution-treated alloy indicating less number of β-phase. The solution treatment reduced the number of β-phase precipitates similar to the reported work in ref [6].

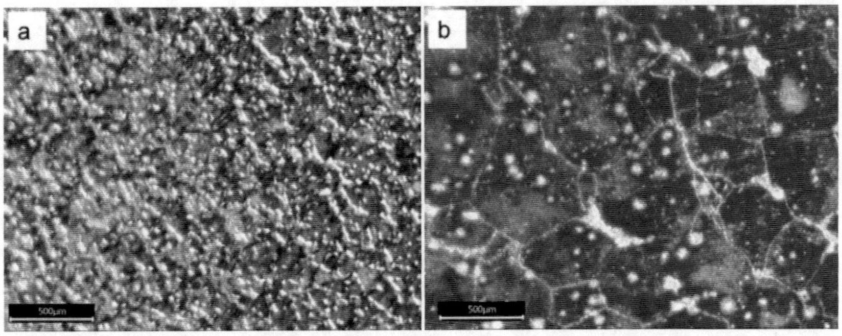

Figure 1. The optical microscope images showing the microstructure of a) as-cast and b) solution-treated Mg-9Al-1Zn alloys.

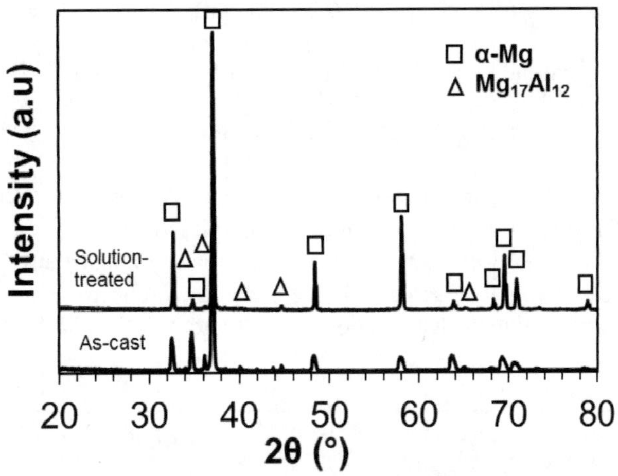

Figure 2. X-ray diffraction pattern of as-cast and solution-treated Mg-9Al-1Zn alloys.

The effect of solution treatment on the mechanical hardness of as-cast Mg-9Al-1Zn alloy was not significant. The hardness decreased slightly as a result of solution-treated as displayed in figure 3. The as-cast alloy exhibited an average hardness of 61.86 HV while the solution-treated alloy showed a hardness value of 60.66 HV. Reduction of hardness as a result of solution treatment was due to the dissolution of intermetallic precipitate β-phase into solid solution matrix. The precipitate played a role as a hardening phase in the alloy. The release of residual stress during solution treatment also contributed to the reduction of the mechanical hardness of the alloy [11].

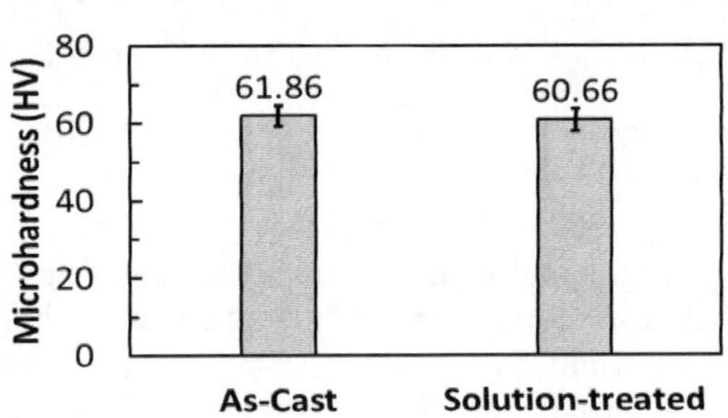

Figure 3. The mechanical hardness of as-cast and solution-treated Mg-9Al-1Zn alloys.

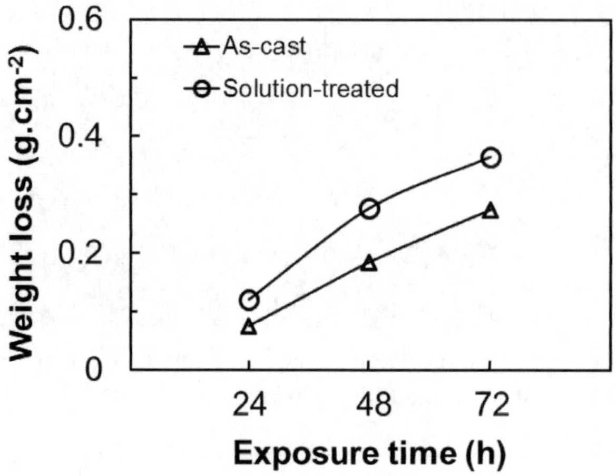

Figure 4. The weight loss plot of as-cast and solution-treated Mg-9Al-1Zn alloys as function of exposure time in 3.5% NaCl solution.

Figure 4 shows the results of the corrosion test by measuring weight loss after 24, 48, and 72 h immersion in 3.5% NaCl solution at 30°C. Both specimens showed a linear increase in weight loss with exposure time but at a different slope. The solution treated alloy consistently showed higher material loss than that of the as-cast alloy. The as-cast alloy exhibited weight loss rate of 0.045 $g.cm^{-2}.h^{-1}$ while the rate for solution-treated specimen was higher at 0.056 $g.cm^{-2}.h^{-1}$ indicating higher dissolution than the as-cast one.

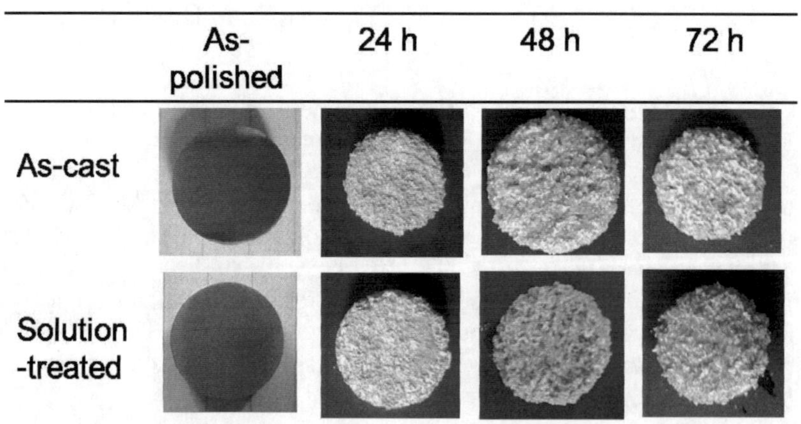

Figure 5. The specimen appearance after immersion test in 3.5% NaCl solution.

Figure 5 shows the specimen appearance after the immersion test for 24, 48, and 72 h in 3.5% NaCl solution at 30°C. The as-polished surface which exhibited a metallic reflection turned to a whitish rough surface after immersion for 24 h and longer. In agreement with the weight loss data presented in figure 4, the specimen already suffered from harsh corrosion attack after 24 h immersion. The material loss increased significantly with extension immersion time as indicated by decreasing in the specimen thickness and diameter. The specimen dimension decreased faster in the solution-treated specimen than in the as-cast one. Significant material dissolution occurred during immersion as indicated by the accumulation of white deposit in the solution at the bottom of the glass. During drying of the specimen after the 72 h test was terminated, the specimen was easily disintegrated into powder. The specimen became spongy after 48 and 72 h immersion. The spongy structure was obtained as a result of selective corrosion of the α-Mg matrix leaving the β-phase. The corrosion test results demonstrated that the Mg-9Al-1Zn was susceptible to micro galvanic corrosion due to the potential gradient between the α-Mg matrix and the β-phase intermetallic precipitate. However, the β-phase precipitate also contributed to providing a barrier for corrosion propagation across the grain when forming a continuous network along grain boundaries. Therefore, the as-cast alloy had a lower weight loss rate relative to the as-cast alloy. The dual role of β-phase precipitates had been observed in AZ series alloys [1].

4. Conclusion

The effect of solution treatment on the microstructure development and the weight loss rate during the corrosion test of as-cast Mg-9Al-1Zn alloy has been investigated. The metallographic observation revealed that the solution treatment enhanced dissolution of the intermetallic precipitate β-phase in the α-Mg matrix. The reticular distribution observed in the as-cast alloy disappeared after solution-treatment leaving a random distribution of the β-phase intermetallic. Since the metallic grain size became significantly larger (10x), the volume fraction of β-phase precipitate decreased significantly as a result of solution treatment. The reticular distribution of β-phase precipitate existed in the as-cast alloy was beneficial in stopping the corrosion propagation across the grain. The weight loss rate of the as-cast alloy was hence lower than that of the solution-treated alloy.

References

[1] Song G 2011 *Woodhead Pub.* 117-260
[2] Pekguleryuz M O, Kainer K U, Kaya A A 2013 *Woodhead Pub.* 125-189
[3] Esmaily M, Svensson J E, Fajardo S, Birbilis N, Frankel G S, Virtanen S, Arrabal R, Thomas S, Johansson L G 2017 *Prog. Mater. Sci.* **89** 92-193
[4] Jin X, Xu W, Li K, Zeng X, Shan D 2018 *Mater. Sci. Eng. A* **729** 219-229
[5] Liu Q, Ma Q, Chen G, Cao X, Zhang S, Pan J, Zhang G, Shi Q 2018 *Corros. Sci.* **138** 284-296
[6] Yuan G, You G, Bai S, Guo W 2018 *J. Alloys and Compd.* **766** 410-416
[7] Li J, Jiang Q, Sun H, Li Y 2016 *Corros. Sci.* **111** 288-301
[8] Barylski A, Kupka M, Aniolek K, Rak J 2017 Vacuum 77-86
[9] Anawati A, Asoh H, Ono S 2016 *AIP conference Proc.* **1725** 020002
[10] Anawati A, Asoh H, Ono S 2018 *International Journal of Technology* **3** 622-631
[11] Wang C, Luo T, Zhou J, Yang Y 2018 Mater. Sci. Eng. A **722** 14-19

Acknowledgement

This work was fully supported by Hibah Publikasi Internasional Terindeks untuk Tugas Akhir Mahasiswa (PITTA) Universitas Indonesia with contract no. 2226/UN2.R3.1/HKP.05.00/2018. The authors would like to thank to Adi Ganda Putra for helping with the casting process.

IOP Conf. Series: Materials Science and Engineering **541** (2019) 012009 doi:10.1088/1757-899X/541/1/012009

Relaxation behavior of natural rubber composites based on mooney stress relaxation and rheometer data

I N Indrajati[1] and I Setyorini[1]
[1] Centre for Leather, Rubber and Plastics, Ministry of Industry Republic of Indonesia, Sokonandi St. No. 9, Yogyakarta, Indonesia, 55166

E-mail: i-novia@kemenperin.go.id

Abstract. Studies on relaxation behavior of natural rubber composites were conducted by Mooney stress relaxation coupled with rheometer data. The effect of filler (carbon black N-330) and sulphur loading was monitored on the relaxation behavior. Elasticity of the composites was also determined in term of elongation at break and hardness, as confirmation to the relaxation behavior. The experiments were conducted by varying the N-330 30, 35, 40 and 45 phr with the sulphur baseline of 2 phr. While sulphur loading was varying from 1, 2, 3 and 4 phr with the N-300 baseline of 30 phr. Mooney experiments were carried out at 100 °C, while rheometer experiments were conducted in a moving die rheometer at 150°C. Results finding showed that the relaxation rate was reduced with N-330 loading or sulphur loading due to restriction of rubber molecule chain mobility by crosslinking formation. N-330 provided higher effect on stress relaxation rate compared to sulphur. Elongation of the composites was found to decrease upon filler or sulphur loading, while hardness was inversely.

1. Introduction

Rubber is widely used in all aspect life because of its elasticity in nature. Designing a durable rubber based product needs to consider many factors. For load bearing application or when rubber is subjected to any short dynamic loading, it is not enough to define the long term behavior by only determining the mechanical properties, such as tensile strength, elongation, hardness, which are time-independent. Since rubber is viscoelastic material, it is important to take account the viscoelastic time-dependent characteristic such as creep, stress relaxation, dynamic mechanical properties, etc. [1,2]. In general, stress relaxation is observed when a fixed strain applied on rubber to a constant deformation, the stress required to retain the strain is found to decay with the time [1,3]. The time spans for relaxation can be described as transition of the system from one stage to another [4], but typically the rate is strongly related to the molecular mobility involving the elastic and viscous elements [5]. Moreover, the surrounding structure of the rubber chain molecules will also contribute to the rate of rubber relaxation [1]. Chain entanglements and crosslinking could alter the structure of the rubber, and in turn will affect the relaxation rate.

The relaxation behaviour becomes more complicated in filled-vulcanized rubber. During stress relaxation, at least three process could be taken place including physical relaxation due to relocation of rubber chains and the fillers, degradation caused by heat, light or chemicals, and crosslinking breaking or rearrangement of crosslink and/or polymer backbone [4]. Several studies had been conducted to investigate the stress relaxation behaviour in filled rubber. Abu-Abdeen (2010) studied single and double-step stress relaxation in carbon black filled natural rubber [4]. Maria et al. (2014) investigate the effect of organically modified montmorillonite on stress relaxation of natural rubber/nitrile rubber nanocomposites. It is concluded that the rate relaxation was increased with filler loading due to the decrease of rubber-filler interaction [1]. Mohammadian-Gezaz and Karrabi (2017) conducts a deep study on stress relaxation behaviour of natural rubber filled by nano silicate/carbon black using rubber processing analyzer (RPA). Obaid et al. (2017) investigates the stress relaxation of polymer reinforced by short elastic fiber. They revealed that the presence of short fibers slowed the relaxation rate, and it increases with fiber loading. But, the underlying phenomenon was not well understood because of its complexity [2]. da Rocha et al. (2018) studied the effect of hybrid filler meta-kaolin/carbon black on stress relaxation of nitrile rubber [6]. Most of them studied the stress relaxation behaviour using rubber processing analyzer or monitoring the stress decay in universal testing machine, either in tensile or compressive force. None of them was applying Mooney stress relaxation and rheometer to predict the stress relaxation behaviour.

Vulcanized rubber is composed by two types of crosslink, namely physical and chemical crosslink. Physical crosslinks are resulted from physical interaction between rubber molecule and filler particles (bound rubber formation). Chemical crosslinks are obtained from vulcanization reaction by the formation of the sulphur bridge that linked one rubber molecule chain to another. These crosslinks formation can be monitored through many ways. One of them is by assessing the torque during rheometer experiment. It is generally accepted that the minimum rheometer torque (ML) represents the physical crosslink, while the torque delta (MH-ML) torque indirectly describes the chemical crosslink [7].

In this present study, the stress relaxation behaviour of NR composites was studied by evaluating the effect of filler (carbon black N-330) and sulphur loading on Mooney stress relaxation. Further discussion would be related to the contribution of physical and chemical crosslink to the stress relaxation behavior. Curing characteristics obtained from rheometer experiment were used as additional data. Also, the elasticity of the NR composites was taken account to confirm the relaxation behaviour.

2. Experimental

2.1. Materials

Natural rubber grades Standard Indonesian Rubber (SIR) 10 was supplied by PTPN IX. Carbon black N-330 and N-774 (Ex. OCI, Korea), zinc oxide active (Indoxide), Aflux 42 processing promotor. Paraffinic oil was purchased from CV. Indrasari, Semarang. Antidegradant system consisted of 2,2,4-Trimethyl-1,2-Dihydroquinoline (TMQ) (Kemai) antioxidant, N-(1,3-Dimethylbutyl)-N'-phenyl-p-phenylenediamine (6PPD) antiozonant (Northeast), and paraffin wax Antilux 654 A (Rhein Chemie). Accelerators N-Cyclohexylbenzothiazole-2-Sulfenamide (CBS) (Northeast) and diphenyl guanidine (DPG) (Shandong Shianxian). Sulphur Midas SP 325 (Miwon) was used as vulcanizing agent. All chemicals were industrial grade.

2.2. Rubber Compound Preparation

The rubber compound formulation was set according to

Table **1**. Carbon black N-330 and sulphur loading were varied to evaluate the effect of crosslink formed onto the stress relaxation behavior, while another ingredient was incorporated in a fixed amount. The detail of the variable taken in this study was listed on Table 2.

All compounding processes were performed in a two roll mill mixing machine laboratory scale. Rubber was first masticated and then followed by additive incorporation in a sequence. After completed, the rubber compound was soaked in the cold water for 5 minutes to stop the pre-vulcanization which might be proceeded. Then, the compound was allowed to rest for 24 h at 25ºC before testing.

Table 1. Compound formulation.

Ingredients	Loading (phr)
SIR 10	100.0
ZnO	5.0
Aflux 42	1.5
N-330	Varied
N-774	30.0
Paraffinic oil	10.0
TMQ	2.0
6PPD	1.0
Paraffin wax	0.5
CBS	1.9
DPG	0.2
Sulphur	Varied

Table 2. Variables of the experiment.

Variable	Baseline	Range
N-330	30 phr	30, 35, 40, 45 phr
Sulphur	2 phr	1, 2, 3, 4 phr

2.3. Mooney Viscosity and Mooney Stress Relaxation (MSR) Measurement

Mooney Stress Relaxation (MSR) was a part of Mooney viscosity measurement. This was carried out using Montech Mooney Viscometer MV 3000. Large rotor was applied with the speed of 2 rpm (0.21 rad/sec). The temperature at upper and lower die were maintained at 100±0.03°C. Sample of unvulcanized compounds were prepared using sample cutter, then placed in lower die. Mooney viscosity measurement was performed according to ASTM D-1646. Total testing time for each rubber sample was six minutes, including one minute preheating, four minutes viscosity testing and one minute relaxation testing. Figure 1 depicts the typical Mooney curve obtained from the instrument. The Mooney viscosity, ML1+4, is noted by point D, while Mooney stress relaxation (MSR) is described by curve DE.

The stress relaxation is often described by Power law model as expressed on equation (1),

$$M = kt^a \tag{1}$$

where M is the Mooney torque during the relaxation test, k is the constant equal to the torque unit 1 s after the rotor has been stopped, t is the relaxation time and a is the exponent determines stress relaxation. Taking the log of both sides, equation (1) has transformed equation (2) which is the linear regression form.

$$log\,M = log\,k + a\,log\,t \tag{2}$$

a is the slope of plot of log M vs log t that represents the relaxation rate. The slope a is obtained from the instrument.

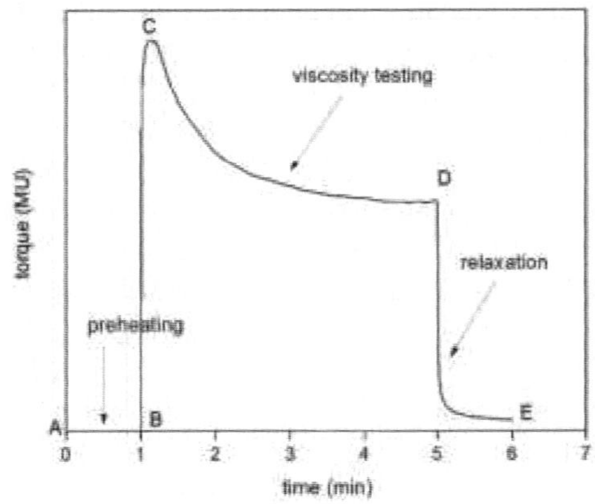

Figure 1. Typical Mooney curve.

2.4. Rheometer Experiment

Moving die rheometer (MDR) Gotech M3000A was used to characterize the curing characteristic of NR composites. A fixed angle of 3° was employed for all experiments. The tests were performed on unvulcanised rubber compound at 150°C. The data recorded from rheometer included elastic minimum (ML) and maximum (MH) torques, scorch time (ts_2), optimum curing time (t_{90}). ML roughly represents the physical crosslink, while MH is the highest torque achieved during testing that represents total crosslink density, both physical and chemical crosslinks. The chemical crosslink formed during vulcanization is determined from the delta of maximum (MH) and minimum (ML) torque, MH-ML. The ts_2 and t_{90} are valuable information regarding vulcanization time. All rheometer tests were performed in triplicate and the average value of each parameter was taken.

2.5. Hardness and Elongation Measurement

Samples were prepared by vulcanizing the rubber compound according to t_{90} obtained from rheometer test. A 2-mm thick slab of elongation test sample and a-12 mm thick round disk of hardness sample were prepared by compression molding in a hydraulic press at 150 kg/cm^2 and at temperature of 150°C. Samples were conditioned for 16 h before testing. Hardness testing was carried out by Shore A durometer. The measurement was performed in three different point of each sample. Elongation testing was conducted in a universal testing machine Tinius Olsen H 50 K. The elongation at which the specimen broken was recorded. All of the samples were tested according to ISO 37. The strain when the sample broken was recorded.

3. Results and Discussion

3.1. Effect of N-330 and Sulphur loading on stress relaxation behavior

Figure 2 and figure 3 depict the torque response after sudden cessation of Mooney rotor for various N-330 and sulphur loading, respectively. All curves exhibit the typical time dependent stress decay of stress relaxation behavior, but with different relaxation degree. The relaxation curves divide into two regions. First region is characterized by rapid drop of Mooney torque (shaded area). This is related to the recovery of elastic motion that takes place at a short time. Second region is noted by irreversibly viscous motion that occurs at longer time, such as chain slippage, disentanglement of rubber molecule chains, breaking down of filler agglomerates or filler-matrix debonding. These govern the flow and causes gradually reduction to very low amount. The degree of the gradual reduction is depended on final configuration [5].

N-330 loading is greatly affected the rate of relaxation as seen on Figure 2 Increasing N-330 loading tend to prolong the relaxation time, indicating that the presence of N-330 slowed the relaxation rate as shown on Figure 4. This finding was agreed to Obaid et al. (2017) [2]. According to Maria et al. (2014) [1], the rate of the stress relaxation is greatly influenced by the filler and its interaction with rubber molecule chains in the matrix. Filler particles interact with rubber molecule chains through weak physical forces such as van der Waals. This interaction is often called by physical crosslinking, and in rubber, it is related to the bound rubber formation. Crosslink will hinder the mobility of the rubber molecule chains [8], thus lower the relaxation rate. Physical crosslink can be roughly estimated from the minimum rheometer torque (ML). It is clearly seen on Table 3 that ML increases upon N-330 loading. It indicates higher physical crosslink formed as a result of better interaction between filler and rubber. Higher physical crosslink is also confirmed by gradually increase of Mooney viscosity (ML1+4) upon N-330 loading. Increasing rubber-filler interaction is responsible to the increase of composite viscosity as observed in the experiments [9].

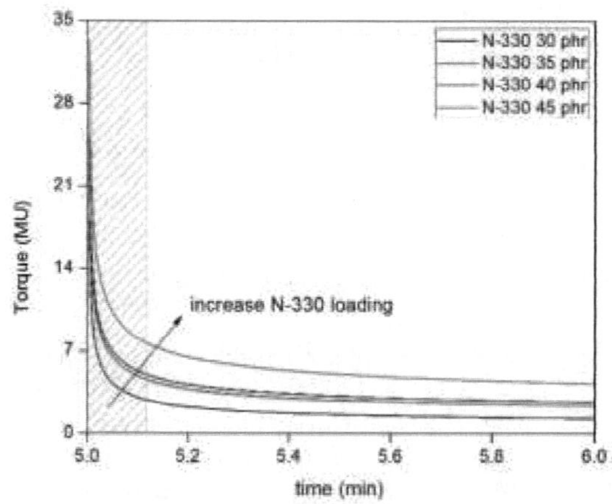

Figure 2. Time dependence of Mooney torque decay of NR composites on various N-330 loading.

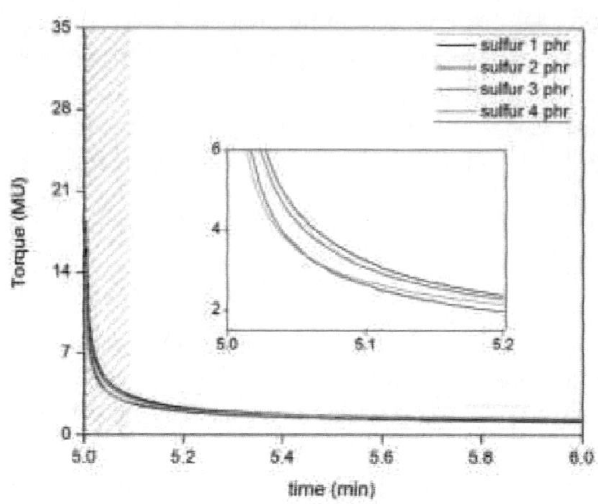

Figure 3. Time dependence of Mooney torque decay of NR composites on various sulphur loading.

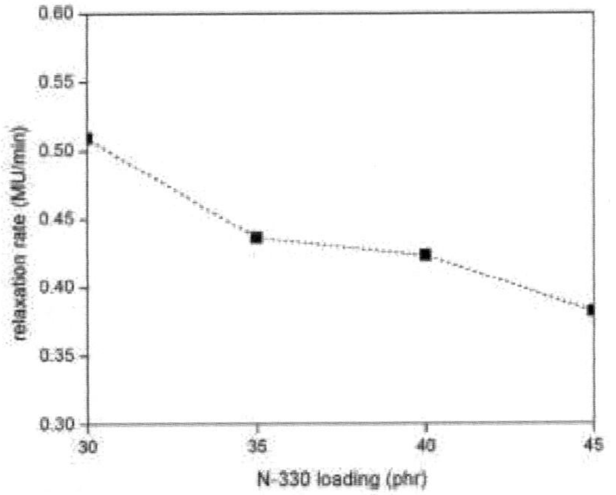

Figure 4. Stress relaxation rate of NR composite on various N-330 loading.

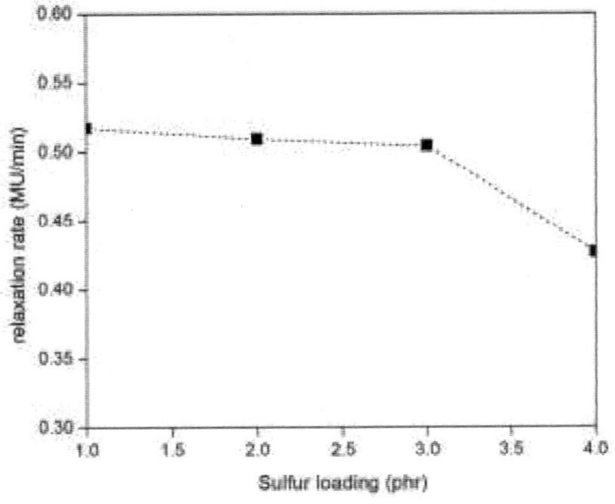

Figure 5. Stress relaxation rate of NR composite on various sulphur loading.

1.1. Sulphur is well-known as crosslinking agent for rubber. When vulcanization reaction proceed, sulphur reacts to the unsaturated sites of rubber molecule chains and forms the chemical crosslink. The density of these crosslinks is proportionally to the sulphur amount in the formula. Chemical crosslinks can be roughly determined by calculating the delta between maximum and minimum elastic rheometer torques (MH-ML) [7]. From Table 3 it is seen that chemical crosslink amount is increased upon sulphur loading, nevertheless it gives less effect on the rate of relaxation compared to N-330 as depicted on

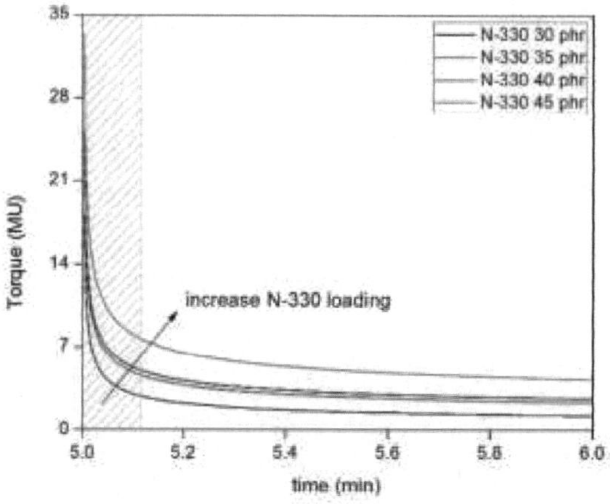

Figure 6. Time dependence of Mooney torque decay of NR composites on various N-330 loading.

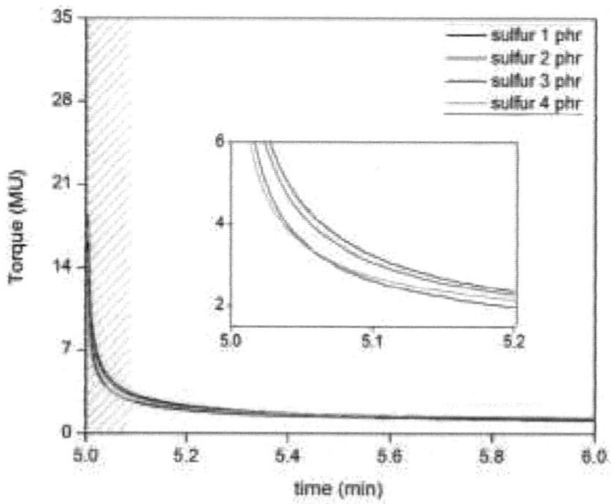

Figure 7. The rate of relaxation also shows little change upon sulphur loading (Figure 5). In this case, chemical crosslink provides less effect than physical crosslink on rate of relaxation of *NR composites.*

Table 3. Curing characteristics and Mooney viscosity of NR composites.

Loading	ts_2	t_{90}	MH	ML	ΔM=MH-ML	ML1+4
	sec	sec	kgf.cm	kgf.cm	kgf.cm	MU
N-300						
30 phr	23	80	56.59	2.99	53.60	26.19
35 phr	22	84	59.88	3.92	55.96	32.82
40 phr	26	94	62.24	3.98	58.26	35.80
45 phr	22	92	68.31	5.04	63.27	46.48
Sulphur						
1 phr	26	47	41.45	3.27	38.18	28.84
2 phr	23	80	56.59	2.99	53.60	26.19

IOP Conf. Series: Materials Science and Engineering **541** (2019) 012009 doi:10.1088/1757-899X/541/1/012009

3 phr	22	97	69.26	4.10	65.16	23.44
4 phr	26	134	80.42	2.60	77.82	17.99

3.2. *Effect of Elasticity of NR Composites on Stress Relaxation*

Elastic properties of rubber composite can be represented by its elongation and hardness. These two properties are depended on the rubber crosslink density [10], and are inversely one to another. Figure 8 displays linearly decrease on elongation upon N-330 loading. This reduction is probably caused by polymer chain stiffening through polymer-filler interaction, i.e. physical crosslinking [11]. At higher crosslink density, the length between two crosslink point becomes shorter, thus hinders of molecular mobility of rubber chain. This restriction brings to a slower relaxation rate as exhibited by Figure 4. and Figure 9 increasing crosslink density, either physical or chemical crosslink. At higher crosslink density, the Low crosslink density provides higher elongation due to free movement of the polymer chain. When filler particles amount increased, it is evident that the more physical crosslinks are formed (noted by higher ML) which restricts the mobility of rubber molecule chains. Similar to N-330, the increasing sulphur loading also leads to a linear decrease of elongation as depicted on Figure 9. The more sulphur amount in the formulation, the more chemical crosslink forms that leads to shorten the intercrosslink length and in turn hinder the molecular mobility. Increasing N-330 and sulphur loading causes linear increases of composite's hardness, as expected. Figure 10 and Figure 11 exhibit increasing of hardness with crosslink density for physical and chemical crosslink, respectively.

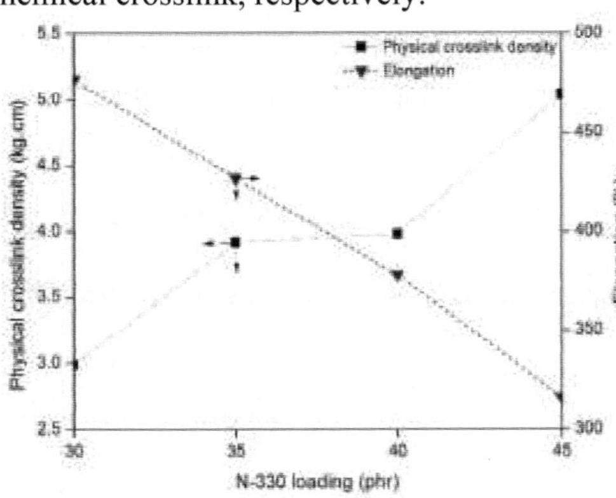

Figure 8. Influence of N-330 loading on physical crosslink density and elongation.

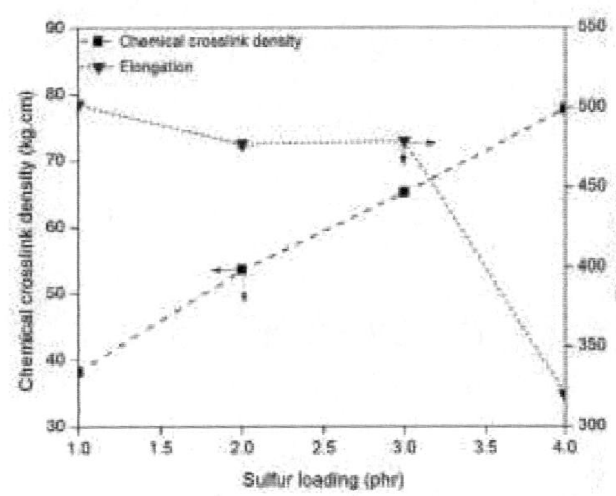

Figure 9 Influence of sulphur loading on chemical crosslink density and elongation.

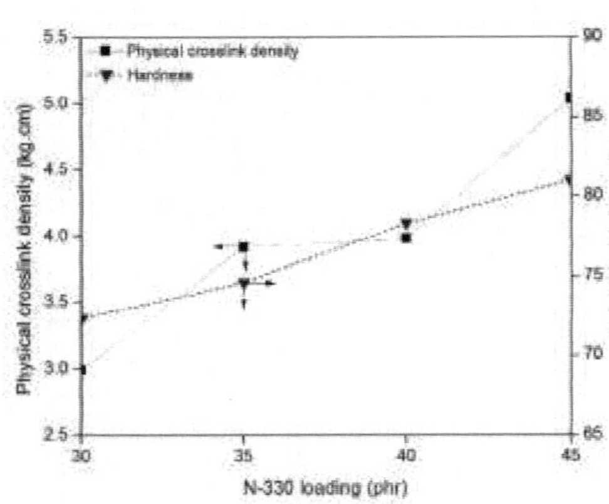

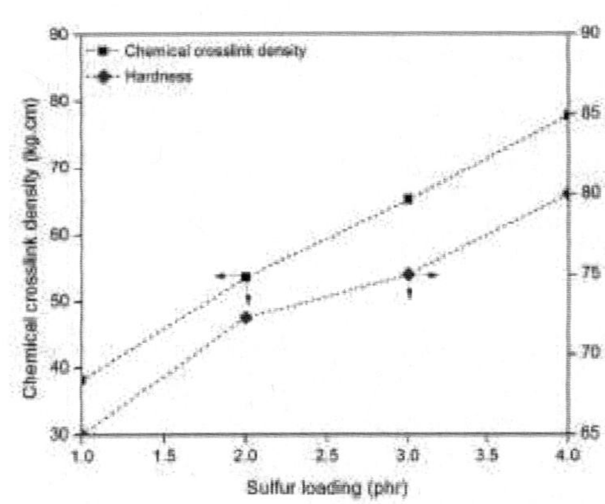

Figure 10. Hardness of NR composites as a function of N-330 loading.

Figure 11. Hardness of the NR composites as a function of sulphur loading.

4. Conclusion

Stress relaxation of NR composites has been studied by Mooney stress relaxation coupled with rheometer data. Vulcanized rubber is usually composed by physical and chemical crosslink. Physical crosslink is resulted from rubber-filler particle, while chemical crosslink is formed during vulcanization reaction between rubber and sulphur. The MSR curve showed typical torque decay in which consisted of two regions as a result of elastic recovery and viscous flow. The relaxation rate of the composites, represented by MSR slope, is strongly related to the crosslink densities, i.e. physical and chemical crosslinks. The relaxation rate decreases upon crosslink densities caused by molecular mobility restriction. Physical crosslink density gives markedly effect on the relaxation rate compared to chemical crosslink. Thus, it should be considered when incorporating filler into the rubber matrix. Crosslink densities also affect the elasticity of rubber composite which represented by elongation and hardness. Elongation was linearly decrease with crosslink densities, while hardness was inversely. This confirms lower of relaxation rate of rubber composite.

References

[1] H. J. Maria, N. Lyczko, A. Nzihou, K. Joseph, C. Mathew and S. Thomas 2014 *Appl. Clay Sci.*, **87** 120–128.
[2] N. Obaid, M. Kortschot and M. Sain 2017 *Materials (Basel)* 10(5) 472.
[3] E. B. D. da Rocha, F. N. Linhares, C. F. S. Gabriel, A. M. F. de Sousa, and C. R. G. Furtado 2018 *Appl. Clay Sci* **151** 181–188.
[4] M. Abu-Abdeen, 2010 *Mater. Des* **31(4)** 2078–2084.
[5] S. Mohammadian-Gezaz and M. Karrabi 2017 *Prog. Rubber, Plast. Recycl. Technol.* **33(4)** 261–280.
[6] E. B. D. da Rocha, F. N. Linhares, C. F. S. Gabriel, A. M. F. de Sousa and C. R. G. Furtado, 2018 *Appl. Clay Sci.* **151(2017)** 181–188.
[7] I. Surya, H. Ismail and A. R. Azura, 2013 *Polym. Test.* **32(8)** 1313–1321.
[8] D. Gabriel, A. Karbach, D. Drechsler, J. Gutmann, K. Graf and S. Kheirandish, 2016 *Colloid Polym. Sci.* **294(3)** 501–511.
[9] J. Gheller, M. V Ellwanger, and V. Oliveira, 2016 *J. Elastomers Plast.* **48(3)** 217–226.
[10] F. Zhao, W. Bi, and S. Zhao, 2011 *J. Macromol. Sci. Part B* **50(7)** 1460–1469.
[11] C. S. Julie Chandra, P. K. Bipinbal, and K. N. Sunil, 2017 *Polym. Test* **60** 187–197.

Acknowledgement

The authors would like to express their thanks to Dr. M. Chalid for intensive discussion related to the stress relaxation in rubber.

An analytical, numerical and experimental study of an octagonal ring force transducer for a 2-axis force measurement

M Rizal[1], Husaini[1], M Abrar[1] and M Y Wiranda[1]
[1]Department of Mechanical Engineering, Faculty of Engineering, Syiah Kuala University (UNSYIAH), 23111, Darussalam, Banda Aceh, Indonesia

E-mail: muh.rizal@unsyiah.ac.id

Abstract. A reliable force transducer is urgently required in engineering design, optimization, monitoring systems, and mechatronic fields. In this paper, characteristic of an octagonal ring as force transducer is studied. This transducer utilized strain gauge which is designed to able measure the external force in 2-axis both vertical and horizontal simultaneously. Strain distribution analysis on the transducer were done to get the accurate position of the strain gauge bonded using theoretical and numerical approaches. In the end, an octagonal ring force transducer is manufactured and the calibration test is conducted. The experimental results show a good performance of sensitivity, linearity, repeatability and low cross coupling error. The result found that developed force transducer is suitable and reliable to measure static and dynamic forces.

1. Introduction

Force measurement is an important task in engineering fields to obtain the information related to mechanical system in various applications. In the past before the 20th century, various methods have been used for measuring forces based on mechanic, pneumatic and hydraulic devices [1]. These nonelectric force measuring methods which not have sensitivity to electric signal as required in the current measurement system to be integrated with control and automation systems. Therefore, force transducer is a key component in the force measurement system which can converting an applied force into a material strain or deformation [2]. This deformation is then can be sensed by resistive, capacitive, inductive and piezoelectric sensors to given the output reading in electrical signals [3].

The force transducer structure have elastic element that can occurs stresses and displacements when external force applied on it. These stresses can be determined by using measurable parameters, such as strain, that vary in response to loading. Strain gauge in one of more strain sensors which is simple, low cost and easy to use. Strain gauges were arranged in the form of a Wheatstone bridge, a means accurately detecting very small changes in resistance of strain gauge during imposed by external force on the transducer structure. The output of the bridge is a measure of the applied force that expressed in the form of an electrical output, like voltage (V) [4].

Recently, many types and various geometrical shape of strain gauge based force transducers have been developed for measuring force. Ştefănescu [1] have classified ten types of force transducers which can be measured force in range 200 N to 50 MN. Ring shaped type force transducer is still populer area of research although many years of research. Typically circular [5], square [4], hexagonal [6] and octagonal [7] shaped rings are approved as the basic geometric structures of a force transducer. Some studies in the past have applied the octagonal ring as dynamometer in turning process [8], milling process [9, 10] and also drilling process [11]. However, the octagonal ring force transducer is multi-purpose 2-axis force transducer that can use in many applications. In previous study, there are still rare studies of strain analysis that compare analytical, numerical and experimental. In addition, an octagonal ring force transducer application as a 2-axis force plate for biomechanics study of human motion is still very rare. Therefore, in this study will focus on design, construction and characterization an octagonal force transducer for two-axis forces measurement that is normal and tangential forces.

2. Design Analysis

An octagonal ring force transducer was designed in this study. The transducer is expected to detect external forces in two directions that are vertical and horizontal forces, or normal and tangential forces. The good model of transducer can be indicated by high sensitivity and low interference error between two directions of external force. An octagonal ring with two force exerted on the top of its structure was shown in Figure 1. The transducer is designed for maximum capability of 750 N, and the other geometries are selected about 15 mm of width (b) and 12.5 mm of radius (r). The maximum stress and strain on the octagonal ring surface due to external force was set as the point to be placed the strain gauge.

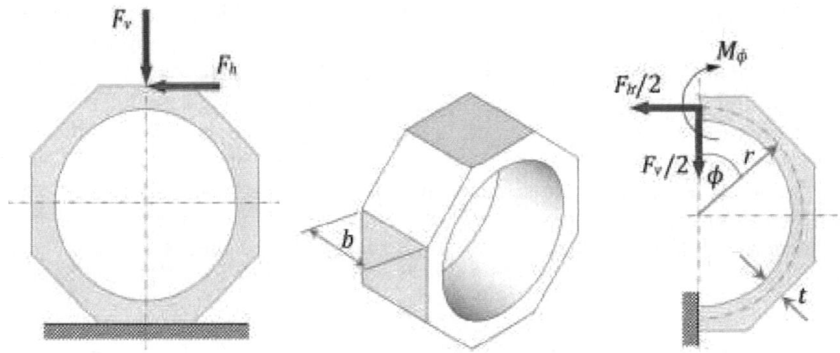

Figure 1. The octagonal ring force transducer model

2.1. Analytical Study

Analytical study was employed to estimate strain and stress of the octagonal ring. Figure 1 shows the octagonal ring is subjected to the vertical or normal force (F_v) and the horizontal or tangential force (F_h). When the bottom of the structure was fixed on the base and the external forces acting on the top of the ring would cause a moment to occur which varied with the angular position (ϕ). Elastic strains ε_{Fv} and ε_{Fh} due to forces F_v and F_h are calculated according to ring theory by using the following equations [10].

$$\varepsilon_{F_v} = \pm \frac{1.09\, F_v r}{Ebt^2} \quad (1)$$

$$\varepsilon_{F_h} = \pm \frac{2.18\, F_h r}{Ebt^2} \quad (2)$$

Since the material structure is stainless steel 304, with yield strength of 215 MPa, then in order to avoid structural failure due to overload, the maximum stress on the transducer must be below the yield strength (S_y) of the structural material. By taking into account the safety factor (sf) of 1.3, the maximum allowable stress can be obtained through the following:

$$\sigma_{max} = \frac{S_y}{sf} \quad (3)$$

IOP Conf. Series: Materials Science and Engineering **541** (2019) 012010 doi:10.1088/1757-899X/541/1/012010

then

$$\sigma_{max} = \frac{215}{1.3} = 165.4 \text{ MPa} \quad (4)$$

Therefore, by taking account the allowable maximum stress 165.4 MPa to equations (8) and (9), it can be obtained the range of allowable thickness of load cell structure that are:

$$t_{max} = \sqrt{\frac{2.32\, F_t r}{\sigma_{max} b}} = \sqrt{\frac{2.18\,(750)(12.5)}{(165.4)(15)}} = 2.87 \text{ mm} \quad (5)$$

$$t_{min} = \sqrt{\frac{1.09\, F_n r}{\sigma_{max} b}} = \sqrt{\frac{1.09\,(750)(12.5)}{(165.4)(15)}} = 2.02 \text{ mm} \quad (6)$$

The allowable thickness of structure is obtained in range 2 mm up to 3 mm, then it is selected with maximum thickness is 3 mm. If the actual force is exerted to the structure with thickness of beam about 3 mm in vertical and horizontal directions, so we obtained the actual stress are:

$$\sigma_{F_h} = \frac{2.18\,(750)(12.5)}{(15)(3)^2} = 151.38 \text{ MPa} \quad (7)$$

$$\sigma_{F_v} = \frac{1.09\,(750)(12.5)}{(15)(3)^2} = 75.69 \text{ MPa} \quad (8)$$

2.2. Numerical Study

Finite element (FE) analysis of the octagonal ring force transducer have been done to validate the analytical method. In the previously, FE method as an effective computational tool was used to identified the optimum location of strain gauge on the cross beam transducer [12, 13]. The octagonal ring has been modelled in three dimensional forms with the dimensions as follows: outer diameter 28 mm, inner diameter 22 mm and width 15 mm. Stainless steel 304 was used for structure material with Young's modulus of elasticity is 200 GPa and Poisson ratio 0.29. The boundary conditions as can be seen in Figure 1, the bottom end of the octagonal ring shaped transducer has been fixed and the top of ring is kept free which the force is applied in vertical and horizontal directions. The external forces are given from 10 N up to 100 N in each direction.

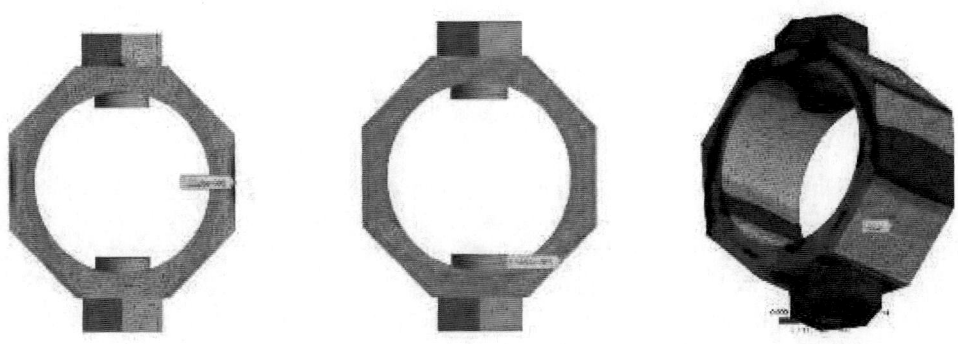

(a) (b) (c)

Figure 2. Strain distribution on the octagonal surface: (a) when vertical force, F_v was given; (b) when horizontal force, F_h acting on the top; (c) stress occurred at angle 39.5° from the bottom point.

Figure 2(a) shows strain occurred due to vertical force (F_v) acting on the top of the ring. It can be observed that the maximum strain occurred at an angle 90° about 12 με or 1.2 x 10^{-5} mm/mm when the external force is 50 N. While the horizontal force was given on the top of the ring, the maximum strain appears at a 39.5° is about 21.5 με or 2.15 x 10^{-5} mm/mm as shown in Figure 2(b).

Figure 2(c) shows stress occurred due to horizontal force (F_h). It can be seen that stress occurred at an angle 39.5° is 165.24 MPa. When this result is compared to the mathematical analysis, it can be found that the result is closely match to each other. The error of stress occurred between analytical and FE is not exceeding 9 % which it can be reasonably acceptable in engineering design study.

2.3. Experimental Study

The point of maximum stress and strain of octagonal ring force transducer was set as mounting of strain gauge. In order to measure strain due to the vertical force, a strain gauge has been mounted at an angle 90° from vertical axis. While, to detect the strain as result of horizontal force, another strain gauge has been mounted at an angle 39.5° from vertical axis as shown in Figure 3(a). A linear strain gauge with the nominal resistance about 350 Ω was used for each channel of the transducer. Its dimension was 3 mm of length, 3.2 mm of width and gauge factor about 2. Before fixing the strain gauge, the surface of structure was polished with sandpaper and cleaned it with acetone. A quarter Wheatstone bridge was used for each channel by utilizing NI-9945 for balancing the circuit. Then, NI-9237 has been used for signal conditioning and data acquisition. The display and data recording program was developed in house using LabVIEW software.

(a)

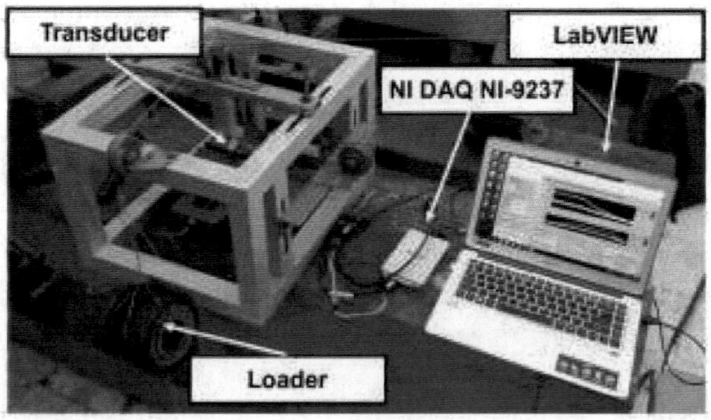

(b)

Figure 3. (a) Strain gauge location on the transducer; (b) experimental setup for static calibration.

The experimental test was performed using a developed calibration frame to determine the relationship between the input load and output of the voltage as shown in Figure 3(b). It is also to investigate the performance and sensitivity of force transducer after design and construction. The calibration was made in two directions or coordinates for F_v (vertical) and F_h (horizontal) for an applied load of up to 100 N, with incremental steps of 10 N. The calibration tests were repeated three times to ensure the consistency, and the average values were recorded.

3. Result and Discussion

Figure 4 shows the output signal of strain gauge during calibration test. It can be seen that the signal increase due to external load is linear calibration curves for normal and tangential force measurements. From the calibration curve we can observed the sensitivities of the octagonal ring force transducer were obtained in vertical force direction is 5.3 x 10^{-4} μV/N, and horizontal force direction is 9.1 x 10^{-4} μV/N. In addition to the curves, their found that the average interference error in vertical direction was 2.9%.

Whereas the transducer was subjected to horizontal force, the interference error was 4.3%. These results are consistent with those of other studies that their found the error about less than 5% [14].

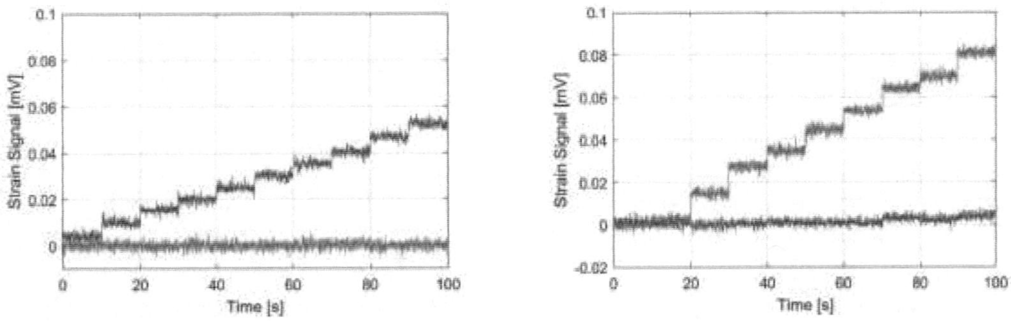

(a) (b)

Figure 4. (a) Strain signal when external force on vertical direction; (b) Strain signal when external force on horizontal direction.

In order to convert the output value in voltage signal (mV) to the strain in theoretical analysis the equation is as follows [13]:

$$\frac{V_o}{V_i} = \frac{1}{4} GF.\varepsilon \quad (9)$$

where V_i is the input voltage of the Wheatstone bridge, V_o is the output voltage, GF is the gauge factor, and ε is the strain. Figure 5 shows the result obtained from analytical, numerical and experimental analysis of strain occurred on the octagonal force transducer. It can be evidenced from the curve that analytical and numerical predictions closely match to each other. The error in vertical and horizontal directions between analytical and numerical is 3.3% and 15%. The present findings seem to be consistent with the results reported earlier regarding the numerical studies of hexagonal ring-shaped force transducer, though the extent of relative deviation is much less between the analytical method and experimental [6, 7]. However, when we observed the experimental result, indicated that the strain measurement on the octagonal ring is successful in following the rising trend of strain values when analysed in mathematical and numerical analysis. Although the value of the strain slightly deviated from the analytical a numerical due to several factors, that are the installation of strain gauge, calibration frame and the environmental condition.

In order to evaluate the performance of the developed octagonal ring force transducer, an experimental test was performed to detect the horizontal force which was conducted in four times. The aim of this test is evaluating the repeatability or consistency of transducer in detecting the fluctuation force during applications. A fixed force 30 N was subjected in the horizontal direction of transducer, but in vertical direction was not given any load.

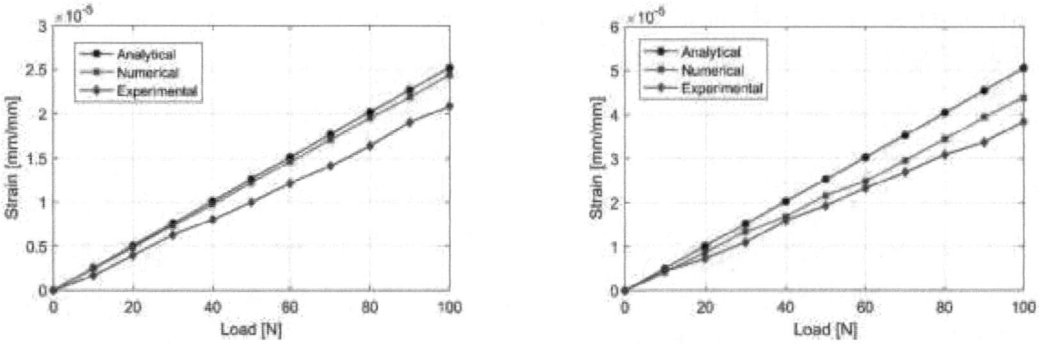

(a) (b)

Figure 5. Comparison strain from analytical, numerical and experimental on the octagonal ring when loading up to 100 N (a) vertical direction; (b) horizontal direction.

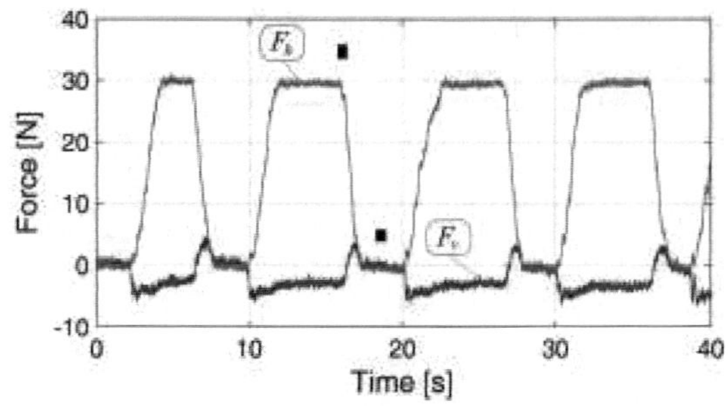

Figure 6. Result of measurement test of developed load cell.

Figure 6 shows the force signal reading in four times loadings. It can be observed that the load cell can be detected horizontal force (F_h) accurately. However, it found any error or disturbance in the vertical direction, but it was not exceed 9% from maximum scale output of developed octagonal ring.

4. Conclusion

An analytical, numerical and experimental study of a transducer for vertical and horizontal force measurements has been presented. The sensor system is developed based on an octagonal type structure which is utilized by strain gauge. The maximum measurement of the external force is designed up to 750 N. The results of calibration and measurement tests showed that its sensitivity was approximately in the range 5.3 x 10^{-4} – 9.1x10^{-4} μV/N and it has cross-sensitivity error of below 4.5%. The differences result between theoretical and experimental analysis were found in range 3.3 – 15%. The developed system gave a satisfactory performance and it has the potential to be used as sensor system in two axis force measurement such as in tribometer and applications.

References

[1] Ştefănescu D M 2011 *Handbook of Force Transducers: Principles and Components* (London: Springer)
[2] Wood R J, Cho K-J and Hoffman K 2009 *Smart Mater. and Structur.* **18** 125002
[3] Rizal M, Ghani J A, Nuawi M Z and Haron C H C 2014 *Eng. Technol. Appl. Sci. Res.* **7** 2083-97
[4] Kumar H, Sharma C and Kumar A 2013 *Meas. Sci. Technol.* **24** 095007
[5] Libii J N 2006 *World Transactions on Engineering and Technology Education* **5** 175–8
[6] Kumar H, Pardeep, Kaushik M and Kumar A 2015 *MAPAN-Journal of Metrology Society of India* **30** 37-47
[7] Soliman E 2015 *Alexandria Eng. J.* **54** 155-62

[8] Yaldız S and Ünsaçar F 2006 *Measurement* **39** 80-9
[9] Korkut I 2003 *Mater. Des.* **24** 631-7
[10] Yaldız S, Unsacar F, Saglam H and Isık H 2007 *Mech. Syst. Signal Process.* **21** 1499-511
[11] Karabay S 2007 *Mater. Des.* **28** 673-85
[12] Rizal M, Ghani J A, Nuawi M Z and Haron C H C 2015. In: *Proceedings of the International Conference on Sensing Technology, ICST,* (Auckland pp 289-94
[13] Rizal M, Ghani J A, Nuawi M Z and Haron C H C 2015 *Mech. Syst. Signal Process.* **52-53** 559–76
[14] Uddin M S and Songyi D 2016 *Measurement* **90** 168–77

Acknowledgements
The authors would like to thank Directorate General of Higher Education and Ministry of Research & Technology for their financial support under Grant 41/UN11.2/PP/SP3/2018.

IOP Conf. Series: Materials Science and Engineering **541** (2019) 012011 doi:10.1088/1757-899X/541/1/012011

Study on Phases Development and Mechanical Properties in a Fe-Ni-Al Carbide Free Bainitic Steel Based on Lateritic Steel After Warm Rolling

Meryanalinda[1], M Ariati[1] and F Citrawati [2]
[1] Metallurgy and Materials Engineering, Faculty of Engineering, Universitas Indonesia, Depok 16424, Indonesia
[2] Research Center for Metallurgy and Materials, Indonesian Institute of Sciences, Tangerang Selatan, Banten 15314, Indonesia

E-mail: meryanalinda.anharmi@gmail.com

Abstract. The abundant reserve of lateritic ores in Indonesia is currently processed and studied to fulfil the national steel demand in several sectors, one of which is for railway application. The development of lateritic steel (Fe-Ni) as carbide free bainitic steel is carried out by adding Al to the Fe-Ni alloy. In this research, the effect of warm rolling and Al addition to the formation of carbide, phases development and mechanical properties was studied. The warm-rolled thermomechanical process (TMCP) was carried out by heating the material at 945°C for 20 minutes followed by second heating to 400°C, 450°C and 500°C with holding time for 30 minutes. The materials then warm rolled with 50% and 70% reductions using 20 tons capacity roller machine and then air cooled outside the furnace chamber. The microstructure of the as-rolled materials was characterized using optical microscope (OM) and scanning electron microscope (SEM), while the phases, the chemical distribution and the possibility of carbide formation were examine using X-Ray Diffraction (XRD) pattern and energy dispersive X-Ray spectroscopy (EDS). The mechanical properties of the material were observed using macro Rockwell hardness test. It was revealed that the addition of Al altered the phases of Fe-Ni lateritic steel significantly. Furthermore, Al addition gives positive effects to the Fe-Ni lateritic steel by increasing hardness. The reductions applied during warm rolling were observed to have effect on the growth of the grains in the Fe-Ni-Al lateritic steel.

1. Introduction

Indonesia is one of countries that has the largest laterite iron ore reserves in the world spread in South Kalimantan, South Sulawesi, Southeast Sulawesi, North Maluku and West Papua. The development of laterite (nickel) steel ore can be used as an alternative to reduce the dependence on steel raw materials, especially raw materials for automotive in Indonesia. One development of Fe-Ni based laterite steel is by adding alloy and microstructural engineering with the thermomechanical control process (TMCP) process. The purpose of this development is to get a new type of steel, which is known as carbide free bainitic steel, in which it is an interesting candidate for automotive and railway industries application[1,2], which exhibits good combination of strength, toughness and ductility, has been designed and manufactured [3-6].

The previous research reported that the microstructure of carbide free bainitic steel consists of fine plates of carbide-free bainitic ferrite and thin film of carbon which is enriched by retained austenite[6]. Fine-scale microstructure is often obtained by the bainite transformation with the diffusion less growth of tiny bainitic ferrite platelets that is mostly gotten at low temperatures, and it is primarily achieved in steels with a higher carbon content, in which does not only decrease the bainite transformation temperature, but also stabilizes retained austenite after bainitic transformation [4-7].

Adding the aluminum in the steel can accelerate the low temperature of bainitic transformation, especially the formation of bainitic ferrite [4,8].C.W. Shao et al [9] studied that the addition of Al does not only increase the stability of α-ferrite but also expands the δ-ferrite phase region. Aluminum has a similar function to suppress the precipitation of cementite to silicon. It also has been observed that adding aluminum to TRIP Steels tends to refine the bainitic lath structure [8,10]. The presence silicon in alloy will affect the microstructure. It was found that silicon can only impede cementite (Fe_3C) formation lower bainite but does not fully suppress it [11]. Producing low carbon bainitic steels in the Fe-Mn-Ni-C was also investigated and found that the platelets of bainitic ferrite were extremely coalesced and the coarsening of microstructure occurred. Thus, it eliminates the beneficial effect of fine scale bainitic microstructure.

Warm rolling has a potential process to get such beneficial microstructure as the structure of the ultrafine-grained ferrite matrix to offer considerable costs savings due to lower reheating temperatures, reduce the roll wear and loss of scale, and allow accurate control of product dimensions [12]. W. Hui et al [11] studied the microstructural evolution and mechanical properties of a 3% Al containing medium Mn steel during warm rolling after intercritical annealing at temperature 750°C and found that the ferrite and retained austenite lath changed gradually to follow rolling direction and refined by increasing the rolling thickness reduction. Rolling effect indicates that the formation mechanism of the lamellar structure is a kind of grain subdivision, the deformation induces grain boundaries subdivide the initial grains to ultrafine regions [13].

Current study is aimed to explore of possibility of formation of the carbide free bainitic microstructure in a lateritic steel alloyed with 2,23 wt. % Al and examine the effect of austempering temperature and warm rolling by reducing the different thickness after austempering on the microstructure and hardness of this steel.

2. Experimental Details

The chemical composition of the Fe-Ni lateritic steel alloyed with Al is given in Table 1. The dimension of the sample used in this study is 75 x 20 x 10 mm were cut to austempering treatment and warm rolled. Before the thermochemical treatments were conducted, the bainitic start temperature (Bs) was calculated to be 579°C for Al-added steel, while the martensitic start temperature (Ms) 418°C. The calculation method occupied was the MUCG83 program developed by Bhadeshia at Cambridge University. During the thermomechanical treatment, the samples were heated to 950°C for 20 mins to obtain full austenite structure, followed by furnace cooling to reach bainitic transformation temperature and hold for 30 mins. Three different bainitic transformation temperatures were used, which are 400°C, 450°C and 500°C. At each of these temperatures, the sample was subjected to warm rolling with 50% and 70% reductions using 20 tons capacity roller machine. Then, the samples were air cooled to room temperature outside the furnace chamber. The schematic image of the thermomechanical schedules is shown in figure 1. The orientation corresponding to warm rolled product are shown in figure 2, in which ND stands for normal direction, TD stands for transverse direction and RD stands for rolling direction.

Table 1. Chemical composition of the experimental steel (in wt.%):

C	Si	Al	P	S	Cr	Ni	Mo	Mn	Fe
0.30	1.38	2.23	0.0005	0.0003	0.28	3.03	0.20	~0.4	Bal

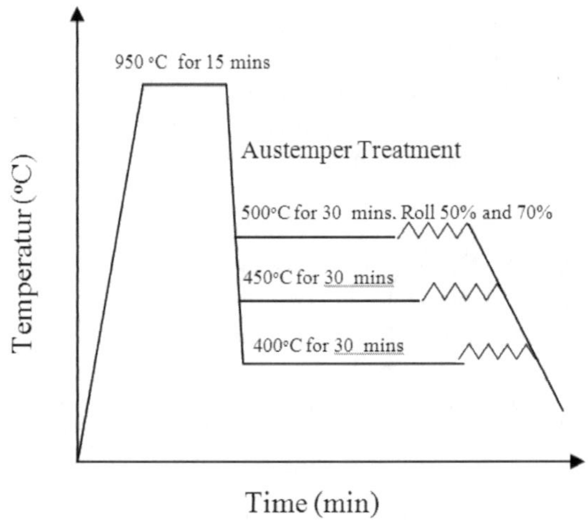

Figure 1. Schematic illustration of the thermomechanical schedules applied to studied samples.

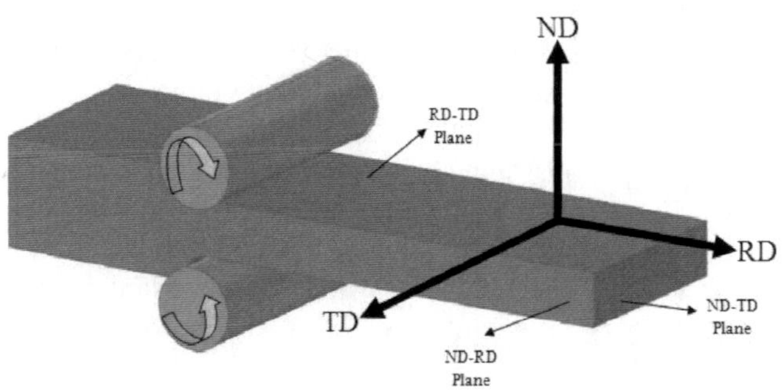

Figure 2. Corresponding directions of the warm rolled product.

The longitudinal sections (along rolling reduction) of the samples were mounted and polished using the standard metallurgical procedure and etched in 4% picral, and their microstructure were observed using an optical microstructure (OM) and a scanning electron microscope (SEM, JEOL 6390). The heat-treated samples were also characterized using X-Ray Diffraction (XRD Rigaku Miniplex 600) pattern diffraction instrument with Cu Kα radiation analyzing between 0° and 125° (2θ), and energy dispersive X-Ray spectroscopy (EDS) to examine the chemical distribution and the possibility of carbide formation. Macro hardness test was carried out on ND of warm rolled samples. Macro hardness measurements were conducted using a load of 150 kgf on a Rockwell hardness testing machine with diamond indenter and the procedural based on ASTM E18. At least six reading were taken for each heat treatment condition and their average is reported here as the hardness value.

3. Result and Discussion

3.1 Microstructure analysis

3.1.1. Microstructure

The SEM micrographs of lateritic steel of different heat treatment are shown in figure 3. Figure 3a depicts the microstructure of as-cast steel with austenization at 950°C for 20 min followed by air cooling. It can be seen that some plate-like transformation product (i.e. bainite) is formed in the sample. The microstructure is composed of acicular grains of ferrite (α), elongated bainite ferrite(αb) and blocky and elongated martensite or retained austenite (M/RA) particles, in which generally surrounded by bainite grains. Isothermal treatment without holding time at 400°C followed warm rolling 50% (figure 3b) shows the microstructure consisted of mainly bainite and ferrite. Holding time is affected by the bainite transformation of microstructure, when the period of holding is short, the amount of bainite formed at the end of the holding period is small. As can be seen, the size of grain is thin and it is discontinue surrounded by acicular ferrite grains. When the isothermal treatment has 30 min holding time the mainly result of microstructure consists of bainite ferrite and ferrite with less or no martensite (figure 3c-3g). By increasing the temperature, the amount of bainite increases, while the amount of ferrite decreases. The bainite ferrite and ferrite are arranged alternatively along the rolling reduction.

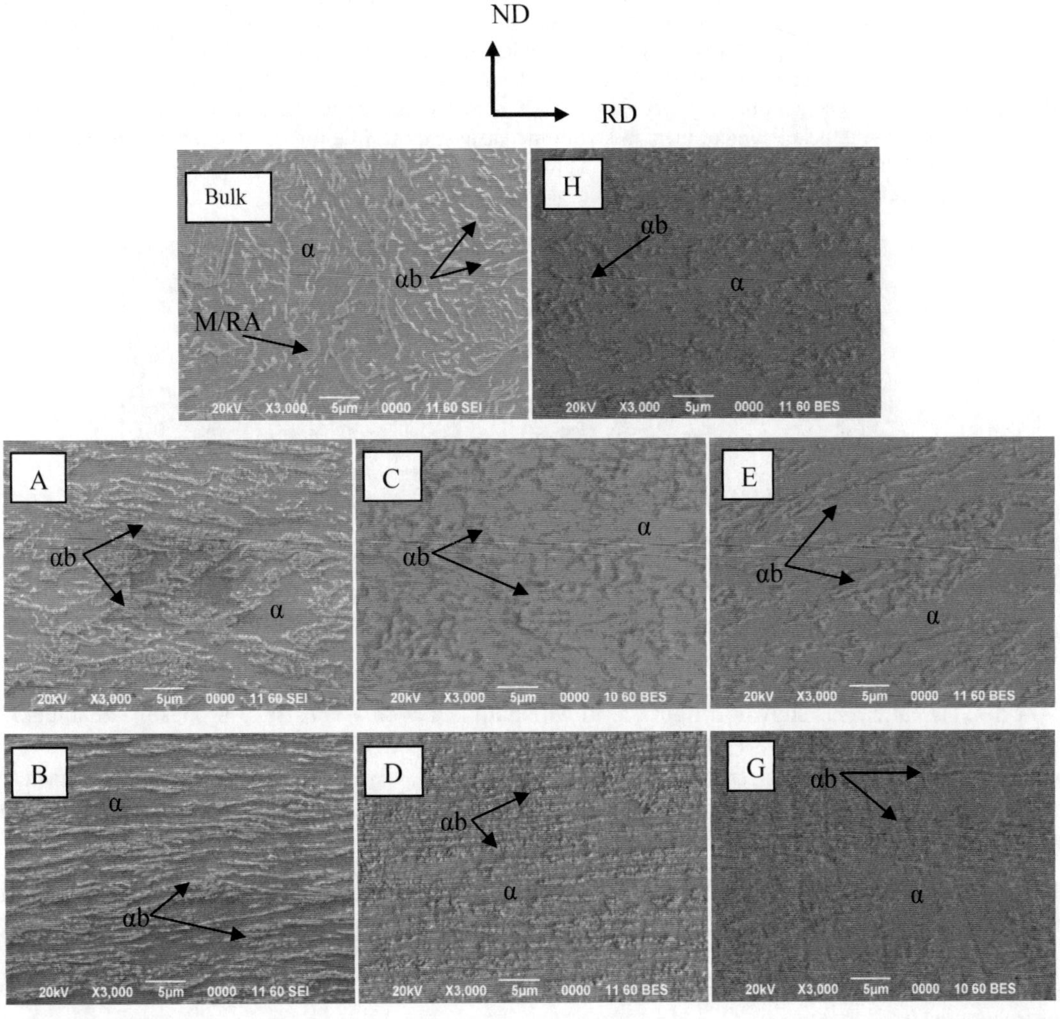

Figure 3. SEM micrographs of the samples, (a) air cooling condition after austenization 945°C; (b) WR 50% at 400°C without holding time; WR 50% after holding time for 30 mins at (c) 400°C, (d) 450°C, (e) 500°C; WR 70% after holding time for 30 mins at (f) 400°C, (g) 450°C, (h) 500°C.

The effect of rolling different rolling reduction on warm rolling microstructure are illustrated in figure 3c-3g. The microstructure will develop lamellar structure with the longitudinal axis along rolling reduction. The bainite ferrite and ferrite phases gradually changed their longitudinal axis to the rolling reduction by increasing the amount of deformation. The microstructure (figure 3c-3g) shows that ferrite changed from figure 3a to be elongated ferrite along the rolling reduction. It is obvious that ferrite phase was significantly elongated and refined with highly rolling reduction at WR 70%. In fact, the microstructure becomes finer with increasing of warm rolling reduction (Figure 3c-3g).

3.1.2. Energy Dispersive X-ray spectroscopy (EDS) Analysis
The microstructure of the SEM micrograph to EDS characterization is shown in figure 4. There are dark particles that indicated the existence of carbide and bainite and ferrite that is masked in grey. The result of EDS characterization shows that dark particle consists of aluminum compound as much as 59% in figure 4a and 61% in figure 4b. The amount of aluminum slightly increases following the rising of warm rolling reduction. The raising of warm temperature rolling will increase the content of aluminum compound up to 92% in dark particle. This is caused by carbide is not being transformed at high temperature and producing the granular microstructure. The grey microstructure consists of mainly Fe compound which consists of ferrite and bainite phase. The amount of carbon contained in the ferrite phase is less than in the bainite phase and increase following the temperature.

Furthermore, the result of EDS analysis depicts that the effect of increasing warm rolling reduction does not only increase aluminum content but also decreases nickel contained in ferrite phase. In bainite phase, aluminum content decreases while nickel content increases with increasing warm rolling reduction. Increasing temperature of warm rolling will decrease the fluctuated aluminum and nickel contained in ferrite and bainite. However, the highest aluminum content is found at the lowest temperature warm rolling.

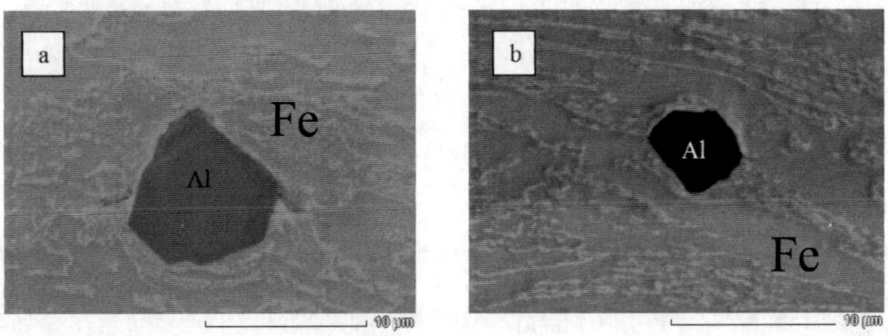

Figure 4. SEM micrographs to EDS characterization of the samples, (a) WR 50% for 30 min at 400°C, (b) WR 70% for 30 min at 400°C

3.1.3. X-Ray Diffraction (XRD) Analysis
Typical XRD patterns is shown in figure 5, in which all peaks only consist of α phase. The integrated intensity ratio of α (110) peaks is much higher than other peaks in the samples. The intensity of α (110) increases when warm rolling temperature decreases. The α (310) peak slightly increases when warm rolling temperature decreases. Furthermore α (200), α (211) and α (220) peak increase fluctuate and tends to be stable without being affected by temperature.

The effect of warm rolling with 50% and 70% reduction on XRD pattern shows the integrated intensity ratios α (110) decreases when the warm rolling reduction increases. The α (200) and α (211) peak fluctuated as the effect of increasing warm rolling reduction. From typical XRD pattern, it can be concluded that there is no carbide form and further confirmation that only bainite and ferrite phases exist in the samples.

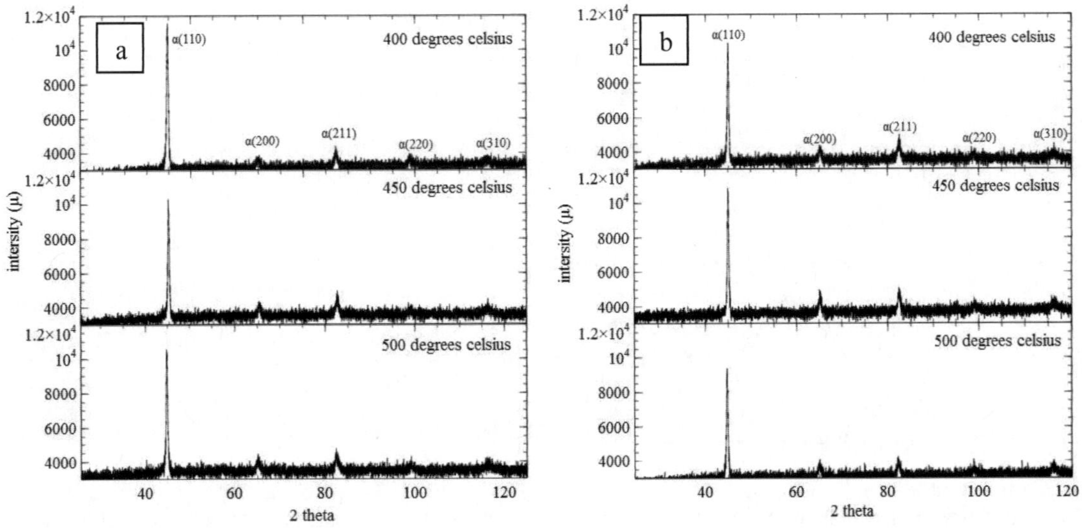

Figure 5. XRD patterns of lateritic steel alloyed with Al sample (a) WR 50% at 400°C, 450°C and 500°C, (b) WR 70% at 400°C, 450°C and 500°C.

3.2 Mechanical Properties

3.2.1. Hardness Measurements

The variety of macro hardness of as cast and warm rolled steels with different reduction are shown in figure 9. Sample with austenization treatment at 945°C followed air cooling and no warm rolling has lower hardness (3.37 HRC) than other samples. This confirms that microstructure of sample No WR consists of M/RA, ferrite and less bainite ferrite. Sample with isothermal treatment without holding time followed WR 50 % has 17.23 HRC. The hardness of sample that is no holding WR 50% increases because the warm rolling treatment and the amount of bainite ferrite increase while the amount M/RA decreases.

Figure 9 also shows the value of hardness increase significantly with warm rolling treatment. The hardness of 70% warm rolling reduction is higher than hardness of 50% warm rolling reduction. The deformation effect increases dislocation density and the material hardening by inhibiting the movements of dislocation. From the results of optic and SEM photographs, shows that warm rolling reduction changes in grain shape from equiaxed to elongated and refined. These internal changes have led to increase the hardness and strength. With the increasing of deformation, the stored energy increases and changes the driving force for recrystallization and form many new nucleation. Furthermore, figure 9 shows that the hardness increases as the transformation temperature reduce. It confirms that hardness increases when the less refinement of bainitic ferrite is detected on microstructure.

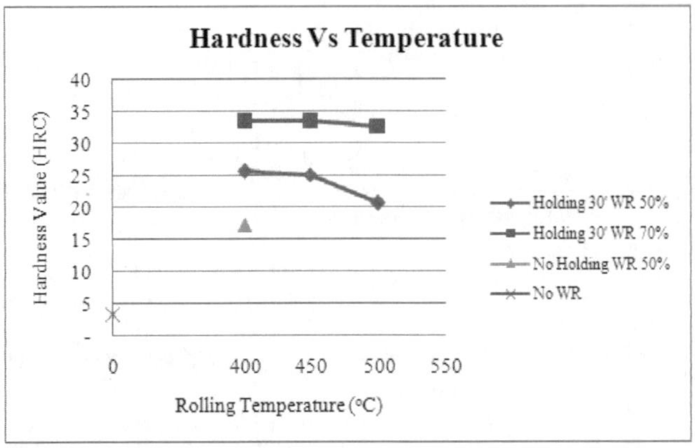

Figure 6. Hardness Value WR versus Rolling Temperature

4. Conclusions

The heat treatment, warm rolling process, microstructure and mechanical property of lateritic steel alloyed with Al was investigated. This studied found that the microstructure of bainite ferrite and ferrite was obtained, and the precipitation of carbide was reduced. This indicates the additionof aluminum effectively suppress carbide formation during bainite transformation.The results of XRD showthat no carbide form and further confirmation that only bainite and ferrite phases are present in the sample.Effect of warm rolling, the microstructure of lateritic steel changes form equiaxed to lamellar structure bainite ferrite and elongated ferrite with the longitudinal axis along rolling reduction. It obvious that the microstructure become finer with increasing warm rolling reduction.The hardness was found to increase with increasing warm rolling reduction and decreasing temperature warm rolling. The maximum hardness value is 33.62 HRC obtained isothermal treatment at 400°C for 30 min followed by 70% warm rolling reduction.

References

[1] P. Pointner 2008 *Wear* vol. 265 (9–10) 1373–1379
[2] S. R. Clayton P, Devanathan R and Jin N 1993 *A review of bainitic steels for wheel/ rail contact. In: Rail quality and maintenance for modern railway operatio*n (Dordrecth: Springer) 41–51
[3] Bhadeshia HKDH 2001 *Bainite in Steels* (London: IOM Commer. Ltd)
[4] F. G. Caballero and H. K. D. H. Bhadeshia 2003 *Isij Int.* **43(11)** 1821–1825
[5] T. S. Wang, J. Yang, C. J. Shang, X. Y. Li, B. Zhang and F. C. Zhang 2009 *Scr. Mater* **61(4)** 434–437
[6] L. Qian, Q. Zhou, F. Zhang, J. Meng, M. Zhang, and Y. Tian 2012 *Mater. Des.* **39** 264–268
[7] D. B. DeMeyer M, Vanderschueren D 1999 *ISIJ Int.* **39** 813–822
[8] C. Shao, W. Hui, Y. Zhang, X. Zhao, and Y. Weng 2017 *Mater. Sci. Eng. A* **682**45–53
[9] T. HOJO, K. SUGIMOTO, Y. MUKAI and S. IKEDA 2008 *ISIJ Int* **48** 824–829
[10] C. Hofer, F. Winkelhofer, J. Krammerbauer, H. Clemens, and S Primig 2015 *Mater. Today Proc.* **2** S925–S928
[11] W. Hui, C. Shao, Y. Zhang, X. Zhao and Y. Weng 2017 *Mater. Sci. Eng. A* **707** 501–510
[12] R. Ueji, N. Tsuji, Y. Minamino and Y. Koizumi 2004 *Sci. Technol. Adv. Mater.* **5(1–2)** 153–162
[13] L. Lv, L. Fu, S. Ahmad and A. Shan 2017 *Mater. Sci. Eng. A* **704** 469–479
[14] F. G. Caballero, M. K. Miller, S. S. Babu and C. Garcia-Mateo 2007 *Acta Mater* **55(1)** 381–390

Acknowledgments
The authors would like to acknowledge financial support from Research Center for Metallurgy and Materials, Indonesian Institute of Sciences.

IOP Conf. Series: Materials Science and Engineering **541** (2019) 012012 doi:10.1088/1757-899X/541/1/012012

Mechanical properties and curing characteristics of shape memory natural rubber

N M Setyadewi[1], I N Indrajati[1] and N Darmawan[1]
[1]Center for Leather, Rubber and Plastics, Ministry of Industry 9 Sokonandi Road, Yogyakarta, 55166, Indonesia

E-mail : nmsetyadewi@gmail.com

Abstract. Natural rubber, which is a polydiene elastomer, is one of the attractive raw materials that could be developed as a shape memory polymer because of its high mechanical and elastic properties. This study aims to determine mechanical properties and curing characteristics of heat-induced shape memory polymer of natural rubber or herein after referred to shape memory natural rubber (SMNR). Natural rubber with different blend ratios of sulphur was prepared by a two-roll mill. The curing characteristics were determined using Moving Die Rheometer at 150°C. The effect of sulphur (0.75; 1.25; 1.5 phr) on mechanical properties such as tensile strength, tear strength and shape memory properties of the stearic acid swollen SMNR vulcanizates were investigated. This research showed that the higher sulphur amount gives the higher torque difference and higher mechanical properties. Shape memory compound swollen stearic acid exhibited shape recovery of 98-100% and shape fixity 59-88.5%.

1. Introduction

The polydiene elastomers (i.e., natural rubber (NR), cis-polybutadiene (PB), synthetic cis-polyisoprene (PI), styrene–butadiene rubber (SBR)) are attractive materials for shape memory polimers (SMPs). They are cheap, have good mechanical properties, have high elasticity and are readily cross-linked using robust chemistries [1]. The used of natural rubber as an elastomer was very broad because the properties of the material formed depend on the degree of cross-linkage. Several studies have provided a brief overview of the extraordinary properties of natural rubber that can be created by the low degree of rubber crosslinking, including the shape memory effect on cold temperatures, the ability to store very large strains, energy, and the ability to sense and remember environmental conditions [2].

SMPs is typically programmed by heating it above a certain temperature, called the trigger temperature T_{trig}, brought into a new shape, and subsequently cooled below T_{trig} to fix this new shape [2]. Shape memory polymers which are a heat-induced form of natural rubber or hereinafter referred to as shape memory natural rubber (SMNR) can be made from various methods, including by making very low cross-link degrees, SMNR is made by manipulating the degree of natural rubber crosslinking. The material is able to return to its original form on critical transverse stress. Transversal stress variations make the trigger temperature range from 45 to 0°C [3]. Another method is to make a program for trigger temperatures (Ttrig), swelling in fatty acids [1],[4], through crosslinking with the Oxa-Michael reaction [5], zinc stearate mixing in EPDM [6].

Refer to Brostowitz et. al., that investigated a commercial rubber band were swollen in molten stearic acid at 75°C (35 wt. % stearic acid loading). Under manual, strain-controlled, tensile deformation the shape memory rubber bands exhibited fixity and recovery of 100% ± 10. In this research, SMNR was made from the compound formula with variations of sulphur addition and swelling stearic acid method (100 wt.%).

2. Experimental Method

2.1. Material

SMNR compound was made from natural rubber (NR) Ribbed Smoke Sheet-1, Zinc oxide (ZnO, Indoxide) and stearic acid (Aflux 42M Rhein Chemie) was used as activator, Carbon black N660 was used as filler, antidegradant 2,2,4-Trimethyl-1,2-Dihydroquinoline (TMQ) and N-(1,3-Dimethylbutyl)-N'-phenyl-p-phenylenediamine (6PPD, Starchem), paraffinic oil (Indrasari), N-Cyclohexyl-2-benzothiazole sulfenamide (CBS) was used as accelerator and sulphur (Miwon) as vulcanizing agent.

2.2. Method

The compounds were prepared by mixing the ingredients for SMNR compounds with a laboratory size two-roll mill as illustrated in Table 1. Formula with amount of sulphur variation 0.75; 1.25; 1.5 phr each given the symbol NRA, NRB and NRC. SMNR compounding was started by mastification of NR approximately 5 min, then followed adding with ZnO and stearic acid, paraffinic oil and carbon black, then 6PPD, TMQ, CBS, and the last is sulphur. After obtained a homogeneous compound, approximately 21 min, the compound was stored for 24 h in room temperature. Curing time of the compound was tested by Moving Die Rheometer Gotech M-3000A at 150°C (Gotech 3000 A). SMNR compounds were vulcanized in a hydraulic press (Toyoseiki A-652) at 150°C with a pressure of 150 kg/cm^2 with time-based on the rheometer data.

Table 1. SMNR compound formulation.

Material	The amount, phr (per hundred rubber)		
	NRA	NRB	NRC
NR	100	100	100
ZnO	0.2	0.2	0.2
Stearic acid	1.4	1.4	1.4
CB N660	3	3	3
TMQ	1	1	1
6 PPD	0.5	0.5	0.5
Paraffinic Oil	3	3	3
CBS	1	1	1
Sulphur	0.75	1.25	1.5

2.3. Curing Characteristics

SMNR compound rheology with various quantity sulphur was acquired by Moving Die Rheometer Gotech M-3000A at 150°C. The curing characteristics such as scorch time (t_{s2}), optimum time (t_{c90}), minimum elastic torque (ML) and maximum elastic torque (MH) can be determined.

2.4. Mechanical Properties

The mechanical properties of SMNR vulcanizates were observed using Universal Materials Testing Machine (UTM) Tinius Olsen.

2.5. Shape Memory Properties

The shape memory properties of the stearic acid swollen SMNR vulcanizates were characterized by modified method refers to Brostowitz et.all and Sun et al. SMNR vulcanizates were swollen in molten stearic acid (100 % w/w) at 75°C for 1 h followed by quenching in a beaker of 10°C water.

For the shape memory testing the SMNR vulcanizates with length L_0 were immersed in a 75°C water bath for 30 s, stretched to 100% strain with a permanent set, and then immersed in a beaker of 10°C water for 30 s. Then the stretched samples were left at permanent set approximately 1 h at temperature room and the length was measured as L_1. After 1 h, SMNR vulcanizate unclamped from a permanent set and the length were measured again as L_2. Shape recovery was initiated by immersing the unclamped SMNR vulcanizate in the 75°C water bath for 30 s, and then immersed in a beaker of 10°C water for 30 s, the final length was measured and taken as L_3. Gauge length was marked on the sample a pen and measured with calipers. The shape memory phenomena were evaluated by calculating the shape recovery (SR) using equation (1) and shape fixity (SF) was defined as the strain fixed by crystalline and calculated with Equation (2).

$$SR = \frac{(L_2 - L_3)}{(L_2 - L_0)} x100\% \quad (1)$$

$$SF = \frac{L_2}{L_1} x100\% \quad (2)$$

3. Results and Discussion

3.1. Cure Characteristics

The cure characteristics as well as scorch time (t_{s2}), optimum time (t_{c90}), minimum elastic torque (ML) and maximum elastic torque (MH) are shown in Table 2. That could be seen from Table 2 if the increased amount of sulphur makes short the scorch time and optimum curing time. Scorch time is a cue of the vulcanization onset, while optimum curing time indicates required time for the cured rubber to get 90% of the maximum accomplishable torque. In the process of crosslinking, heat energy is needed, so the shorter optimum curing time (tc90) is preferable.

Table 2. Cure Characteristics

Compound	ts_2 (min)	t_{90} (min)	MH kgf.cm	ML kgf.cm	MH-ML kgf.cm
NRA (S=0.75 phr)	1.05	1.49	14.58	2.28	12.30
NRB (S=1.25 phr)	0.53	1.30	15.14	1.02	14.12
NRC (S=1.5 phr)	0.56	1.39	17.45	2.27	15.18

NRA compound provides the highest minimum elastic torque (ML) followed by the NRC compound. ML values were related to the viscosity and ability to process compounds [7]. This indicates that the NRA and NRC compounds have better viscosity and process capability than NRB. While the highest maximum elastic torque (MH) value was obtained by NRC compound. The MH value represents the cross-linking process during the vulcanization process. From Table 2 it can also be seen that the NRC compound has the highest torque difference (MH-ML). Torque difference illustrates the number of cross-linkages that occur and was related to the shear modulus of the compound [8]. The rheometric curve with different sulphur rate can be shown in figure 1.

Figure 1 shows rheometric cure curve, it is known that compound maturation is divided into 3 stages. That is induction, which is the stage where the right compound starts to enter the maturation phase. In this induction phase cross-linking between rubber and the filling material has not occurred, so the compound is still plastic and can still be formed. The next phase is the curing phase, which is the phase where the crosslinking is formed, in this phase the rubber compound hardens as a sign of the crosslinking and mature rubber. The next phase is overcure, which is the phase after the compound has matured.

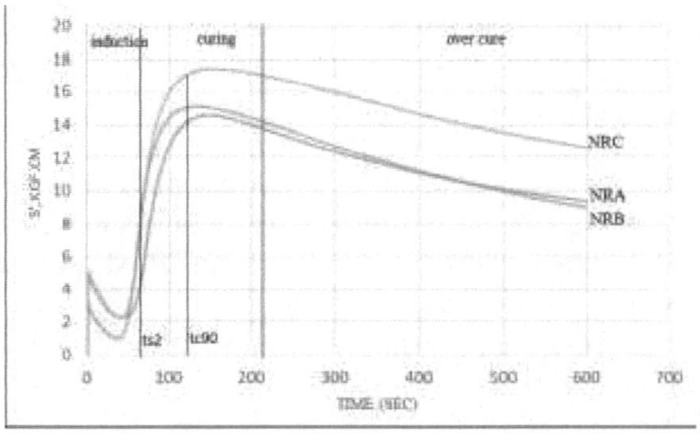

Figure.1. Rheometric cure curve of SMNR compound

3.2. Mechanical properties

Mechanical properties of SMNR is depicted by tensile strength, total elongation and tear strength as shown in Figure 2 and Figure 3. The lowest tensile strength has resulted from NRA blank sample as well as NRA swollen in stearic acid. Generally, tensile strength increases with the increasing of sulphur amount and the crosslink density. Figure. 3 shows the highest tear strength has resulted from NRC (blank) compound and NRB (SA100) compound.

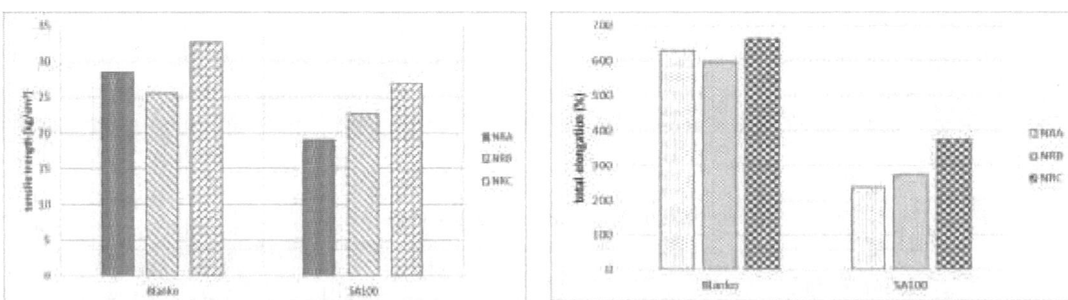

(a) (b)

Figure 2. Effect of a sulphur amount on Tensile Strength (a) and Total Elongation (b)

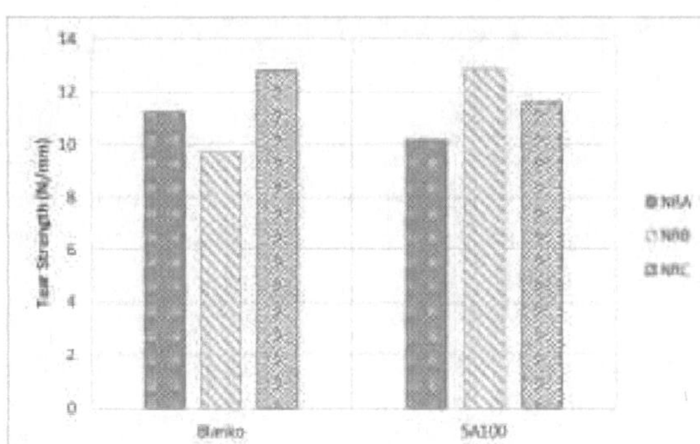

Figure. 3. Effect of a sulphur amount on tear strength.

3.3. Shape Memory Effect

Natural rubber is amorphous at room temperature and changes to crystalline when given strain, so as to produce a self-reinforcing effect. When the stretching force is removed, the rubber will return to its original shape [2]. Table 3 showed the shape recovery (SR) and shape fixity (SF) of SMNR cured with different amount of sulphur (blank) and swollen with stearic acid 100% (SA100).

Table 3. Shape Memory Properties

	Blank			SA100		
	NRA	NRB	NRC	NRA	NRB	NRC
Shape Fixity (%)	8.3	3.5	7.7	88.5	73.4	59.9
Shape Recovery (%)	44.3	57.3	77.2	98.9	98.8	100.1

Immersion treatment with stearate acid makes microscopic crystalline is able to form a temporary network with a permanent crosslinking of a rubber compound. The key to the shape memory effect is the melting crystallization of the stearic acid network and the entropic recovery force of the cross-linked NR [1]. The shape recovery of SMNR vulcanizates comes from the elastic strain and elastic recovery of the SMNR network. Related with data from Table 3, it can also be seen that the NRC compound has the highest torque difference (MH-ML), so is the SR value of NRC compound. The higher the crosslink, the higher the elastic recoverability. So, the higher the sulphur amount, the higher the torque difference, the higher the shape recovery (SR) and the less the shape fixity was.

4. Conclusion

The sulphur amount had an influence on cure characteristics of SMNR compound, mechanical and shape-memory properties. The higher the sulphur amount, the shorter the scorch and optimum time were. The higher the sulphur amount, the higher the torque difference and the higher value the strain recovery degree and the less the shape fixity was.

References

[1] Brostowitz N R, Weiss R A, and Cavicchi K A 2014 *ACS Macro Letters* **3(4)** 374–377
[2] Katzenberg F, and Tiller J C 2016 *Journal of Polymer Science, Part B: Polymer Physics* **54(14)** 1381–1388
[3] Heuwers B, Quitmann D, Katzenberg F, and Tiller J C 2012 *Macromolecular Rapid Communications* 1–6
[4] Huang W M, Zhao Y, Wang C C, Ding Z, Purnawali H, Tang C, and Zhang J L 2012 *Journal of Polymer Research* **19(9952)** 1–34
[5] Lin T F, Ma S W, Lu Y and Guo B C 2014 *Acs Applied Materials & Interfaces* **6(8)** 5695–5703
[6] Dong J and Weiss R A 2013 *Macromolecular Chemistry and Physics* **214(11)** 1238–1246
[7] Jovanovic S, Markovic G, Budinski J and Marinovic M 2013 *Composites: Part B.* **45** 333–340
[8] Jovanović V, Budinski-Simendić J, Samardžija-Jovanović S, Marković G and Marinović-Cincović M 2009 *Chemical Industry and Chemical Engineering Quarterly* **15(4)** 283–289

Acknowledgment

The authors would like to acknowledge the Center of Leather, Rubber, and Plastics, Ministry of Industry, Indonesia for financial support.

IOP Conf. Series: Materials Science and Engineering **541** (2019) 012013 doi:10.1088/1757-899X/541/1/012013

Synthesis of immobilized nanostructured ZnO grown on glass and its photocatalytic activity for cyanide degradation

R S Mohar[1], **Isnaeni**[2], **and G Hadiko**[1]
[1] Center of Technology for Mineral Resources Development, Badan Pengkajian dan Penerapan Teknologi (BPPT), Gedung 820 Kawasan Puspiptek, Tangerang Selatan, Banten 15314, Indonesia
[2] Research Center for Physics, Indonesian Institute of Sciences, Gedung 422 Kawasan Puspiptek, Tangerang Selatan, Banten 15314, Indonesia

E-mail: rahmat.setiawan@bppt.go.id

Abstract. Nanostructured ZnO has been synthesized by using two simple steps, i.e. ultrasonic spray pyrolysis (USP) and chemical bath deposition (CBD) methods on glass substrates. The crystal structure, morphology, photoluminescence (PL) emission, and photocatalytic activity for cyanide degradation were studied. From the X-ray diffraction (XRD) examination, it was observed that the ZnO had the hexagonal wurtzite structure with the preferred orientation of (100). The field emission scanning electron microscopy (FESEM) image displayed that the ZnO had grass-like and bush-like nanostructures. The PL spectrum exhibited the near band-edge emission (NBE) peak at ~380 nm. The photocatalytic activity of the nanostructured ZnO for cyanide degradation in aqueous solution was evaluated under warm-white lamp (color temperature of 2700 K) and cool-daylight lamp (color temperature of 6500 K) irradiations, resulting enhanced cyanide degradation rates in presence of the nanostructured ZnO.

1. Introduction

Zinc oxide (ZnO) is a versatile semiconductor material that nowadays attracts much attention. With its wide band gap of ~3.37 eV, high exciton binding energy of ~60 meV, and native crystal defect properties, ZnO has a potential application as photocatalysts for degrading various hazardous chemicals, such as acetaldehyde [1], nitrophenols [2], organic dyes [3–8], and cyanide [8–12].

The photocatalytic activity of ZnO nanoparticles, which are considered highly mobile once immersed in non-stagnant aqueous solutions, has been reported in many scientific articles [1,2,8–12]. Other articles have also reported studies concerning the photocatalytic activity of immobilized ZnO grown on and adhered to fixed surfaces, such as silicon [13], indium tin oxide [7], sapphire [14], and glass [3–5]. However, none of those articles have discussed the use of immobilized ZnO adhered to fixed surfaces for cyanide photodegradation under daylight irradiation.

Some methods have been developed to produce ZnO nanostructures, such as pulsed laser ablation [10], pulsed laser deposition [14], metalorganic vapor-phase epitaxy [15], sol-gel [3,5,11], ultrasonic spray pyrolysis (USP) [4,7,16], sonochemical impregnation [12], combustion [17], chemical bath deposition (CBD) [18,19], and hydrothermal [3–7,16,20]. Among these methods, USP and CBD were chosen to be conducted in this study as they offer simplicity and efficiency in chemical use, and can be carried out at relatively low temperature [4,7,18,19].

In this report, we demonstrate the synthesis of immobilized nanostructured ZnO grown on glass substrates using two simple steps, i.e. USP and CBD methods, and its photocatalytic activity in degrading cyanide ions under warm-white lamp and cool-daylight lamp irradiations.

2. Experimental Details

2.1. Synthesis of Nanostructured ZnO

The first step was the USP method to seed glass substrates with ZnO seed layers before the nanostructured ZnO could be grown on the substrates. Prior to the process, the substrates were cleaned in acetone, ethanol, and distilled water, respectively. Zinc nitrate tetrahydrate ($Zn(NO_3)_2.4H_2O$, 99.99%, Merck Millipore) dissolved in distilled water with the concentration of 0.2 M was placed into an ultrasonic nebulizer to be vaporized by ultrasonic vibration (1.65 MHz) at room temperature. The vapor then flew through a tube (flow rate set at 1.5 ml/min) and exposed the glass substrates placed on a hotplate at the set temperature of 500°C for 30 minutes without any vacuum system. The seeded substrates were then annealed at 400C for an hour on the hotplate.

In the next step, nanostructured ZnO was grown on the seeded glass substrates using the CBD method similar to the growth method in the aqueous solution developed by Vayssieres [21] with some modifications. Growth solution was prepared by dissolving $Zn(NO_3)_2.4H_2O$ and hexamethylenetetramine ($C_6H_{12}N_4$, 99.99%, Merck Millipore) in distilled water with the equimolar concentration of 0.05 M. The growth process was performed by immersing the seeded glass substrates into the growth solution in an uncovered glass beaker on a hotplate at ~85°C for 3 hours. After the growth process, samples were rinsed in distilled water and dried out on a hotplate set at 170°C for 10 minutes.

2.2. Characterization

The crystal structure was investigated using Rigaku MiniFlex 600 X-ray diffractometer with CuKα radiation (1.5406 Å). The morphology of the sample was observed using JOEL JIB-4610F Multi-Beam System field emission scanning electron microscope (FESEM). The photoluminescence (PL) emission was examined using Maya Ocean Optics detector under the emitting laser's wavelength of 325 nm, measured at room temperature.

2.3. Photocatalytic activity evaluation

The photocatalytic activity for cyanide degradation was evaluated through exposing the cyanide solution in basic condition to the light from warm-white lamps (color temperature of 2700 K, intensity ~4800 lux) and cool-daylight lamps (color temperature of 6500 K, intensity ~4800 lux). The cyanide solution (200 ppm) was prepared by dissolving sodium cyanide (NaCN, 99%, technical grade) in distilled water, in which sodium hydroxide (NaOH, technical grade) had been added to raise pH until it reached 12.5 beforehand. Afterward the cyanide solution was irradiated with warm-white lamps and cool-daylight lamps, separately, in the presence and absence of the nanostructured ZnO grown on glass substrates with a total surface area of 12 square inches. The concentration of residual cyanide ions ($[CN^-]$) was determined every 15 minutes until two hours using the argentometric titration method with a few drops of 5-(4-dimethylaminobenzylidene)-rhodanine (≥98%, Merck Millipore) dissolved in acetone as the indicator and silver nitrate ($AgNO_3$, 99.99%, Merck Millipore) dissolved in distilled water as the titrant.

3. Results and discussion

Figure 1 shows the X-ray diffraction (XRD) pattern of the nanostructured ZnO grown on the glass substrate. The pattern represents diffraction peaks attributed to (100), (002), (101), (102), (110), (103), and (112) planes, corresponding to the hexagonal wurtzite crystal structure of ZnO (ICDS 98-016-5002), and no phase other than ZnO was detected from the pattern. Since spray pyrolysis methods in general characteristically produce rough surfaces [20], the ZnO seed layer produced by the USP was quite rough and had many orientations, then affected the growth orientation of the nanostructured ZnO. The preferred orientation of the ZnO crystals was (100), determined by the extremely prominent peak at 2θ of 31.78°. This result, thus, suggests that most of the ZnO nanorods had the lateral growth orientation along the a-axis (parallel to the substrate's surface).

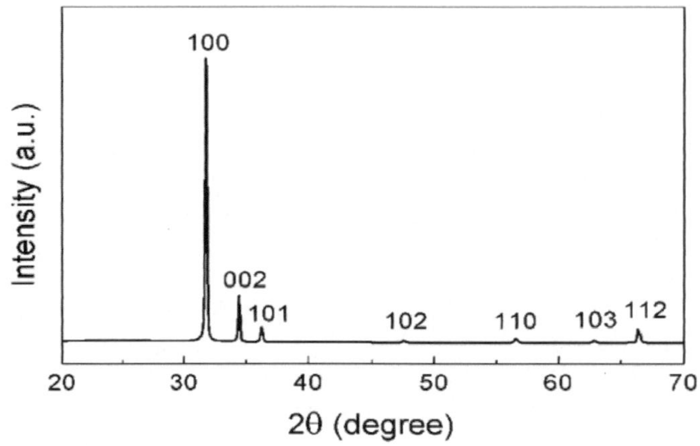

Figure 1. XRD pattern of the nanostructured ZnO grown on glass.

Figure 2. FESEM images of the nanostructured ZnO grown on glass.

IOP Conf. Series: Materials Science and Engineering **541** (2019) 012013 doi:10.1088/1757-899X/541/1/012013

Figure 2 displays the FESEM images taken from some points on the sample's surface. The images represented the typical morphology of ZnO nanostructures grown on the seed layer fabricated using spray pyrolysis methods, which was comprised of numerous disarrayed ZnO nanorods grown along with many directions [20]. This morphology might benefit the photocatalytic activity since it obviously increases the surface area. Most of the nanorods were oriented parallel to the substrate, rendering the predominant grass-like nanostructure, and some of the nanorods randomly tilted and were perpendicular with respect to the substrate, rendering the flower-like nanostructure. This also confirms the result from XRD examination that the nanostructured ZnO had some plane orientations and the preferred orientation was along the a-axis.

Figure 3 shows the PL emission of the nanostructured ZnO grown on glass substrates in logarithmic scale. The near band-edge emission (NBE) of the ZnO, which attributes the excitonic recombination [4,7], is exhibited ranging from ~360 nm to ~420 nm with its peak in the ultraviolet region at ~380 nm. The bandgap obtained from the peak, determined by the equation E = hv, is ~3.26 eV. A deep level emission (DLE) generated by defects appeared extremely weak in the visible region, inferring that the ZnO crystals had very low defect density and excellent crystal quality [15,22].

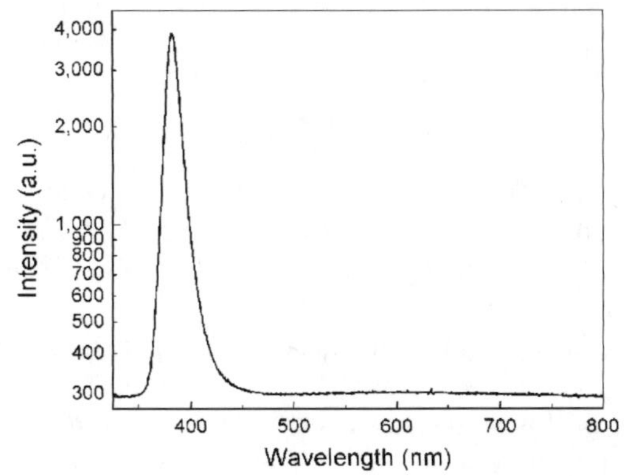

Figure 3. PL emission of the nanostructured ZnO grown on glass in logarithmic scale.

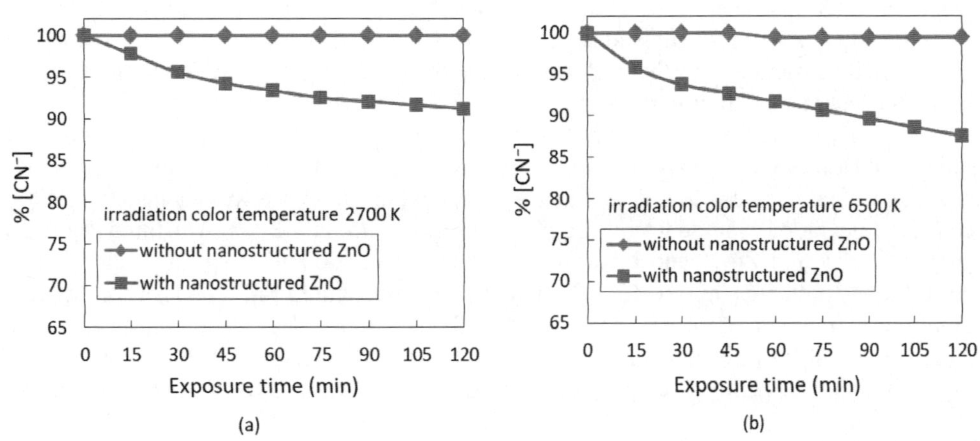

Figure 4. Cyanide degradation under (a) warm-white lamp (color temperature of 2700 K) and (b) cool-day light lamp (color temperature of 6500 K) irradiations.

IOP Conf. Series: Materials Science and Engineering **541** (2019) 012013 doi:10.1088/1757-899X/541/1/012013

The evaluation of photocatalytic activity for cyanide degradation is presented by figure 4. It is shown that the cyanide degradation rates were slightly enhanced in presence of the nanostructured ZnO grown on glass substrates. Without the nanostructured ZnO immersed in the cyanide solution, the warm-white irradiation for two hours seemed to have a negligible effect on cyanide degradation. With the nanostructured ZnO, the warm-white lamp irradiation for two hours could degrade 8.79% of cyanide ions as shown in figure 4(a). On the other hand, the cool-daylight exposure seemed to render tenuous cyanide degradation up to 0.52% in absence of the nanostructured ZnO. However, with the nanostructured ZnO, the degraded cyanide ions reached 12.44% in two hours of cool-daylight lamp irradiation as shown in figure 4(b).

The mechanism behind the photocatalytic degradation of cyanide by ZnO can be explained as follows. When ZnO is irradiated by photons possessing enough energy, the electrons excite from the valence state to the conduction state generating free electrons (e^-) and holes (h^+). The electron–hole pairs (excitons) then react with oxygen and water, producing hydroxyl radicals ($OH^•$). In general, the mechanism for the photocatalytic degradation of cyanide by ZnO can be described as follows [8–11]:

$$\mathrm{ZnO} + h\nu \rightarrow \mathrm{ZnO}(e^- + h^+) \tag{1}$$

$$e^- + \tfrac{1}{2}\mathrm{O_2} + \mathrm{H_2O} \rightarrow \mathrm{OH^-} + \mathrm{OH^\bullet} \tag{2}$$

$$h^+ + \mathrm{H_2O} \rightarrow \mathrm{H^+} + \mathrm{OH^\bullet} \tag{3}$$

$$\mathrm{CN^-} + 2\mathrm{OH^\bullet} \rightarrow \mathrm{OCN^-} + \mathrm{H_2O} \tag{4}$$

In the case when ZnO has a fairly high amount of crystal defects, the above mechanism not only depends on the interband excitation. The existence of crystal defects in ZnO also plays an important role in photocatalytic degradation, as reported by some authors [4,5,8,11]. The defects provide some deep level energy states between the valence and conduction bands that can be occupied by electrons and allows photons lower than the bandgap to excite electrons and generate excitons [8,11]. Moreover, the defects can act as traps for electrons and holes, hence inhibiting recombination and improving the photocatalytic activity [4,5]. In this study, however, it can be inferred that the above photocatalytic degradation mechanism depends thoroughly on the interband excitation and does not quite rely on the deep level states since the PL emission (figure 3) indicated that the nanostructured ZnO contained a very slight quantity of crystal defects [10,12].

4. Conclusions

Nanostructured ZnO has been successfully synthesized using two simple steps, i.e. USP and CBD methods, immobilized on glass substrates. It was investigated that the nanostructured ZnO had the hexagonal wurtzite structure with the preferred orientation of (100), suggesting that most of the ZnO crystals had the lateral growth orientation along the a-axis. The nanostructure of the ZnO was made up of grass-like and bush-like formations of ZnO nanorods. From PL examination, a very strong NBE was exhibited at ~380 nm (~3.26 eV) and an extremely weak DLE appeared in the visible region, indicating that the ZnO crystals had very low defect density. The photocatalytic activity evaluation showed the enhancement of degraded cyanide from negligible decrease (without the nanostructured ZnO) to 8.79% (with the nanostructured ZnO) under warm-white lamp irradiation for two hours and from 0.52% (without the nanostructured ZnO) to 12.44% (with the nanostructured ZnO) under cool-daylight lamp irradiation for two hours. It is inferred that the photocatalytic degradation mechanism depends thoroughly on the interband excitation as the PL examination indicates that the nanostructured ZnO contained only a slight quantity of crystal defects.

References

[1] Chen Y, Katsumata K, Chiu Y, Okada K, Matsushita N and Hsu Y 2014 *Appl. Catal. A, Gen.* **490**
[2] Qamar M T, Aslam M, Ismail I M I, Salah N and Hameed A 2015 *ACS Appl. Mater. Interfaces* **7** 8757–69
[3] Nasr-Esfahani M and Nekoubin A 2011 *AIP Conference Proceedings* **1341** 309–14
[4] Putri N A, Fauzia V, Iwan S, Roza L, Umar A A and Budi S 2018 *Appl. Surf. Sci.* **439** 285–97
[5] Ali A, Zhao X, Ali A, Duan L, Niu H, Peng C, Wang Y and Hou S 2015 *Superlattices Microstruct.* **83** 422–30
[6] Faisal M, Ismail A A, Ibrahim A A, Bouzid H and Al-Sayari S A 2013 *Chem. Eng. J.* **229** 225–33
[7] Aisah N, Gustiono D, Fauzia V, Sugihartono I and Nuryadi R 2017 *IOP Conference Series: Materials Science and Engineering* **172** 012037
[8] Maya-Treviño M L, Guzmán-Mar J L, Hinojosa-Reyes L and Hernández-Ramírez A 2018 *Mater. Sci. Semicond. Process.* **77** 74–82
[9] Bagabas A, Alshammari A, Aboud M F and Kosslick H 2013 *Nanoscale Res. Lett.* **8** 516
[10] Saleh T A, Gondal M A and Drmosh Q A 2010 *Nanotechnology* **21** 495705
[11] Salinas-Guzmán R R, Guzmán-Mar J L, Hinojosa-Reyes L, Peralta-Hernández J M and Hernández-Ramírez A 2010 *J. Sol-Gel Sci. Technol.* **54** 1–7
[12] Karunakaran C, Gomathisankar P and Manikandan G 2010 *Mater. Chem. Phys.* **123** 585–94
[13] Azam A and Babkair S S 2014 *Int. J. Nanomedicine* **9** 2109–15
[14] Kenanakis G, Pervolaraki M, Giapintzakis J and Katsarakis N 2013 *Appl. Catal. A Gen.* **467** 559–67
[15] Park W I, Kim D H, Jung S-W and Yi G-C 2002 *Appl. Phys. Lett.* **80** 4232
[16] Mohar R S, Iwan S, Djuhana D, Imawan C, Harmoko A and Fauzia V 2016 *AIP Conference Proceedings* **1729** 20024
[17] Wang Y, Zhao X, Duan L, Wang F, Niu H, Guo W and Ali A 2015 *Mater. Sci. Semicond. Process.* **29** 372–9
[18] Tian F and Liu Y 2013 *Scr. Mater.* **69** 417–9
[19] Kuriakose S, Satpati B and Mohapatra S 2015 *Phys. Chem. Chem. Phys.* **17** 25172–81
[20] Breedon M, Rahmani M B, Keshmiri S H, Wlodarski W and Kalantar-zadeh K 2010 *Mater. Lett.* **64** 291–4
[21] Vayssieres L 2003 *Adv. Mater.* **15** 464–6
[22] Miao L, Ieda Y, Tanemura S, Cao Y G, Tanemura M, Hayashi Y, Toh S and Kaneko K 2007 *Sci. Technol. Adv. Mater.* **8** 443–7

Acknowledgements

This study was funded by Center of Technology for Mineral Resources Development, BPPT and supported by the research collaboration with Research Center for Physics, Indonesian Institute of Sciences.

IOP Conf. Series: Materials Science and Engineering **541** (2019) 012014 doi:10.1088/1757-899X/541/1/012014

Mechanical and microstructure properties of the Ni-Cr-Mo modified steel by heat treatment process

S Herbirowo[1], M Syahrum[2], M Y Hasbi[1], S A Chandra[1], F M Ridlo[1], B Adjiantoro[1]
Research Center for Metallurgy and Materials – Indonesian Institute of Sciences, 470 building PUSPIPTEK. Tangerang Selatan, Indonesia
STT-PLN, Jalan Duri Cengkareng, Jakarta Barat, Indonesia

E-mail : satrio.herbirowo@lipi.go.id

Abstract. The development of Iron and steel is the fundamental essential needs for a construction. Indonesia should seek local raw materials for steelmaking production to cooperate with industrial markets. In this research, a new innovation in the manufacturing of bucket teeth components using laterite steel by optimizing the nickel alloy content reaches 1.5-3% and other alloys with special heat treatment process. The best results of optimization abrasion wear resistance and hardness properties on 1Cr1Mo steel with quenching oil treatment resulted in hardness of 43.60 HRC and had a good abrasive resistance of 0.16 gr/cm^2 weight loss. In morphological analysis of microstructure showed that the grain boundaries become more flattened and sharpened into martensite phase transformation.

1. Introduction

Iron and steel manufacturing industries are very rapidly developing and also material innovation has been carried out in the development of bucket teeth product using laterite steel by optimizing the nickel alloy content which reaches 1.5-3%[1]. It is expected that laterite steel can be a new alternative in the production of bucket teeth with a value of mechanical properties and a longer lifetime. In this study will be discussed about the mechanical study of materials especially on the nature of hardness and wear resistance abrasion through a tempering process using metallographic observation methods, abrasion testing, chemical composition testing, and hardness testing which is useful to determine the price of hardness in bucket teeth.

Laterite steel is steel that has much better formability and mechanical properties compared to carbon steel and stainless steel because it has a Ni content of 0 to 3.0% and Cr is 0 to 2.5%. In addition to Fe and Ni content, laterite steel also contains other elements, namely Cr and Mn, where both elements will have an impact on the properties of steel weldability, surface hardness, and corrosion resistance[2]. Low alloys steel is widely used as the main material in piping systems, hydroelectric power plants, petrochemical industries and nuclear power because of its long service life, corrosion resistance, and economical[3].

Abrasion wear is a mechanism of hard material collides with a soft surface and usually leaves wear fragments. A heat treatment approach that can modify phase and microstructures and also modify steel surfaces by applying coatings is a way to increase abrasion wear resistance. Tempering process with the addition of Mo-Cr has no effect on the carbide morphology at low tempering temperatures but can increase wear resistance when the temperature is above 350°C[4].

This goal will be achieved as desired to the factors that influence, such as tempering temperature and cooling media also variations in the material composition. Heat treatment has an influence on the microstructure and mechanical properties of variations in cast steel of Ni-Cr-Mo alloy, tempering temperature can change the microstructure and abrasion resistance and surface hardness of laterite steel.

2. Experimental Method

This experiment starts with cutting the test sample with various composition (Ni, Cr, Mo) to get the specimens ready for abrasion, hardness and metallography testing so as not to be affected by the machining process after heat treatment. The heating process at 950°C with 60 minutes holding time then quench using oil media and tempering at a temperature of 500°C.

Furthermore, heat treatment samples were tested for hardness using rockwell hardness method, and abrasion testing using ASTM G 132-96 standard, ADAMEL LHOMARGY abrasion rotary on cylindrical sample with 2-3 cm^2 surface area where the sample was placed under sandpaper grid size 120 rotating at a speed of 100 rpm as many as 1000 rounds with a load of 1 kg at room temperature of about 30 °C. The results of abrasion testing are weight loss in units of mg / cm^2 which shows the abrasion resistance. Metallographic testing with an optical microscope with grinding preparation, polishing to etch with 2% nital concentrate to analyze the shape, grain boundary and phase formed.

3. Results and Discussion

Chemical composition testing is carried out to determine the content of chemical elements contained in laterite-based alloy cast steel resulting from the casting process with material balance calculations.

Table 1. Chemical composition of the experimental steel

Type	**Chemical Element**								
	C	Si	Mn	P	S	Cr	Ni	Mo	Fe
As-Cast	0.2	0.6	0.47	0.0005	0.0003	0.02	2.4	0.006	Bal.
1Cr1Mo	0.2	0.4	0.5	0.0005	0.0003	1.074	2.2	0.981	Bal.

The chemical composition testing of alloy steel, it can be seen that the main chemical composition contained in the laterite steel samples of As-Cast is (Ni) = 2.483 %, Ni is the main constituent element. From the results of this test the carbon content (C) = 0.1556 % with this carbon content can be concluded that this steel is categorized in medium carbon steel and low alloy steel, medium carbon steel that is steel containing carbon elements from 0.10 % - 0.30 % approached the modified steel standard AISI 4140[5], [6].

Abrasive resistance testing aims to determine the effect of variations of Ni-Cr-Mo alloy in laterite steel and variations in heat treatment of abrasive resistance. Bucket teeth have the main function as soil and rock crushing with wear resistance and toughness to obtain bucket teeth have an optimum service life.

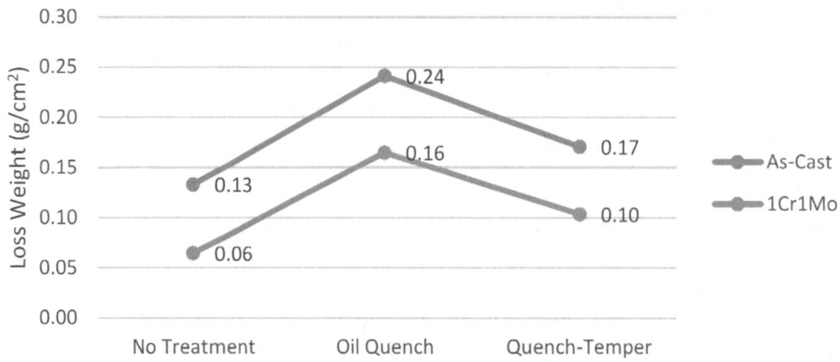

Figure 1. Abrasion test result

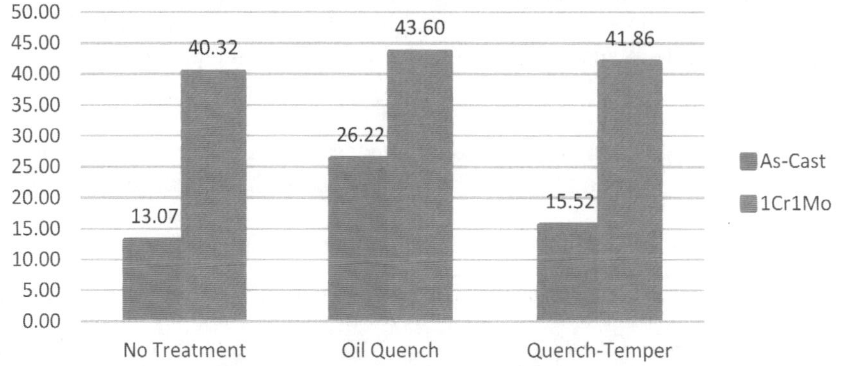

Figure 2. Hardness Rockwell test result

In Figure 1 and 2 showed that the effect of heat treatment on material variations with abrasive resistance values and hardness properties. It can be seen in Figure 1 and 2 on the graph that weight loss and hardness on surface area shown in the untreated 1Cr1Mo sample was 0.06g/cm^2; 40.32 HRC, then at the oil quench the abrasion resistance oil increased by 0.16g/cm^2; 43.60 HRC and when at quench-temper the abrasion resistance was stable at 0.10 g / cm^2 ;41.86 HRC respectively.

Abrasive resistance is strongly dependent on the value of hardness where the higher hardness value also the good abrasive resistance value. But the graph above where different abrasive resistance values are due to the influence of the value of hardness and microstructure formed during the heat treatment process in variety of material[7][8].

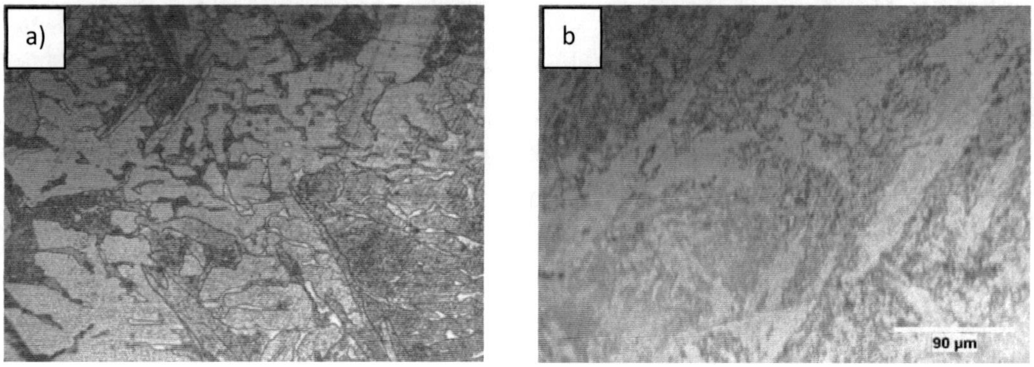

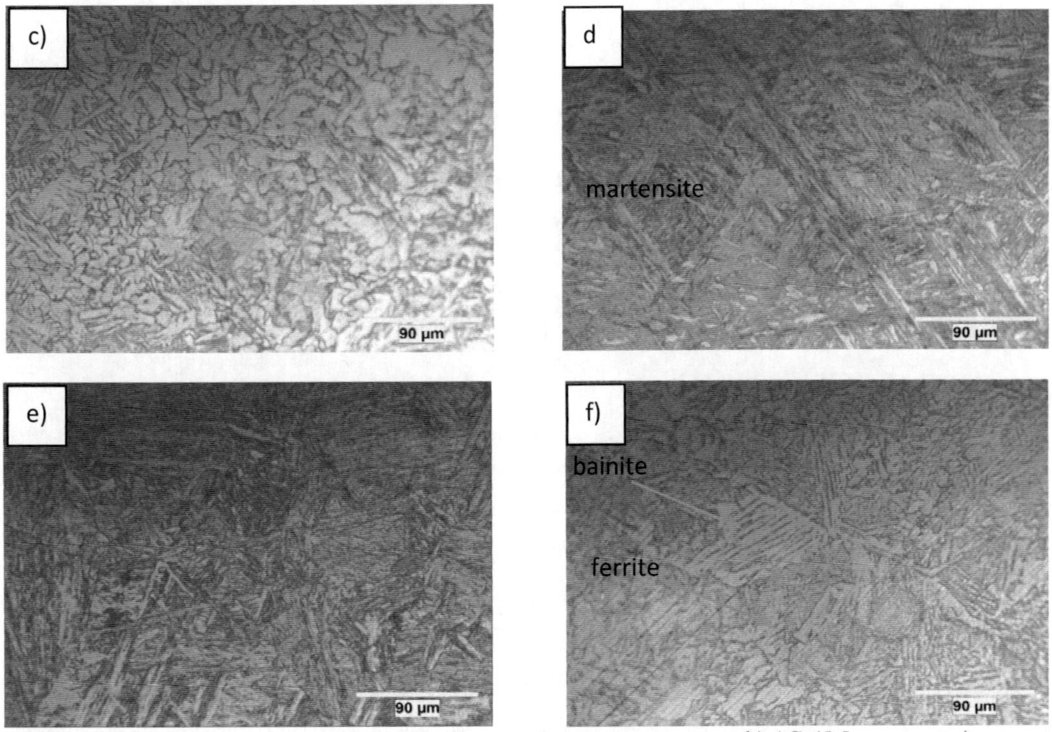

Figure 3. The microstructure of lateritic alloy steel a) as-cast no treat b) 1Cr1Mo no treat c) as-cast quench oil d) 1Cr1Mo quench oil e) as-cast quench-temper f)1Cr1Mo quench-temper

Figure 3 show the microstructure of 1Cr1Mo material obtained from heating at a temperature of 950°C and cooling using oil media showed that the microstructure were increasingly flat and the phase transformed into martensite with visible residual austenite. The martensite phase will made the sample stronger and harder but tend to be brittle. The microstructure of sample which heated at 950°C follow with quench temper showed quench tempering after heat treatment the formation of tempered martensite phase, bainite and ferrite phase with the success of eliminating retained austenite. This increasingly tenacious microstructure supported by the presence of Cr-Mo alloys[9], [10] .

4. Conclusions

1Cr1Mo treated laterite steel has the good characteristics compared as-cast or non-treatment, the highest hardness properties in this study had 1Cr1Mo material with quenching oil treatment produces hardness of 43.60 HRC and has a wear resistance of 0.16 g/cm^2 significantly. It can be projected that Ni-Cr-Mo alloy cast steel can be applied for the development of bucket teeth material which refers to AISI 4140 specifications with 40 HRC hardness. In metallographic testing, the grain boundaries are increasingly flat and the phase is transformed into martensite-tempered and with a pearlite phase.

References

[1] Herbirowo S, Adjiantoro B, and Citrawati F 2018 *Maj. Metal* **1** 35
[2] Jena P. K. *et al* 2010 *Int. J. Impact Eng.* **37**,3, 242
[3] Herbirowo S, Adjiantoro B, Romijarso T. B, and Pramono A W 2018 *AIP Conf Proceedings* **20043**
[4] Haryadi G D 2014 *Rotasi* pp. 1–8
[5] Hadi S, Adi S, and Syariati A 2018 **4**, no. **2** 294
[6] Suryo S H, Bayuseno A P, Jamari J, and Rahmat M A 2018 *AIP Conf. Proc.* **30022** 1977

[7] Apriania E. M. A S, Anwar M S, Rusnaldy 2018 *Maj. Metal* **3** 115
[8] Liu F. *et al* 2016 *J. Alloys Compd.* **689** 225
[9] Yanuar Y P and Nugroho S 2016 7
[10] Herbirowo S and Adjiantoro B 2016 *Widyariset* **2** 153

Acknowledgement
The authors would like to acknowledge financial support from Research Center for Metallurgy and Materials, Indonesian Institute of Sciences.

IOP Conf. Series: Materials Science and Engineering **541** (2019) 012015 doi:10.1088/1757-899X/541/1/012015

Effect of Punch Parameters on Springback for Mild Carbon Steel in A V-Shape Bending Process

M A Suyuti, M Iswar, and R Nur
Mechanical Engineering Department, Politeknik Negeri Ujung Pandang
Center for Materials and Manufacturing, Politeknik Negeri Ujung Pandang

Email: rusdinur@poliupg.ac.id

Abstract. The springback is variably sensitive to materials and process parameters. Considering springback occurred in a formed part is significant for designing tools and dies. In this paper, experiments of a V-shape bending test were performed for investigating the springback effect. The punch parameters variation has been observed in ASTM A-36 mild carbon steel. Punch radius used and punch angle). The V-shape bending of mild steel was performed using Universal Testing Machine (UTM) at various punch radius (2, 4, and 6 mm) and punch angle (80°, 85°, and 90°) and at constant of die radius (10 mm) and die width (50 mm). The experimental result indicates that the punch radius has a most significant impact on springback behaviour of these sheet metal. This is verified through our experimental work using V-shape bend testing.

1. Introduction

In metal forming techniques, bending is one of the most widely used methods, such as the automotive industry and the production of home appliances. The raw material in the form of a plate is formed using bending dies. Bending is a uniform stretching process around a straight axis of flat sheet material. At the point under review, the bending will produce a voltage that is directionally normal to the cross-section of the field. There are two types of bending process namely type V and type U. In the bending process, the elastic limit of the material will be exceeded but not exceed the yield limit [1].

The bending process for plates is usually accompanied by the springback phenomenon. Springback is a problem that cannot be avoided or prevented. The more widespread use of high strength steel and aluminum plates, the springback deviation problem will be even greater. However, springback can be minimized in many ways. One example is to use a smaller punch radius and angle compensation [2-4]. Therefore, the understanding of the springback phenomenon needs to be improved by various studies aimed at overcoming the springback phenomenon.

Various studies to investigate the phenomenon of springback on bending techniques have been widely practiced. Garcia-Romeu et al. [5] investigating springback for air bending on aluminum and stainless steels with a thickness of 2 mm and 3 mm, and a bending angle 22° to 90°. The results obtained from the springback regression model as a function of the end angle. Osman et al [6] developed a theoretical model for water bending along with experimental tests of V-bending annealed and pure commercial aluminum plates. Yilamu et al. [7] studied springback on aluminum clad aluminum on V-shaped water bending with die angle 90°. Simulation with finite element method is done to predict springback. Reche et al. [8] performed a two-dimensional mechanical analysis and simulation of 1.6 mm steel bending test plate with bending angles up to 150°.

IOP Conf. Series: Materials Science and Engineering **541** (2019) 012015 doi:10.1088/1757-899X/541/1/012015

This study investigates the punch parameters (i.e. punch radius and punch angle) on springback when V-shaped bending of mild carbon steel.

2. Experimental Details

This bending experiment is conducted by using die set tool in the Mechanical Laboratory of State Polytechnic of Ujung Pandang. The bending tool is installed on the Universal Testing Machine (UTM) Galdabini Type PM 100. The bending experiment can be seen in the following Figs. 1 and 2.

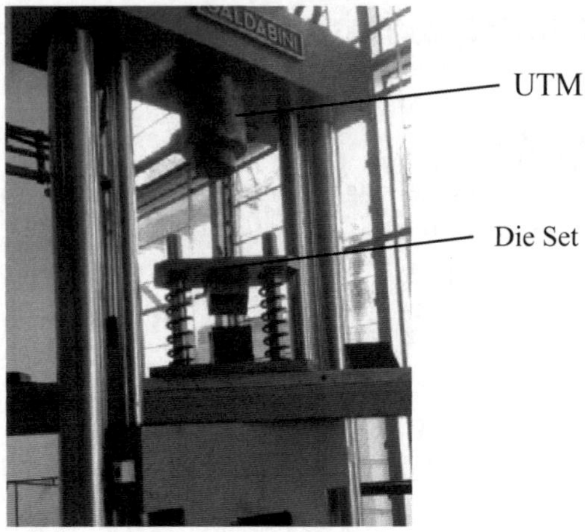

Figure 1. Die Set Installed on Universal Tensile Machine (UTM)

Figure 2. Bending Test on Universal Tensile Machine (UTM)

The process of bending experiment is performed by varying variables, namely punch angle of 80°, 85°, and 90°, and punch radius of 2, 4, and 6 mm. The dies are constant with the angle of 90° and radius of 10 mm. This experiment had been performed by Arsyad et al [9] for a thickness of 6 mm. The material used is carbon steel material ASTM A-36 with a thickness of 2 mm. The mechanical properties of ASTM A-36 are shown in Table 1.

IOP Conf. Series: Materials Science and Engineering **541** (2019) 012015 doi:10.1088/1757-899X/541/1/012015

Table 1. Mechanical properties of ASTM A-36

Tensile strength (MPa)	Yield strength (MPa)	Elongation (%)
416,062	365,000	35.86

3. Result and Discussion

3.1. Punch Radius on Springback

Based on a bending test of steel ASTM A36, it was obtained the punch radius data on springback as shown in figure 3.

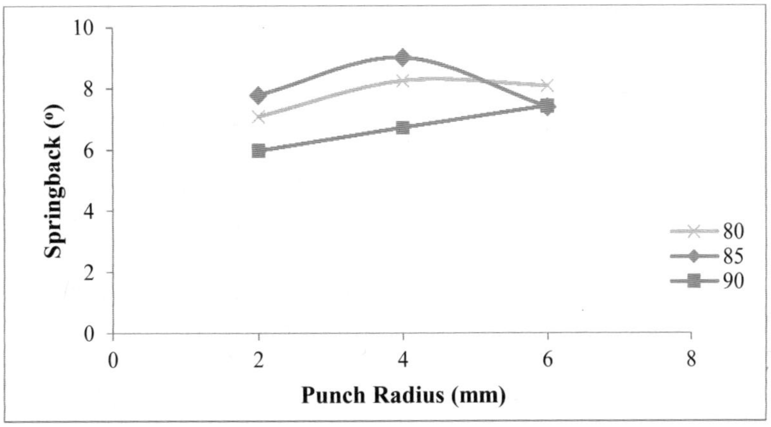

Figure 3. Bending Test on Universal Tensile Machine (UTM)

Figure 3 shows the relationship between punch radius and springback. The smallest springback value is obtained at 5.97° at a 2 mm punch radius, while the largest is 8.25° and 4 mm punch radius. This figure also shows that the increase in springback to increase the punch radius, except at a radius of 6 mm for the corner of punch 85°. Vice versa, the smaller the punch radius and the greater the angle of punch, the resulting springback tend to be small.

3.2. Punch Angle on Springback

Similarly, for the punch angle results, it was obtained the punch radius data on the springback as shown in figure 4.

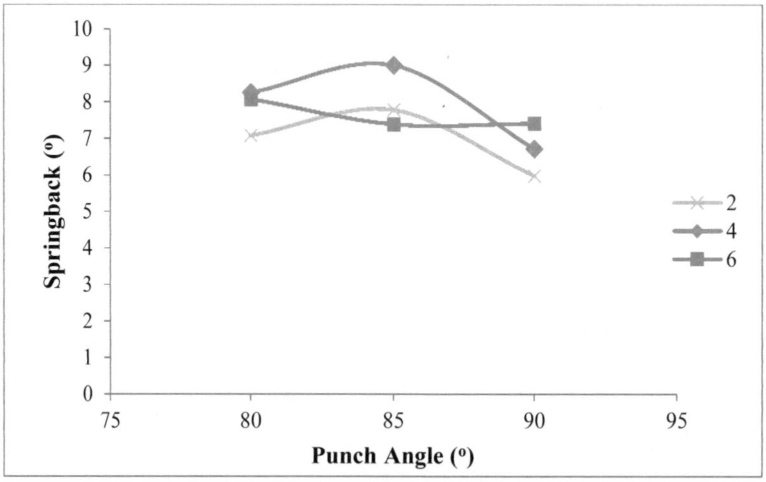

Figure 4. Bending Test on Universal Tensile Machine (UTM)

Figure 4 shows the relationship between the punch angle and springback. The smallest springback value is obtained at 5.97° at punch angle of 90°, while the largest is 8.25° at the punch angle of 85°. This figure also shows that springback looks quadratic curve for increased radius angle, except on 6 mm of punch radius and 85° of punch angle. Vice versa, the smaller the punch radius and the greater the punch angle and the resulting springback tend to be small. Based on the picture above, the smaller the punch radius and the greater the punch angle then the resulting springback tends to be small.

The reasonably small springback angle in this experiment is predictable since there has been a permanent deformation [1, 9]. Springback involves a small strain equivalent to a strain on the elastic deformation of the metal [10]. However, if the permanent deformation is quite dominant, it will reduce the elastic deformation required by the springback phenomenon. Plastic deformation was located in a bending region that changes shape due to pressure from the punch. Chongthairungruang et al. (2013 were investigated springback effects of a mild steel grade JSC270C using method of experiments and simulations. It found that the material strength considerably affected the springback magnitude [11]. Oya et al. (2010) were studied the V-bending process on all of the samples to investigate their bending formability with a springback angle of 3.16, 4.86, 5.99 and 7.14. They stated that the usefulness of the multilayered steel sheet due to its improved plastic formability caused by the remarkable effect of the ductility enhancement [12].

4. Conclusion

The bending test of ASTM A-36 carbon steel was performed by using the universal tensile machine. The punch radius and punch angle were varied at 2, 4, and 6 mm and 80°, 85°, and 90°, respectively. The bending response evaluated was springback. It found that the smaller the punch radius and the greater the punch angle then the resulting springback value tends to be small.

References

[1] Tekıner Z 2004 *Journal of Materials Processing Technology* ***145*****(1)** 109-117
[2] Banabic D 2010 Sheet Metal Forming Process: Constitutive Modelling and Numerical Simulation, Springer
[3] Slota Ján and Miroslav Jurčišin 2012 *Mechanika* **84 (284)** 55-68
[4] Chikalthankar S B, G. D. Belurkar, and V. M. Nandedkar 2014 *International Journal of Engineering and Advanced Technology (IJEAT)* **3**

[5] Garcia-Romeu M L, Ciurana J, and Ferrer I 2007 *Journal of Materials Processing Technology* ***191*(1-3)** 174-177
[6] Osman M A, Shazly M, El-Mokaddem A, and Wifi A S 2010 *Journal of Achievements in Materials and Manufacturing Engineering* ***38*(2)** 179-186
[7] Yilamu K, Hino R, Hamasaki H, and Yoshida F 2010 *Journal of Materials Processing Technology* ***210*(2)** 272-278
[8] Reche D, Besson J, Sturel T, Lemoine X, and Gourgues-Lorenzon A F 2012 *International Journal of Mechanical Sciences* ***57*(1)** 43-53
[9] Suyuti, Muhammad Arsyad, and Rusdi Nur 2015 *Advanced Materials Research* ***1125*** 157-160
[10] Wagoner R H, Lim H, and Lee M G 2013 *International Journal of Plasticity* ***45*** 3-20
[11] Chongthairungruang B, Uthaisangsuk V, Suranuntchai S, and Jirathearanat S 2013 *Materials & Design* ***50*** 253-266
[12] Oya T, Tiesler N, Kawanishi S, Yanagimoto J, and Koseki T 2010 *Journal of Materials Processing Technology* ***210*(14)** 1926-1933

IOP Conf. Series: Materials Science and Engineering **541** (2019) 012016 doi:10.1088/1757-899X/541/1/012016

Pitting corrosion resistance of CA6NM as geothermal turbine blade material in simulated artificial geothermal brine

A Nikitasari[1,2], M A Prasetyo[1], R Riastuti[2], and E Mabruri[1]
[1]Research Center for Metallurgy and Material, Indonesian Institute of Sciences, Indonesia
[2]Department of Metallurgy and Material Engineering, FT UI, Indonesia

E-mail: arininikitasari89@gmail.com

Abstract. Fatigue failure of turbine blades originating from pitting corrosion is a critical issue for power plant. Herein, we study the pitting corrosion resistance of cast CA6NM as turbine blade material in simulated geothermal environment with and without CO_2 gas by using cyclic polarization measurement. Three different molybdenum and nitrogen content of CA6NM were used for experiments: (i) CA6NM with a molybdenum content of 1 wt.% and without addition of nitrogen, (ii) CA6NM with a molybdenum content of 2 wt.% and without addition of nitrogen, (iii) CA6NM with a molybdenum content of 2 wt.% and nitrogen content 0.1 wt.%. The measurement reveals CA6NM with molybdenum 2 wt.% and nitrogen 0.1 wt.% is the most resistance to pitting corrosion in simulated artificial geothermal brine with and without CO_2 gas.

1. Introduction

Geothermal energy can contribute a significant portion of electricity to the future energy mix because it has characteristics of low cost, wide distribution, and low pollution [1-2]. However, geothermal steam contains corrosive chemical species that can cause corrosion which can result in high costs associated with maintenance, materials, and loss in production [3]. In some geothermal fields, the geothermal steam contains ammonium chloride (NH_4Cl) and hydrogen chloride (HCl) gas that is generally regarded as originated from the volcanic system. If condensation and/or re-boiling of the steam with HCl and NH_4Cl, risk of localized corrosion arises on blades of the wet section of steam turbine. Nowadays, corrosion mitigation is mainly achieved through steam scrubbing system and chloride abatement with alkaline sodium hydroxide solution. Besides an increase in operating costs, such treatment reduces the fluid temperature leading to relevant loss of power about 0,5 MW on 15 MW turbine. Thus, the use of corrosion resistance turbine material can be a suitable solution [4].

Steam turbine blades are critical components in power plants, which convert the linear motion of high temperature and high-pressure steam flowing down a pressure gradient into a rotary motion of the turbine shaft. The low pressure (LP) turbine blades are designed to extract the final remnant of energy from the passing steam flow. Statistics show that low pressure (LP) turbine blades are generally more susceptible to failure than those of the high pressure (HP) and intermediate pressure because trace amounts of corrosive chemical species present in the steam are easily concentrated in condensed water droplet and accumulate inside gap regions in the phase transition zone due to alternative dry and wet phenomenon [5-6].

Almost 50% of the turbine blades failures are related to fatigue, stress corrosion cracking, and corrosion fatigue [5]. Fatigue failure of turbine blades originating from pitting corrosion can cause catastrophic damage. Fatigue failure of steam turbine blades caused by pitting corrosion occurs preferentially in low pressure (LP) blades where early condensate develops. Activation of pitting is not expected during standard operational condition because the condensate on the turbine blades is oxygen free. But unexpected occurrences – e.g. leaking of the condenser – may create corrosive environments with critical amounts of corrosives such as chloride. When combined with off-load aeration, corrosion pits can form that can act as stress raisers. If a corrosion pit exceeds a critical size, the increase of the local stresses can initiate a fatigue crack [7]. Therefore, pitting corrosion resistance becomes crucial properties in the material selection of turbine blades to avoid failure.

Material selection guidelines for geothermal turbines mainly recommend the use of stainless steel because of its ability to passivate in a corrosive environment. Cast CA6NM is a low carbon 13%Cr-4%Ni martensitic stainless steel, which is used specifically when applications require a high resistance to fraction, creep, and fatigue[8–10]. This alloy displays improved weldability, strength and corrosion resistance compared to typical martensitic SS such as 410 SS, thereby CA6NM is used as turbine blade material in this experiment. Effect of alloyed elements on pitting corrosion resistance have been investigated and beneficial effect of molybdenum (Mo) has been widely reported [11-12].

Nitrogen (N) is another alloying element that is known to improve the resistance of stainless steels to pitting corrosion. As such, it participates in the empirical pitting resistance equivalent numbers (PREN) equation with a large weight [13].

$$PREN = \%Cr + 3.3\ \%Mo + 16\ \%N \quad (1)$$

The effect of Mo and N seem to synergistically interact when both of them are present in the stainless steel, but this issue has not been clearly understood yet. The purpose of present work is to elucidate the role of alloyed molybdenum and nitrogen on pitting corrosion resistance of CA6NM in geothermal environment using electrochemical measurement.

2. Experimental Details

2.1. Materials

Three types of cast CA6NM with composition as in table 1 were used for the experiments. Cast CA6NM as specimen cut to rectangular coupon and heat treated. All spesimen austhenized at 1050°C for 3 hours and then tempered at 670°C for 3 hours. After heat treated, CA6NM mounted with resin until the exposed area was 1 cm^2. The mounted CA6NM abraded with abrasive paper from 120 cw until 1200 cw. Microstructure of all spesimen observed using optical microscope. All test material immersed into artificial geothermal brine with or without CO_2 gas during measurement.

Table 1. Materials for experiment.

Samples	C	Mn	Si	P	S	Cr	Ni	Mo	N	Fe
CA6NM1	0.04	0.40	0.35	0.018	0.008	12.83	4.17	1.02	-	Bal
CA6NM2	0.04	0.43	0.55	0.017	0.009	13.02	4.08	1.96	-	Bal
CA6NM3	0.04	0.35	0.35	0.019	0.009	13.07	4.09	2.14	0.11	Bal

2.2. Artificial Geothermal Brine

These experiment were performed using simulated artificial geothermal brine with and without CO_2 gas. To simulate geothermal environment, artificial geothermal brine synthesized from chemicals in table 2. The synthesized artificial geothermal brine contains chloride ions that cause pitting corrosion.

2.3. Measurement Techniques

Pitting corrosion resistance of all samples evaluated using electrochemical measurement. The glass cell used for the experiments contains a three-electrode setup with the test material as the working electrode, a platinum counter electrode, and a saturated calomel electrode (SCE) reference electrode. It has a volumetric capacity of 500 ml. A scanning electron microscope (SEM, JEOL Model JSM-5400) was employed for the pitting corrosion size and an EDAX attachment was used for qualitative elemental chemical analysis.

The electrochemical measurement was conducted by cyclic polarization techniques at ambient pressure, room temperature, and pH neutral in artificial geothermal brine. Cyclic polarization was applied using Gamry Instruments G750 Series. The polarization experiments were started within 1 h after OCP (Open Circuit Potential) measurement. The measurement of cyclic polarization referred to ASTM G-61. The potential of the sample was swept from the -500mV to +1500mV where a current density limit of 10mA/cm^2 was applied in the specimen. After reaching current density limit, reverse polarization was started to find susceptibility to pitting corrosion. The forward scan rate was 5mV/s and reverse scan was 2.5 mV/s.

Table 2. Artificial geothermal brine composition[14].

Salt used for synthesis	g/l
NaCl	224.6
NaSO4.10H2O	12.074
MgCl.6H2O	10.620
CaCl2.2H2O	6.452
KCl	0.390
KOH	0.321
NaHCO3	0.048

3. Result and Discussion

3.1. Open Circuit Potential Analysis

Open circuit potential (E_{oc}), also referred to as the free corrosion potential (E_{corr})is a measure of a tendency of the investigated, i.e. working electrode to dissolve (loose electrons) and undergo oxidation. It is measured versus a stable reference electrode, when no potential or current is being applied to the electrochemical cell[15]. OCP is the mixed electrode potential in the given corrosion environment. The evolution of the OCP for the test materials in artificial geothermal brine with and without CO_2 gas is given in figure 1. Based on figure 1, the OCP increases with increase in molybdenum content and excistance of nitrogent. Table 3 showed the maximum and minimum electrode potential and OCP value of test materials after 1 hours immersion in artificial geothermal brine with and without CO_2 gas. The highest OCP value is CA6NM alloy with the highest molybdenum and nitrogent content. From these results, it can be concluded that molybdenum and nitrogent content give effect on the reactivity of CA6NM alloys. The higher the molybdenum and nitrogent content, the less reactive and more noble of CA6NM alloy and vice versa. From figure 1, CA6NM immersed in artificial geothermal brine without CO_2 gas have more positif value of OCP than CA6NM immersed in artifial geothermal brine with CO_2 gas. This happens because artificial geothermal brine with CO_2 gas is more aggressive environment.

Table 3. Maximum and minimum electrode potential and open circuit potential value for 1h.

Samples	Vmax (mV vs SCE)	Vmin (mV vs SCE)	OCP (mV vs SCE)
CA6NM1 with CO_2 gas	-470.3	-532.8	-532.7
CA6NM2 with CO_2 gas	-439.9	-527.1	-527.1
CA6NM3 with CO_2 gas	-472.7	-519.3	-519.3
CA6NM1 without CO_2 gas	-483.8	-510.5	-510.2
CA6NM2 without CO_2 gas	-476.2	-496.9	-489.6
CA6NM3 without CO_2 gas	-403.7	-450.1	-450.1

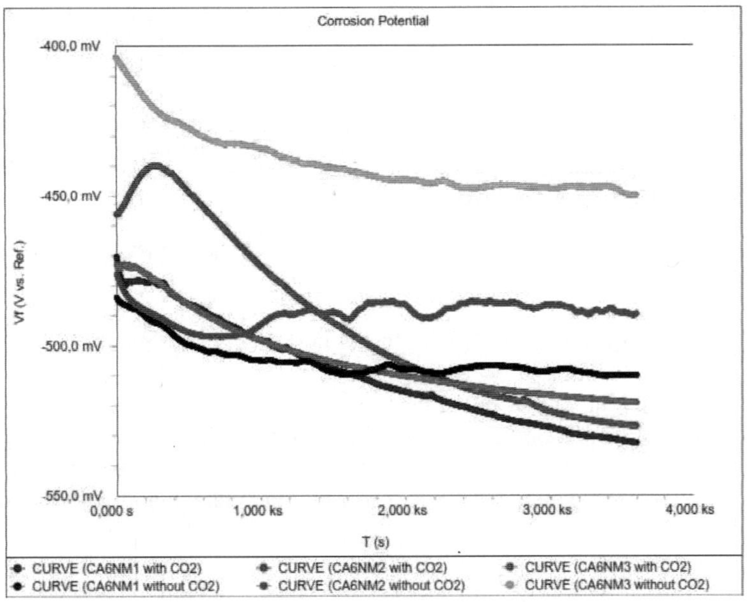

Figure 1. Open circuit potential measurement in artificial geothermal brine with and without CO_2 gas.

3.2. Cyclic Polarization Analysis

Figure 2 and figure 3 depict the cyclic polarization curves of all specimen obtained in the artificial geothermal brine with and without CO_2 gas, respectively. The parameters determined from the cyclic polarization of all test materials, such as pitting potential (E_{pit}), protection potential (E_{prot}), and corrosion potential (E_{corr}) were presented in table 4. Based on table 4, all test material immersed in artificial geothermal brine with CO_2 gas have higher protection potential value than pitting potential. If the protection potential value is more positive (anodic) than the pitting potential, there will be no tendency to pit [16]. From table 4, CA6NM materials immersed in artificial geothermal brine without CO_2 gas have higher pitting potential value than protection potential. It means, CA6NM immersed in artificial geothermal brine without CO_2 was susceptible to pitting corrosion [17]. However, in figure 3, only small hysteresis loop is found in all CA6NM immersed in the artificial geothermal brine without CO_2 gas. This indicates that CA6NM can quickly repassivation of the attacked local sites [15].

The pitting potential of different molybdenum and nitrogent content of CA6NM follow the sequence: CA6NM1 < CA6NM2 <CA6NM3. The lower E_{pit} value indicates more chances of easily breakdown of passive film and early start of pitting corrosion[18]. CA6NM immersed in artificial geothermal brine with CO_2 gas have lower E_{pit} value than CA6NM immersed in artificial geothermal brine without CO_2 gas. This proves OCP measurement that artificial geothermal brine with CO_2 gas is a more aggressive

environment. Nevertheless, E_{prot} value of CA6NM in artificial geothermal brine with CO_2 gas higher than CA6NM in artificial geothermal brine without CO_2 gas. This result shows that CA6NM has ability to protect its passive film in artificial geothermal brine with CO_2, so pitting corrosion does not occur.

In cyclic polarization curves the probability of localized corrosion is related to the difference between the protection potential, E_{prot}, and corrosion potentials, E_{corr}, thus the higher the difference the lower the probability of pitting corrosion [19].Overall, the experimental results obtained on CA6NM with varying content of molybdenum and nitrogent immersed in artificial geothermal brine showed that in absence of nitrogent and lower molybdenum content the pitting corrosion activity on the CA6NM surface strongly increases leading to low pitting potentials as observed in table 4. However, the addition of molybdenum and nitrogent content to the alloys, once corrosionwas initiated, leads to a repassivation ofthis activity. Thus the results of electrochemical measurements suggested that molybdenum and nitrogent are able to reduce pit initiation. The results of the electrochemical measurements for predicting pitting corrosion resistance of CA6NM in artificial geothermal brine are certified and confirmed by the results of the calculations of PREN value listed on table 5. The higher the PREN value, the more resistant the pitting corrosion.

To study the synergistic interaction between Mo and N the pitting potential direct increase, ΔE_{pit}, which appeared as a result of the addition of Mo or both is listed on table 5. In this sense, ΔE_{pit}, represents individual effects of only Mo being added to the alloy, that is, ΔE_{pit}, Mo = E_{pit} CA6NM2 – E_{pit} CA6NM1 (1).

In the sameway, the synergistic effect should be related to the concomitant presence of Mo and N, expressed by

$$\Delta\ E_{pit},\text{Mo and N} = E_{pit}\ \text{CA6NM3} - E_{pit}\ \text{CA6NM1} \quad (2)$$

Results of the calculation of the pitting potential direct increaseclearly shows that the combined effect of Mo and N represented by equation (2) is largely beyond the sum of Mo individual effects (equation (2)). Hence the existence of a synergetic positive effect of Mo and N on the pitting potential values is confirmed for CA6NM. This experiment result is in good agreement with Carole Loable, et.al experiment result[13].

Table 4. Parameters from cyclic polarization.

Samples	E_{pit} (mV vs SCE)	E_{prot} (mV vs SCE)	E_{corr} (mV vs SCE)
CA6NM1 with CO_2 gas	-548.2	-335.9	-528.7
CA6NM2 with CO_2 gas	-462.7	-265.4	-528.1
CA6NM3 with CO_2 gas	-454.2	-256.9	-439.2
CA6NM1 without CO_2 gas	-435.1	-462.5	-530.5
CA6NM2 without CO_2 gas	-311.0	-331.0	-486.2
CA6NM3 without CO_2 gas	-258.8	-308.1	-463.9

IOP Conf. Series: Materials Science and Engineering **541** (2019) 012016 doi:10.1088/1757-899X/541/1/012016

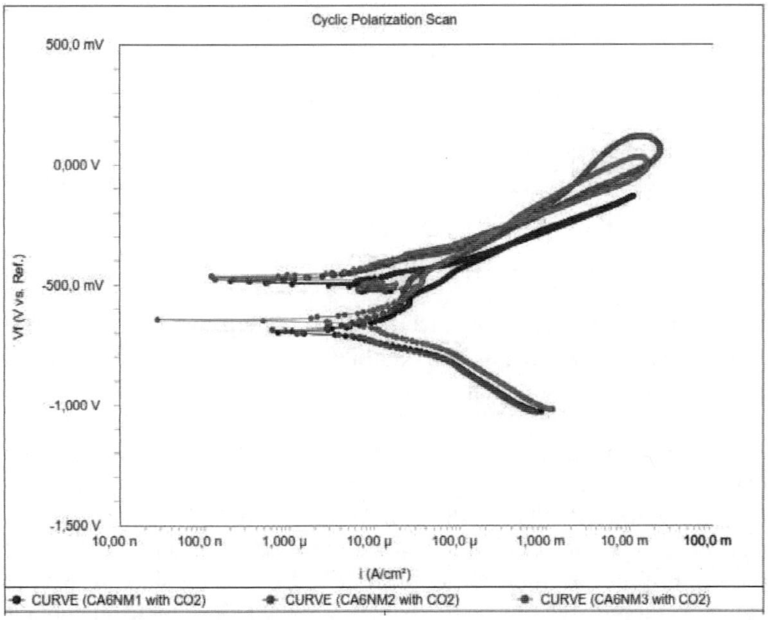

Figure 2. Cyclic polarization curve CA6NM in artificial geothermal brine with CO_2 gas.

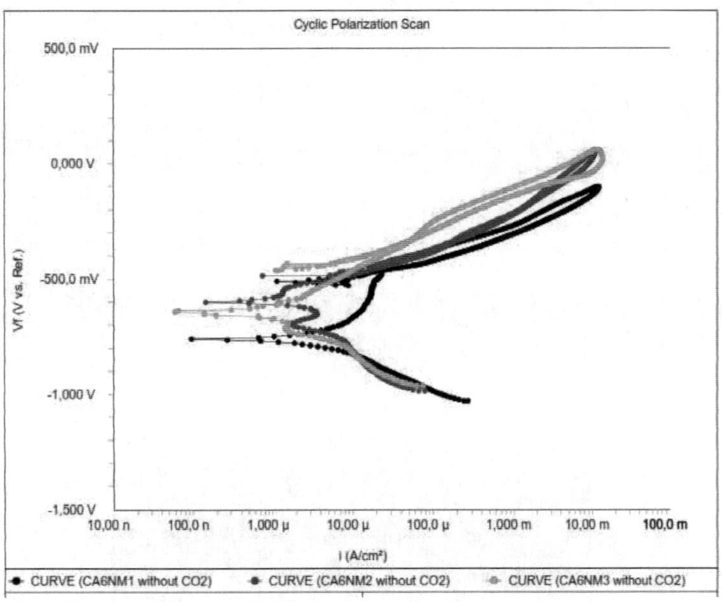

Figure 3. Cyclic polarization curve CA6NM in artificial geothermal brine without CO_2 gas.

Table 5. PREN value of CA6NM.

Samples	PREN
CA6NM1	16.3
CA6NM2	19.6
CA6NM3	21.2

3.3. Microstructure Analysis

Figure 4(a)-4(c) were microstructure photos using optical microscope of CA6NM1, CA6NM2, and CA6NM3, respectively that austhenized at 1050°C for 3 hours and tempered at 670°C for 3 hours. The result showed all samples have same microstructure that consists most of lath martensite, small portion of ferrite and austenite.

After conducting cyclic polarization measurement, all samples were characterized using scanning electron microscope (SEM) to observe pitting size. Figure 5(a)-5(c) show the SEM images of CA6NM1, CA6NM2, and CA6NM3, respectively in artificial geothermal brine without CO2 gas.Based on figure 5, pitting corrosion exist in test materials immersed in artificial geothermal brine without CO_2 gas at different diameter size of pit. The smallest pit size is CA6NM3 which has an average pit diameter of 3μm. To prove that pitting corrosion exist was carried by using EDAX analysis. Figure 6 shows EDAX result in pit area and not corroded area of CA6NM1. The presence of O element in pit area proves that there is product corrosion in pit area, so it is true that the pit is pitting corrosion.

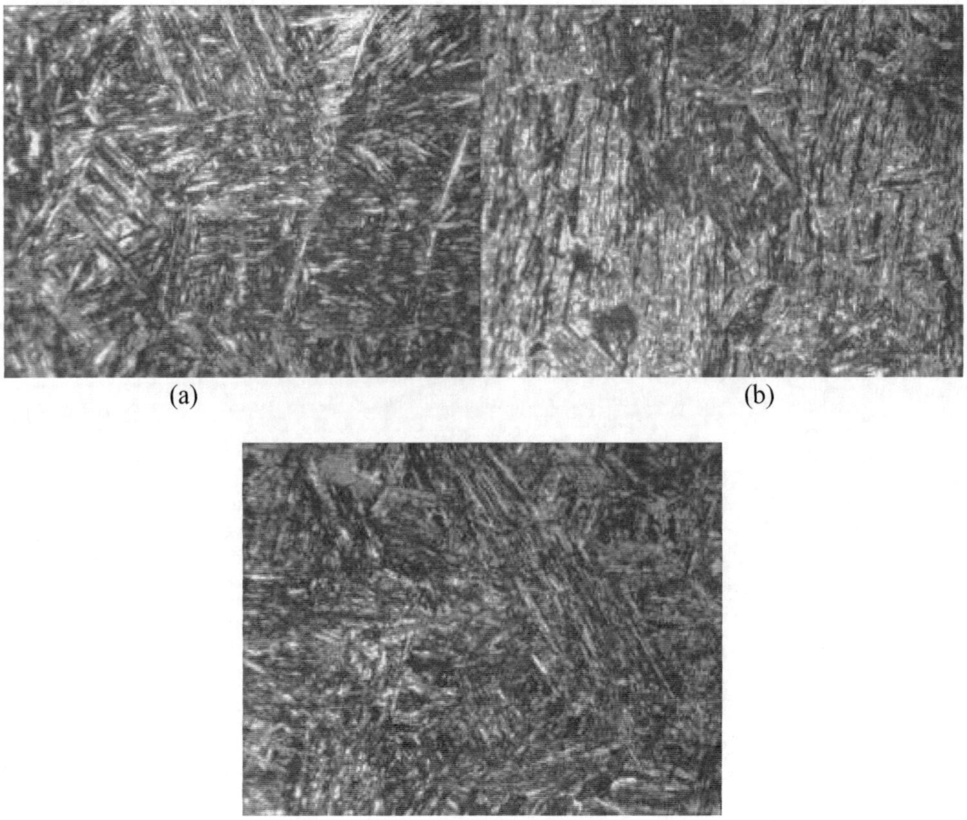

(a) (b)

(c)

Figure 4. Microstructure of (a)CA6NM1, (b)CA6NM2, (c)CA6NM3.

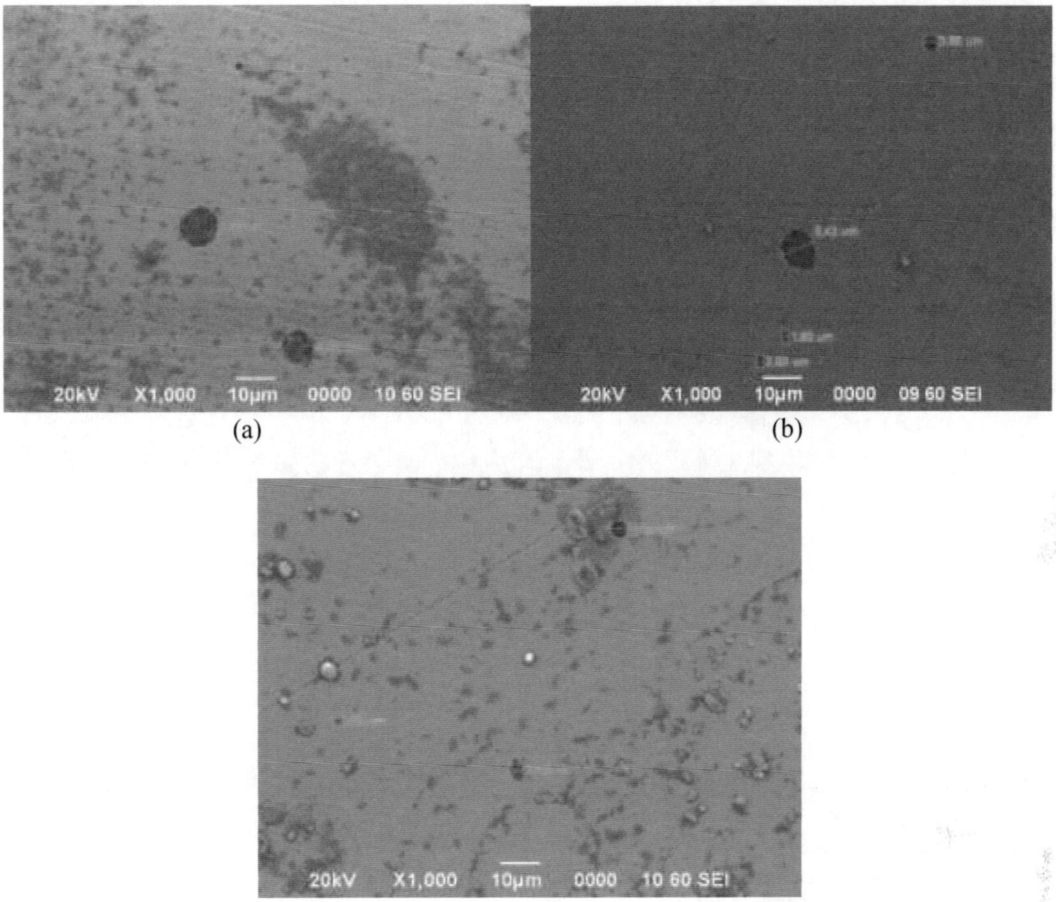

Figure 5. SEM micrograph of (a) CA6NM1, (b) CA6NM2, (c) CA6NM3 after conducting cyclic polarization in artificial geothermal brine without CO_2 gas.

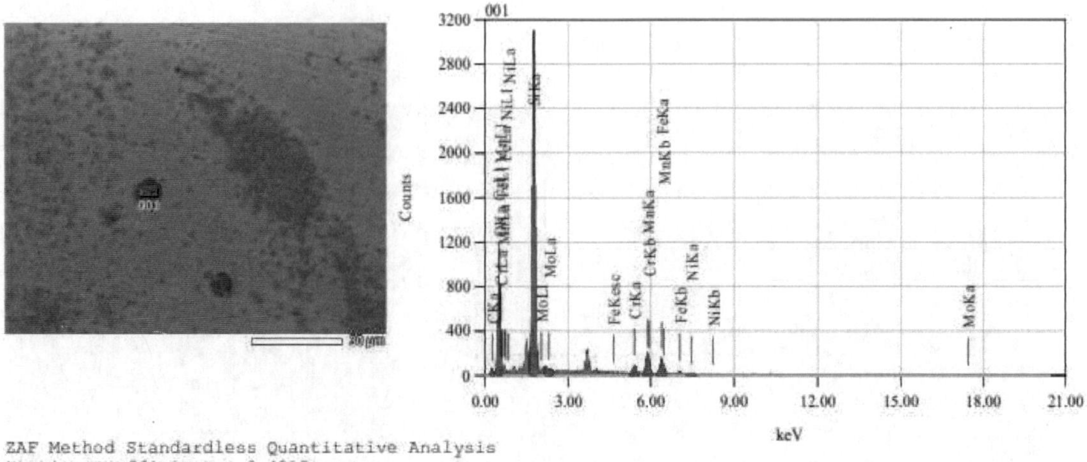

ZAF Method Standardless Quantitative Analysis
Fitting Coefficient : 0.4212

Element	(keV)	Mass%	Error%	Atom%	Compound	Mass%	Cation	K
C K*	0.277	11.50	0.86	20.98				1.6383
O K	0.525	30.66	0.57	42.01				32.1814
Si K	1.739	36.56	0.28	28.53				40.0757
Cr K	5.411	2.88	0.70	1.21				3.6181
Mn K	5.894	9.65	0.86	3.85				11.7226
Fe K	6.398	8.20	0.89	3.22				10.1195
Ni K*	7.471	0.46	1.40	0.17				0.5602
Mo L*	2.293	0.10	0.91	0.02				0.0842
Total		100.00		100.00				

Figure 6. EDAX analysis in pit area of CA6NM1.

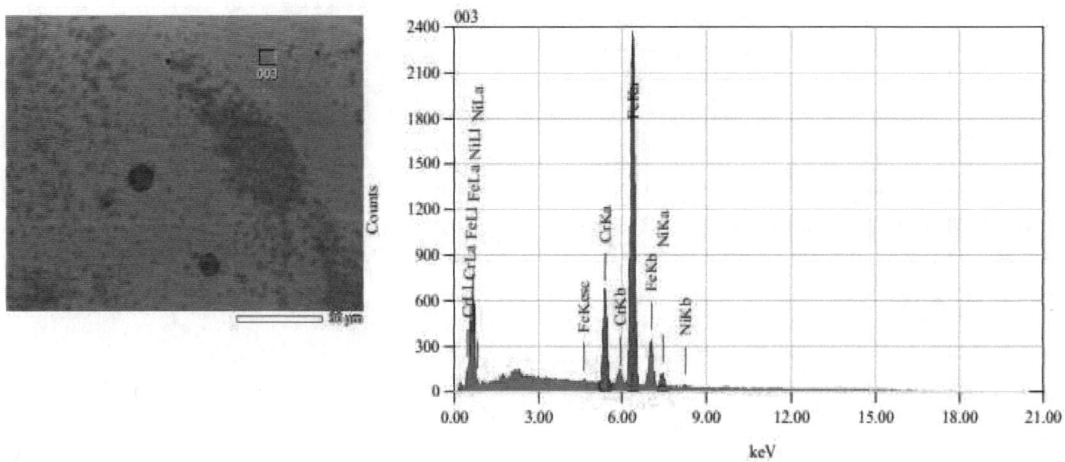

ZAF Method Standardless Quantitative Analysis
Fitting Coefficient : 0.2942

Element	(keV)	Mass%	Error%	Atom%	Compound	Mass%	Cation	K
Cr K	5.411	12.00	0.30	12.80				13.8234
Fe K	6.398	83.46	0.44	82.91				82.1016
Ni K	7.471	4.54	0.74	4.29				4.0750
Total		100.00		100.00				

Figure 7. EDAX analysis in not corroded area of CA6NM1.

IOP Conf. Series: Materials Science and Engineering **541** (2019) 012016 doi:10.1088/1757-899X/541/1/012016

4. Conclusion

The following conclusions can be drawn from the results of the present work with regard to the performance of CA6NM stainless steel material at different content of Molybdenum and Nitrogent in the artificial geothermal brine with and without CO2 gas. CA6NM in artificial geothermal brine without CO2 gas have susceptibility to pitting corrosion while CA6NM in artificial geothermal brine with CO2 gas have no tendency to pit even though the environment is more aggressive. CA6NM have ability to lead repassivation when pitting corrosion initiated so CA6NM resistant to pitting corrosion. This resistance to pitting increases with increasing Mo content in the alloy. Pitting corrosion resistance of CA6NM in artificial geothermal brine was significantly intensified when added both Mo and N as we clearly showed that the combined effect ofMo andNis largely beyond the sum of their individual effects in artificial geothermal brine. Pitting corrosion considerably in the presence of O element in pit area as product corrosion. Average diameter of pit decreases with increasing Mo and N content in CA6NM. Therefore, Mo and N content give positive effect in the pitting corrosion resistance of CA6NM in artificial geothermal brine.

References

[1] Yangshuhan X and Mingyan L 2015 *Geothermics* **70** 339–350
[2] N Mundhenk, P Huttenloch, R Bäßler, T Kohl, H Steger and R Zorn 2014 *Corros. Sci* **84** 180–188
[3] S N Karlsdottir, K R Ragnarsdottir, I O Thorbjornsson and A Einarsson 2015 *Geothermics*. **53** 281–290
[4] M Cabrini, S Lorenzi, T Pastore, M Favilla, R Perini and B Tarquini 2017 *Geothermics*. **69** 139–144
[5] D Ziegler, M Puccinelli, B Bergallo, and A Picasso 2013 *Biochem.* Pharmacol. **1** 192–199
[6] L Niu and K Nakada 2015 *Corros. Sci.* **96** 171–177
[7] Bernd M S *et al.* 2014 *International Journal of Fatigue* **65** 33–43
[8] F Mirakhorli, X Cao, X Pham, P Wanjara and J L Fihey 2017 *Mater. Charact.* **123** 264–274
[9] A Trudel, M Lévesque and M Brochu 2014 *Eng. Fract. Mech.* **115** 60–72
[10] L B Winck, J L A Ferreira, J A Araujo, M D Manfrinato and C R M. Silva 2013 *Surf. Coat. Technol.* **232** 844–850
[11] M Monnot, V Roche, R Estevez, M Mantel and R P Nogueira 2017 *Electrochim. Acta*. **252** 58–66
[12] A Tomio, M Sagara, T Doi, H Amaya and N Otsuka 2015 *Corros. Sci.* **98** 391–398
[13] C Loable *et al* 2017 *Material Chemistry and Physic*. **186** 237–245
[14] A Pfennig, R Wiegand, M Wolf and C Bork 2013 *Corros. Sci.* **68** 134–143
[15] A Keserovich 2014 "*Geothermal systems in Indonesia – Influence on the corrosion resistance of stainless steel materials*." Dissertation (Germany: TU Berlin)
[16] A Poursaee 2010 *Concrete Research Letter* **1** 90–97
[17] Mochammad S A, Boby F, Arini N, Soesaptri O and Efendi M 2017 *Procedia Engineering,* **171** 517–525
[18] A Malik and S Ahmad *Swcc.Gov.Sa* 1805–1855
[19] E Rakanta, T Zafeiropoulou, and G Batis 2013 *Constr. Build. Mater.* **44** 507–513

Acknowledgement
Gratitude is expressed to the Research Center for Metallurgy and Materials, Indonesian Institute of Sciences for financial support during this project in fiscal year 2018.

Green simple microwave-assisted extraction (MAE) of cellulose from *Theobroma cacao* L. (TCL) husk

S T C L Ndruru[1], D Wahyuningrum[2], B Bundjali[1] and I M Arcana[1]
[1]Physical and Inorganic Chemistry Divisions, Faculty of Mathematics and Natural Sciences, Bandung Institute of Technology, Bandung, Indonesia,40132
[2]Organic Chemistry Divisions, Faculty of Mathematics and Natural Sciences, Bandung Institute of Technology, Bandung, Indonesia, 40132

E-mail: arcana@chem.itb.ac.id ; stcln07chem@gmail.com

Abstract. This work had successfully produced Theobroma cacao L. (TCL) husk-based cellulose used efficiently and friendly environmentally simple microwave-assisted extraction (MAE) method. The microwave-assisted heating time optimization of alkaline-treatment was carried out for 10, 20, 30 and 40 minutes. Microwave-assisted bleaching process was also conducted using green bleaching agent H_2O_2, and also utilizing microwave heating for 60 minutes. The followed treatment was filtered, neutralized, washed and overnight freeze-dried. Fourier transform infrared (FTIR) analysis confirmed common cellulose functional groups of TLC, are β-(1,4)-glycosidic bonds at ~897 cm^{-1}, O-H (hydrogen bond) at 3412 cm^{-1}, C-H vibration at 2902 cm^{-1}, C-O-C asymmetric at 1161 cm^{-1}, and C-OH out-of-plane bending at 665 cm^{-1}. FTIR analysis, in addition, also studied crystallinity ratio (*CrR*), hydrogen bond energy (E_H) and hydrogen bond distances (R_H), while particle size analyzer (PSA), X-ray diffraction (XRD) and thermogravimetry analysis (TGA) were carried out to confirm particles sizes, crystallinities and thermal stability properties of TCL husk, cellulose of TCL husk and commercial-microcrystalline cellulose (MCC), respectively.

1. Introduction

Searching and utilization of cellulose in industrial and research activities are the answer to issue in reducing environmentally unfriendly fossil fuels. Cellulose is appealing ones due to easily and costly obtained, renewable, biodegradable, biocompatible, good mechanical strength, and good thermal stability properties [1].

Cellulose is the most abundant and renewable biopolymer in nature. Cellulose is a main component of plants, strengthening the cell wall [2], forming one-third to one-half of plants tissue and photosynthesis reproduced [3]. Biosynthetically, cellulose is as much as 10^{11} tons produced every year [4]. In its chemical structure, cellulose is a homopolymer composed of units of D-glucopyranose bound by β (1 → 4) glycoside bonds [5].

Cellulose is widely used in industries such as papermaking, food as stabilizer, anti-caking agents, suspending agents and packaging, pharmaceutical applications as a binders, absorbents, flowability and innovative composite applications [6]. Natural cellulose fibers are used as a reinforcing material in thermoset and thermoplastic cellulose matrix especially glass fibers due to unique characteristics, such as renewability, low density and high specific strength [3].

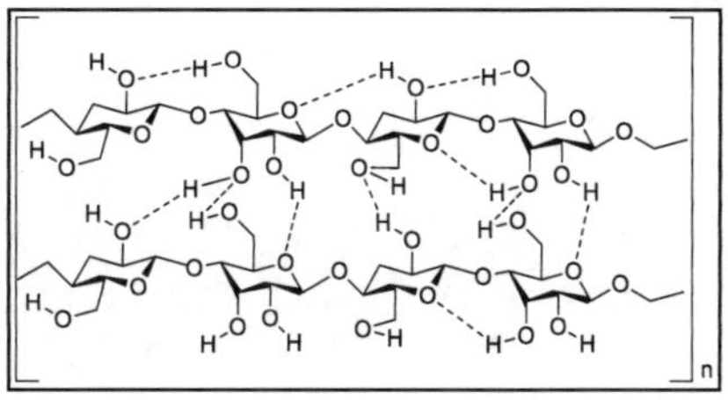

Figure 1. Supramolecular hydrogen bond of cellulose [7].

A reason to be widely applied in some applications such as an alternative to synthetic materials of electrochemical devices caused cellulose has excellent characteristics, are strength and modulus, large surface area, high aspect ratio, high strength, and stiffness, inexpensive, lightweight, recyclable and biodegradable [6]. Cellulose material has excellent properties and meets green friendly environmentally and sustainable requirements of lithium-ion batteries (LIBs) separator membranes compare with others [8,9], such as polyethylene (PE), polypropylene (PP), polyethylene oxide (PEO), polyvinyl alcohol (PVA) [10], and polyvinylidene fluoride (PVDF) [11]. Several reports have been published as a source of cellulose such as sugarcane bagasse [12,13], seaweed [6], durian rind [3], pineapple peel [14], cassava peel [15], and others [16].

Indonesia is the third highest *Theobroma cacao* L. (TCL)-producing country in the world, after Ghana and Republic of Cote d'Ivoire. In Indonesia TCL is often popular called as *kakao*, *cocoa* and *buah* (*pohon*) *cokelat* [17]. Every year Indonesia produces 800,000 tons of TCL, and contribute 16% of world TCL production. TCL has 73-75% husk, 2-3% placenta and 22-24% seed. TCL husk contains 32-45% crude fiber in the form of lignin and cellulose. Unfortunately, TCL husk waste was only used as animal feed [18,19]. However, there were not many reports to explore lignocellulose of TCL husk for last two decades.

Cellulose extraction consists of two important stages, commonly, are an alkaline treatment (NaOH/KOH) and bleaching process (bleaching agent: NaClO/ $NaOCl_2$). Conventional alkaline treatment aims to break bonds between lignin and holocellulose, to obtain lignin, hemicellulose, and cellulose. The alkaline treatment using strong bases such as NaOH and KOH requires neutralizing process the pH of the extracted product. The filtration process is then followed by a process of washing with excess distilled water, then using acetic acid in neutralizing the extraction product. This product is usually called holocellulose. Holocellulose consists of hemicellulose and cellulose. Bleaching process aims to separate the cellulose-bonded hemicellulose, and also to make the product brighter (white). The bleaching agent which usually used is acidified sodium chlorite ($NaOCl_2$) after alkaline treatment in breaking lignin completely. The conventional cellulose extraction long takes 4-6 hours interval for alkaline treatment and then the bleaching process needs 1-2 hours [3,12,13,15].

This study utilized a new method *green simple microwave-assisted extrac*tion (MAE), applied of electromagnetic radiation to a material [20], which can make more efficient so that cellulose extraction process can be done in the short duration (between 10-60 minutes). MAE is an interesting technique due to able it can accelerate biomaterials isolation method [21]. Microwave irradiation can reduce the dewaxing process in cellulose extraction. Microwave irradiation increases heating perfectly to alkaline, bleaching, and hydrolysis treatment. And also, it needs environmentally friendly hydrogen peroxide (H_2O_2) bleaching agent to replace $NaClO_2$ [6].

A microwave is a form of electromagnetic energy which exists in the electromagnetic spectrum (300-300,000 MHz). There are two basic mechanisms of energy transfer of microwaves in the matter, through dipole rotation and ionic conduction. Dipole rotation is the interaction between polar molecules as a result of the rapidly changing electric field of microwaves. The ion conduction mechanism includes a momentary superheating of ionic substances due to the ion motion generated by an electric field. As temperatures increase, energy transfer becomes more efficient [22].

Previous studies stated that microwave irradiation convert electromagnet heating into energy, a familiar method to organic synthesis with many advantages such short duration time, need small amount of solvent and produce high yields [23]. Santos [24] and Smiderle [23] described the effectiveness of microwave-assisted extraction of cellulose. Likewise, in synthesizing cellulose derivatives, the use of microwave heating is practically highly preferred. Biswas [25] reported that alkyl cellulose (methyl cellulose and ethyl cellulose) have been successfully synthesized using microwave-assisted heating.

Based on reasons above, so this work studied and developed a green simple microwave-assisted extraction of cellulose from TCL husk. We investigated effect of alkaline treatment with various time using this method, continued bleaching process. Bleaching product, as cellulose product, was characterized with Fourier transform infrared (FTIR), particle size analyzer (PSA), x-ray diffraction (XRD) and thermogravimetry analysis (TGA), and this result was compared with TCL husk and commercial microcrystalline cellulose (MCC).

2. Experiment

2.1. Material
TCL (*Theobroma cacao L.*) husk was obtained from Hilisimaetano Village (Lahusa Subdistrict, South Nias Regency, Province of North Sumatera, Indonesia), hydrogen peroxide (H_2O_2), sodium hydroxide (NaOH), distilled water and ethanol were commercially obtained from Merck.

2.2. Characterization Instrument
Functional groups of samples were carried out with fourier transform infra-red, FTIR (type: Prestige 21 Shimadzu), crystallinities of samples were carried out with polymer electrolytes membranes were carried out with diffractometer (type XRD Rigaku Smartlab), particle size of sample was measured with PSA Delsa™ Nano C (Merk Beckman Coulter), morphology images of samples was carried out with SEM instrument, Type JEOL-JSM-6510LV, and thermal analysis of samples were carried out with thermal gravimetry analysis (TGA, merk Linseis, type STA PT 1600).

3. Methods

3.1. Klason lignin content
Klason lignin was investigated using Viera *et al.*-developed method [26], with some modifications. A 1 gram of TCL husk was subjected into round bottomed flask 100 mL, and then added 7.5 mL sulfuric acid 72% under stirring in the room temperature for 1 hour. After previous treatment, the solution was diluted with 280 mL distilled water. In addition, the system was stand for residue precipitation, next it was filtered dan washed with 250 mL distilled water. The product was dried in the vacuum freeze-dry, weighed and analyzed by FTIR.

3.2. Holocellulose content
Holocellulose was also investigated using Viera *et al.*-developed method [26], with some modification. 2.5 g of TCL husk was added to 125 mL beaker, containing 0.75 g of sodium chlorate ($NaClO_3$), 0. 5 mL glacial acetic acid, 50 mL of distilled water was added and the collision was agitated until chlorate homogenized. The mixture was than heated and stirred at 75°C for 1 h. Each 1 hour for a total duration time of 3 h, the reagents added into beaker containing mixture. The mixture was subjected into ice water and next filtered. The bright product was washed with ethanol and dried in the vacuum freeze dry, weighed and FTIR analyzed.

3.3. Extraction of cellulose
The method of cellulose extraction in this study refers to a method developed by Santos *et al.* [24] and Singh *et al.* [6] with some modification. The ratio of TCL husk and 2% NaOH solution is 1: 20 (m/v). Microwave CEM reactor settings were power: 100 watts, temperature: 50°C, stirring: medium, time variations: 10, 20, 30 and 40 minutes. The product of the NaOH treatment, separated from the mixture with 2% NaOH, was neutralized with aquadest, followed by ethanol. The separated product was dried in vacuum freeze dry and then weighed.

Figure 2. (a) ground TCL husk; (b) microwave-assisted alkaline treatment; (c) product of alkaline treatment; (d) after bleaching process; (e) cellulose product.

Alkaline treated products were then bleached by green bleaching agent H_2O_2 using heating with microwave. Microwave CEM reactor settings were power: 100 watts, temperature: 50°C, stirring: medium, time: 1 hour. The bleaching treatment product, separated from the mixture using a glass filter, was then neutralized with distilled water until the odor of the bleaching agent removed, then again washed with ethanol, so that the bleaching product can be terminated by water disturbance. The filtered product was then dried in freeze dry and after drying it was milled again and weighed.

3.4. Yields calculation

Weight of both alkaline-treatment and bleaching process product were calculated using formulation below.

$$\%\ yields = \frac{weight\ of\ product}{weight\ of\ sample} x\ 100\ \% \qquad (1)$$

Weight of product = weight dried product (alkaline treatment/bleaching process)
Weight of samples = weight of TCL husk initially.

3.5. Crystallinity Index (C.I) Calculation

C.I calculation used a Segal *et al.*-proposed formulation [27,28].

$$C.I. = \frac{I_{200} - I_{am}}{I_{200}} x\ 100\% \qquad (2)$$

where I_{200} = maximum intensity of lattice diffraction; I_{am} = amorph region of diffraction intensity.

3.6. Crystallinity rasio (CrR) calculation

$$CrR = \frac{A_{1372}}{I_{2900}} \qquad (3)$$

where A_{1372} is absorption peak around 1372 cm^{-1} and A_{2900} is absorption peak around 2900 cm^{-1}.

3.7. Hydrogen bond energy (E_H)

$$E_H = \frac{1}{k}[\ \frac{(v_0 - v)}{v_0}\ ]x\ 100 \qquad (4)$$

where v_0 is standard wave number relate to free OH group (3650), v is wave number of bonded and observed -OH group to cellulose samples FTIR spectra, dan k is constant ($1/k = 2.625 \times 10^2$ kJ).

3.8. Hydogen bond distances (R_H)

$$\Delta v(cm^{-1}) = 4430\, x\, (2.84 - R_H) \quad (5)$$

$\Delta v = v_0 - v$, v_0 is OH^- stretching wave number at 3600 cm^{-1} and v is observed-OH stretching wave number to cellulose samples FTIR spectra.

4. Results and Discussion

The microwave-assisted extraction method of cellulose from TCL husk meets green chemistry principle. Designing energy efficiency is mainly principle approved in this method [29]]. Short time duration of cellulose isolation is promising for many applications. A hydrogen peroxide (H_2O_2) was bleaching agent also fulfill one of green chemistry principle, that is designing safer chemicals [29], if we compare to common bleaching agent of sodium (hypo)chlorite ($NaOCl/NaOCl_2$) which not friendly to environment.

4.1. FTIR analysis

Theobroma cacao L. (TCL) husk is one of most lignocellulose biomass, mainly comprised of lignin and holocellulose. TCL husk component analysis is very important employed quantitively. In this work, klason lignin and holocellulose extraction were also carried out. Fig 4 shows the differences of the main functional groups FTIR absorption of lignin and holocellulose, whereas Table 1 performs their main functional groups FTIR absorption differences compare to isolated cellulose product. The wavenumber of 1521.81 and 1253.73 cm^{-1} are lignin-typical peaks, C=C stretching of aromatic ring and C-O stretching ether linkage. Those bands are disappeared on holocellulose and cellulose (see figure 3).

Differences of TCL husk to cellulose by FTIR analysis, are mainly shown to wave number at 898 (commercial MCC) and 896.89 (~897) cm^{-1} (bleaching product of NaOH-treatment). The wave number is a C-O-C bond between the anhydroglucose unit (AGU), known as β- (1.4) -glycosidic bonds. We concluded that the bleaching product of NaOH-treatment is isolated cellulose product.

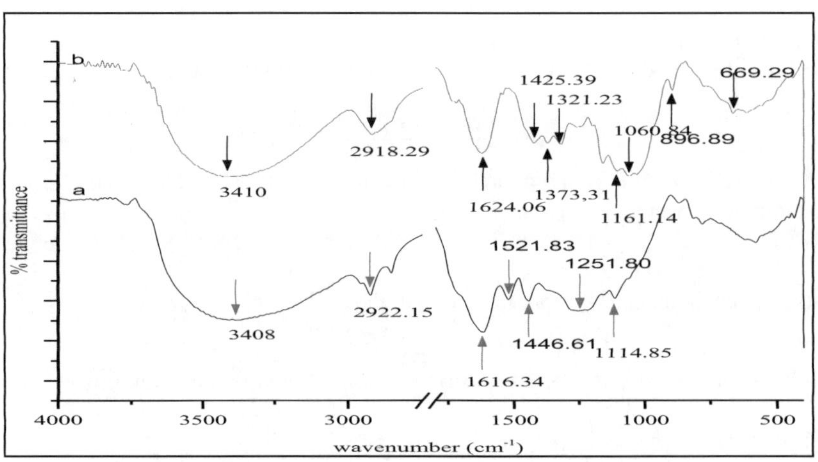

Figure 3. FTIR spectra: (a) klason lignin; (b) holocellulose.

Holocellulose and cellulose have significantly similar of FTIR absorption, but they are some different on wavenumber shift. Holocellulose and cellulose product expose the β- (1.4) -glycosidic bonds absorption at 896.89 (~897) cm^{-1}.

The complete assignment of main functional groups absorption of lignin, holocellulose and cellulose are depicted in Table 1 below.

Table 1. Assignment of main functional groups absorption of lignin, holocellulose and cellulose.

Functional group absorption	Klason lignin	Holocellulose	Cellulose product of TCL husk
OH stretching	3408	3410	3412
CH stretching	2922.15	2918.29	2902
H-O-H bending of absorbed water	1616.34	1624.06	1641
C=C stretching of aromatic ring (lignin)	1521.83	-	-
CH_2 bending	1446.61	1425.39	1431
C-H deformation	-	1373.31	1373
C-O stretching of ether linkage	1251.80	-	-
C-O-C antisymmetric bridge stretching	-	1161.14	1161-1163
C-O symmetric stretching of primary alcohol	-	1060.84	1080.91
β- (1.4) -glycosidic bonds	-	896.89	896.89 (~897)

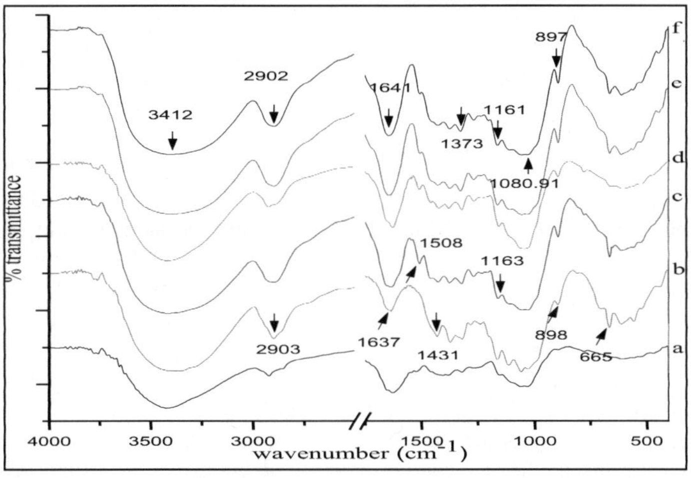

Figure 4. FTIR spectra: (a) TCL husk; (b) commercial MCC; (c) bleaching product of NaOH-treatment at 10 minutes; (d) bleaching product of NaOH-treatment at 20 minutes; (e) bleaching product of NaOH-treatment at 30 minutes; (f) bleaching product of NaOH-treatment at 40 minutes.

The time optimization of the NaOH-treatment produces different yields. Table 2 shows the yields of the isolated cellulose product with time optimization. The longer the NaOH-treatment, the lower the product yields. This is because of the amount of lignin and hemicellulose attached to cellulose decreases. In figure 3, it can be explained that existence of lignin is appeared at 1508 cm^{-1} on the duration of NaOH-treatment to the TCL husk at 10 minutes and 20 minutes. However, the typical peak of the lignin is decreased, even disappeared in the NaOH-treatment time of 30 minutes and 40 minutes. With considering its purity and yields, so bleaching product of NaOH-treatment time of 30 minutes that was next followed to analysis.

Table 2. NaOH 2%-treatment product and TCL husk bleaching yields (setting Microwave reactor: Power = 100-watt, T = 50°C, stir = average).

Samples	**Long time of NaOH-treatment**	**% weight of NaOH-treatment product**	**% weight of bleaching process product**	**Typical lignin group (~1500 cm^{-1})**
Sample 1	10 minutes	79.450	35.044	appeared
Sample 2	20 minutes	73.150	31.180	appeared
Sample 3	30 minutes	68.080	25.220	disappeared
Sample 4	40 minutes	65.240	11.810	disappeared

Based on the obtained results, the use of cellulose extraction methods from TCL husk with microwave heating compared to conventional extraction methods (reflux with heating using hotplate) is more efficient especially in the use of time and temperature of heating. A very short duration of time in the treatment of NaOH with the best condition of 30 minutes, greatly accelerates the process of propagation, further work, and other analyzes. While based on several previous reports, the process of treatment of NaOH needed time duration between 4 and 6 hours.

Table 3. Crystallinity ratio (CrR), hydrogen bond energy (E_H) and hydrogen bond ranges (R_H).

Samples	CrR	E_H (kJ)	R_H (A°)
commercial MCC	1.082	17.25	2.797
Sample 1	1.010	17.25	2.797
Sample 2	1.031	16.97	2.798
Sample 3	1.048	17.25	2.797
Sample 4	1.048	17.25	2.797

Some absorption peaks can be used to study the polymorphic types of cellulose samples. The absorption peaks around 1420, 1160, 1110 and 890 cm^{-1} can be used to study cellulosic polymorphic types. From the analysis of these wavelength numbers, cellulose of TCL husk is identified as cellulose I. It is exposed that wavenumber shifts around 1420 to 1431 for all conditions, confirmed to trend of cellulose I, whereas cellulose II content is further decreased [2,30]. Similarly, the absorption peak around 890 will shift to about 896 cm^{-1} confirming the presence of cellulose I, which is not possessed by TCL husk.

Calculation of CrR depict that commercial MCC crystallinity as shown in table 3 is higher that all samples. While among of cellulose products of TCL, the bleaching products of alkaline-treatment for 30-40 minutes have higher CrR than the bleaching products of alkaline-treatment for 10-20 minutes. E_H and R_H calculation confirm that similarity of all samples, indicating that cellulose properties are displayed.

4.2. *Particle size analyzer* (PSA) *and Scanning Electromotive Microscopy* (SEM*)*

Based on particle size analysis data, we obtained mean diameter of isolated cellulose product of TCL is 2,439.7 nm (~2.4397 µm). The particles were expected as micrometer-sized, and not as nanometer-sized.

IOP Conf. Series: Materials Science and Engineering **541** (2019) 012017 doi:10.1088/1757-899X/541/1/012017

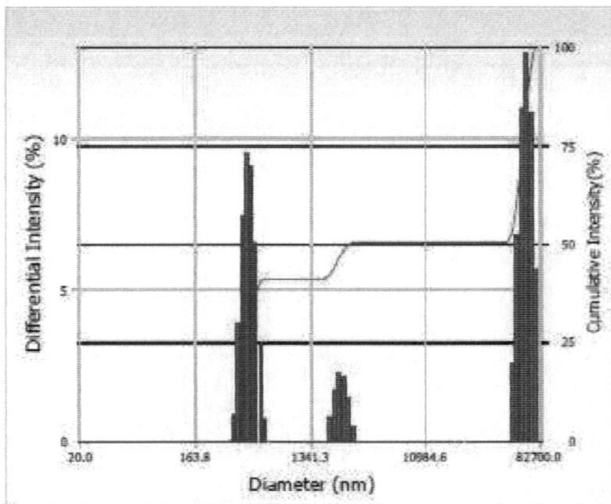

Figure 5. Intensity distribution of isolated cellulose product particle size.

From the histogram above, depicted that there are three peaks of distribution results at 431.2, 2306.9 and 69147.1 nm.

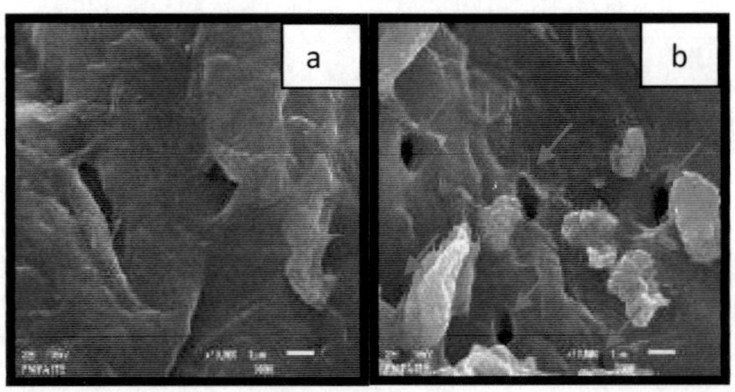

Figure 6. SEM analysis of (a) Commercial MCC and (b) Isolated cellulose of TCL husk.

Scanning electromotive microscopy (SEM) analysis was performed to both commercial MCC and isolated cellulose of TCL husk for 10,000x. Figure 6 explains the different of morphology appearance of the samples. Commercial MCC shows dense morphology than isolated cellulose of TCL husk. Some pores of isolated cellulose of TCL husk are depicted by arrow in figure 6b. Pores presence confirm crystallinity index of isolated cellulose of TCL husk decrease, it means increase amorph region as effect of molecular weight decreasing, if we compare to commercial MCC. Molecular weight decreasing weakens hydrogen bond interaction intra- and interchain of cellulose polymer chain.

4.3. XRD analysis

Figure 7 below shows X-ray diffraction patterns for each sample (a) TCL husk, (b) isolated cellulose and (c) commercial MCC cellulose respectively. The commercial MCC X-ray diffraction pattern shows three major peaks located at 2θ: 22.48; 14.29 and 34.36. The X-ray diffraction pattern for cellulose of TCL husk shows a similar pattern a commercial MCC. The cellulose of TCL husk shows the main peaks at shift 2θ: 22.40; 15.20 and 34.7.

The X-ray diffraction pattern of the commercial MCC is appeared having a strong intensity with three major peaks, indicating the crystal structure. While the cellulose of TCL husk is appeared weaker intensity and with the wide area at the shift of the main peaks, indicating lower crystallinity than commercial MCC. XRD pattern of TCL husk shows amorphous structure, there is no strong peak appeared. Crystallinity Index (*C.I.*) calculation of TCL husk, isolated cellulose, and microcrystalline cellulose are shown in Table 4 below. The calculation refers to *Equation* 2.

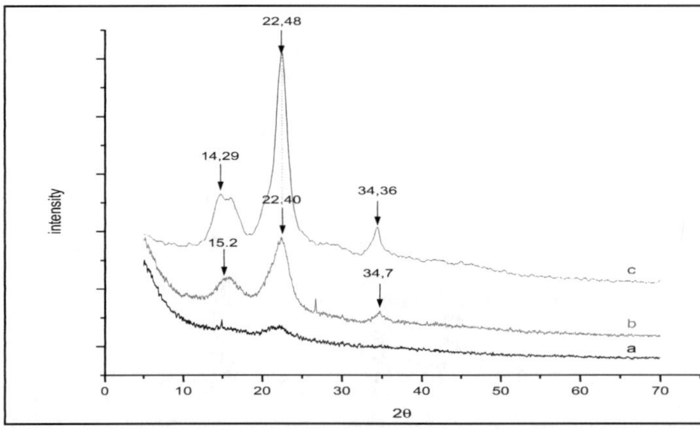

Figure 7. X-ray Diffraction Pattern: (a) TCL husk; (b) cellulose product dan (c) commercial MCC.

Table 4. Crystallinity Index of Samples.

Sample	I_{200}	I_{am}	*C.I. (%)*
TCL Husk	-	-	-
Cellulose product of TCL husk	335	199.36	40.48
Commercial MCC	1029	320	68.90

We didn't calculate *C.I. (%)* of TCL husk, because it's totally amorphous structure, no crystalline peak appeared. TCL husk's amorphous fraction is contribution of lignin content. While, the low crystallinity index, *C.I. (%)* of cellulose of TCL husk compared with commercial MCC, is predicted caused of lower molecular weight. However, this prediction needs some studies forward. The low molecular weight cellulose from biopolymer provides advantages, fortunately, such as cellulose is more easily modified. This statement is similar to previous report (Hutomo, et al. In press) and also confirmed by preceding SEM analysis. The low molecular weight indicates hydrogen bonding inter/intra-chain weak that relate to crystallinity index, *C.I. (%)* of cellulose of TCL husk reducing.

The XRD pattern provides information of crystalline allomorphs structure of cellulose. There are four crystalline allomorph are celluloses I, II, III and IV [28]. As we know, the cellulose XRD pattern of the cellulose of TCL husk is significant similar to commercial MCC, refers to cellulose-I allomorph structure. It approves that the product is native cellulose, an abundant biopolymer in nature.

4.4. TGA analysis

The thermal stability of cellulose products from TCL husk was analyzed by Thermogravimetry analysis (TGA). The following thermogram shows the differences in thermal stability of the sample. Based on Moran's report [31], hemicellulose and cellulose degradation began at temperature intervals of 220 and 250 °C respectively, whereas lignin was degraded at lower temperatures, at 200 °C. At higher temperatures, heat-resistant lignin is higher than hemicellulose and cellulose, due to the slow rate of degradation. Weight loss between 200 and 300 °C is primarily due to decomposition of hemicellulose and slow parallel decomposition of lignin, whereas weight loss between 250 and 500°C is caused by cellulose decomposition (250-350°C) and lignin (200-500°C). At temperatures higher than 400 °C, oxidation and breakdown of residues take place into gas products with lower molecular weight [24].

According to Trache *et al.* [16], degradation processes of cellulose, commonly, consist of both releasing of water (60-140°C) and dehydration, decarboxylation, depolymerization and decomposition of glycosyl unit of cellulose (250-450°C). Water releasing of polysaccharide is detected with a little weight loss appeared for samples TGA thermogram [32]. From TGA thermogram above showed that TCL husk exhibits higher weight than commercial MCC and cellulose product of TCL husk indicating TCL husk absorbed higher water than another both samples.

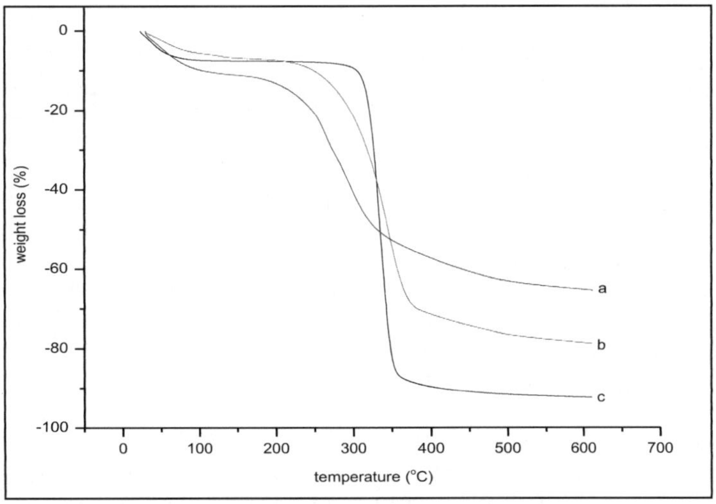

Figure 8. Thermogram of: (a) TCL husk; (b) Cellulose product of TCL husk and (c) commercial MCC.

Table 5. Initial and final decomposition temperatures of samples

Samples	Initial decomposition temperature (°C)	Weight loss of initial (%)	Final decomposition temperature (°C)	Weight loss at final (%)	Weight remains (%)
TCL HUSK	203.5	13.35	332.9	50.72	49.28
Commercial MCC	308.0	9.7	352.6	86.60	13.40
Cellulose product of TCL husk	248.7	9.8	375.5	69.60	30.40

By comparing the obtained data from the table 5, can be depicted that isolated cellulose products from TCL husk have good thermal stability compared to their original TCL husk. The weight loss of isolated cellulose product occurs at higher temperature temperatures (248.7°C) while TCL husk first undergoes weight loss at lower temperatures (203.5°C). The mass that is lost too rapidly on TCL husk is the component of lignin and hemicellulose, whereas in the first weight loss of isolated cellulosic unit is first estimated from the cellulose itself.

When compared with MCC, the thermal stability is better compared with the isolated cellulose. This is due to the molecular weight of MCC is much greater than that of isolated cellulose from TCL husk. From the above data has been obtained information that the mass remaining to temperatures above 600°C. As mentioned before, after temperatures above 400 °C, there is a decrease in the decomposition rate of lignin. Consequences of the rate, then the largest residual gain is on TCL husk is 49.28% whereas on the cellulose of TCL husk and the commercial MCC are lower at 30.40% and 13.40% respectively. The high char residue of TCL husk showing no significant lignin component remains else degrade at the final decomposition and the next. Cellulose of TCL husk and commercial MCC, however, are stable at the high temperature, just remain lower residue of char.

5. Conclusion

Green simple microwave-assisted extraction (MAE) of cellulose from TCL husk had succeeded to employ short duration of alkaline-treatment, with best condition of 30 minutes. Applying of low concentration of alkaline solution (NaOH 2%) and H_2O_2 as a bleaching agent indicated that the extraction method was greener compared to conventionally extraction methods. Designing energy efficiency and chemical safer are mainly approved of green chemistry principle from the MAE method. Delignification of TCL husk had produced lignin and hemicellulose-free cellulose product, which depicted to FTIR peaks absorption differences. The cellulose products of TCL husk were confirmed having pure cellulose-typical important groups by FTIR analysis, they are –OH group at 3000, CH_2/CH_3 group at 2902, C-O-C group at 1161-1163 and β-(1,4)-glycosidic bond at ~897 cm^{-1}. PSA analysis shows that isolated cellulose of TCL husk has average sizes 2,439.7 nm (~2.4397 μm). Morphology study depicts some pores of isolated cellulose of TCL husk, while commercial MCC have dense morphology and tends has no pores. XRD analysis indicated that cellulose product has cellulose-I crystalline structure with 2θ: at 21.89; 14.20 dan 34.64 and has crystallinity index ~40.48%. Cellulose product of TCL husk has good thermal stability, with decomposition temperature between 248.7 and 375.5°C, which was analyzed using TGA.

References

[1] Jabbour L, Bongiovanni R, Chaussy D, Gerbaldi C and Beneventi D 2013 **20** 1523–45
[2] Fan M, Dai D and Huang B 2012 Fourier transform infrared spectroscopy for natural fibres *Fourier transform-materials analysis* (Intechopen)
[3] Penjumras P, Rahman R B A, Talib R A and Abdan K 2014 *Agric. Agric. Sci. Procedia* **2** 237–43
[4] Singh R K 2013 *J. Therm. Anal. Calorim.* **114** 809–19
[5] Bono A, Ying P H, Yan F Y, Muei C L, Sarbatly R and Krishnaiah D 2009 *Adv. Nat. Appl. Sci.* **3** 5–12
[6] Singh S, Gaikwad K K, Park S-I and Lee Y S 2017 *Int. J. Biol. Macromol.* **99** 506–10
[7] Brandt A, Gräsvik J, Hallett J P and Welton T 2013 *Green Chem.* **15** 550–83
[8] Chen D, Yang X, He Z and Ni Y 2016 *J. Bioresour. Bioprod.* **1** 18–21
[9] Pérez-Madrigal M M, Edo M G and Aleman C 2016 *Green Chem.* **18** 5930–56
[10] Kanimozhi K, Basha S K and Kumari V S 2016 *Mater. Sci. Eng. C* **61** 484–91
[11] Xiao S Y, Yang Y Q, Li M X, Wang F X, Chang Z, Wu Y P and Liu X 2014 *J. Power Sources* **270** 53–8
[12] Kumar A, Negi Y S, Bhardwaj N K and Choudhary V 2012 *Carbohydr. Polym.* **88** 1364–72
[13] Liu C-F, Ren J-L, Xu F, Liu J-J, Sun J-X and Sun R-C 2006 *J. Agric. Food Chem.* **54** 5742–8
[14] Anwar B, Bundjali B and Arcana I M 2015 *Procedia Chem.* **16** 279–84
[15] Leite A L M P, Zanon C D and Menegalli F C 2017 *Carbohydr. Polym.* **157** 962–70
[16] Trache D, Hussin M H, Chuin C T H, Sabar S, Fazita M R N, Taiwo O F A, Hassan T M and Haafiz M K M 2016 *Int. J. Biol. Macromol.* **93** 789–804
[17] Kristanto A 2010 Panduan budidaya kakao
[18] Hutomo G S, Marseno D W and Anggrahini S 2012 *African J. Food Sci.* **6** 180–5
[19] Nisa D and Putri W D R 2013 *J. Pangan dan Agroindustri* **2** 34–42
[20] Kumar S, Sivakumar M and Ruckmani K 2016 *Int. J. Biol. Macromol.* **92** 682–93
[21] Han D and Row K H 2010 R **15** 2405–26
[22] Rana K K and Rana S 2014 *Open Access Libr. J.* **1** 1
[23] Smiderle F R, Morales D, Gil-Ramírez A, de Jesus L I, Gilbert-López B, Iacomini M and Soler-Rivas C 2017 *Carbohydr. Polym.* **156** 165–74
[24] dos Santos D M, de Lacerda Bukzem A, Ascheri D P R, Signini R and de Aquino G L B 2015 *Carbohydr. Polym.* **131** 125–33
[25] Biswas A, Kim S, Selling G W and Cheng H N 2013 *Carbohydr. Polym.* **94** 120–3
[26] Viera R G P, Rodrigues Filho G, de Assunção R M N, Meireles C da S, Vieira J G and de Oliveira G S 2007 *Carbohydr. Polym.* **67** 182–9
[27] Poletto M, Ornaghi H and Zattera A 2014 *Materials (Basel).* **7** 6105–19
[28] Park S, Baker J O, Himmel M E, Parilla P A and Johnson D K 2010 *Biotechnol. Biofuels* **3** 10
[29] Farrán A, Cai C, Sandoval M, Xu Y, Liu J, Hernáiz M J and Linhardt R J 2015 *Chem. Rev.* **115** 6811–53

[30] Oh S Y, Yoo D Il, Shin Y, Kim H C, Kim H Y, Chung Y S, Park W H and Youk J H 2005 *Carbohydr. Res.* **340** 2376–91
[31] Morán J I, Alvarez V A, Cyras V P and Vázquez A 2008 *Cellulose* **15** 149–59
[32] Rodrigues Filho G, de Assunção R M N, Vieira J G, Meireles C da S, Cerqueira D A, da Silva Barud H, Ribeiro S J L and Messaddeq Y 2007 *Polym. Degrad. Stab.* **92** 205–10

Acknowledgements
This research was supported by Doctoral Program of Chemistry, Faculty of Mathematics and Natural Sciences, Institut Teknologi Bandung (ITB), Bandung, Indonesia. The authors are very thankful to Bioorganic and Organic Synthesis Laboratory, which have supported this works. The authors are also thankful to technician of FTIR of Analytical Chemistry Laboratory (ITB), XRD of Hydrology and Hydrogeochemistry Laboratory (ITB), PSA of Instrument Laboratory (ITB), SEM of Basic Science Centre A (ITB) and TGA of Universitas Lampung.

IOP Conf. Series: Materials Science and Engineering **541** (2019) 012018 doi:10.1088/1757-899X/541/1/012018

Synthesis and characterization of mesoporous silica nanoparticles (MSNp) MCM 41 from natural waste rice husk

B Purnawira[1], H Purwaningsih[1], Y Ervianto[1], V M Pratiwi[1], D Susanti[1], R Rochiem[1], A Purniawan[1]
[1]Institut Teknologi Sepuluh Nopember, Surabaya

Email: purwaningsihhariyati@gmail.com

Abstract. Rice husk contains silica up to 86.9-97.3%, it potentially as precursor silica in the synthesis process of mesoporous silica MCM 41. First, use acid leaching as pre-treatment of rice husk extraction, followed by heat treatment at 600°C in the furnace, continued with sol-gel process to get amorphous silica. Mesoporous silica was synthesizing using CTAB as surfactant to assist porous of silica MCM 41 (CTAB-templated silica). CTAB concentration were 1, 1.25 and 1.5%. Mesoporous silica nanoparticle MCM 41 synthesized by sol-gel method, followed by hydrothermal and reflux methanol + HCl. X-ray diffraction pattern showed that silica has amorphous liked-structure, indicated by peak broadening at 2-theta around 22-24°. Fourier Transformed Infrared-FTIR shown functional group of Si-O, Si-O-H, Si-O-Si, -OH, Si-H, and Si-OH bond have been identified. Scanning electron microscope (SEM) shown particle agglomeration and particle sizes range between 311-482 nm. The desorption adsorption nitrogen analysis (BET analysis) showed that surface area and pore diameter at 1%, 1.25% and 1.5% CTAB have surface area 552.429 m^2/g, 768.947 m^2/g and 705.501 m^2/g. In hence pore diameters are 3.4178 nm, 3.0517 nm and 3.4098 nm respectively.

1. Introduction

The development of mesoporous silica technology has been very developed. Since the discovery of new families of mesoporous molecular sieves such as MCM-41, they have attracted much attention in the filed of catalysis, separation media and host materials for inclusion compounds [1,2]. Many research have been reported, such as various synthesis and formation mechanisms for the M41S family (MCM-41, MCM-48, and MCM-50) [3-16]. These materials serve as an excellent choice for such applications, since they offer high specific surface areas, large and defined pore sizes, defined surface acidity, and excellent mechanical and thermal stability. They are synthesized with surfactant used as template and can be functionalized with different organic groups later. These textural properties should allow fast extraction processes [17–19].

Rice is Indonesia's main agricultural commodity. From paddy to rice, it produces secondary product in the form of rice husks. The use of rice husk itself is still limited to burning brick media. The rice husk ash has a silica content of 86.9-97.3% [20]. Based on previous study, the burning rice husks are expected to replace TEOS as a precursor in the synthesis of MCM 41 mesoporous silica. TEOS is a commercial product that has high prices while rice husks are abundant and inexpensive. Thus, rice husk charcoal is a potential source for producing silica.

Silica that produced from rice husk ash combustion is amorphous silica [21]. Amorphous silica in various conditions is considered more reactive than crystalline silica and has a complicated spherical structure. This complex structure causes a high surface area, usually above 3 m^2/g [22]. The natural specific surface area of silica can still be maximized by using surfactants as templates. Mesoporous silica can be synthesized by modifying the Stöber process with additional surfactants (for example, cetyl trimethyl ammonium bromide-CTAB [23].

In this paper, we report the synthesis of MCM 41 silica mesoporous from rice husk and its characterization. The detailed phase, crystal structure, porous structure, and their possibility as drug delivery material were explored.

2. Experimental

2.1. Reagents

Rice husk from Blitar, Cetyltrymethyl Ammonium Bromide (CTAB) was purchased from HiMedia Laboratories Pvt. Ltd. hydrochloric acid 37% (HCl), Acetic Acid glacial (CH_3COOH), Sodium hydroxide (NaOH), Methanol 100% (CH_3OH), 96% Ethanol (CH_3CH_2OH) and aquadest.

2.2. Synthesized mesoporous silika MCM 41 from rice husk

Rice husk leached with HCl at 80°C for 2 hours, washed and burned at 600°C for 6 hours to produced rice husk ash (RHA). Dissolved 6 grams of RHA into 100 ml of 2M NaOH and filtered using filter paper. The filtration solution is titrated with 2M HCl until gel formed, the titration process continued to pH 7. Strained the gel using filter paper and dried it at 100°C for 6 hours to produced silica-extracted rice husk. Mixed 6 grams SiO_2 and 100 ml of 2M NaOH and heated at 80°C in oil bath for 24 hours.. Dissolved CTAB 1%, 1.25%, and 1.5% into 100 ml aquades according to the weight ratio and stirred it to produced Na_2SiO_3. Followed by acetic acid titration until pH 10 and stirred for 6 hours. Continued hydrothermal process at 100°C for 24 hours and aging treatment for 24 hours. Then washed using a mixture of distilled water and ethanol. Aging at room temperature for 12 hours. Finally, reflux with solution 9 ml HCl-37% and 160 ml methanol for 12 hours. Separated the sediment using centrifuge and dry it at 80°C for 2 hours.

2.3. Characterization

Fourier transform infra-red (FTIR-Thermo Scientific Nicolet IS10) was employed to determine the chemical bonding, x-ray diffraction (XRD-PanAnalytical X'Pert Pro) was conducted to determine the crystalline structure. Microstructure analysis was observed with microscop electron (SEM-FEI Inspect S50 and TEM). The specific surface area, pore size and pore distribution were determined by Brunauer-Emmett-Teller (BET) Nitrogen adsorption-desorption method using Quantochrome surface analyzer.

3. Result and Discussion

Figure 1 shown FTIR spectrum of pure CTAB and CTAB-templated silica mesoporous. FTIR spectrum of pure CTAB has identi1fied wavenumber at 2915.75 cm^{-1}, 2848.39 cm^{-1}, 1472.87 cm^{-1}, 960.31 cm^{-1}, 911.15 cm^{-1} and 718.79 cm^{-1}. This peak corresponded to vibrational bands of CH, CH_2, $N(CH_3)_2$ and CH functional group. According to Becerra (2012), he had reported IR spectra of the surfactants CTAB. The symmetric (υ_s-(CH_2), d_+) and asymmetric (υ_{as}(CH_2), d_-) stretching vibrations of pure CTAB indicate equivalent gauche defects which lie at 2849 and 2918 cm^{-1}. The peaks at 3017 and 1487 cm^{-1}, and at 1473 and 1462 cm^{-1} were attributed to the asymmetric and symmetric C-H scissoring vibrations of CH3-N^+ moieties and to the CH_2 scissoring mode, respectively [25]. FTIR spectrum of CTAB-templated silica various composition 1%, 1.25% and 1.5% had shown wavenumber at 3388.31 cm^{-1}, 1636.71 cm^{-1}, 1070.46 cm^{-1}, 956.95 cm^{-1}, 795.09 cm^{-1} and 456.15 cm^{-1}. According to the Aldrich handbook library, these wavenumbers indicated functional groups of -OH, Si-O-Si, Si-OH and Si-O. Compared to spectra of pure CTAB, CTAB-templated silica did not show C-H vibrational band and others as showed at pure CTAB spectrum. CTAB-templated silica treatment followed by methanol reflux was able to remove the surfactant template due to the absence of pure CTAB functional groups on the silica reflux results.

Adding CTAB in extracted silica did not show the wave number for the -CH_2-, N $(CH_3)_2$, -CH- groups. Many more surface hydroxyl groups can also be kept, enhancing the hydrophilic property and modifying the reactive ability of pore channels forming.

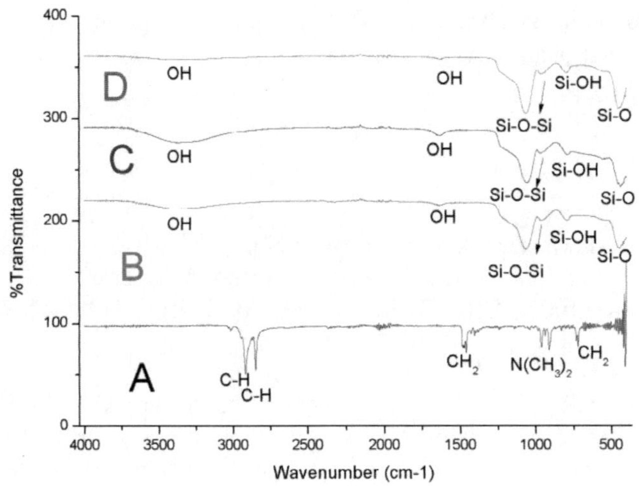

Figure 1. FTIR result of (A) pure CTAB, silica reflux results in concentration (B) 1% CTAB, (C) 1.25% CTAB and (D) 1.5% CTAB.

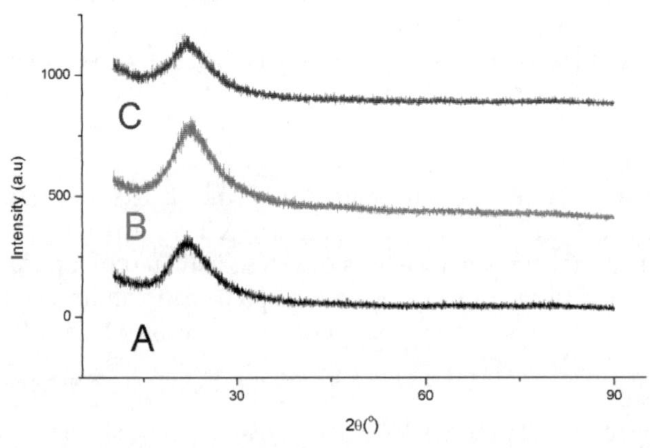

Figure 2. X-Ray diffraction curve of MCM 41 silica mesoporus with surfactant concentration (A) 1% CTAB, (B) 1.25% CTAB and (C) 1.5% CTAB.

Figure 2 shown x-ray diffraction pattern of various CTAB-templated silica for 1% CTAB (figure 2.A), 1.25% CTAB (figure 2.B) and 1.5% CTAB (figure 2.C). XRD pattern of 1% CTAB-templated silica indicated peak diffraction at $2\theta = 22.0155°$. XRD pattern of 1.5% CTAB and 1.5% CTAB-templated silica shown similar curve as 1% CTAB. All XRD pattern correspond to PDF card standard for silica-cristobalite phase ICDD # 00-001-0424. Diffraction patterns showed similarities in profile (figure 2), peaks diffraction broadened at around $2\theta = 22–24°$ as occur in the diffraction pattern of silica amorphous phase.

Amorphous silica is usually present in living things such as diatoms, radiolarians, silicoflagellates, and some sponges. Non-crystalline or amorphous silica has an arrangement of atoms and molecules in the form of random and irregular patterns [22]. Another study had been reported that peak broadening also indicated porous structure forming. This hypothesis will prove with another characterization measurement such as TEM and BET surface analyzer.

Based on figure 2, the diffraction peaks broadened further as CTAB concentration increases. The sample 1.5% CTAB has the widest diffraction peak compared to other samples. The widening of the sample diffraction peaks indicates the formation of non crystalline structures due to pore formation by the CTAB template. According to Wanyika, et.al (2011) had explained that wide angle x-ray diffraction (WAXRD) have amorphous-like pattern that corresponded to existence of large porous in material. The same pattern had shown in figure 2, this profile could be assumed that silica have porous structure [7]. It concludes that the peak broadening of peaks diffraction in the CTAB surfactant – templated silica is due to the pores formed in silica.

Figure 3 shows scanning electron microscope images of various wt. % CTAB-templated silica, SEM image of 1 % CTAB (A), 1.25% CTAB (B) and 1.5% CTAB (C). Fig 3 shows CTAB-templated silica resulted coral-liked structure According to the Figure 3.A, silica particles have the smallest size of 476 nm and the largest is 627 nm, while the average particle size is 482 nm. Figure 3.B shows silica particles have the smallest size of 343 nm and the largest is 542 nm and average particle size is 445 nm. Figure 3.C shows silica particles have the smallest size of 234 nm and the largest is 431 nm, so the average particle size is 311 nm. It was found that the particle size is getting smaller as the concentration of CTAB increases.

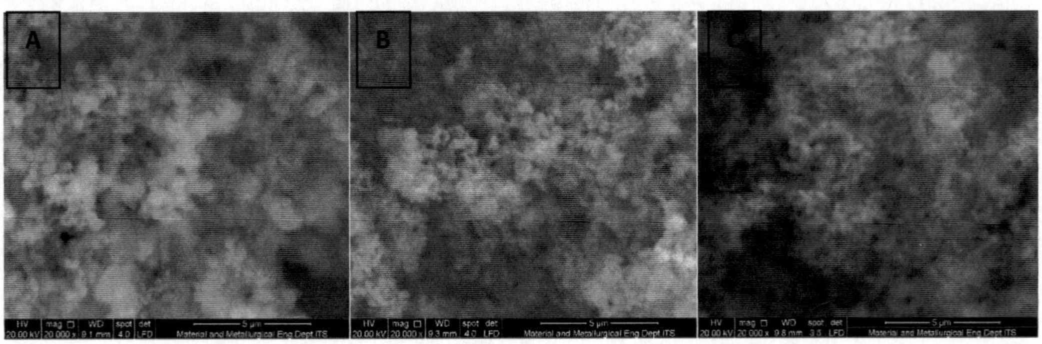

Figure 3. SEM images of silica with (A) 1% CTAB, (B) 1.25% CTAB, and (C) 1.5% CTAB.

Increasing the concentration of CTAB causes the formation of micelle-structures, in hence increasing the bonds between silica and surfactants. Whereas at lower CTAB concentrations, silica that do not bind to surfactants would bind Si-O-Si group itself which causes greater silica particle size. Zhao, et al. (2013) explain the mechanism for the formation of mesoporous material, in which part of the R_n-$N^+(R)_3$ group would bind to silica on the entire surface. Silica formed a long Si-O-Si chain that become a particle [26].

Figure 4 shown TEM images of various concentration CTAB-templated silica, TEM images showed that CTAB induced pores structure in silica particles. Figure 4.A (1% CTAB) shown that silica has a pore diameter 3.58 nm, silica-1.25% has a pore diameter 3.13 nm (B) and silica-1.5% has identified pore size 3.12 nm (C). TEM images also represented pore structure such as honeycomb or hexagonal for all various CTAB, in accordance with the material of the mesoporous silica MCM 41 in other research [24].

Table 1 presented N_2 adsorption-desorption analysis-BET analysis of all various CTAB. It appears that surface area tends to be wider. The total pore volume shows a tendency to increase along with the increasing concentration of CTAB, this is consistent with the results of the study [11] and [12]. An increase in CTAB causes an increasing number of templates so that it can bind more Si-O-H functional groups. When the template removal process is carried out, the number of pores formed increases according to the increase in concentration of the surfactant. Yun-yu, et al. (2012) reported that the increase in CTAB concentration would increase the surface area of the mesoporous silica produced [27].

Table 1 showed that surface area data in this study conformity with Yun-yu et al (2012), because the addition of CTAB concentration has a tendency to increase the surface area of the mesoporous silica produced except at 1.5% CTAB-templated silica. This requires more intensive research to determine the effect of adding CTAB above 1.5%. TEM image and BET surface analyzer for 1.5% CTAB had strongly mentioned that 1.5%CTAB-templated silica had classified as mesoporous material according to pore size ~3.12 nm from TEM analysis and 5.814 nm form BET surface analyzer.

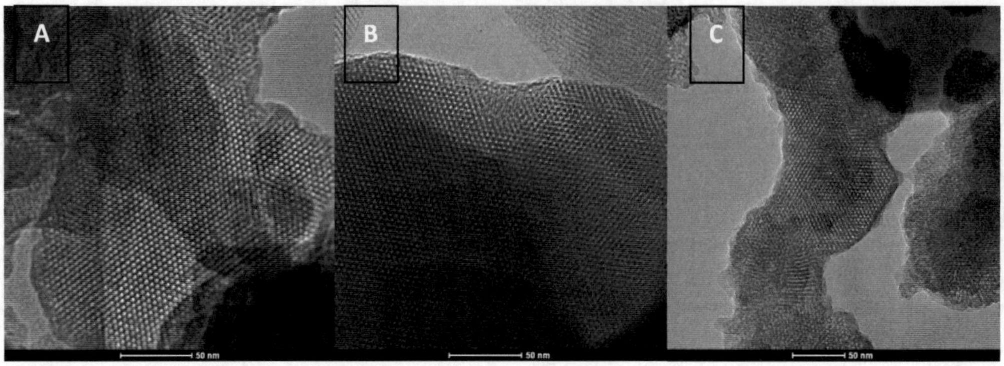

Figure 4. TEM results from silica with concentration (A) 1% CTAB, (B) 1.25% CTAB and (C) 1.5% CTAB.

Table 1. Surface area, average diameter pore size, and total pore volume of silica samples with the addition of 1%, 1.25%, and 1.5% CTAB

No	Sample	Surface Area (m^2/g)	Average diameter pore size(nm)	Total pore volume (cc/g)
1	1%	552.429	7.58248	1.047
2	1.25%	768.947	6.07107	1.167
3	1.5%	705.501	5.81420	1.025

4. Conclusion

MCM 41 silica mesoporus had been successfully extracted from rice husk ash. MCM-41 silica has functional groups -OH, Si-O-Si, Si-OH and Si-O. It has the smallest particle size at a concentration of 1.5% CTAB which is 311 nm with a pore diameter of 3.12 nm and has a honeycomb or hexagonal pore structure. The largest surface area at the greatest concentration of 1.25% CTAB was 768.947 m^2/g, the average pore diameter was 6.07107 nm and the pore volume was 1.167 cc/g. CTAB surfactant successfully as pores-templated silica and resulted mesoporous silica nanoparticles.

References

[1] C T Kresge, M E Leonowics and W J Roth, J C Vartuli, J S Beck 1992 *Nature* **359** 710
[2] L Mercier and T J Pinnavaia, 1997 *Adv. Mater.* **9** 500
[3] Kresge C T, Leonowicz M E, Roth W J, Vartuli J C and Beck J S 1992 *Nature* **359** 710
[4] Beck J S, Vartuli, J C, Roth W J, Leonowicz M E, Kresge C T, Schmitt K T, Chu C T-W, Olson D H, Sheppard E W, McCullen S B, Higgins J B and Schlenker, J L J 1992 *Am. Chem. Soc.* **114** 10834
[5] Vartuli J C, Schmitt K D, Kresge C T, Roth W J, Leonowicz M E, McCullen S B, Hellring S D, Beck J S, Schlenker J L, Olson D H and Sheppard E W 1994 *Chem. Mater.* **6** 2317
[6] Vartuli J C, Kresge C T, Leonowicz M E, Chu A S, McCullen S B, Johnson I D and Sheppard E

W 1994 *Chem. Mater.* **6** 2070
[7] Beck J S, Vartuli J C, Kennedy G J, Kresge C T, Roth W J and Schramm S E 1994 *Chem. Mater.* **6** 1816
[8] Huo Q, Margolese D I, Ciesla U, Feng P, Gier T E, Sieger P, Leon R, Petroff P M, Schu¨th F and Stucky G D 1994 *Nature* **368** 317
[9] Huo Q, Margolese D I, Ciesla U, Demuth D G, Feng P, Gier T E, Sieger P, Chmelka B F, Schu¨th F and Stucky G D 1994 *Chem. Mater* **6** 1176
[10] Tanev P T and Pinnavaia T J 1995 *Science* **267** 865
[11] Firouzi A, Kumar D, Bull L M, Besier T, Sieger P, Huo Q, Walker S A, Zasadzinski J A, Glinka C, Nicol J, Margolese D, Stucky G D and Chmelka B F 1995 Science **267** 1138
[12] Monnier A, Schu¨th F, Huo Q, Kumar D, Margolese D, Maxwell R S, Stucky G D, Krishnamurty M, Petroff P, Firouzi A, Janicke M and Chmelka B F 1993 *Science* **261** 1299
[13] Stucky G D, Monnier A, Schu¨th F, Huo Q, Margolese D, Kumar D, Krishnamurty M, Petroff P, Firouzi A, Janicke M and Chmelka B F 1994 *Mol. Cryst. Liq. Cryst.* **240** 187
[14] Chen C, Li H and Davis M E 1993 *Micropor. Mater.* **2** 17
[15] Chen C, Burkett S L, Li H and Davis M E 1993 *Micropor. Mater.* **2** 27
[16] Cheng C F, Luan Z, Klinowski J 1995 *Langmuir* **11** 2815
[17] J S Beck, J C Vortuli, W J Roth, M E Leonowicz, C T Kresge, K D Schmitt, C T W Chu, D H Olson, E W Sheppad, S B McCullen, J B Higgins and J L Schlenker 1992 *J. Am. Chem. Soc.* **114** 10834
[18] S R Hall, C E Fowler, B Lebeow and S Mann 1999 *Chem. Commun.* **201**
[19] M Algarra, M V Jimenez, E Rodriguez-Castellon, A Jimenez-Lopez and J Jimenez-Jimenez 2005 Chemosphere **59** 779
[20] Y Shiraishi, G Nishimura, T Hirai and I Komasawa 2002 *Ind. Eng. Chem. Res.* **41** 5065
[21] Danarto Y C, Nur A, Setiawan D P Kuncoro and N D 2010 *Prosiding Seminar Nasional Teknik Kimia Pengembangan Teknologi Kimia untuk Pengolahan Sumber Daya Alam Indonesia*, Yogyakarta 1-2
[22] Chandrasekhar S, Pramada P N and Praveen L 2005 *J. Mat. Sci.* **40** 6353-6544
[23] Kirk R E and Othmer 1984 *Encyclopedia of Chemical Technology 4th Edition* (New York: John Wiley and Sons, Inc.)
[24] Cai Q, Lin W-Y, Xiao F, Pang W, Chen X and Zou B 1999 *Microporous and Mesoporous Materials* 1-15
[25] Ghorbani F, Younesi H, Mehraban Z and Celik M S 2013 *J. the Taiwan Institute of Chemical Engineers* 821–828
[26] Alberto G B, M B Soto, V Soto, J A C, N Casillas, S Prévost, L Noirez, M Gradzielski and José I E 2012 *Nanoscale Research Letters* **7(1)** 83
[27] Zhao D, Wan Y and Zhou W 2013 *Ordered Mesoporous Materials* (Singapore: Markono Print Media Pte Ltd)
[28] Yun-yu Z, Xiao-xuan L and Zheng-xing C 2012 *Powder Technology* **226** 239–245

IOP Conf. Series: Materials Science and Engineering **541** (2019) 012019 doi:10.1088/1757-899X/541/1/012019

Effect of austempering treatment time on strength, ductility, hardness, and abrasive wear resistance of JIS FCD 700 ductile iron

H Rinaldo, M Januar*, M M Sinaga
Department of Mechanical Engineering, Atma Jaya Catholic University of Indonesia

E-mail: moch.januar@atmajaya.ac.id

Abstract. The effects of austempering treatment time on the strength, ductility, hardness, and abrasive wear resistance of JIS FCD 700 were studied. The austempering was carried out at a temperature of 315 °C and treatment times of 19, 30, 50, and 77 mins. The results indicate that the variations of the austempering time alternate the mechanical properties of the material, depending on the retained phases in the microstructure. A higher retained austenite induces a higher abrasive wear resistance and strengths, while the presence of acicular ferrite produces a lower hardness. Moreover, the large number of losses in martensite and retained austenite phases deteriorate the overall mechanical performances, as exhibited in the austempering time longer than 50 min. It is found that the 50 min of austempering time yields the optimum performance in terms of strengths (tensile strength of 1058 MPa and yield strength of 859 MPa), ductility (4,08 %), hardness (37 HRC), and abrasive wear resistance (0,21 x 10^{-6} mm^3/mm at 6,32 kg load). This optimum performance was achieved mainly due to the relatively larger amounts of retained austenite in the JIS FCD 700.

1. Introduction

Nowadays, ductile irons are greatly considered as replacement materials of steels for automotive components, such as camshaft, transmission gear, truck axle, and so on [1,2]. Ductile iron generally has better strength, ductility, hardness, and abrasive wear resistance compared to steel [3]. Fortunately, these are the characteristics required for the automotive components with regard to the ability to sustain external loads, machinability, and component lifetime [1,4,5]. Moreover, the melting temperature of ductile iron is lower than steel, which makes the casting process of ductile iron relatively simpler and less expensive [6].

JIS FCD (*Japanese International Standard Ferro Casting Ductile*) 700 is one of widely used ductile irons for automotive component material due to its good machinability along with its superior tensile and yield strengths [7,8]. However, the mechanical properties of the JIS FCD 700 are still needed to be developed to offer higher performances of the automotive components as demanded by the industry. Austempering is one effective way to enhance the ductile iron mechanical properties [6,9]. A ductile iron subjected to the heat treatment, namely austempered ductile iron (ADI), will retain the composition of nodular graphite, martensite, retained austenite, and acicular ferrite [10]. The presence of these microstructures allows the increment of strength, ductility, hardness, and abrasive wear resistance in the corresponding ductile iron [1,6].

In the process, the microstructure and mechanical properties of ADI are strongly affected by the austempering treatment temperature and time. As reported by Sohi *et al.* [11], the retained austenite phase increases as the austempering time increases, leading to an improvement in tensile and yield strengths. However, while the retained austenite increases, the martensite phase tends to decrease. Although the ductility can be improved, the hardness might be deteriorated. Moreover, the reduced amount of martensite has an impact on the loss of hardness, leading to a decrease in abrasive wear resistance [1,10]. These findings were also corroborated by the work of Suratman *et al.* [6], implying a decrease in hardness as austempering treatment time increases specifically on JIS FCD 700 material. Unfortunately, these trade-offs aggravate the practical use of ADI in automotive components.

Considering the trade-off between hardness/abrasive wear resistance and ductility in JIS FCD 700, investigating parameters involved in the austempering time should be performed. This research was carried out to determine the austempering time which produces the optimum strength, ductility, hardness, and abrasive wear resistance for JIS FCD 700.

2. Experimental details

JIS FCD 700, with chemical composition listed in Table 1, was cast into a block-shaped mould with a dimension of 560 mm x 350 mm x 20 mm. For abrasive wear test, hardness test, and metallography, samples with 200 mm x 10 mm x 5 mm dimension were machined from the block. For the tensile test, samples were machined from the block, in accordance with ASTM E8 standard. Before all the samples were subjected to austempering, phase homogenization was carried out to all samples by normalizing at 900 °C for 30 min. After normalizing, austempering was carried out in a continuous furnace. All samples were austenitized at 900 °C for 30 min, and then quenched into a salt bath (50% $NaNo_3$ + 50% KNO_3) at 315 °C for 19 min, 30 min, 50 min, and 77 min. Prior to metallography, all samples were ground using sand papers ranging from 180 mesh to 1500 mesh, polished with aluminium oxide, and then etched with Nital 5%. The metallographic testing was subsequently performed by using Olympus BX53MRF-S optical microscope. Rockwell hardness test (HRC) was conducted using Mitutoyo AR-10 hardness tester. Each sample was indented at 10 different points along the surface. Vickers hardness test (0,1 HV), conducted to characterize the phases present in metallography, was carried out using Willson-VH1102 hardness tester. The mean hardness of one phase was obtained from indentation at 5 different points. Tensile tests were carried out on each sample at room temperature, using Hung Ta tensile testing machine. Ogoshi High-Speed Universal Wear Testing Machine was used to conduct abrasive wear test. Abrasive ring 3 mm in width and 15 mm in radius span at 1,97 m/s for a sliding distance of 1 m. The applied load was increased from 3,16 kg to 6,32 kg, to 12,64 kg. Each load was applied 3 times for each sample. All three wear marks were measured using a Nikon measurement microscope. The mean wear mark was obtained from the measurement, and wear rate of each sample was calculated using $W_s = \frac{B \times b^3}{12 \times r \times l}$, where W_s is the wear rate (mm^3/mm), *B* is the width of the abrasive ring (mm), *b* is the mean wear mark (mm), *r* is the radius of abrasive ring (mm), and *l* is the ring sliding distance (mm).

Table 1. The chemical composition of JIS FCD 700 (wt. %).

C	Si	Mn	P	S	Cr	Cu	Mg
3,678	2,406	0,021	0,021	0,022	0.061	0.427	0.048

3. Results and Discussion

The phase hardness of JIS FCD 700 after normalizing and austempering treatments are presented in Table 2. The obtained phases agree with the phase hardness data as stated in Iron Castings Handbook [12]. According to the handbook, the hardness ranges of martensite, acicular ferrite, pearlite, ferrite, and nodular graphite are > 550 HV, 260 - 350 HV, 217 - 269 HV, 149 - 187 HV, 42 - 52 HV, respectively.

The microstructure of JIS FCD 700 after normalizing treatment is shown in figure 1(a). Ferrite, pearlite, and nodular graphite are found in this sample. Upon cooling to room temperature, carbon atoms diffuse away from austenite to nodular graphite, the carbon solubility of austenite starts to decrease at this point. This will create carbon depleted zones, which are ferrite nucleation zones. Pearlite nucleates at ferrite/austenite boundary as the eutectoid temperature is reached [3].

Table 2. Phase hardness of JIS FCD 700 after normalizing and austempering.

Heat Treatment	Phase Hardness (HV)				
	Nodular Graphite	Ferrite	Pearlite	Martensite	Acicular Ferrite
Normalized	47,5	170,3	252,0	-	-
Austempered 19'	47,7	-	-	792,2	-
Austempered 30'	46,4	-	-	651,3	328,9
Austempered 50'	47,7	-	-	638,3	322,0
Austempered 77'	47,9	-	-	628,8	315,1

Figure 1(b) shows the microstructure of JIS FCD 700 after austempering for 19 min, which consists of martensite and nodular graphite. The martensite presents in the microstructure because of the 19 min of treatment time is inadequate for the residual austenite to decompose into acicular ferrite and high carbon austenite [13]. Consequently, the residual austenite will not be able to reach enough carbon contents required to stabilize at room temperature. Thus, it transforms to martensite upon cooling process to room temperature [14].

The microstructure of JIS FCD 700 after austempering for 30 min is presented in Figure 1(c). The treatment time of 30 min generates nodular graphite, martensite, and acicular ferrite. Acicular ferrite containing carbide is formed due to a fair amount (2,406 %) of silicon alloying content. The silicon increases carbon activity in ferrite and inhibits cementite formation. Thus, the ferrite will reach a carbon supersaturation, and carbides will precipitate within the ferrite [15]. Although retained austenite cannot be seen directly from the metallographic and Vickers hardness tests, its formation is highly possible in the 30 min of austempering. According to Panneerselvam *et al.*, this is probably due to the formation of high carbon austenite as a result of residual austenite decomposition [16]. Austenite with high carbon content (approximately 1,7 %) will cause the M_s to drop below room temperature. Therefore, the high carbon austenite will be retained at room temperature [17]. Some amounts of residual austenite, which are not decomposed into acicular ferrite and high carbon austenite, transform into martensite during the cooling process to room temperature [14].

Figure 1(d) shows the microstructure of JIS FCD 700 after austempering for 50 min. Similarly, the 50 min of treatment time generates microstructure consists of nodular graphite, martensite, and acicular ferrite. Moreover, retained austenite might also be produced in the 50 min treatment time, but possibly in a larger amount compared to the 30 min of treatment time [15,18]. The microstructure of JIS FCD 700 after austempering for 77 min is shown in Figure 1(e), which also consists of nodular graphite, martensite, and acicular ferrite. Although carbides cannot clearly be detected by the metallographic and Vickers hardness tests, the decomposition of the high carbon austenite into carbides is highly possible to occur at such longer treatment time [16].

Figure 2(a) shows the tensile and yield strengths of the normalized and austempered JIS FCD 700. The normalizing generates the lowest tensile and yield strengths value due to the presences of ferrite and pearlite [6,19]. Austempering for 19 min increases the tensile and yield strengths due to the existence of martensite [6]. However, a high amount of martensite leads to extreme brittleness of the sample, resulting in lower tensile and yield strengths compared to the other austempered samples [20]. Austempering for 30 min causes the tensile and yield strengths to increase, because the formed acicular ferrite absorbs some of the strain, reducing the brittleness effect caused by martensite [1]. The increase in the tensile strength can be attributed to the formation of retained austenite that transforms into martensite phases during plastic deformation, as elaborated by Aranzabal *et al.* [13]. Moreover, the 50 min of austempering also increases the tensile and yield strengths, which are possibly due to the formation of higher amounts of retained austenite [18]. Furthermore, we also argued that decreasing amounts of martensite and retained austenite are the cause for the reduction of the tensile and yield strengths in the 77 min of treatment time [11].

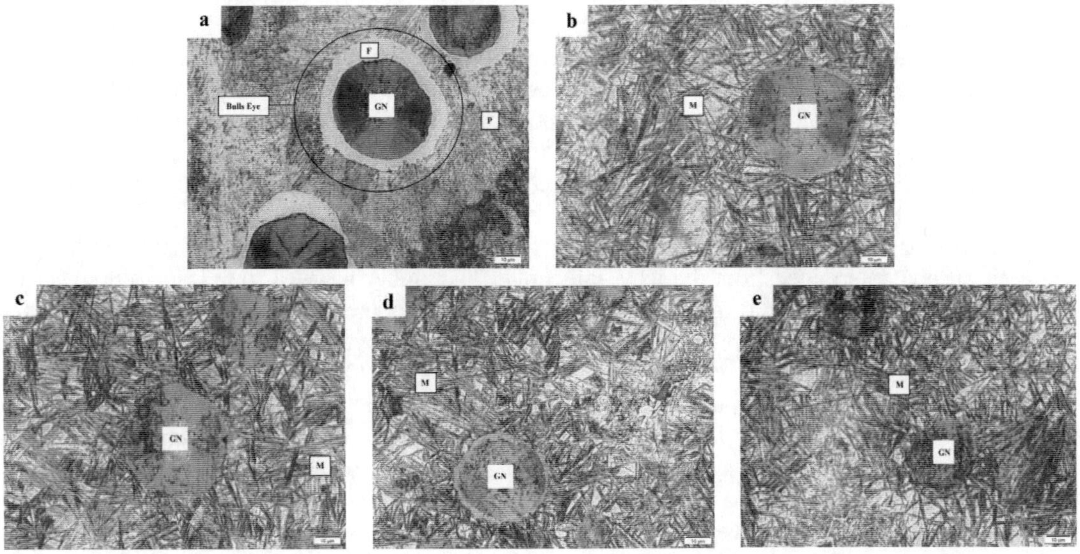

Figure 1. Microstructures of JIS FCD 700 after (a) normalizing, (b) austempering for 19 min, (c) austempering for 30 min, (d) austempering for 50 min, and (e) austempering for 77 min.

Figure 2(b) shows the elongation of normalized and austempered JIS FCD 700. Normalized JIS FCD 700 possesses a fair elongation value because it is dominated by ferrite and pearlite [6,19]. Austempering time of 19 min produces JIS FCD 700 with the lowest elongation value, as the brittle martensite dominates the microstructure [6]. Austempering treatment for 30 min generates the highest value of elongation, which can be associated with the formation of acicular ferrite [21]. Sample austempered for 50 min generates lower elongation compared to that of 30 min. The presumed higher amount of retained austenite produced by austempering for 50 min causes elongation value to decrease, as it would transform into martensite during plastic deformation [13]. Austempering for 77 min also causes elongation value to decrease. The decrease of elongation value is caused by the formation of carbides [18].

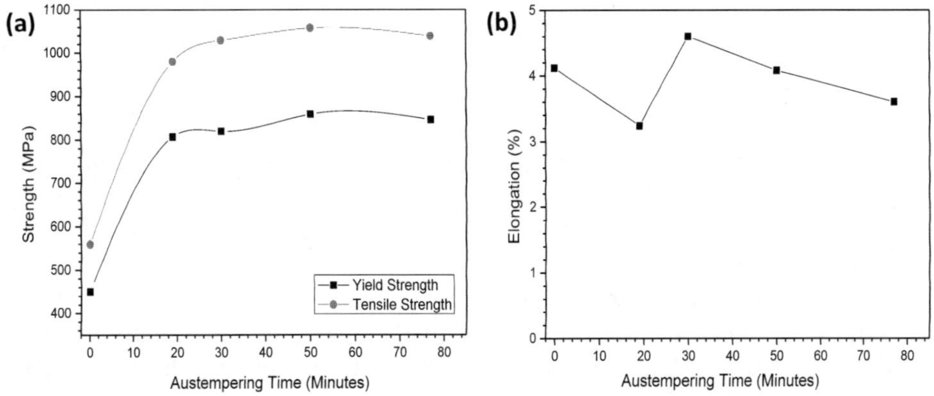

Figure 2. (a) Tensile strength and yield strength of JIS FCD 700 after normalizing and austempering. (b) Elongation of JIS FCD 700 after normalizing and austempering.

Figure 3(a) shows the hardness of normalized and austempered JIS FCD 700 in HRC. The normalizing treatment generates the lowest hardness value since ferrite and pearlite exist in the microstructure [12]. The highest hardness value is obtained by 19 min of austempering time, which is dominated by martensite [11]. The increase in austempering time leads to hardness deterioration. The decreasing amount of martensite formed and the formation of softer phase (acicular ferrite and retained austenite) as the austempering time increases are responsible for the deterioration of hardness [13,18]. Although some amounts of retained austenite are possibly formed during the austempering of 30 min, 50 min, and 77 min, these phases may not transform into martensite when the samples were indented. Based on the work of Sohi *et al*, the presence of retained austenite does not lead to an increase in the hardness value [11].

Figure 3(b) shows the abrasive wear rate of normalized and austempered JIS FCD 700. The abrasive wear resistance is determined from the wear rate value. The higher the wear rate value, the lower the abrasive wear resistance. The normalized JIS FCD 700 possesses the lowest abrasive wear resistance. Both ferrite and pearlite microstructures contribute to the lower level of hardness compared to other samples. This leads to the highest wear rate value of all samples [12,22,23]. The 19 min of austempering produces an improved abrasive wear resistance compared to the normalized JIS FCD 700. The high amount of martensite leads to an increase in hardness, which in turn leads to an increase in abrasive wear resistance [12,23]. As the austempering time increases, the abrasive wear resistance tends to decrease due to a larger loss of martensite. Consequently, the hardness and abrasive wear resistance of JIS FCD 700 decline [23]. However, the sample with 50 min austempering opposes the trend, as exhibited in its lowest abrasive wear rate value. The high amount of retained austenite is assumed to be the reason behind the low wear rate value since retained austenite will transform into martensite when it was subjected to wear [10]. Moreover, the fair hardness of this sample, due to the presence of acicular ferrite, also contributes to the superior abrasive wear resistance [1]. The abrasive wear resistance of ductile iron, particularly JIS FCD 700, appears to be affected by two variables, *i.e.*, the hardness and the amount of retained austenite. This result is in line with the work of Sahin *et al.* [1].

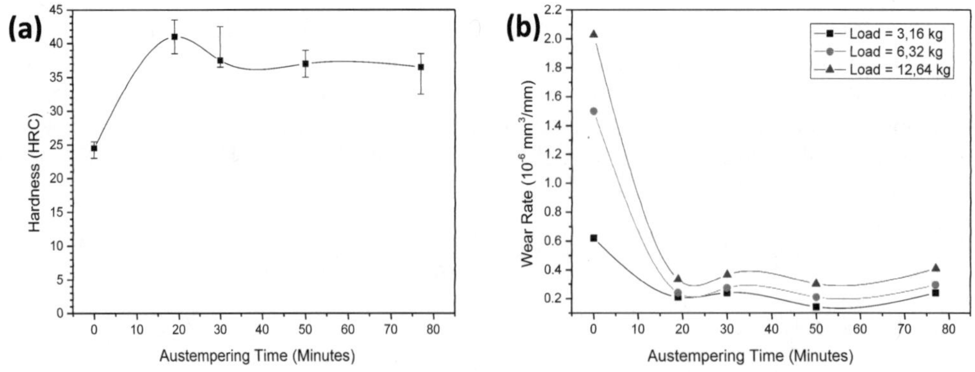

Figure 3. (a) Hardness of JIS FCD 700 after normalizing and austempering. (b) Abrasive wear rate of JIS FCD 700 after normalizing and austempering.

The comparison of normalized and austempered JIS FCD 700 performances are shown in Table 3. It can be inferred that the optimum strengths (tensile and yield), ductility, hardness, and abrasive wear resistance of JIS FCD 700 are obtained through the 50 min of austempering treatment time. The large amounts of retained austenite could possibly be attributed to the reason behind the higher level of strengths (tensile and yield) [13]. Moreover, the existence of acicular ferrite, which is moderate in amount, results in a fair ductility of this sample [21]. The fair hardness of this sample is due to the presence of acicular ferrite [11]. The presence of retained austenite, which is large in amount, is assumed to be the reason behind the exceptional abrasive wear resistance possessed by this sample [10]. In addition, the fair hardness of this sample is also assumed to be contributing to the superior abrasive wear resistance this sample exhibits [1].

Table 3. Comparison of JIS FCD 700 performances.

Heat Treatment	σ_U	σ_Y	ε	Hardness	Abrasive Wear Resistance
Normalized	5	5	2	5	5
Austempered 19'	4	4	5	1	2
Austempered 30'	2	2	1	2	3
Austempered 50'	1	1	3	3	1
Austempered 77'	3	3	4	4	4

*1 = best, 5 = worst.

4. Conclusion

In summary, strengths (tensile and yield), ductility (elongation), and hardness of JIS FCD 700 increase along with the treatment time up to 50 min, 30 min, and 19 min, respectively, but then decrease afterward. On the contrary, the abrasive wear resistance of JIS FCD 700 tends to be deteriorated as the austempering time increases. Nevertheless, 50 min of the austempering time yields an enhancement in the abrasive wear resistance. We argued that this is due to the relatively higher retained austenite contents and a high enough hardness. Hence, the optimum strength, ductility, hardness, and abrasive wear resistance of JIS FCD 700 can be obtained through the 50 min of austempering treatment time.

References

[1] Sahin Y, Erdogan M and Kilicli V 2007 *Mater. Sci. Eng. A* **444** 31-38
[2] *ASM Speciality Handbook, ASM Handbook: Cast Irons* (Ohio: ASM International)
[3] Gagne M and Labrecque C 1988 *Can. Metall. Q.* **37** 343-378
[4] Keough J R and Hayrynen K L 2000 *Journal of Materials and Manufacturing* **109** 344-354
[5] Fatahalla N, Bahi S and Hussein O 1996 *J. Mater. Sci.* **31** 5765-72
[6] Suratman S and Bandanadjaja B 2002 *Jurnal Teknik Mesin* **17** 2
[7] Yanda H, Ghani, J A, Rodzi M N A M, Othman K and Haron C H C 2010 *International Journal of Mechanical and Materials Engineering* **5** 182-190
[8] 1981 *JIS Handbook 1981: Ferrous Materials and Metallurgy* (Tokyo: Japanese Standard Association)
[9] Fuller A G 1985 *Mater. Des.* **6** 127-130
[10] Shepperson S and Allen C 1988 *Wear* **121** 271-287
[11] Sohi M H, Ahmadabadi, M N and Vahdat A B 2004 *J Mater. Process. Trch.* **153** 203-208
[12] Walton C F and Opar T J 1981 *Iron Castings Handbook: Covering Data on Grey, Malleable, Ductile, White, Alloy, and Compacted Graphite Irons* (Michigan: Iron Castings Society, Inc.)
[13] Aranzabal J, Gutierrez I, Rodriguez-Ibabe, J M and Urcola J J 1997 *Metall. Mater. Trans* **28A** 1143-56
[14] Erić O, Rajnović D, Zec S, Sidjanin L and Jovanović M T 2006 *Mater. Charact.* **57** 211-217
[15] Bhadeshia H K D K 1992 *Bainite in Steels: Transformations, Microstructures, and Properties*, (London: The Institute of Materials)
[16] Panneerselvam S, Martis C J, Putatunda S K and Boileau, J M 2015 *Mater. Sci. Eng. R Rep.* **626** 237-246
[17] Aranzabal J, Gutierrez I and Urcola J J 1994 *J. Mater. Res. Technol.* **10** 728-737

[18] Blackmore P A and Harding R A 1984 *Journal of Heat Treating* **3** 310-325
[19] Iacoviello F, Di Bartolomeo O, Di Cocco V and Piacente V 2008 *Mater. Sci. Eng. A.* **478** 181-186
[20] Kaczorowski M and Krzynska A 2008 *Arch. Foundry Eng.* **8** 87-92
[21] Erić O, Jovanović M, Šidanin L, Rajnović D and Zec S 2006 *Mater. Des.* **27** 617-622
[22] 2001 *ASM Metals Handbook* Vol. 18*: Friction, Lubrication, and Wear* (Ohio: ASM International)
[23] Khruschov M M 1974 *Wear* **28** 69-88

IOP Conf. Series: Materials Science and Engineering **541** (2019) 012020 doi:10.1088/1757-899X/541/1/012020

Preliminary Synthesis of Slow Release Fertilizer Material by Using Obsidian Rock

Solihin[1], A D Handoko[1], and R R Noviardi[1]
Indonesian Institute of Sciences Research Center for Geo Technology
Komplek LIPI, Jl. Sangkuriang, Bandung, Jawa Barat, Indonesia

E-mail: solihin@lipi.go.id

Abstract. Slow release fertilizer is a new kind of fertilizer in which the release rate of elements from fertilizer can be controlled. The controlling of nutrient component release rate lead to the increasing of fertilizer efficiency, decreasing agricultural cost, and decreasing ground water pollution. Obsidian from West Java, Indonesia is one the mineral that can be used to synthesize slow release fertilizer. By using this method, ordinary fertilizer is placed in the matrix of polymerized obsidian. The release test to the slow release fertilizer material shows that the material exhibits slow release property.

1. Introduction

Fertilizer is the common material that is used to assist plant growth [1]. Fertilizer supplies nutrient elements for the plants by releasing the nutrient elements in fluids that flow through soil cavity [2]. Plant root absorbs the nutrient elements from the fluid and distributes the elements to the vital parts of the plants. The most efficient utilization of fertilizer can be achieved if the rate of nutrient elements being absorbed by root plant is the same with the rate of nutrient elements being released by fertilizer [3]. Unfortunately, the ordinary fertilizer material is easily dissolved in water. Thus, the rate of nutrient release is very much higher than that of absorbed by plant root, which obviously leads to the decreasing of fertilizer efficiency. Another effect of the unbalance rate between nutrient elements release and absorption of the nutrient elements by plant root is the pollution of ground water [4- 6]. The nutrient elements that are not absorbed by plant eventually reach the ground water reservoir and contaminate the water. The fertilizer that has low rate of nutrient elements release is needed. Therefore, the aim of our research work is to synthesize the fertilizer that has low rate of nutrient elements release. The raw material that is used in this slow release fertilizer synthesis is the obsidian rock, which is the natural material that will not pollute the environment after being used. Obsidian is well known as the raw material that was commonly used in ancient structure, which indicates that obsidian is a quite environmental friendly material [7,8].

2. Experiment

Obsidian rock was obtained from West Java Provence Indonesia. The obsidian rock was crushed and ground. The size of ground obsidian rock is 100 mesh. Obsidian rock was mixed with ammonium di-hydrogen phosphate ($NH_4H_2PO_4$), sodium hydroxide (NaOH) and sodium silicate (Na_2SiO_3) prior to heating at 60°C for 24 hours. The minerals that construct obsidian rock was revealed through X-Ray Diffraction (XRD) by using Cu-Kα radiation. The elemental composition of obsidian mineral was analyzed through wet analysis method by using Atomic Absorption Spectroscopy (AAS), Spectrophotometry, and Gravimetric method. Obsidian mineral was dissolved in aqua regia and filtered.

The filtrate was analyzed by using AAS to determine quantity of Al, K, Na, Fe, Ti, Mg, and Mn. Meanwhile, the quantity of P in the filtrate was determined through spectrophotometry. The quantity of Si and S in the residue was determined through gravimetry method. Since obsidian is the oxide mineral, the quantity of every element in the obsidian is converted into oxide form. The result of wet analysis is combined with X-Ray Diffraction (XRD) to identify the minerals that compose the obsidian rock. The material that represents fertilizer is potassium di-hydrogen phosphate (p.a. Merck). The raw material and method used in our experiment refers to the common procedure to synthesize silica based slow release fertilizer [9,10].

3. Result and Discussion

The composition of obsidian from West Java Indonesia, based on oxide, is shown in Table 1. The weight percentage of total SiO_2, Al_2O_3, K_2O, Fe_2O_3, Na_2O and CaO in the ore are 77.38, 12.13, 4.15, 2.11, 1.57, and 1.21 %, respectively. The oxide form is chosen due the nature of obsidian rock, which is classified as oxide rock. The exact mineral that was naturally constructed in obsidian rock was analyzed through XRD. The result of XRD analysis is shown in Figure 1. The XRD analysis shows that obsidian rock is composed by sodium aluminum silicate ($NaAlSi_3O_8$), Crystobalite and Quartz (SiO_2). Sodium aluminum silicate and crystobalite is a reactive mineral. They are more active than quartz. Quartz is obviously a dense crystalline mineral, in which it is the final stage of SiO_2 crystal transformation according to Bowen series. These oxides construct minerals within obsidian.

Table 1. Composition of Obsidian from Garut,West Java, Indonesia.

Component	% Weight
SiO_2	77.38
Al_2O_3	12.13
K_2O	4.15
Na_2O	1.57
Fe_2O_3	2.11
CaO	1.21
TiO_2	0.147
MgO	0.0637
MnO	0.0460
P_2O_5	0.0178
SO_3	0.0105
LOI	0.39

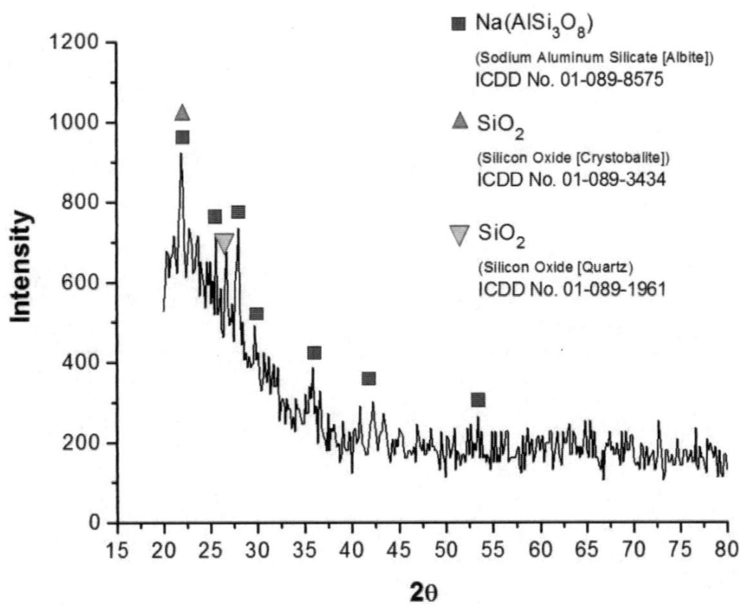

Figure 1. XRD Pattern of obsidian from West Java, Indonesia.

(a) (b)

Figure 2. Morphology of Slow Release Fertilizer at magnification (a) 100x and (b) 500x.

Figure 3 shows the morphology of the synthesized slow release fertilizer at magnification 100x and 500x. The surface of the sample seems to be quite dense. The cotton-like material can be seen on the surface of the obsidian. The EDS analysis to this material has revealed that this material mainly consists of 21, 26, and 34% weight of Silicon, Sodium, and Oxygen, respectively. This material possibly is the excessive geo-polymer material that has been resulted during the formation of slow release fertilizer. Geopolymerization reaction is the most important stage in the slow release fertilizer synthesis. It takes place rapidly on the interface of obsidian with other raw materials. The formation of polymer network of Na-Si-O contributes to the decreasing of nutrient release from the fertilizer.

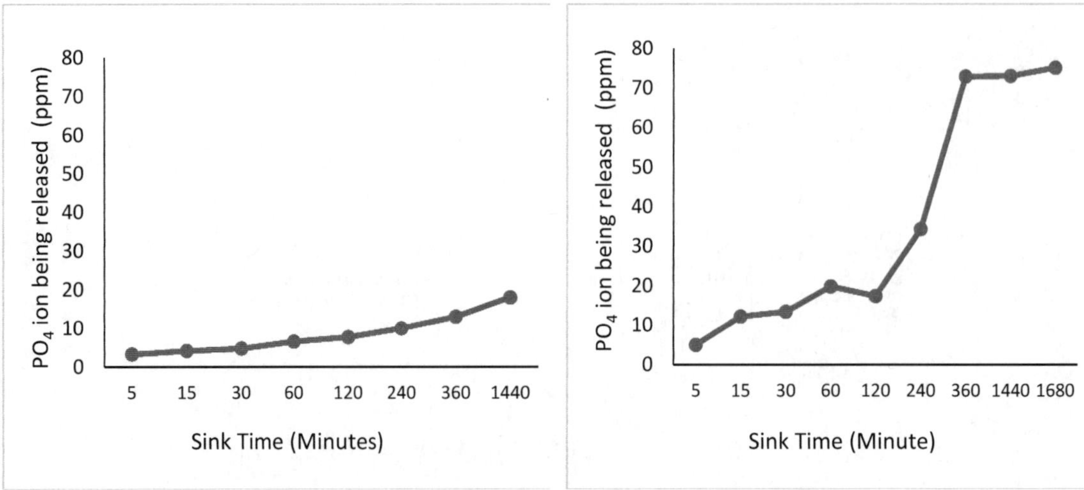

(a) (b)

Figure 3. PO_4 Release Profile of (a) low SiO_2 and (b) high SiO_2

The particular effect of geo polymerization reaction on slow release fertilizer synthesis is the reduction of nutrient release rate. Figure 3 represents the PO_4, which is one of nutrient, release rate at different amount of SiO_2 in the samples. Both of the figures show that the release of PO_4 can be very much decreased. The difference is only in the release rate value. This implies that during the application of this kind of fertilizer, the nutrients release rate can be controlled. The amount of SiO_2 can be used as a tool to control the release rate.

4. Conclusion

Slow release fertilizer is a new kind of fertilizer capable to increase fertilizer efficiency and decrease environmental impact caused by fertilizer. Obsidian consists of silica-based mineral, which can be used as raw material in slow release fertilizer synthesis. The synthesized sample shows slow release property, in which the nutrient release rate can be decrease significantly. The amount of SiO_2 can be used as tool to control the release rate value.

Reference

[1] Food Agricultual Organization, 2000 *Fertilzer and Their Use*. (Rome: Food Agricultual Organization)
[2] ABarke AV and Pilbeam DJ, 2007 *Handbook of Plant Nutrition* (New York: Taylor & Francis)
[3] Zhang Q, Tongamp W, Saito F, 2010 *Ind. Eng. Chem. Res.* **49** 2213–2216
[4] Jana IW, Sudarmanto IG, Rusminingsih NK, 2014 *Skala Husada* **11(10)** 34–40
[5] Djuwansah MR, Suriadarma A, Suherman D, Rusydi AF, Naily W, 2009 *J. Ris. Geol. dan Pertamb.* **19(2)** 109
[6] Food and Agriculture Organizaton, 1972 *Effects of intensive fertilizer use on the human environment* (FAO United Nations)
[7] Hughes RE, 2017 *Quat. Int* **482** 67–82
[8] Brown KA, Tykot RH, 2018 *J. Archaeol. Sci. Reports* **20** 284–292
[9] Solihin, Mursito AT, Dida EN, Erlangga BD, Widodo, 2017 *IOP Conf. Ser. Mater. Sci. Eng.* ***214*(1)**
[10] Solihin, Erlangga BD, Yusianita N, Saepulloh A, Santoso EB, 2014 *Forum Tahunan Pengembangan Iptek dan Inovasi Nasional IV*

IOP Conf. Series: Materials Science and Engineering **541** (2019) 012021 doi:10.1088/1757-899X/541/1/012021

Experimental study on crack propagation of stainless steel 304 material under amplitude overloads

Andreas Hartanto[1)], Roziq Himawan[2)], Hadi Sutanto[3)]
[1,3] Department of Mechanical Engineering, Atma Jaya Catholic University, Indonesia
[2] Center for Nuclear Reactor Technology and Safety, National Nuclear Energy Agency, Indonesia

E-mail: hadi.sutanto@atmajaya.ac.id

Abstract. Fatigue mechanisms caused by dynamic loadings starts with cracks that occurred in a material. Under conditions such as transient conditions, an earthquake or instantaneous external loads, some parts of nuclear reactor that initially receive operating loads can be overloaded. The purpose of this research is to investigate the change of crack propagation in stainless steel 304 material related with amplitude overloads. The crack propagation test used the axial dynamic testing with the maximum stress (σ_{max}) of 145 MPa, stress ratio R= 0.1 and over loading ratio (OVR) of 1.5, 1.75, 2. The results showed that the amplitude overloads caused the crack to propagate slowly as the crack tip formed a branch. The higher crack growth rate also showed the higher stress intensity factors in the crack tip.

1. Introduction

Nowadays, extensive research studies have been carried out contributing to the understanding of the mechanisms of fatigue damage along with stages of fatigue in the structures of metallic materials. For many fatigue-critical parts of structures, fatigue crack propagation under service conditions generally involves variable amplitude rather than constant amplitude loading conditions. If these structures contain fatigue cracks and the crack propagation rate depends on the loading sequence, it is difficult to persistently understanding the fatigue life. The reason for this is that many factors affect the initiation and the growth of fatigue cracks. Much works has been done on the subject related with the following aspects such as (i) geometrical effects like stress concentrations, (ii) microstructural effects: grain size and shape, second phase particles, precipitation and texture, (iii) surface effects: surface roughness, surface damage or surface treatment, (iv) environmental effects: corrosion and temperature, and (v) overload effects [1].

Fatigue crack growth could be retarded after applying a tensile overload while a constant tensile load is being repeated. However, the crack growth rate can be accelerated after applying an overload at a critical level while a tension-compression load is being repeated [2]. Figure 1 shows graphs of overload cycle (top graph) and crack length (bottom graph) related with number of cycles [3]. On the beginning of the graph, the loads are constant, however on a certain cycle the loads given are exceeding the normal value causing the crack closure, such as plasticity, crack tip blunting, branching or deflection. This crack closure will also start the crack retardation. As seen in the Figure 1 (bottom graph) this overload effect has two different crack lengths on the same cycles. The crack with constant load is not retarded and the crack with overload is retarded. Crack retardation after overload caused by crack closure such as plasticity, crack tip blunting, branching or deflection on crack tip. Crack closure first introduced onto the fatigue crack growth analysis by Wolf [4]. After that, many of researches have been done concerning the crack closure using experimental investigation, numerical analysis and theoretical studies [5-14].

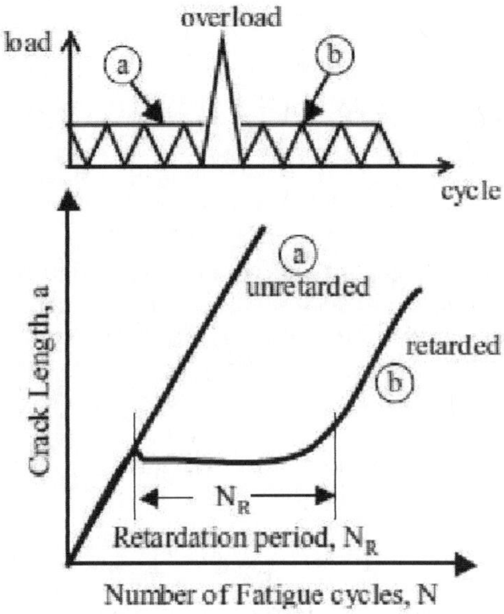

Figure 1. Illustration of the overload effect [3].

Paris and Erdogan proposed a fracture mechanics in 1860s using a method of fatigue life prediction related with the fatigue crack rate to the applied of stress intensity factor range [15]. Many modifications Paris law after that which one of the most important is the inclusion of the crack closure concept. A general fatigue crack growth model was showed as (1) by Wolf [4]. The fatigue crack growth rate is determined by the effective stress intensity factor range,

$$\frac{da}{dN} = C\left(\Delta K_{eff}\right)^m \tag{1}$$

$$\Delta K_{eff} = K_{max} - K_{op} \tag{2}$$

where dA/dN is the crack growth rate, ΔKeff is the effective stress intensity factor, Kmax is the stress intensity factor of peak load and Kop is the crack closure level. C and m are the calibration parameters. The objective of this research is to investigate the change of crack propagation for stainless steel 304 material related with amplitude overloads. Stainless steel 304 material is commonly used for many parts of nuclear reactor in power plants.

2. Experimental Procedure

Experiments used stainless steel 304 (SS 304) materials as testing specimens, because of its usage as nuclear reactor's components. Results from chemical composition test using Dyna-tech spectrometer are shown in Table 1. The mechanical properties SS 304 material are tensile strength = 621 MPa, yield strength = 290 MPa and elongation (%) = 55 [16].

The specimen's first form is a strip plate. Then, for crack propagation test, the specimen M (T) is formed following the ASTM E647 standard as seen in Figure 2 [17]. The EDM machine is used to form a notch on the specimen. The 2a value on the crack propagation test is the length of crack initiation.

Table 1. Chemistry composition of SS304 material.

Element	C	Mn	P	S	Si	Cr	Ni	Fe
%	0.0283	1.62	0.038	0.0126	0.36	18.8	7.58	balance

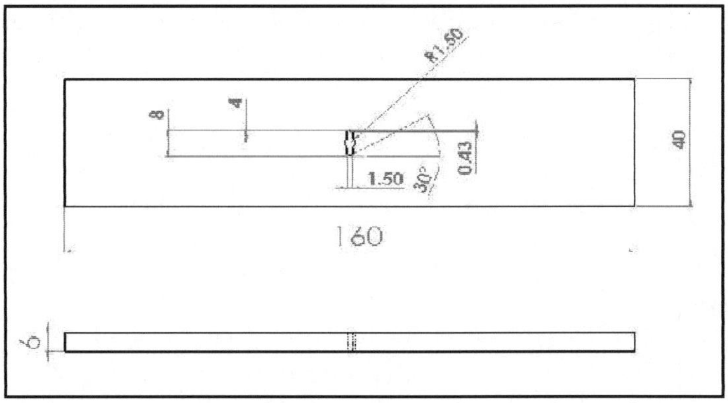

Figure 2. Specimen of SS304 material for experiment (in mm).

Crack propagation test was done using 250 kN axial dynamic test machine. Figure 3 shows the fixture for specimen on the dynamic test machine. The tests were done gradually, due to observing and measuring the crack's length were done manually. Also, for each specimen, the test was done until 50.00 cycles and the propagation was observed using optical microscope with 10x magnification. On this test, load ratio used is 0.1 and the assumed value of maximum stress (σmax) as normal load is 145 MPa, which came from the half of the SS 304's yield strength. There are three overload ratios (OVR) which are 1.5; 1.75; and 2.

$$OVR = \frac{\sigma_{OVL}}{\sigma_{max}} \tag{3}$$

Figure 3. Fixture of dynamic test machine for specimen

3. Experimental Results

Crack propagation tests used specimens of SS304 material with three different overload ratios OVR 1.5; 1.75 and 2 are shown in Figure 4. On Figure 4, there is a similarity on all of the specimens. Before the overload cycle, the crack growth is increasing, however, after the overload cycle, the crack growth on specimens are starting for retardation.

For crack propagation test, the purpose of deciding da/dN value is to see how big the propagation rate value related with stress intensity factor. Analyzing the value of da/dN could give an insight on how the crack growth after overload is given. Strain intensity factor used in fracture mechanics to predict the strain intensity near the crack tip that is caused by residual stresses. Results from tests of crack propagation rate are shown on Figure 5. For specimen with OVR = 1.5 showed that the crack propagation rate is different but it has similarity on how the crack propagation test is retarding after overload cycle. Other experiments with OVR = 1.75 and 2 also showed retardation after overload cycles as in Figure 5.

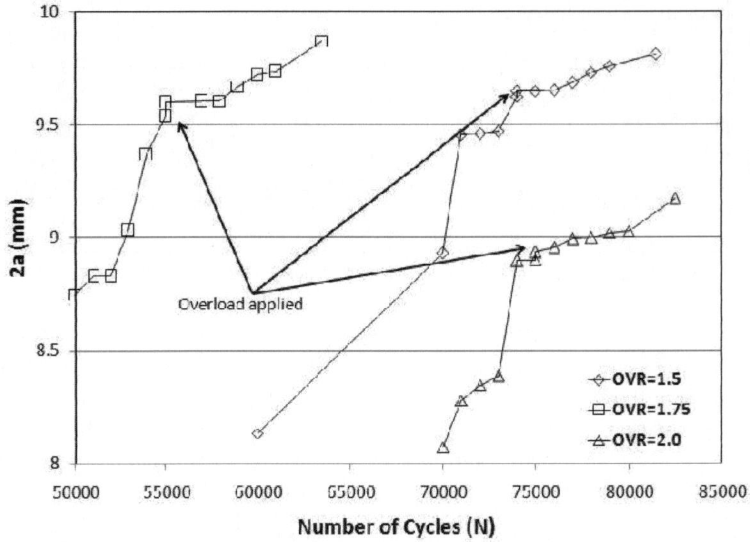

Figure 4. Graph of crack length 2a vs number of cycles N for OVR 1.5; 1.75 and 2 with OVL position.

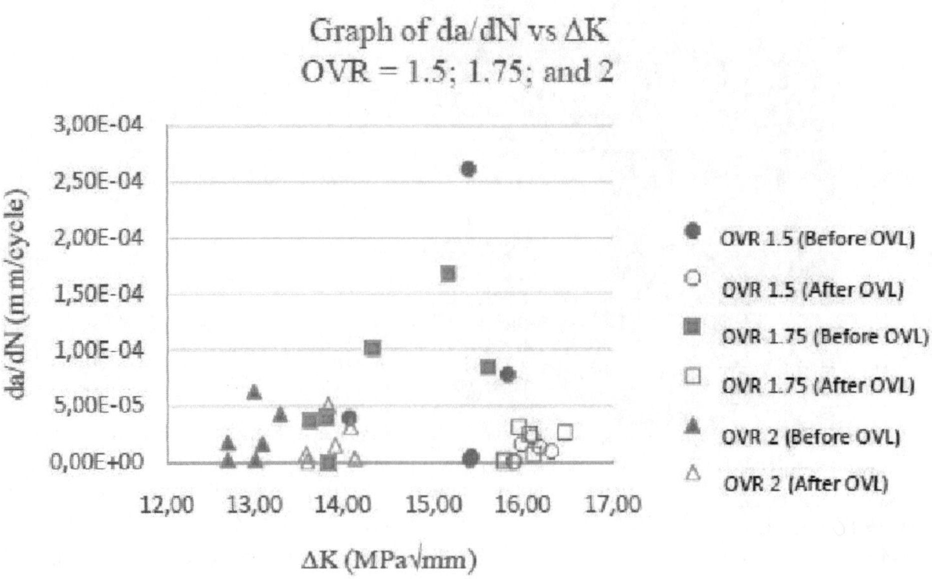

Figure 5. Comparison of crack propagation rate (da/dN) with stress intensity factor for overload ratio OVR 1.5, 1.75 and 2 before and after OVL.

Branched crack tip is one of the factors that causing crack growth retardation. The observation of crack from the notch point to the crack tip is seen in Figure 6 a. On crack propagation test, after the overload cycles, the observed crack tip is become branched. As seen in Figure 6 b, c and d the crack propagations for OVR 1.5, 1.75 and 2 are shown with the crack tip after overload (see the arrow on the picture). The crack tip after overload will cause crack growth retardation. The driving forces on that crack tip are distributed that causing crack growth retardation.

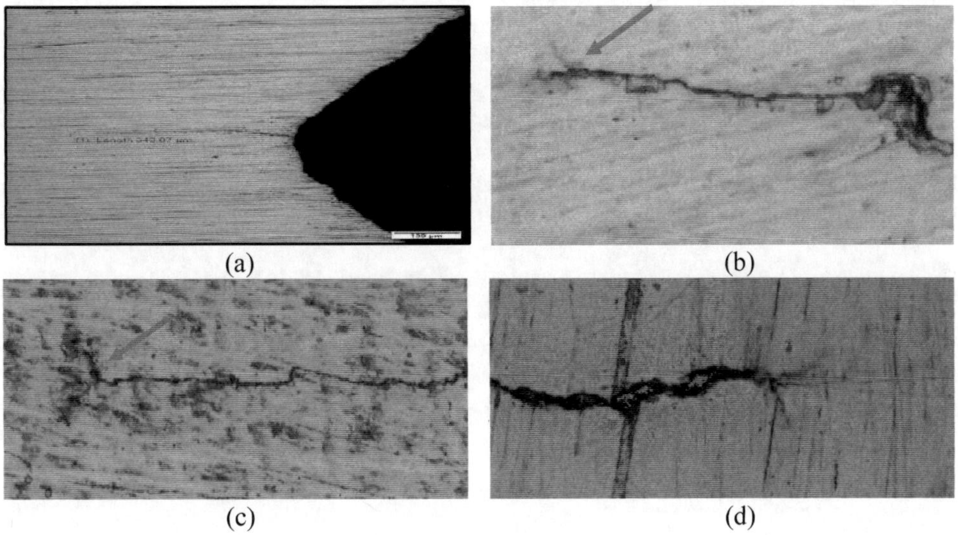

Figure 6. (a) Sample of specimen test, (b) Crack tip after overload with OVR = 1.5, (c) Crack tip after overload with OVR 1.75, (d) Crack tip after overload with OVR 2.

4. Conclusions

From this study concerning the influence of overload with differential overload ratio as 1.5, 1.75 and 2 using crack propagation test of σmax = 145 MPa in stainless steel 304 (SS304) material, the following conclusions can be drawn.

Graphs of 2a vs. N and da/dN vs. ΔK shows the results of crack propagation test for all specimens with the results that overload caused crack propagation retarding. On crack observation using optical microscope, after overload, the crack tip will become branched. Bigger value of crack propagation rate also makes strain intensity value on crack tip become higher.

References

[1] Mlikota M, Schmauder S, Bozic Z and Hummel M 2017 *Fatig. Fract. Eng. Mater. Struct* **00** 1-9
[2] Yamauchi A, Miyahara H, Makabe C and Miyazaki T 2012 *Int. J. Modern Physics: Conference Series* **6** 239-244

IOP Conf. Series: Materials Science and Engineering **541** (2019) 012021 doi:10.1088/1757-899X/541/1/012021

[3] Schijve J, Rijk D and Broek J 1961 Technical Report NLR-TN M. *Amsterdam, National Aero- and Astronautical Research Institute*
[4] Wolf E 1970 *Eng. Mech.* **2** 37-45
[5] Elbert W 1971 *The significance of fatigue crack closure in Damage Tolerance in Aircraft Structure: A Symposium Presented at the Seventy-Third Annual Meeting ASTM,* pp 486 ASTM International, Toronto, Canada
[6] Zhang W and Liu Y 2012 *International Journal of Fatigue* **43** 188-196
[7] Newman J C 1984 *International Journal of Fracture* **24** 131-135
[8] Zhang J Z and Bowen P 1998 *Engineering Fracture Mechanics* **60** 341-360
[9] Budiansky B and Hutchinson J W 1978 *Journal of Applied Mechanics* **45** 267-276
[10] Antunes F V, Chegini A G, Branco R and Camas D *Fatig. Fract. Eng. Mater. Struct* **27**
[11] Llorca L and Galvez V S 1990 *Fract. Mechanics* . **37** 185-196
[12] Kim J H and Lee S B 2001 *International Journal of Fatigue* **23** S247-S251
[13] Daneshpour S, Kocak M, Langlade S and Horstmann M 2009 *International Journal of Fatigue* **31** 1603-1612
[14] Himawan R and Susmikanti M 2016 J. Tek. Reaktor. Nukl. **18** 155-164
[15] Paris P and Erdogan F 1963 *Journals of Fluid Engineering* **85** 528-533
[16] ----- 2007 *Product Data Sheet 304/304L Stainless Steel* AK Steel Corporation
[17] ASTM E647 2004 *Metal Test Method and Analytical Procedures* (United States: ASTM)

IOP Conf. Series: Materials Science and Engineering **541** (2019) 012022 doi:10.1088/1757-899X/541/1/012022

The Effect of deposition times on preparation of SnO_2:F conductive glass by Indonesian local stannic chloride precursors

Tri Arini[1*], Latifa Hanum Lalasari[1], F. Firdiyono[1], Lia Andriyah[1], Gennady Fahmi[2], Erlina Yustanti[2], Achmad Subhan[3], Akhmad Herman Yuwono[4]

[1] Research Center for Metallurgy and Material, Indonesian Institute of Sciences (LIPI), Kawasan Puspiptek Serpong, Tangerang Selatan 15314, Indonesia

[2] Department of Metallurgy Engineering – Universitas Sultan Ageng Tirtayasa

[3] Research Center for Physics, Indonesian Institute of Sciences (LIPI), Kawasan Puspiptek Serpong, Tangerang Selatan 15314, Indonesia

[4] Department of Metallurgy and Materials Engineering, Faculty of Engineering, Universitas Indonesia, Kampus UI Depok, Depok 16424, Indonesia

E-mail: arinitri82@gmail.com

Abstract. The performance of fluorine-doped tin oxide (FTO) as an alternative semiconductor besides indium tin oxide (ITO) has been investigated. Currently, the precursors often used in the manufacture of fluorine-doped tin oxide (FTO) are commercial precursors on the market. In the present work, a thin film FTO will be fabricated using a pyrolysis spray technique with an ultrasonic nebulizer using Indonesian local precursors of stannic chloride with the variation of deposition times (5, 10, 15, 20 and 25 minus). Thin films were characterized by using x-ray diffraction (XRD), scanning electron microscope (SEM), ultraviolet-visible (UV-Vis) spectroscopy, and four-point probes. The optimum result obtained by pyrolysis technique with a deposition time of 20 minutes and substrate temperature at 300 °C has a resistivity of 1.059 x 10^{-4} Ω.cm and transmittance 89.614%. Fluorine-doped tin oxide (FTO) fabricated by using these local Indonesian precursors promise for further development as a glass of dye-sensitized solar cells (DSSC).

1. Introduction

The current energy crisis is beginning to remind people to find or even create new energy sources that can be utilized globally. Renewable energy sources are needed in large quantities and could be used for a long time. Sunlight is an energy source that suitable for human because it is environmentally friendly and can be obtanined in large quantities. There are tools that can store sunlight and convert into electrical energy such as Direct Sensitized Solar Cell (DSSC) [1].

Currently, most solar cells still use Indium Tin Oxide (ITO) as Transparent Conductive Oxide (TCO). The reserves of indium mineral is small found in the world so the price is very expensive to be used as a doping ingredient. Therefore, the role of indium as doping for Tin Oxide is replaced by fluorine which has conductive properties that are not inferior to indium. The reserves of fluorine mineral is also founded greater than indium. In addition, FTO coated glass substrates can reduce about 40% of material costs [2]. Fabrication of FTO by using many different methods such as thermal evaporation [3] [4], sputtering [5], spray pyrolysis [6], spin coating [7].

A. H. Yuwono et al. [8] have successfully conducted researches in the manufacture of fluorine-doped tin oxide (FTO) by comparing the precursors and solvents. The precursors used are tin (II) chloride dihydrate ($SnCl_2.2H_2O$) and anhydrous tin (IV) chloride ($SnCl_4$) pro analysis (p.a). Precursor pro analysis (p.a) is a commercial precursor that has a fairly expensive price and limited availability. Indonesia is one of the countries that have a large potential for producing tin.

In this study, a thin glass conductor coating was made by ultrasonic spray pyrolysis nebulizer method using stannic chloride ($SnCl_4$) local Indonesia as a precursor in the manufacture of fluorine-doped tin oxide (FTO) conductive glass.

2. Experimental Method

The materials used in this research were glass substrate microscopes of soda lime, stannic chloride ($SnCl_4$) local Indonesia, NH_4F (98%, Merck Ltd., Germany) and methanol (Merck Ltd., Germany). The research began with the substrate glass cleaning by using commercial detergent, then the substrate glass was soaked with acetone then ultrasonication for 15 minutes.

The preparation of the solution was carried out by mixing 8.18 grams of $SnCl_4$ with 9.82 ml of methanol. The solution was stirred thoroughly with a magnetic stirrer at a moderate rate for 30 minutes. The next step was an addition of doping as much as 0.2 gram NH_4F then the solution was stirred homogeneous for 30 minutes.

The deposition was performed by ultrasonic spray pyrolysis nebulizer method (GEA Medical 402A1) with 10 cm distance and velocity ± 30 ml / 10 min and substrate glass placed on hot plate with varying deposition times of 5, 10, 15, 20 and 25 minutes at a fixed substrate temperature of 300 °C. The characterizations of the fabricated FTO glass were performed by scanning electron microscope-energy dispersive spectroscopy (JEOL-JSM 6390A), x-ray diffraction/XRD (Shimadzu XRD-7000), UV-Vis spectroscopy (Thermo UV-Vis Genesys 10s) and four-point probe (FPP5000).

3. Results and Discussion

The x-ray diffraction pattern of SnO_2 thin film that grows on a glass substrate with a variation of deposition times of 5, 10, 15, 20, 25 minutes is shown in Figure 1. The XRD measurement results show that the peaks show the SnO_2 material phase with the plane crystals (110), (101), (200), (211), (310), (301) and tetragonal crystal structures according to the SnO_2 rutile pattern [5].

At 5 minutes deposition time (Figure 1.a), the crystalline structure looks amorphous, this is probably due to the short deposition time so that the droplet of solution is not able to be dispersed evenly to all parts of the glass surface [9]. With a crystal structure that is still amorphous, its conductivity is small. The grain boundary consists of an amorphous phase where the amorphous phase will absorb the energy that is passed so that the material conductivity will decrease.

Crystal growth begins to occur at 10 minutes of deposition time. Increasing deposition time from 10 to 25 minutes causes a change in intensity on the crystal plane. The crystal field (200) has the most dominant structure compared to the intensity of other structures. The research conducted by G. Gordilo, et.al [10] explained that the most dominant crystal plane (200) indicates that the quality of the crystal is getting better and also if the size of the crystal is getting larger. The increase of deposition time will cause an optimal balance between electrical resistance and optical transmittance.

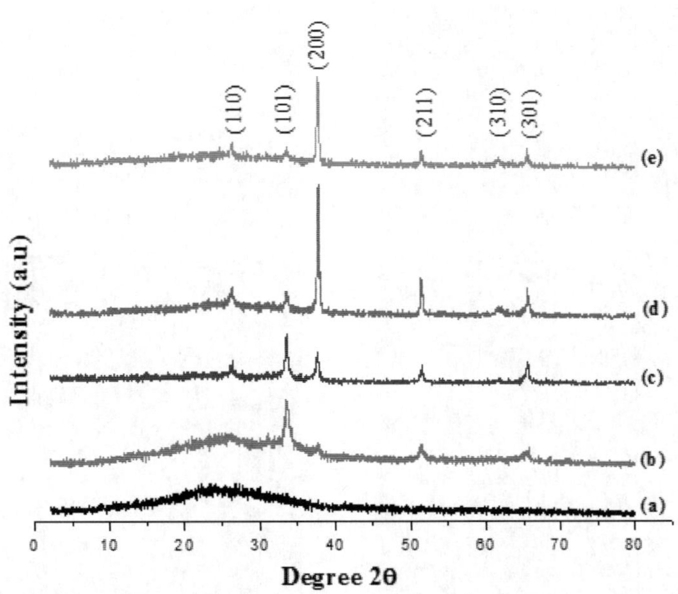

Figure 1. XRD patterns of FTO thin layers heated at 300 °C during the pyrolysis process with a variation of deposition times : (a) 5; (b) 10; (c)15; (d) 20 and (e) 25 minutes.

The longer deposition time causes the crystallite size to expand because the longer the deposition time, the SnO_2 particles grow denser thus reducing porosity and electron transfer becomes easier and thus decreases the reactivity value [3]. This is supported by research conducted by D. Tatar, et. al [11] which said that the higher crystallinity values will make the lower resistivity value.

Table 1. Semi-quantitative EDS-SEM analysis of FTO at 300 °C with a variation of deposition times.

Element	% mass				
	5 min	10 min	15 min	20 min	25 min
O	22.34	19.86	17.47	16.20	15.44
Na	3.75	1.43	-	-	-
Mg	1.53	-	-	-	-
Si	28.59	15.12	13.21	3.49	1.60
Ca	6.74	5.02	3.66	2.01	1.19
Sn	37.05	58.57	65.66	78.30	81.77

The results of semi quantitative analysis of SnO_2 thin layers with variations of deposition times are shown in Table 1. From the results of semiquantitative analysis of EDS, it can be seen that the chemical composition of the thin layer shows major elements, namely O and Sn derived from precursor solutions of Indonesian local precursors of stannic chloride whereas Na, Mg, Si, and Ca are obtained from soda glass [12]. Lalasari et.al [13] in the previous research using pro-analytic precursors indicated Sn and O elements without any other impurities.

IOP Conf. Series: Materials Science and Engineering **541** (2019) 012022 doi:10.1088/1757-899X/541/1/012022

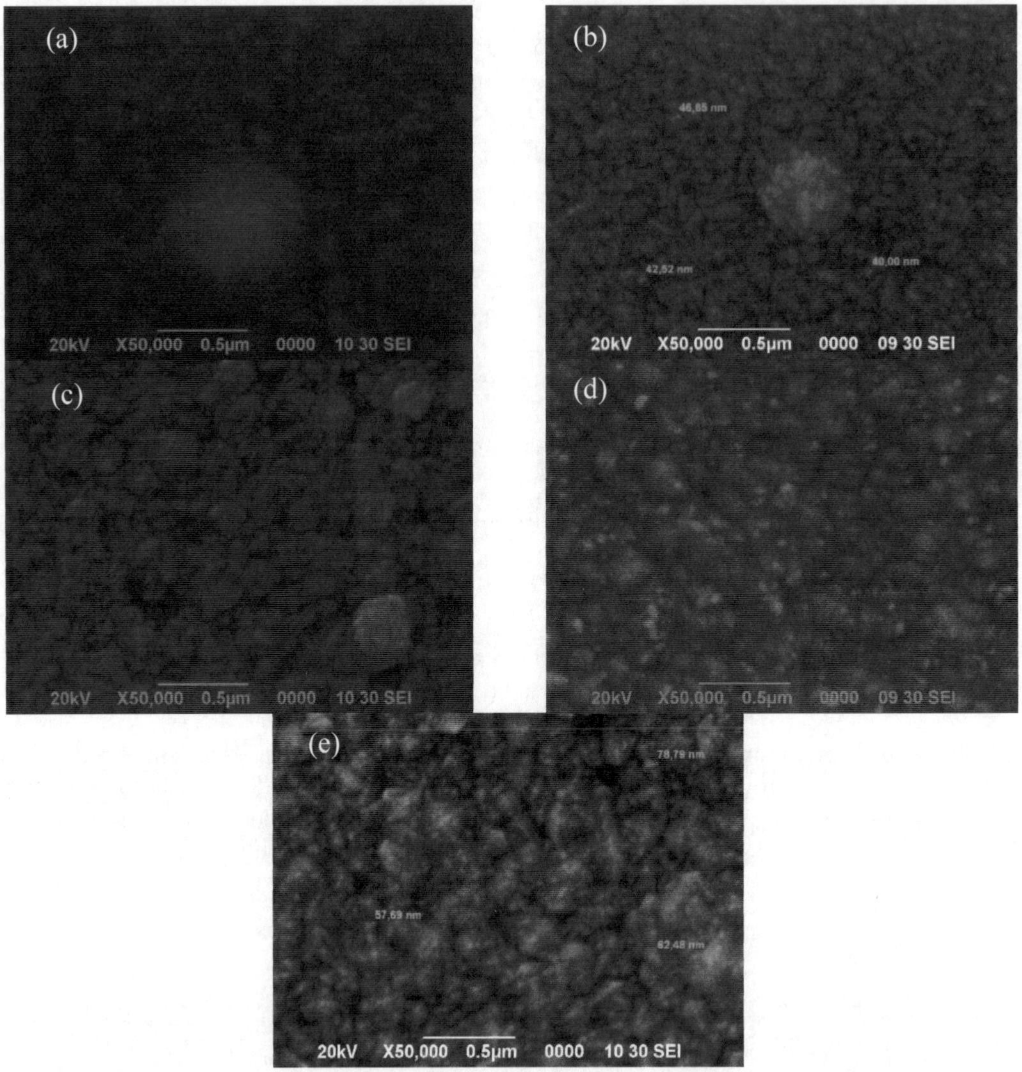

Figure 2. SEM images of FTO thin layers at a substrate temperature of 300°C and with variations of deposition times : (a) 5, (b) 10; (c)15; (d) 20 and (e) 25 minutes.

Figure 2 a-e shows grain growth trends with variations of deposition times. The particle morphology at 5 minutes deposition time is shown in Figure 2.a, thin films still show some void porosity between granules, this is probably due to the short deposition time, droplet solution is not able to be dispersed evenly to all parts of the glass surface[9]. With a magnification of 50,000x, at a deposition time of 10 minutes, the average grain size is approximately 43.56 nm. The average grain size for thin layers in a row for 20 and 25 minutes deposition time is 56.53 and 66.32 nm as shown in Figure 2 d-e.

With the addition of deposition time, the grain becomes more compact and interconnected where porosity has been significantly reduced. The SnO_2 thin film with a deposition time of 25 minute looks denser and has a larger particle size compared to the deposition times of 10, 15 and 20 minutes (Figure 2 b-e)With increasing deposition time, the grain grows larger and the crystal interface decreases. Meeting the arrangement of particles causes lower electrical resistance. This is in line with previous studies that FTO with many grain boundaries causes a high value of electrical resistance [14][3].

IOP Conf. Series: Materials Science and Engineering **541** (2019) 012022 doi:10.1088/1757-899X/541/1/012022

Table 2. The resistivity of FTO thin layers with deposition time variations of 5, 10, 15, 20 and 25 minutes at 300 °C during the pyrolysis process.

Deposition time (min)	Resitivity (Ω.cm)
5	3.930×10^{-4}
10	3.020×10^{-4}
15	1.580×10^{-4}
20	1.059×10^{-4}
25	4.960×10^{-5}

From the results that can be seen in Table 2, it can be seen that the longer the deposition time the smaller the resistivity value. The resistivity val ues for thin layers for deposition times 5, 10, 15, 20 and 25 minutes are 3.930×10^{-4}; 3.020×10^{-4}; 1.580×10^{-4}; 1.059×10^{-4} and 4.960×10^{-5} Ω.cm as shown in Table 2.

Nafiseh Memarian, et. al said this scattering mechanism affects the electrical properties and increases the resistivity of the layer. In addition, the effect of grain boundaries also influence in quality of resistivity [16]. At 5 minutes deposition time, the granules in the glass appear small and there are inter-grain holes. The grain boundary appears when the particles are small and also because of incomplete bonds between atomic layers. Therefore, carrier mobility in thin films is low and resistance becomes high [17]. This is supported by the XRD results, at 5 minutes deposition time, the crystal structure formed is still amorphous so the conductivity is small [5].

With an increase in deposition time, the SnO_2 particles grow denser thus reducing porosity and interconnects are more connected, electronic transfer between grains becomes easier and resistivity has been significantly reduced [3]. From the results obtained from Table 3, it is found that the longer the time of deposit, the grain size has increased so that the sensitivity value becomes smaller, this is supported by research conducted by D. Tatar, et.al [11] which says that as high as the crystallinity value, the lower the resistivity value.

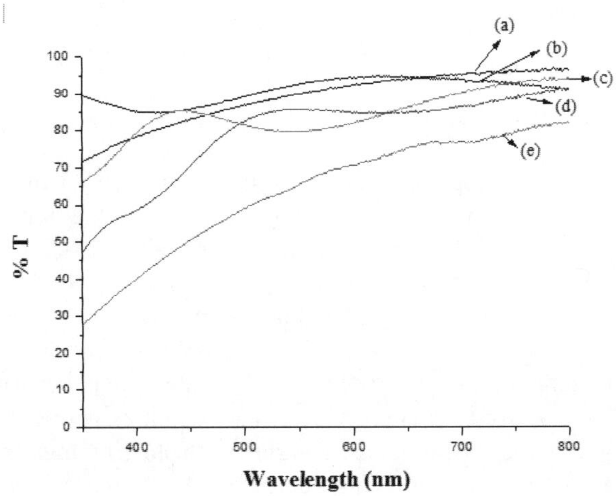

Figure 3. The optical transmittance of FTO thin layers at a constant deposition temperature of 300°C during the pyrolysis process with deposition time variation of (a) 5; (b) 10; (c)15; (d) 20 and (e) 25 minutes.

The observation result of SnO_2 thin layer transmittance spectrum with deposition time variation is shown in Figure 3. Transmittance values respectively at deposition times 5, 10, 15, 20 and 25 minutes were 96.881; 94.829; 94.294; 89.614 and 82.228%. This shows that the longer the coating time, the thicker the layer is formed so that the light passed is less and less [18]. Latifa, et al. [13] have carried out research using the same method but the precursor $SnCl_4$ pro analytic, in their research produced the best value for FTO produced a 15 minutes deposition time, which provides transmittance of 80,786%

and a resistivity of $4{,}01 \times 10^{-5}$ Ω.cm. This is in line with the longer deposition time, the smaller the resistivity value and the less transmittance value [9][15].

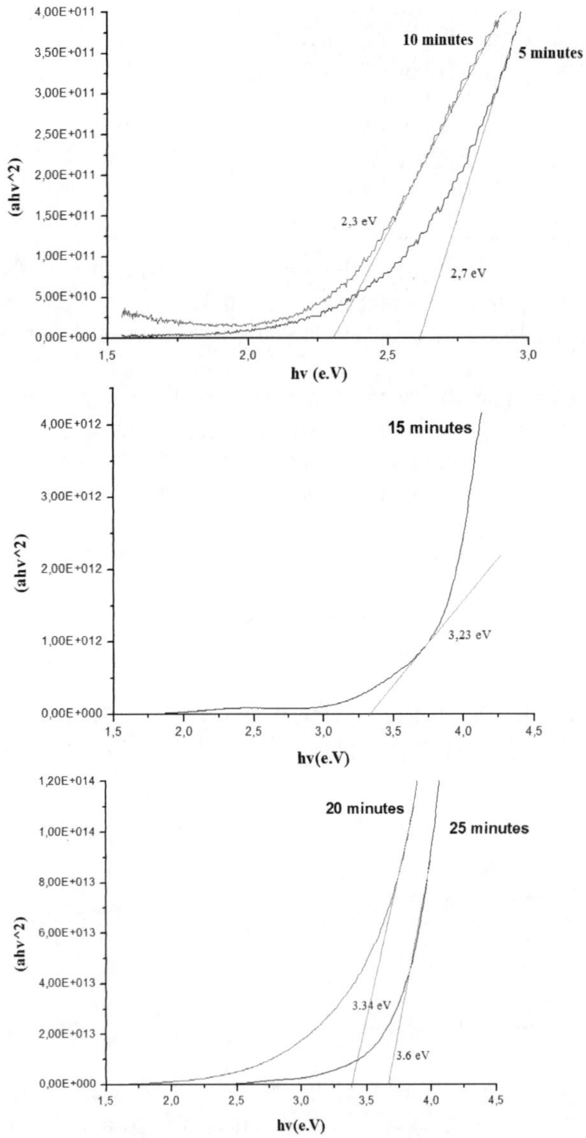

Figure 4. Energy band gap of FTO thin layers at a constant deposition temperature of 300 °C during the pyrolysis process with deposition time variation of (a) 5; (b) 10; (c)15; (d) 20 and (e) 25 minutes.

An energy band is a collection of lines at the same energy level which will coincide and form a ribbon. Band gap energy can be known as the distance from the energy band that coincides with each other. The greater the value of the band gap energy obtained is expected to be slower in electron transfer due to the mobilization of electrons that are farther away. Figure 4 is a graph of band gap energy from the pyrolysis spray method at a temperature of 300°C.

Based on Figure 4, it is known that band gap energy from the pyrolysis spray method with the variation of deposition time 5, 10, 15, 20, 25 minutes obtained the band gap energy values are 2.7, 2.3, 3.23, 3.34, 3.6 eV, respectively. Bandgap results show a decrease in the deposition time of 10 minutes and then increase again from the deposition time of 15 minutes to 25 minutes. This result shows that the band gap energy is the smallest at 10 minutes deposition time so that it can be said that at 10 minutes deposition time has the highest energy density compared to other samples of deposition time.

4. Conclusion

Based on the investigation, it has been found that the use of Indonesian local stannic chloride ($SnCl_4$) can be used as an alternative precursor material in the manufacture of FTO thin films on a laboratory scale using pyrolysis spray technique using an ultrasonic nebulizer. The longer deposition time will increase the thickness of the FTO thin film so that electrical resistance decreases. However, layers that are too thick create an adverse effect an optical transparency. Therefore there must be a balance between electrical resistance and optical transparency. The optimum result obtained by pyrolysis technique with a deposition time of 20 minutes and substrate temperature 300°C has a resistivity of 1.059×10^{-4} Ω.cm and transmittance 89.614%.

References

[1] Choubey P K, Kim M S, Srivastava R R, Lee J C and Lee J Y 2016. *Miner. Eng.* **89** 119–137.
[2] Bu I Y Y 2014 *Ceram. Int.* **40 (1)** 417–422
[3] Muruganantham G, Ravichandran K and Saravanakumar K 2011 *Superlattices Microstruct.* **50 (6)** 722–733
[4] Hammad T M, Hejazy N K, Strip G and Branch G 2011 *Int. Nano Lett.* **1 (2)** 123–128
[5] Banyamin Z Y, Kelly P J, West G and Boardman J 2014 *Coatings* **4** 732–746
[6] Luangchaisri C, Dumrongrattana S and Rakkwamsuk P 2012 *Procedia Eng.* **32** 663–669
[7] Battal A, Tatar D, Koçyiğit A and Düzgün B, 2014 *J. Ovonic Res.* **10** 23–34
[8] Yuwono A H *et al.* 2017 *Int. J. Technol.* **8 (7)** 1336–1344
[9] Arini T *et al.* 2016 *Int. J. Technol.* **8** 1335–1343
[10] Gordillo G, Moreno L C, De la Cruz W D L and Teheran P 1994 *Thin Solid Films* **252 (1)** 61–66
[11] Tatar D and B. Duzgun B 2012 *J. Phys.* **79 (1)** 137–150
[12] Karlsson S, Jonson B and Wondraczek L 2012 *Glas. Technol. J. Glas. Sci. Technol. Part A* **53 (1)** 1–7
[13] Lalasari L H, Arini T, Andriyah L, Firdiyono F and Yuwono A H 2018 *AIP Conference Proceedings* **20001** 1–7
[14] Acosta D R, Estrada W, Castanedo R, Maldonado A and Valenzuela M A 1998. *Thin Solid Films* **375** 147–150
[15] Elangovan E, Singh M P and Ramamurthi K 2004 *Mater. Sci. Eng.* **113** 143–148.
[16] Memarian N, Rozati S M, Elamurugu E and Fortunato E 2010 *Phys. Status Solidi Curr. Top. Solid State Phys.* **7 (9)** 2277–2281.
[17] Zhao H, Liu Q, Cai Y and Zhang F 2008 *Mater. Lett.* **62** 1294–1296
[18] Tatar D, Turgut G and Duzgun B 2013 *J. Phys.* **58** 143–158

Acknowledgement

The authors would like to thank the Ministry of Research and Higher Education of the Republic of Indonesia for funding this research through an Indonesian Institute of Sciences (LIPI) Research Center for Metallurgy and Materials 2016 DIPA Grant and PT Timah Industri for its support.

IOP Conf. Series: Materials Science and Engineering **541** (2019) 012023 doi:10.1088/1757-899X/541/1/012023

The effect of first step holding time of low high austempering heat treatment to the mechanical properties of Austempered Ductile Iron (ADI)

B Bandanadjaja[1,*], D Idamayanti[1], N Widarmansyah[1] and A Lestarianto[2]
[1] Foundry Engineering Department of Politeknik Manufaktur Bandung, Jl. Kanayakan 21 Dago, Bandung, Indonesia
[2] Student of Foundry Technology of Politeknik Manufaktur Bandung, Jl Kanayakan 21 Dago, Bandung, Indonesia

E-mail: benybj@polman-bandung.ac.id

Abstract. A low-high austempering process can be referred to as the application of undercooling condition in the austempering process of Austempered Ductile Iron (ADI). Undercooling condition promote more grain nucleation and producing finer grains with good mechanical properties. This study aims to investigate the influence of the first stage of lower temperature austempering holding time on its mechanical properties and find the best holding time which can produce the highest toughness of ADI materials. The experiment began with the heating of the nodular cast iron sample up to 927 °C and held for 120 minutes. Then it quenches to a liquid salt bath with a temperature of 260 °C as a lower temperature of the first austempering stage with a variation of holding time at 30, 60 and 90 minutes. Then it was transferred to a second salt bath medium of 400 °C as the second austempering stage with a holding time of 120 minutes then it was given the air cooling. The sample was tested by hardness, tensile, impact, and metallographic examination. The results show that the best toughness value in this research is obtained at 30 minutes holding time on first austempering stage with mechanical properties is tensile strength 1273 MPa, yield strength 1162 MPa, elongation 4,87% and impact value at 54.15 J. Value modulus of toughness is $5.93x10^7$ J/m^3.

1. Introduction

Austempered Ductile Iron (ADI) has long been used as an engineering material. If SG Iron is given an austenitization process and then followed by an austempering process, it will produce a mixed microstructure consisting of ausferrite, ferrite, and austenite with graphite nodules dispersed in it. This unique microstructure produces mechanical properties in ADI equivalent to steel forging [1]. ADI also has excellent fatigue strength, [2] high fracture toughness [3], and excellent wear resistance [4]. The interesting properties of ADI are related to its unique microstructure consisting of ferrite (α) and high carbon austenite (γHC). The ADI microstructure is different from forged steel where the bainite microstructure of steel consists of ferrite and carbide. Because of these differences, austempering products in SG iron are often referred to as ausferrite rather than bainite. The ausferrite is free from carbide which makes it has better toughness than bainite structure. The substantial amount of silicon present in SG iron suppresses carbide formation during the austempering reaction and retains large amounts of stable high carbon austenite (γHC) [5–7].

1.1. Conventional Austempering Process

ADI's conventional austempering process can be implemented first with the SG Iron austenitizing process to a temperature of 871-927 °C and held for 2 hours; then SG Iron is given a quench process to temperatures between 260 and 400 °C. The SG Iron is then isothermally held at this temperature for about 2 hours and finally cooled by air. During the austempering process, a phase transformation reaction occurs in two stages [8,9]. In the first stage, austenite (γ) decomposes into ferrite (α) and high carbon austenite (γHC):

$$\gamma \rightarrow \alpha + \gamma_{HC} \quad (1)$$

If the casting is held at austenitizing temperature for too long, then a second reaction occurs, where the high carbon austenite (γHC) further decomposes into ferrite (α) and carbide (ε)

$$(\gamma_{HC}) \rightarrow \alpha + \varepsilon \quad (2)$$

This excessive reaction is undesirable because it produces brittleness in ADI because the presence of carbides can eliminate the ductility and toughness. Therefore, for the successful production of ADI, SG Iron must be processed austempering in the bainite transformation region before the end of the bainite transformation.

The other problem that some alloying element such as Mn and Mo make the austempering reaction becomes slow, and as a result, some unreacted austenite remains, which may turn into martensite during the final cooling process. This unreacted austenite leaves a continuous film layer in martensite in the intercellular region and consequently can trigger crack propagation through this region, and this may result in poor ductility and lower fracture toughness in ADI [10].

1.2. Two Step Austempering

Due to the limitation of the austempering results above and to obtain a better combination of tensile strength and toughness, the two-step austempering process has been proposed by several researchers [11–15]. The process which is used in this research is the step-up austempering method. In this process (often called a low-high austempering), SG iron, after austenitizing, is given the quench process to a lower austempering temperature and held there for a predetermined period. The second austempering step carried out at higher temperatures either by raising the first salt bath temperature or transferring the casting to another salt bath which is maintained at a higher temperature than the first salt bath. The sample is austempered at this second temperature for the period needed and finally cooled in air. The austempering step-up process scheme is shown in figure 1.

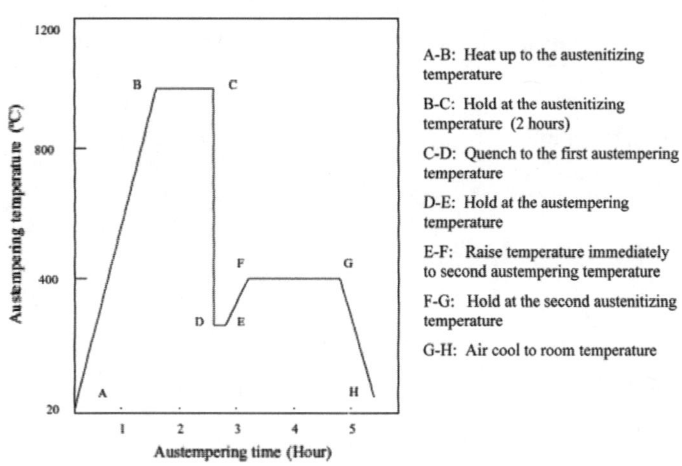

Figure 1. Step-up austempering process [10].

Many researchers for ADI material development have applied the step-up austempering process methods. It was found that the step-up austempering process has a significant improvement in the properties of ADI due to improved ferrite properties and higher austenitic carbon content in ADI avoiding the formation of iron carbide bracket structures [16,17]. The step-up austempering process produces a combination of lower and upper ausferrite. The first step of the austempering process is carried out at temperatures above the initial temperatures of martensite (MS) to 330 °C resulting in lower ausferrite, SG Iron has a high strength/hardness but low toughness, while in the second step of austempering holding time is done at 330 °C to 425 °C produce upper ausferrite with opposite behavior. [18,19]. A low-high austempering process can also be referred to as the application of undercooling condition in the austempering process of Austempered Ductile Iron (ADI) [20]. With undercooling, more grain nucleation can be promoted finer grains which are producing good mechanical properties. This study aims to investigate the influence of holding time of the first (lower) temperature of the low–high austempering process to the mechanical properties of ADI. How long is the best time for the first austempering process which give the best combination of lower and upper ausferrite structure and also give the finer structure. It was expected to find the good toughness of ADI materials among the variable used.

2. Methodology

2.1. Material

The sample for the experiment was made from SG Iron material which was cast in the form of keel Block using ASTM A897 standard with a thickness of 13-38 mm. Keel block castings are cut in a square shape with a size of 22x15x15 mm for hardness test and microstructure examination, 13 x 13 x 58 mm for impact test and cylinders with dia. 10 x 50 mm for the tensile test. The number of samples each is arranged to the variations needed. The chemical composition of SG Iron as-cast material is verified using optical emission spectroscopy and the results as shown in the following table 1:

Tabel 1. Chemical composition of SG iron as-cast.

C	Si	Mn	P	S	Mg	Cu
3.300	3.550	0.260	0.010	0.004	0.047	0.850

2.2. Heat Treatment Process

The heat treatment process was carried out by heating the material to the austenitizing temperature at 927 °C with holding time for 120 minute, then continued with quenching to the first salt bath medium with a temperature of 260 °C with a holding time variation of 30 minute, 60 minute and 90 minute. Then each sample was rapidly transferred to a second salt bath medium which has a higher temperature of 400 °C with a holding time of 120 minute after the hold time is reached then the sample is taken to be air cooled. Figure 2 shows a graph of the heat treatment experiment on the sample which is arranged in this study:

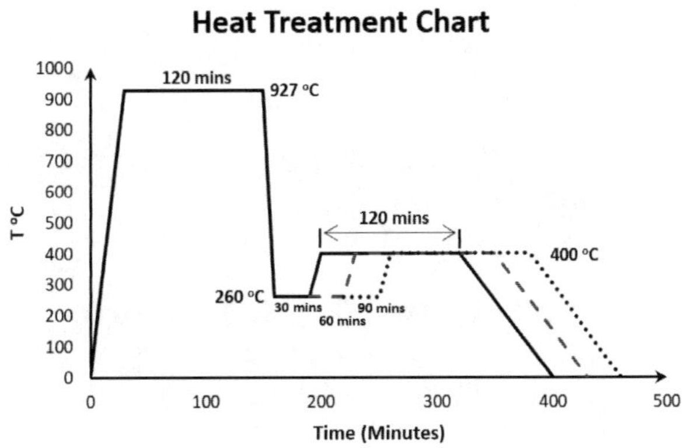

Figure 2. Heat treatment cycle.

2.3. Salt Bath

A salt bath is used for the austempering process. Liquid salt temperature is determined according to the required temperature. Two units of the salt bath are used. The first salt bath for the first austempering process uses salt with a composition of 50% $NaNO_2$ + 50% $NaNO_3$. This salt bath is maintained at 260 °C. The second salt bath is used for austempering the second uses salt with a composition of 95% $NaNO_3$ + 5% NaCl and the temperature set is 400 °C.

2.4. Mechanical Test and Metallographic Examination

All sample variations are tested by tensile testing using a universal tensile testing machine which refers to ASTM E-8M standard and hardness test with Rockwell C with ASTM E-18 standard. The impact test was also carried out with Charpy impact machine and refer to ASTM E-23 standards with unnotches Charpy sample specifications. The metallographic examination is also carried out to analyze the microstructure resulting from each process. Metallographic samples are etched using 3% Nital and observations are done using optical microscopy. To ensure repeatability of the process, each variation was carried out on three samples given similar heat treatment. Thus the average test results from the three samples were obtained.

3. Result and Discussion

3.1. Microstructures

The SG Iron as-cast material has a microstructure with the majority of Pearlite (90%) with a small amount of ferrite and spheroidal graphite. Figure 3 shows the microstructure of SG Iron as-cast material.

Metallographic examination of low-high austempering results are shown in figure 4 (a), (b) and (c). The matrix of microstructure consists of lower ausferrite, upper ausferrite, ferrite needles, and light-colored austenite with dispersed globular graphite. The first austempering process is producing lower ausferrite. The ausferrite is consist of ferrite and austenite. Ferrite needles are getting more and more and also look finer or look closer and sharp with longer of holding time on the first austempering. This shows that the longer austempering holding time at 260 °C result in the amount of lower ausferrite much more. The two-step austempering process generally produces finer ferrite [20] which increase the strength and hardness of the material, where the low-high process produces greater supercooling for ferrite nucleation which is increasing and can produce finer ferrite.

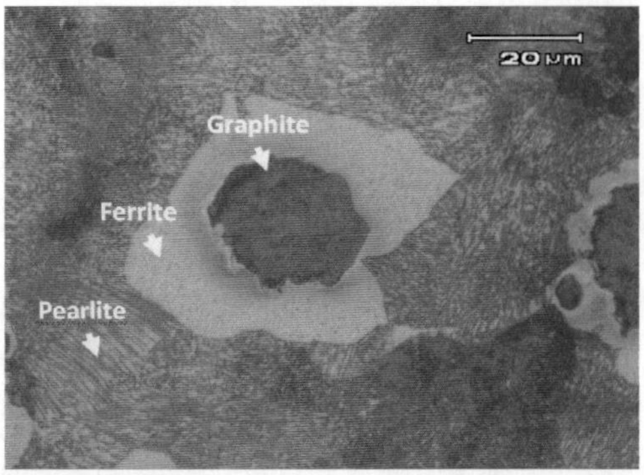

Figure 3. The microstructure of SG iron as-cast, nital 3% etchant.

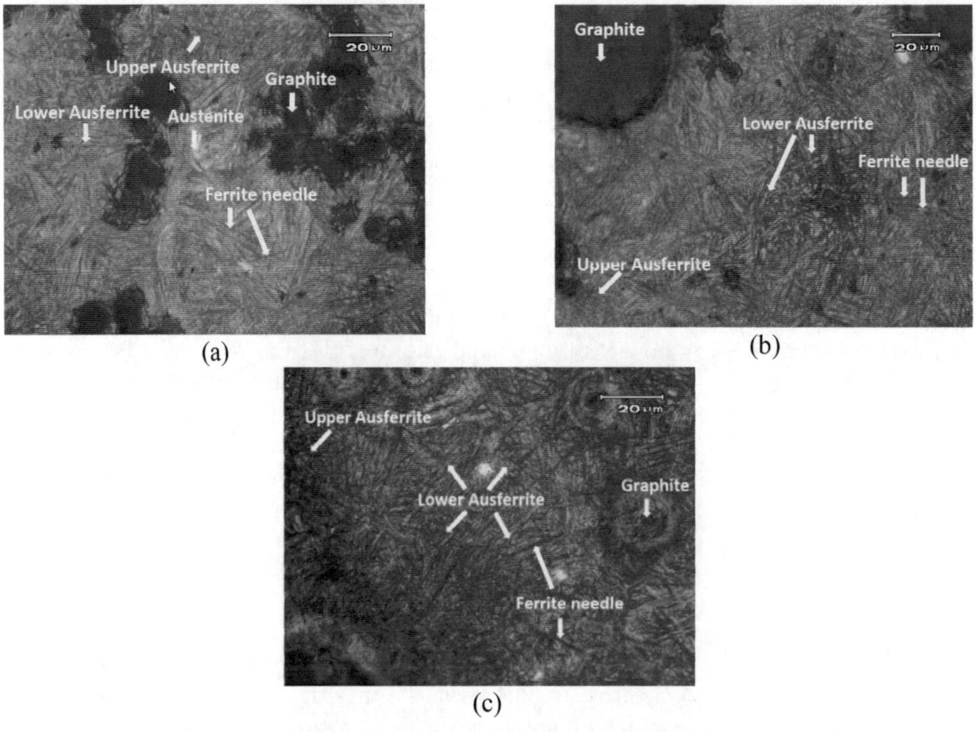

Figure 4. Microstructure heat-treated sample (a) first austempering holding 30 minute (b) first austempering holding 60 minute (c) first austempering holding 90 minute, nital 3% Etchant.

3.2. Mechanical Properties

The tensile test results are shown in figure 5 Shows that the longer holding time at first austempering the tensile strength and yield strength increases. At holding time 30 minute obtained tensile strength 1,274 MPa and yield strength 1,163 MPa, elongation of 4,87%. Whereas for the third sample for holding austempering time for 90 minute obtained tensile strength 1,450 Mpa increased by 13.8% compared to the first sample, the yield strength of 1,291 Mpa increased by 11%. Elongation for holding time 90 minute decreases to 1.22%. The increase in tensile strength and yield strength is related to the formation of lower ausferrite which has higher tensile strength and the finer form of ferrite needles. The finer form of microstructure influence the dislocation movement becomes more difficult as formulated by Hall-Petch where the finer the grain will result in higher tensile strength and yield strength.

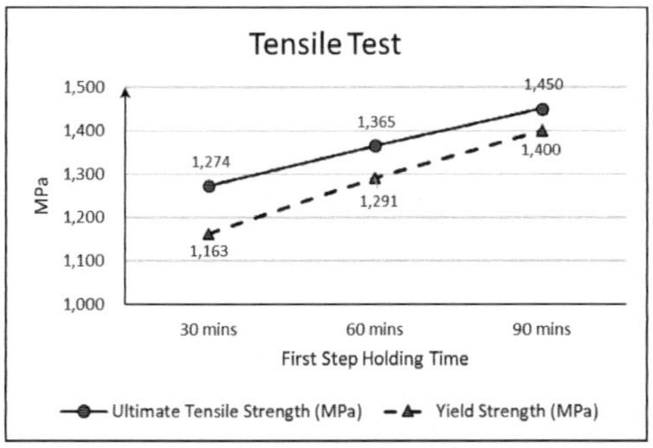

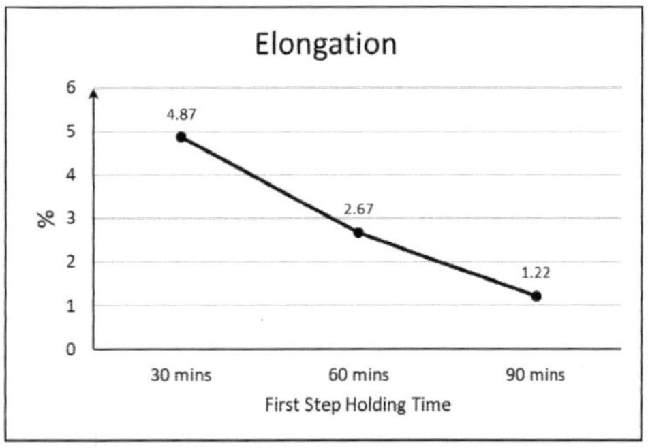

Figure 5. Tensile test result of all specimens.

From the tensile test results, the modulus of toughness for the tensile test can be calculated. The calculation of the area under the tensile test curve can be approximated by using the formula:

$$U_T = \frac{So+Su}{2}.\varepsilon f$$, where So: Yield Strength, Su: UTS and εf: elongation

The result showed that the longer holding time is given at the first austempering, then the value of the modulus of toughness is decreased. The highest value of modulus of toughness is shown by the first sample that is $5.93x10^7$ J/m^3. This modulus of toughness value is correlated with the impact test result. Figure 6 shows the results of impact testing. In the graph, it shows that the highest impact energy is produced by the first sample with a value of 54 J and decreases with increasing holding time on the first austempering.

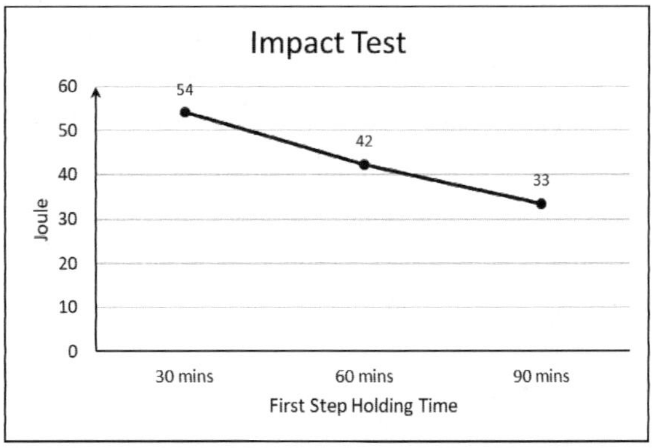

Figure 6. Impact test result of all specimens.

Hardness testing was carried out on the sample, and the results can be seen in figure 7 it shows that the hardness of the holding time at 30 minute is 38 HRC and increases to 40 HRC at the holding time of 90 minute.

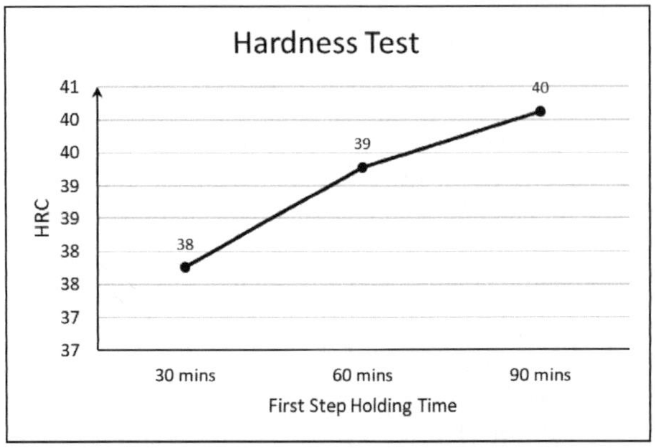

Figure 7. Hardness rockwell C test of all specimen.

The increase in hardness value followed by a decrease in the value of elongation is related to the holding time condition that is too long. If the sample is held at austenitizing temperature for too long, then it is possible to a second reaction takes place; where is the high carbon austenite (γHC) can further decompose into ferrite (α) and carbide (ε) [10]. Thus, if the desired material is obtained that has good ductility, the ADI material must be austempered in the process window. The process window is defined as the time interval between the completion of the first reaction and onset of the second reaction (figure 8). Thus the formation of carbides can be avoided.

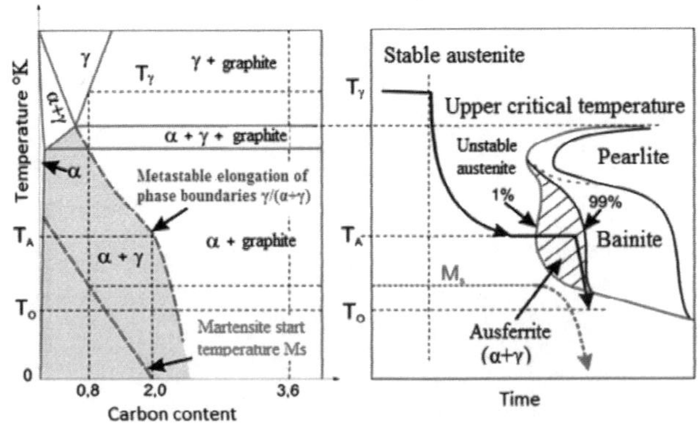

Figure 8. Process window of ADI [21].

4. Conclusion

The experimental results done in this study can be concluded that the longer holding time on first austempering affects ADI's material with increasing of tensile strength, yield strength, and hardness. But on the other hand elongation and impact values decrease. Low-high austempering can produce finer ferrite needle structures that contribute to the increase in ADI material's tensile strength and yield strength. This is also related to the formation of a combination of lower and upper ausferrite structures. The longer the holding time is given at, the lower temperature then, the lower ausferrite is formed more, this can increase the material brittleness. It is also analyses that the formation of carbide (ε) can be possible occurs due to long of holding time which can exceed of the process window in the ADI material (Figure 8).

As the combination of tensile and elongation then the toughness can be calculated to find the best value among the experimental variable. The highest value of modulus of toughness is shown by the holding time of 30 minutes that is 5.93×10^7 J/m^3. To obtain tougher material, the total holding time at austempering process must not be too long, and it is better not exceed the process window in the ADI material. Thus the formation of a brittle carbide structure can be avoided.

Reference

[1] Janowak J F and Norton P A 1985 *AFS Trans* **88** 123–35
[2] Shanmugam P, Rao P P, Udupa K R and Venkataraman N 1994 *J. Mater. Sci.* **29** 4933–40
[3] Rao P P and Putatunda S K 1998 *Metall. Mater. Trans. A* **29** 3005–16
[4] Schmidt I and Schuchert A 1987 *Z. Met.* **78** 871–5
[5] Bandanadjaja B, Lilansa N, Harianton I and Ruswandi A 2016 *American Institute of Physics* **030019** 3-4
[6] Bandanadjaja B 2004 *Pengembangan Material Bainitik As-Cast Ductile Iron Dengan Pengendalian Laju Pendinginan Dan Komposisi Nikel Molibden* (Indonesia: Intitut Teknologi Sepuluh Nopember)
[7] Bandanadjaja B 2004 *Gelar Ilmiah Teknologi dan Industri Dies Natalis ke-27 Polman Bandung*
[8] Janowak J F and Gundlach R B 1983 *Trans. Am. Foundrymen's Soc.* **91** 377–88
[9] Moore D J, Rouns T N and Rundman K B 1985 *J. Heat Treat.* **4** 7–24
[10] Putatunda S K 2010 *Mater. Manuf. Process.* **25** 749–57
[11] Hafiz M 2003 *Mater. Sci. Eng. A* **340** 1–7
[12] Pereloma E V and Anderson C S 2006 Microstructure and properties of austempered ductile iron subjected to single and two step processing *Mater. Sci. Technol.* **22** 1112–8
[13] Putatunda S K and Yang J H 2003 *Materials Science Forum* **426** 913–8
[14] Bayati H and Elliott R 1997 *Mater. Sci. Technol.* **13** 117–24
[15] Hsu C H and Chuang T L 2001 *Metall. Mater. Trans. A* **32** 2509–14
[16] Grech M, Bowen P and Young J M 1991 *World Conference on Austempered Ductile Iron.* **2** 338–74
[17] Putatunda S K 2001 *Mater. Sci. Eng. A* **315** 70–80
[18] Lee S C, Hsu C H, Feng H P and Chang C C 1998 *Metall. Mater. Trans. A* **29** 2511–21
[19] Hamid Ali A S and Elliott R 1996 *Mater. Sci. Technol.* **12** 679–90
[20] Yang J and Putatunda S K 2004 *Mater. Des.* **25** 219–30
[21] Čatipović N, Živković D and Dadić Z 2018 *Teh. Vjesn.* **25** 635–42

Acknowledgment

We would like to show our gratitude to the Kemenristekdikti and POLMAN Bandung which has supported the research and also to Ari Lestarianto with his contribution to the research data from his final project.

IOP Conf. Series: Materials Science and Engineering **541** (2019) 012024 doi:10.1088/1757-899X/541/1/012024

Effect of samarium on microstructure and intermetallic formation in Al-5Zn-0.5Si alloy

D Ferdian[1]*, J R Pratama[1], Y Pratesa[1]
[1]Department of Metallurgy and Materials Engineering, Universitas Indonesia, Depok 16424, Indonesia

Email: deni@metal.ui.ac.id

Abstract. Aluminium-Zinc-Silicon alloy is a material candidate for low voltage sacrificial anode. Nevertheless, iron-rich intermetallic phase that formed in the interdendritic region can reduce its efficiency due to the self-corrosion phenomenon. Samarium addition is expected to modify the formation of iron-rich intermetallic phase. The alloying process was conducted with melting of aluminium and other additional element in mini electric resistance furnace. Furthermore, characterization of the alloys was performed using optical emission spectrometer, differential scanning calorimetry, optical metallography and scanning electron microscopy equipped with energy dispersive spectrometer. The result showed that samarium can refined the secondary dendrite arm spacing (SDAS) from 89.6 μm to 61.4 μm. Samarium also shifted the solidification temperature of the alloy which affected the size of SDAS and formation of Al_2Si_2Sm intermetallic phase.

1. Introduction

The most common type of corrosion protection in offshore structural applications is by cathodic protection method, such as sacrificial anode, impressed current, or coating. Sacrificial anodes method commonly used is attributed to their easy to apply and can be installed without any requirement of external energy. Aluminium Sacrificial Anodes are widely used because they have a good efficiency in the offshore environment. Aluminium Sacrificial Anode has an efficiency up to 25.000 Ah / Kg and driving voltage of -1.1 V vs. SCE.

The first generation alloy used for aluminium sacrificial anode is Al-5Zn-Hg. However, this alloy has been replaced because it is considered not environmentally friendly due to the release of mercury ions during the protection. Current generation of sacrificial anode that widely used is Al-Zn-In alloy. Sakano et al. revealed that indium-activated aluminium alloy anodes with composition is Al- 5Zn-0.03In increase anode efficiency [1]. Furthermore, J.U Chavarin showed that chloride is also important in Al-Zn-In activation [2]. Indium ions made highly polarizing phenomena that promote chloride migration and adsorption, so it increases aluminium dissolution by forming complex chlorides. In general, Al-Zn-In alloy has a good efficiency, nevertheless one of weaknesses is the low potential protection value that could reach -1.1V vs. SCE. As a result, it could causing a stress corrosion cracking (SCC) phenomena on the protected material [3].

New generation of a sacrificial aluminium anode that being develop is low driving voltage anode. This type of sacrificial anode is considered safer to use for steel materials that have high mechanical strength. Ma Li et al. found working potential of Al-Ga alloy was -802 mV ~ -818 mV (vs. Ag / AgCl seawater) respectively [4]. Furthermore, Pratesa et al. [5] is used copper and silicon to substitute gallium for low voltage properties. It showed that copper and silicon increased potential protection of Al-Zn alloy. However, the presence of iron element triggers the formation of iron-rich intermetallic phase within the product. Thus, this intermetallic phase could initiate more selective corrosion which gives a detrimental effect for effectivity [5].

Samarium (Sm) additions have shown the ability to act as a grain refinement in solidification process [6,7]. With the addition of Samarium element, there is an alteration morphology of α-Al precipitates from dendritic into equiaxed forms [8]. The addition of Sm elements is also expected to improve the shape of the eutectic Si precipitates. Therefore, the objective of this present work is to characterize the effect of Sm on microstructure of Al-5Zn-0.5Si alloy as candidate for low voltage sacrificial anode.

2. Experimental Method

The alloying process was conducted using a mini electric resistance furnace. The alloying material consist of high purity Al ingot (99.5% from Inalum), Al-Si alloy (made from Al ingot and 443 Si grade), Zinc (99% purity) which were used to synthetize Al-Zn-Si alloys under normal ambient atmosphere at 850°C. Samarium element was added using a high purity binary Al-Sm alloy to achieve the desired content of 0.3, 0.6 and 1 wt. % Sm. The result of the process is characterized by differential scanning calorimetry (DSC) with a cooling rate of $10°C \cdot min^{-1}$. Moreover, optical emission spectrometer (OES) was used to analyse the chemical composition of the alloy while energy dispersive spectrometer (EDS) was used to investigate the micro chemical composition of the formed phase. Microstructure examination by optical microscope is conducted using metallographic techniques to observe the intermetallic phase and other feature in the matrix.

3. Results And Discussion

The chemical composition test result showed in Table 1. Unfortunately, Sm addition could not be measured due to limitation with the apparatus that was used. Therefore, the addition of Sm is given based on the weight amount of Al-Sm added to the melt. The chemical composition showed no significant decrease in zinc and silicon levels that indicating fading process due to melting processes.

Tabel 1. Composition of the primary element in Al-5Zn-0.5Si-xSm

	Al	Zn	Si	Fe	Cu	In
Al-5Zn-0.5 Si	94.8	4.55	0.561	0.0345	<0.001	<0.01
Al-5Zn-0.5Si-0.1Sm	94.5	4.74	0.542	0.0690	0.0354	< 0.01
Al-5Zn-0.5Si-0.3Sm	94.7	4.55	0.521	0.0653	<0.001	<0.01
Al-5Zn-0.5Si-0.5Sm	94.8	4.32	0.491	0.0726	0.0029	<0.01

DSC thermograms upon cooling as illustrated in Figure 1 showed the solidification reaction at decreasing temperature of nucleation and growth of primary (Al) dendrites and the existence of second phase (samarium-rich intermetallic phase) formation. It seems that, the precipitation of iron-rich intermetallic phase (β-$Al_9Fe_2Si_2$) is over shadowed by the peak of primary α-(Al). Formation of Al_2Si_2Sm phase successfully prevented the formation of iron- rich intermetallic phase (β-$Al_9Fe_2Si_2$ phase). This change is visible at 0.3 Sm addition while at 0.1 Sm is not seen clearly. DSC thermograms also observed a shifting of thermal arrests correspond to the second phase to a higher temperature as the Sm addition increases.

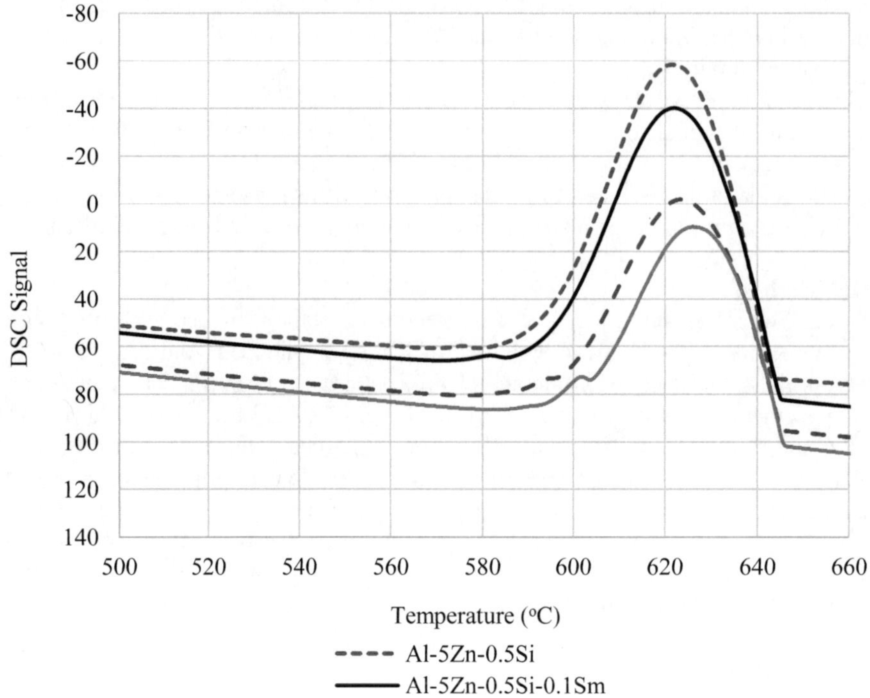

Figure 1. DSC thermograms upon cooling at 10°C·min^{-1} for various samarium additions.

Microstructure of the as-cast samples observed by optical microscopy showed the presence of the (Al) matrix, β-Zn and intermetallic phases in the interdendritic region as seen in figure 2. The presence of undissolve β-Zn within the aluminum matrix causing the damage of passive layer of anode surface [9]. The zinc addition to the alloys also shift the anode potential to more negative value, therefore increases the anode efficiency [10]. Moreover, the result showed a reduction of the SDAS due to Sm addition. This result proved that samarium acted as a grain refinement in aluminium. Measurement showed that the SDAS was successfully reduced from 89.6 μm to 61.4 μm when the alloy was added with 0.5Sm as seen in figure 3. Moreover, the optimum reduction of SDAS was achieved at 0.3Sm additions. Further increase of samarium addition did not give sufficient effect as with 0.3 Sm. At 0.5Sm addition, the ability of samarium to reduce the SDAS is 5% lower compared with 0.3Sm addition. This showed that samarium is no longer acted only as grain refinement but probably served as an intermetallic modifier. This result is confirmed with the DSC result when the new peak occurs at temperature 600°C which belongs to samarium-rich intermetallic phase (Al_2Si_2Sm) formation [11,12].

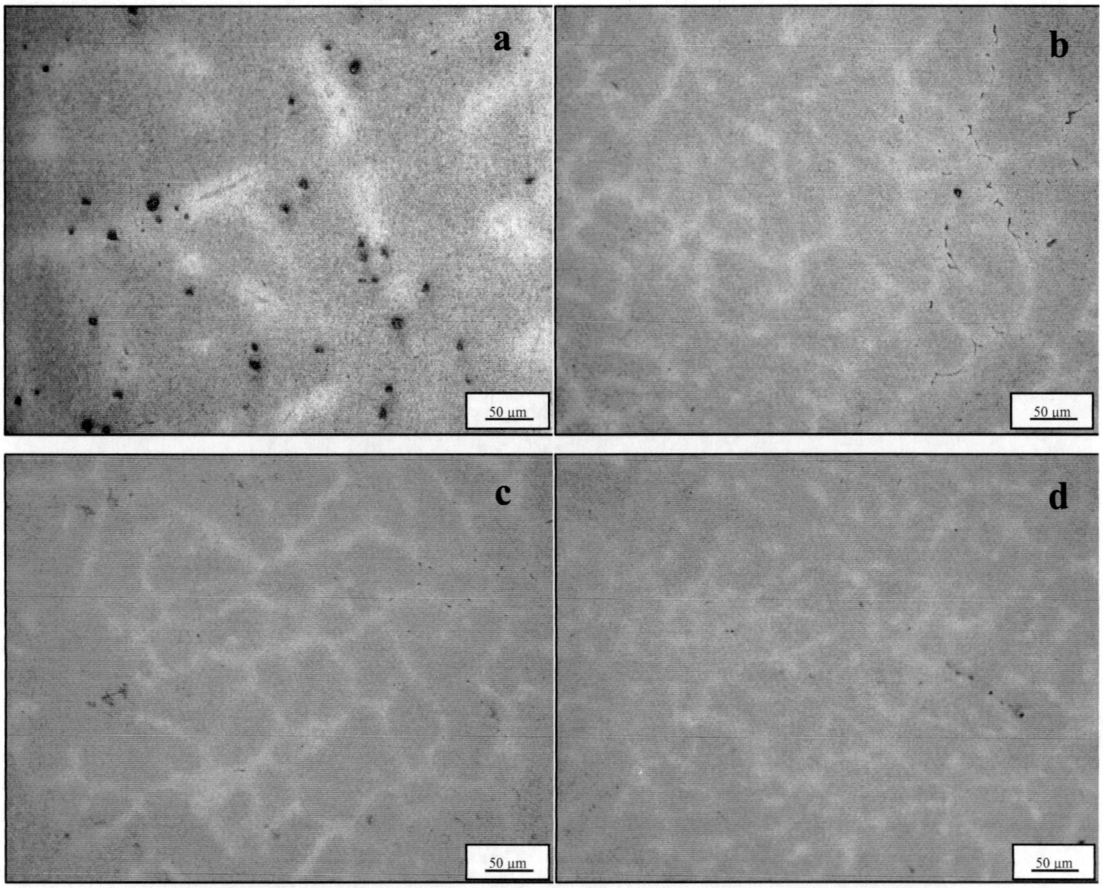

Figure 2. Optical micrograph of as cast Al-5Zn-0.5Si-Sm samples with various samarium additions: (a) 0%, (b) 0.1%, (c) 0.3%, and (d) 0.5%.

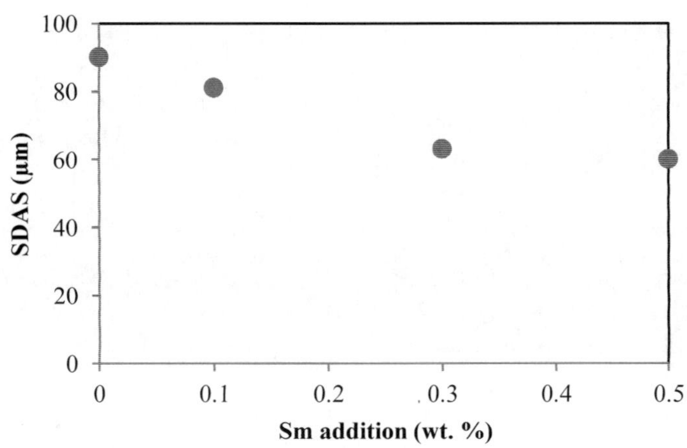

Figure 3. Effect of samarium in SDAS size.

EDS mapping and micro chemical analysis indicate the formation of an intermetallic compound containing samarium. Element mapping as seen in figure 4 showed dispersion of Zn element in the Al matrix while samarium is showed in the interdendritic region along with silicon and other minor element.

IOP Conf. Series: Materials Science and Engineering **541** (2019) 012024 doi:10.1088/1757-899X/541/1/012024

Further examination with EDS was conducted at selected area (white circle in figure 4a). The composition of samarium intermetallic compound (spectrum 1) as seen in figure 5 could relate to Al_2Si_2Sm phase [12].

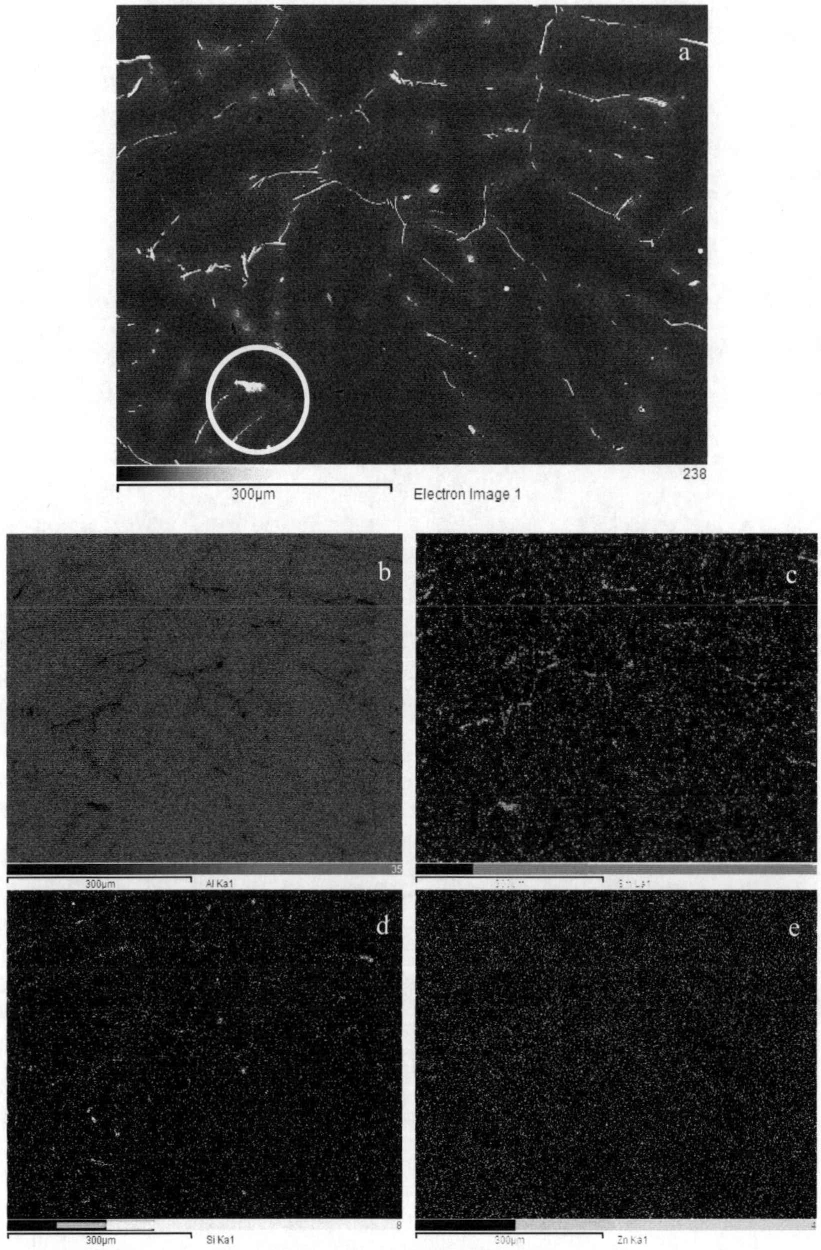

Figure 4. Backscattered electrons SEM micrograph of Al-5Zn-0.5Si-0.5Sm alloy (a) and related EDS mappings of aluminium (b), silicon (c), samarium (d) and zinc (e).

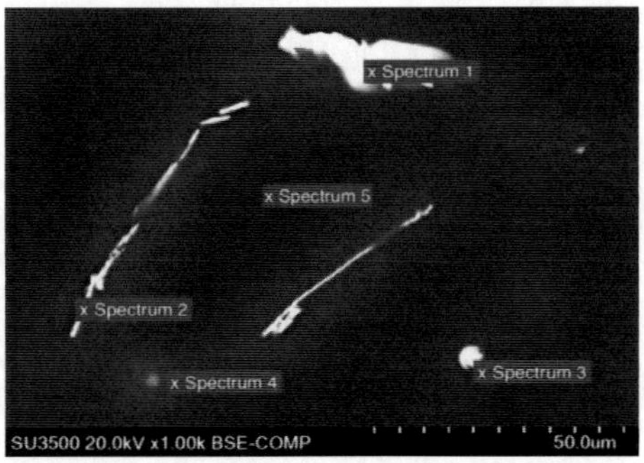

	Composition (*at.* %)					Phase identification
	Al	Sm	Si	Zn	O	
Spectrum 1	58.76	13.35	12.68	3.09	12.12	Al_2Si_2Sm
Spectrum 2	76.83	1.04	10.62	2.45	3.45	
Spectrum 3	62.34	3.46	12.99	3.36	6.24	

Figure 5. EDS micro-analysis of the phases in Al-5Zn-0.5Si-0.5Sm alloy taken from BSE image.

4. Conclusion

Sm addition to Al-5Zn-0.5Si form an Al_2Si_2Sm intermetallic precipitate that shift the solidification process as Sm addition increases. Sm addition also reduced SDAS from 89.6 μm to 61.4 μm in 0.5 Sm with 0.3 Sm additions as the optimum reduction.

References

[1] Sakano M H T and Toda K 1966 *Mater. Prot.* **5** 45
[2] Chavarin JU 1991 *Corrosion* **47 (6)** 472
[3] Garverick L 1994 *Corrosion in petrochemical* (United States: ASM)
[4] Ma L, Li W, Zeng H, Yan Y and Huo B 2010 *J. Chinese Soc. Corros. Prot.* **4** 2
[5] Pratesa Y, Ferdian D and Togina I 2017 *IOP Conf. Ser. Mater. Sci. Eng.* ***204*** 12026
[6] Wu D, Yan S, Wang, Zhang Z, Miao R, Zhang X and Chen D 2014 *J. Rare Earths* **32** 663
[7] Sun M, Hu X, Peng L, Fu P and Peng Y 2015 *J Mat Sci Eng A* **20** 89
[8] Yuansheng H Z R and Hong Y 2013 *J. Rare Earths* **31** 916
[9] Salinas D R, Bessone J B, Garcia S G 1999 *J. Appl. Electrochem.* **29**
[10] Lameche-Djeghaba S, Benchettara A, Kellou F 2013 *Arab. J. Sci. Eng.* **39 (1)** 113
[11] Raghavan V 2007 *J. Phase Equilibrium* **28** 556
[12] Markoli B, Spaic S and Zupanic F 2001 *Z. Metallkd* **92 (9)** 1098

Acknowledgement

The authors are grateful to PT Inalum for providing the aluminium ingot and Direktorat Riset dan Pengabdian Masyarakat (DRPM-UI) through PITTA 2017 grant for the financial support.

IOP Conf. Series: Materials Science and Engineering **541** (2019) 012025 doi:10.1088/1757-899X/541/1/012025

Optimizing performance of $Li_4Ti_5O_{12}$ nanorod doped C@ZnO by hydrothermal synthesis as half-cell lithium-ion battery anode

B Priyono[1], **M R Nugraha**[1], **A Z Syahrial**[1], **Faizah**[1], **A Subhan**[2]
[1]Department of Metallurgy and Materials, Faculty of Engineering, Universitas Indonesia, Depok, 16424, Indonesia
[2]Center for Research of Physics, LIPI, PUSPITEK, Tangerang, Banten, 15310, Indonesia

E-mail: bambang.priyono@ui.ac.id

Abstract. Lithium titanate, $Li_4Ti_5O_{12}$ (LTO) is a promising candidate as lithium ion battery anode material. In this investigation, LTO/C@ZnO was synthesized with LTO nanorod by hydrothermal method using TiO_2 xerogel that prepared by the sol-gel method, lithium hydroxide (LiOH), Activated carbon, and Zinc Oxide (ZnO) nanorod. Three variations of ZnO content addition in weight %, i.e., 4, 7 and 10%, labelled as sample LTO/C@ZnO-4, LTO/C@ZnO-7 and LTO/C@ZnO-10, respectively. The characterizations were made using XRD, FE-SEM, and BET testing. These were performed to observe the effect of ZnO addition on structure, morphology, and surface area of the resulting samples. Result showed that the optimum discharge capacity from each samples was 32.84 mAh/g in LTO/C@ZnO-4 with the crystallite size of 11.86 nm and the surface area of 348.736 m^2/g. In cyclic voltammetry testing, it shows a shift in reaction voltage and reduction in capacity that caused by the addition of C@ZnO and the lack of $Li_4Ti_5O_{12}$ that are formed.

1. Introduction

One of the most advanced technology development in secondary rechargeable batteries are Li-ion batteries. Li-ion batteries has one of the highest energy efficiency that allows it to be used in many applications that may increase energy sustainability [1]. Accordingly, Li-ion batteries is in high demand from industry and government funding agencies. About 96% of hybrid electric vehicles in the world market in present time use nickel metal hydride batteries. However, within 10 years, 70% of hybrids and 100% of plug-in hybrid and other electric vehicles are expected to use lithium-ion batteries [2].

The fact about lithium-ion battery is that the battery is relatively light and compacted which also give the properties of being able to have high energy output. In the application, the specific energy of lithium-ion batteries with the voltage of 4 V can be done and ranged between 100 Wh/kg to 150 Wh/kg [1]. This is also the consideration of choosing the lithium-ion batteries for future development and applications. Material development for Lithium-ion batteries has seen a bright future with LTO ($Li_4Ti_5O_{12}$ or Lithium Titanate) as base anode. LTO has been successful on the market because it can exhibit high rate and high cycle life despite the higher cost of Titanium, the reduced cell voltage, and lower capacity. This leads to LTO having high stability during charging and discharging process.

Development of other material for LIBs is promising, with LTO as base anode such as lithium titanate ($Li_4Ti_5O_{12}$). $Li_4Ti_5O_{12}$ has high lithium insertion/extraction voltage, approximately 1.55 V during intercalation, which has high stability during charging and discharging process [3]. Also compared with graphite which has tendency to pulverization, $Li_4Ti_5O_{12}$ has higher safety concern due to their zero-strain property and proposed to be used in electric vehicle. During intercalation in $Li_4Ti_5O_{12}$, every formula can accommodate three lithium ions, so theoretically it has the capacity of 175 mAh g^{-1}, compared than graphite (372 mAh g^{-1}) [4]. Despite their superiority, it has some drawbacks of low capacity and also low conductivity (10^{-9} S cm^{-1}). Researches has been conducted to improve its capacity and rate capability by conducting morphological optimization, doping other element, and nano-structuring of $Li_4Ti_5O_{12}$ [5].

This research is conducted by fabricating spinel $Li_4Ti_5O_{12}$/C@ZnO, with C@ZnO as doping of other element, with sol-gel and hydrothermal method for TiO_2 nanorod, solid state method for $Li_4Ti_5O_{12}$, and mixing the carbon with ZnO. Difference of ZnO concentration is observed in this research and its effect to the material.

2. Materials and Methods

Preparation and synthesis of $Li_4Ti_5O_{12}$/C@ZnO is done through several processes before it is ready to be characterized under XRD, BET, SEM, and EDS. Processes that will be done consists of sol-gelling, hydrothermal, calcination, ball-milling, sintering, and mixing.

2.1. Synthesis of LTO nanorod

It begins by preparing primary solution which is 20 ml ethanol (pH 3) with the addition of 3.40-gram of Titanium (IV) Butanate, and secondary solution which is 5 ml ethanol (pH 3) + H_2O which were later go through titration process with RW 3.5.

Gelling process is done using magnetic stirrer and magnetic spin bar until gel is produced from the process. The formed gel will be dried overnight to evaporate the moisture inside the gel and proceed the evaporation process for one week to evaporate the moisture content inside the gel. After evaporation process is done with no moisture content inside the gel left, it proceeds to the next process which is calcination. Calcination process is done inside a furnace with 300 °C temperature and is done in 2 hours. Calcination process is done to obtain TiO_2 anatase, to evaporate moisture content inside the gel, and enhance crystallinity.

Next process will be hydrothermal to achieve the TiO_2 nanorod structure. First it begins with mixing the TiO_2 with NaOH 10M solution and 10ml of water in beaker glass, then stirred in the magnetic stirrer for 30 minutes long. After the mixing is done, it will be put inside the autoclave for heating process in the furnace at temperature of 180 °C for 24 hours. The use of autoclave here is to have the highest purity result. Then the precipitated need to be filtered with membrane filter and dried in hot plate at 80 °C temperature until all the moisture evaporated. Sintering process was done in tube furnace with temperature 750 °C for 3 hours to increase the crystallinity, achieve the spinel structure and remove the impurities.

Next process is ball-milling where the lithium hydroxide will be mixed with TiO_2 nanorod using high energy ball-mill (HEBM). Ball-milling process is aimed to obtain optimum surface area.

2.2. Synthesis of ZnO nanorod

Zinc oxide nanorods synthesis process involves several stages before being mix with the activated carbon. First, Zinc oxide (ZnO) nanorods that will be produced in this research will have 0.05 M concentration obtained from 0.05 M concentration of Zinc nitrate tetrahydrate ($Zn(NO)_3.4H_2O$, Merck) and Hexamethylenetetramine (HMTA) ($C_6H_{12}N_4$, Merck). Both of them are dissolved in cold water in a beaker glass by putting the beaker glass inside the cooler box with crushed ice blocks covering all the beaker glass, then keep the cooler box closed for about 1 hour. After being cooled from the cooler box, the beaker glass will be moved into the oven for heating process in 90°C temperature for about 3 hours.

After the heating was done, the precipitated need to be strained using membrane filter paper and dried it in petri dish for approximately 1 day. Then it needs to be heated again in 80 °C for 30 minutes to ensure that all the water content inside the powder already disappeared.

2.3. Synthesis of activated carbon

The activation process of carbon is carried out to produce carbon with specific surface area. In this study, carbon activation was performed using NaOH because NaOH is more advantageous and can produce high specific surface area. The carbon used in this study is commercial carbon from PT Brataco.

Commercial carbon sources that used are still in the form of granules with ununiform size, so it need to be reduced into smaller size to achieved uniform grain. Reducing the size of carbon is done by using high energy ball mill and involves the use of zirconia balls. The fine carbon of the milling results is then heated to the furnace. Carbon is placed on a combustion boat located on a metal buffer and inserted into a tube furnace. Heating is carried out in temperature of 500° C with a heating rate of 20 °C min^{-1} and holding time about 2 hours. The heating carbon is then mixed with NaOH pellets. The ratio of the amount of NaOH pellets with carbon is 3:1 (3 grams of pellet NaOH : 1 gram of carbon) followed by the addition of 10 mL of distillated water. Mixing between carbon and NaOH pellets is done using the help of magnetic stirrer for 2 hours. First, NaOH pellet was mixed with 10 mL of water until it homogenized in a beaker glass and then it added with carbon. The mixture which has been stirred for 2 hours then will be heated in an oven at 130° C for about 4- 24 hours until the carbon is completely dried.

The next stage in the process of carbon activation is heating the carbon in the furnace with nitrogen gas flow. The heating is carried out in temperature of 700 °C at a 20 °C min^{-1} heating rate and a nitrogen gas flow rate of 100 cm^3 min^{-1}. The result of this step is to produce carbon with very dry conditions and minimize the volatile substances that still exist from the previous stage.

Then the heated carbon will be washed using a 0.1 M HCl solution and hot distillated water until it achieved 6.5 pH. The purpose of this washing is to remove residues or inorganic species formed during the process. Sufficient quantity of carbon is fed into a beaker glass and poured 0.1 M HCl solution then stirred. The pH measurement of the solution was performed using pH indicator paper. This step is repeated until the pH of solution achieved 6.5 then washed with hot distillated water. Filtered the solution are necessary with the membrane filter paper to get the carbon precipitate. The heating is done in the oven with the filtered carbon placed inside the beaker glass. The purpose of this heating is to remove or evaporate water that may still be contained in carbon.

2.4. Synthesis of LTO/C@ZnO composite

The concept of this LTO/C@ZnO composite is to have the ZnO nanorod particle encapsulate into the porous of activated carbon. This phenomena called ballistic effect [6]. In order to have that, the composite mixing need to be done step by step from the ZnO nanorod and the carbon first. After that, the C@ZnO powder being mixed again with LTO nanorods. The mixing of these substances is carefully carried out with certain composition that shows in table 1. The percentage of ZnO addition in this research are 4%, 7%, and 10% followed by 5% of activated carbon in all variables.

Table 1. Weight of each material for composite mixing.

No	Sample	LTO (g)	Carbon (g)	ZnO (g)
1	LTO/C@ZnO-4	1.82	0.1	0.08
2	LTO/C@ZnO-7	1.76	0.1	0.14
3	LTO/C@ZnO-10	1.70	0.1	0.2

2.5. Coin cell fabrication

In the battery fabrication process, the initial stage which had to be done is slurry making. In this stage, the active material which obtained from previous synthesis process was prepared with other substances of PVDF and acetylene black. The ratio of each substance of active material, PVDF and acetylene black respectively are 8:1:1. All material mixed together along with DMAC 5 gram using a magnetic stirrer. The mixture was stirred for 3 hours with hot plate temperature 80°C. Slurry results were coated to Cu-foil and dried for 1 hour in 80°C with doctor blade as the media. The coated material then assembled inside a glove box with a argon atmosphere.

The electrochemical analyses were performed using using a CR2032-type coin cell in which lithium metal chip was the counter electrode and $LiPF_6$ as an electrolyte. EIS (Electrochemical impedance spectroscopy) was conducted in a range from 0.10 Hz to 20 kHz using the HIOKI RM3544 instrument. The CV (cyclic voltammetry) test was performed between 0.01 and 3.0 V with the scan rate of 0.1 mVs-1. The charge–discharge (CD) performance was conducted from a low-current (C/5) to a high-current (12C) discharge rate using the WonATech WBCS 2000 instrument.

3. Results and Discussion

3.1. Brunauer–Emmett–Teller

The pores formed in the sample would make the gas flow easily through it. The higher the pores present in the sample, the higher the surface area value of the sample. From BET result on the Table 2, It shows with the addition of 4% ZnO, the surface area in the compound was 348.736 m^2/g which is higher than the other variables. Along with the addition of the ZnO composition, the surface area has decreased due to the pores that blocked by the substances [7].

Table 2. Result of BET examination of different variables.

Sample	Surface Area (m^2/g)
LTO/C@ZnO-4	348.736
LTO/C@ZnO-7	98.786
LTO/C@ZnO-10	88.246

3.2. FE-SEM

3.2.1. LTO nanorod

Figure 1. FE-SEM result of LTO nanorod powder with 10,000 X magnification.

From figure 1, the morphology of the sample has a rough morphology and has a non-uniform particle size, the structure of the sample slightly different from the structure of nanorod, the structure of the sample is more similar with damaged nanorod samples in the process, although the basic form of the sample still appears to look like a rod despite the damage, this may be due to mechanical treatment in this study in the form of a ball mill or due to process temperature. From the particle size measurement, the particle have a size 200-400 nm for diameter, and has a length from 700 nm – 1 μm [8]. This proves that the rod structure still exists even if it is damaged, on generally the nanorod structure has a diameter of nanometer and length micrometer size, this indicates that the mechanical treatment to mix TiO_2 nanorod, damages the structure of nanorod so that the length of the micron-sized particles becomes nano sized. On the sample observation with 500x magnification as in the picture will be seen a lot of agglomerates from nanorod LTO samples. Agglomeration is common in nano-sized particles because of their high surface energy, so to reduce this surface energy, the particles will agglomerate.

3.2.2. ZnO nanorod

Figure 2. FE-SEM result of ZnO nanorod powder with 10,000 X magnification.

From figure 2 it shows hexagonal crystal structure of ZnO in good accordance with the reference ZnO nanorod (JCPDS no. 01-080-0075) [9]. The structure of ZnO nanorod that formed are still very few and can be seen there are still many agglomeration in it. It is supported by XRD data that showing the low peak of ZnO nanorod. Therefore, it can be said that the ZnO nanorod purity are still low.

3.2.3. Activated Carbon

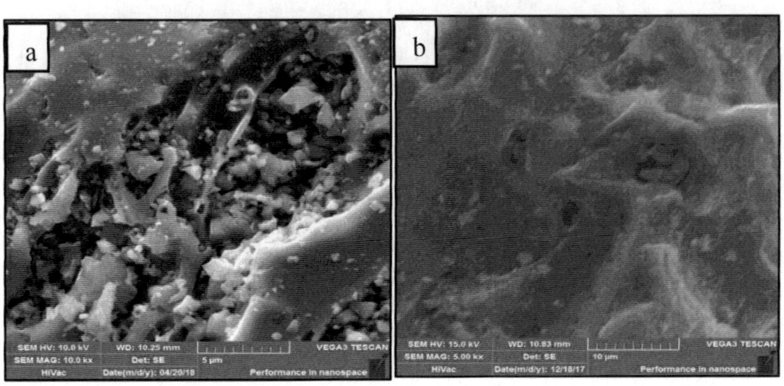

Figure 3. SEM result of carbon (a) before activation (b) after activation.

From the figure 3, the morphology of the carbon before activation was not have enough porous compare to carbon after activation. The porous in the structure can greatly enhance the surface area of the carbon [7].

3.2.4. LTO/C@ZnO

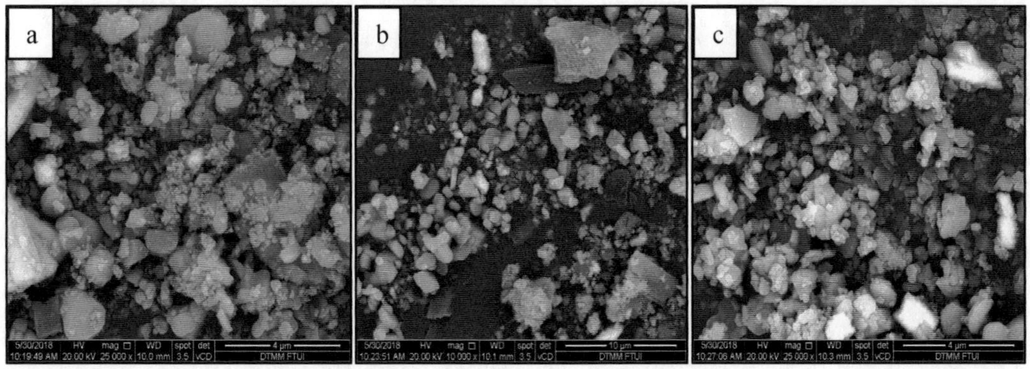

Figure 4. FE-SEM result of LTO/C@ZnO (a) 4% (b) 7% (c) 10% powder with magnification 25.000X.

3.3. X-Ray Difragment

From the figure 5, it shows the XRD patterns of LTO/C@ZnO with the 4%, 7% and 10% composition of ZnO addition. The presence of the $Li_4Ti_5O_{12}$ compound in the sample are indicated by 3 highest peaks at an angle of 18.509° in plane (111), 35.708° in plane (311), and 66.486° in plane (440) (JCPDS No. 00-049-0207). This is has slightly different with reference [10], where the second highest peak is the plane (400) and (111) is the third highest. This is due to the presence of Li_2TiO_3 in the sample which also has a high peak at an angle of 18.514°, 43.676°, and 35.967° (JCPDS 00-033-0831), the result of the Scherrer equation shows that the crystallite size of $Li_4Ti_5O_{12}$ nanorod have a size of 11.86nm (LTO/C@ZnO-4), 29.82nm (LTO/C@ZnO-7) and 29.82nm (LTO/C@ZnO-10). This shows that the synthesized $Li_4Ti_5O_{12}$ nanorod spinel have small size of crystallite size but low in degree of crystallinity because the intensity in xrd peak was still low.

For ZnO nanorods in the samples are also indicated by 2 highest peaks at an angle of 36.47° and 56.71°. It confirmed from the intensity of 3 diffraction graphs that, along with the addition of ZnO, the intensity in each sample are slightly increased. For the activated carbon, the phase are shows in angle 15.95° and 28.97°.

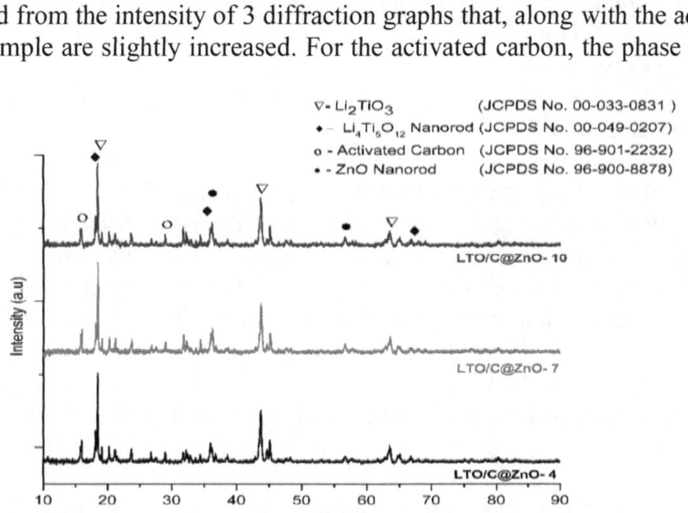

Figure 5. Phases indicated of samples (a) LTO/C@ZnO-4, (b)LTO/C@ZnO-7, (c) LTO/C@ZnO-10.

3.4. Electro Impedance Spectroscopy

Based on the EIS test, it could be seen in figure 6 that the sample of LTO/C@ZnO-4 has the lowest charge transfer resistivity (Rct) compare to other samples. In this case, the optimal ZnO content is 4% in which the conductivity is better than other three samples. As seen on the table 3, the LTO-K/C@ZnO-10 having the moderate resistivity among the other samples.

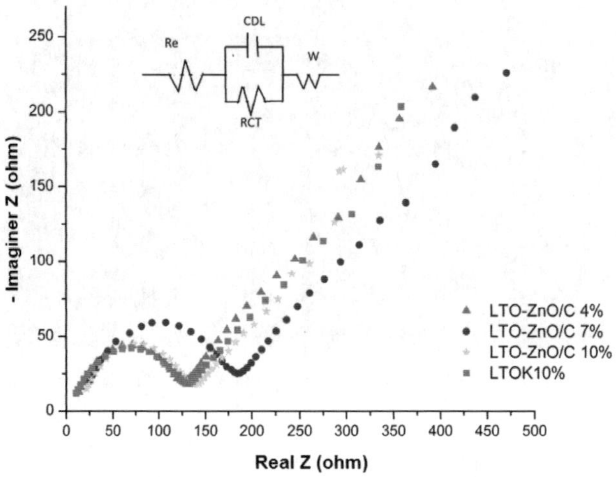

Figure 6. Graph of EIS result and the equivalent circuit.

The equivalent Randle's circuit was obtained by fitting the EIS graph using the software EISSA with the component containing Rs for solution resistance with series connection to the two component in parallel components containing Rct (charge transfer resistance) and Cdl (Capacity double layer) representing the impedance in the porous electrode. The Warburg impedance in semi-infinite diffusion then connected to said parallel system.

Table 3. Rs, Rct, Cdl and W impedance values of the experimental with variation ZnO content.

Sample	Rs (ohm)	Rct (ohm)	Cdl (F)	W
LTO/C@ZnO-4	13.40	103.95	1.00 E-06	221.62
LTO/C@ZnO-7	26.83	145.87	7.19 E-07	247.87
LTO/C@ZnO-10	23.15	107.76	9.54 E-07	174.5
LTO-K/C@ZnO-10	11.37	105.07	1.21 E-06	175.12

The corelation between charge transfer resistance and active surface area in terms of heterogeous electron transfer constant (Ks) is that the higher active surface area will facilitate the electron transfer and make the Ks value become higher and thus lowering the charge transfer resistance (Rct) . It can be seen in sample LTO/C@ZnO-4 which has the highest surface area, the Rct value of this sample is the lowest. According to Zeng, et. all, the impedance of porous electrode is inversely proportional to the interfacial surface area of the particle per unit volume of electrode [11].

Further, the Rct value of sample LTO/C@ZnO-7 is the highest among other sample which seems correlate to the result in the specific capacity value obtained in the Cyclic Voltammetry result below.

3.5. Cyclic Voltammetry

The discharge capacity which could be observed on Table 4, the sample of LTO/C@ZnO-4 has the capacity of 32.84 mAh/g. For LTO/C@ZnO-7 has 28.54 mAh/g and LTO/C@ZnO-10 for 31.55 mAh/g. As seen on the figure, the trend of the discharge capacity of every samples decrease with the addition of

ZnO nanorods 7 wt. % concentration to the LTO. It has a significant decrease compared with the other samples. This could indicate the agglomeration and pulverization which happened in the LTO/C@ZnO-7 and LTO/C@ZnO-10 samples having capacity fade. Compare with the samples using the commercial LTO (LTO-K/C@ZnO-10), it has the highest discharge capacity of 123.4 mAh/g. The use of non-nanorod commercial LTO can achieved higher capacity than the LTO nanorods. This also means the LTO nanorods that have been synthesized is still not successful enough because there are not enough $Li_4Ti_5O_{12}$ that formed in the process, which can be seen from low peak of LTO in CV graph result.

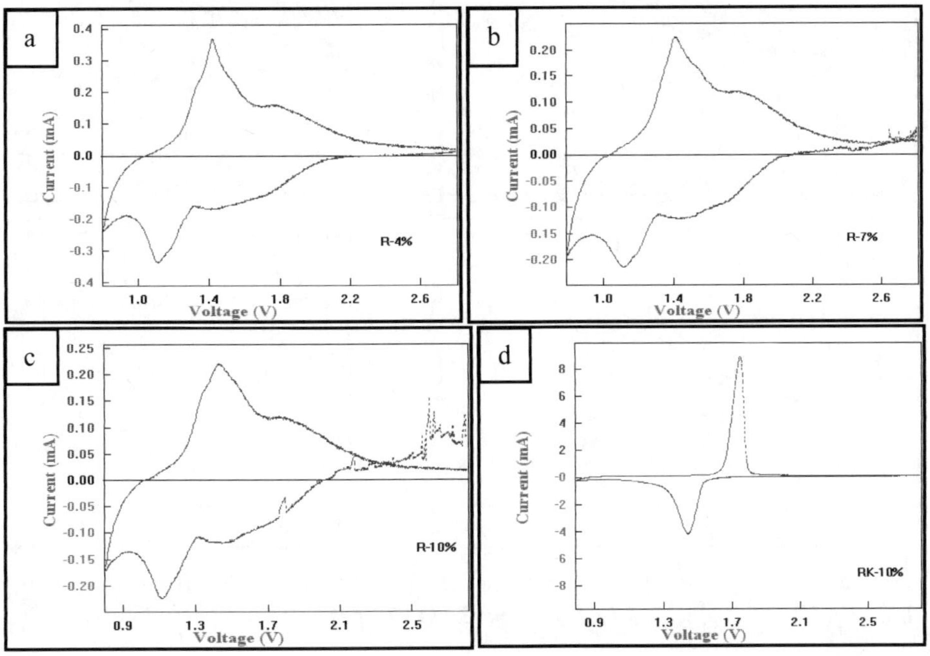

Figure 7. Result of cyclic voltammetry of the samples with different variables (a) LTO/C@ZnO-4, (b) LTO/C@ZnO-7, (c) LTO/C@ZnO-10.

Table 4. Result of CV of different samples.

No.	Sample	Charge - Discharge Capacity (mAh/g)	
		C+	C-
1	LTO/C@ZnO-4	26.54	32. 84
2	LTO/C@ZnO-7	27.06	28.54
3	LTO/C@ZnO-10	4.24	31.55
4	LTO-K/C@ZnO-10	120	123.4

3.6. Charge-Discharge

It can be seen in figure 8, for all the variables of ZnO, the charge and discharge capacity for each variable has shown decrement of capacity. As seen on the current rate from C/5 to 20C the LTO/C@ZnO-4 the charge capacity is 27.45 mAh/g into 2.75 mAh/g and 29.42 mAh/g into 2.67 mAh/g for the discharge capacity. For the LTO/C@ZnO-7, the charge capacity is 25.6 mAh/g into 4.88 mAh/g and the discharge capacity from 25.81 mAh/g into 4.72 mAh/g. In the LTO/C@ZnO-10 having the same decrement for the charge and discharge capacity with the charge capacity of 26.12 mAh/g decrease to be having the lowest compared to other variables to be 1.91 mAh/g. For the discharge capacity from 27.87 mAh/g and decrease to be 3.01 mAh/g but it only reach up to 15C. For the sample with commercial LTO (LTO-

IOP Conf. Series: Materials Science and Engineering **541** (2019) 012025 doi:10.1088/1757-899X/541/1/012025

K/C@ZnO-10) it reach the highest charge capacity of 113.17 mAh/g into 6.28 mAh/g. for the discharge it has the capacity of 115.4 mAh/g into 14.38 mAh/g in 15C current rate.

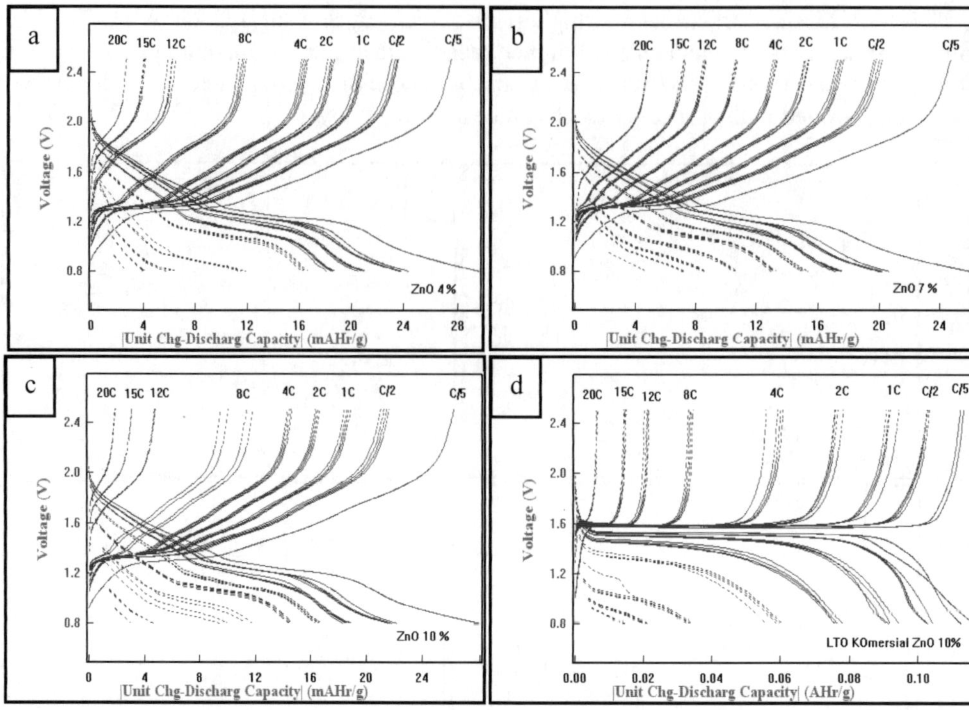

Figure 8. Result of charge-discharge of the samples with different variables (a) LTO/C@ZnO-4, (b) LTO/C@ZnO-7, (c) LTO/C@ZnO-10.

4. Conclusions

For the active material of $Li_4Ti_5O_{12}$ nanorod, it could be concluded that by using hydrothermal and solid-state methods, the synthesis of lithium titanate is not quite successful, in which the result of XRD showed the small amount of $Li_4Ti_5O_{12}$ in every samples. The impurities that appears such as Li_2TiO_3 and salt can lower the conductivity performance. The surface area obtained from the BET examination has shown relatively high surface area.The result for ZnO with 4% addition through 7% and 10% addition respectively 348.736 m^2/g, 98.786 m^2/g and 88.246 m^2/g.

Conductivity of each samples varied which could be compared from EIS examination. The highest conductivity was 4% of ZnO addition. The resistivity of LTO/C@ZnO-4 was 100.4 Ω. The agglomeration that happened in other samples make the conductivity decrease. The highest capacity that resulted from charge-discharge achieved by LTO/C@ZnO-4 was 32.84 mAh/g. As comparison, the sample that using commercial LTO (LTO-K/C@ZnO-10) has the capacity of 123.4 mAh/g.

References

[1] Harks P P R M L, Mulder F M and Notten P H L 2015 *J Power Sources* **288** 92–105
[2] Safari M 2018 *Energy Policy* **115** 54–65
[3] Nordh T *Lithium titanate as anode material in lithium- ion batteries* (Uppsala University : Ångström Advanced Battery Centre)
[4] Wang L, Chen B, Meng Z, Luo B, Wang X and Zhao Y 2016 *Electrochim. Acta* **188** 135–144
[5] Chauque S, Oliva F Y, Visintin A, Barraco D, Leiva E P M and O R Cámara 2017 *J Electroanal Chem* **799** 142–155
[6] Li Q, Zhang H, Lou S, Qu Y, Zuo P, Ma Y and Cheng X 2017 *Ceram Int* **43** 11998–12004
[7] Chen Z, Liu K, Liu S, Xia L, Fu J, Zhang X, Zhang C and Gao B 2017 *Electrochim. Acta* **237**

102–108
[8] Rajamanickam G, Narendhiran S, Pandian S, Mukhopadhyay S and Perumalsamy R 2017 *Chem Phys Lett* **689** 19–25
[9] Suryavanshi U, Iijima T, Hayashi A, Hayashi Y and Tanemura M 2012 *Chem Eng J* **179** 388–393
[10] Wang L, Wang F, Zhu J, Zhang X, Tang Y and X Wang 2017 *Ceram Int* **44** 1–8
[11] Zeng Z Y, Tu J P, Wang X L and Zhao X B 2008 *J Electroanal Chem* **616** 7–13

Acknowledgements
The authors would like to thank the Directorate Research and Public Services Universitas Indonesia for their financial support to do this research under HIBAH PITTA 2018 with contract PITTA/1062/FT/2018.

Erosion behaviour of alumina ceramic coating on mild steel by the modified composition of phosphate binder

D Idamayanti[1], D Ginanjar[1], B Bandanadjaja[1], W Purwadi[1], and N Lilansa[2]
[1] Foundry Engineering Department, Politeknik Manufaktur Bandung, Jl. Kanayakan no. 21 Bandung, West Java, Indonesia
[2] Automation Engineering Department, Politeknik Manufaktur Bandung, Jl. Kanayakan no. 21 Bandung, West Java, Indonesia

E-mail : idamayanti79@gmail.com

Abstract. The Pulverizer pipe made of mild steel had erosion failure due to coal dust impacting, thus its service life also reduces. The ceramic coating overlay on the surface of mild steel is one of the appropiate ways to protect the mild steel from erosion. This research is aimed to perform a ceramic coating over the surface of the mild steel using a dipping method to improve its erosion resistance by using the alumina-phosphate ceramic coating. The coating layer is formed by the reaction between monoaluminum phosphate (MAP) as a binder and Al_2O_3 particles. It transforms into berlinite phase when heated at an elevated temperature. The observation is carried out with the variation of the MAP binder composition Al:P 25:75, 28:72, 30:70 and the Al_2O_3/MAP slurry is given at 40/60. Scanning electron microscopy is used to characterize the coating morphology. X-ray diffraction is applied to investigate the ceramic coating phases. The gas erosion jet measures erosion resistance of the ceramic coating. From the test result, it can be concluded that the binder composition influenced the erosion behaviour of alumina ceramic coating, the binder with Al:P (30:70) showed the erosion resistance increasing four times compared to the condition without coating.

1. Introduction

The pulverizer pipe is a part of the coal mill pulverizer system which has a function to deliver coal dust to a combustion chamber. Mild steel is commonly used for the pulverizer pipe material due to its mechanical strength, machinability, weldability, formability and reasonable cost. Many energy projects also rely on the amount of steel for pipelines and electric power turbine component [1]. Unfortunately mild steel does not have good erosion resistance against solid particles such as coal dust. During coal grinding at the working chamber of a coal mill pulverizer, the process releases a certain amount of heat which raises the temperatures to 300ºC. Pulverizer pipe and inner cone which made of mild steel is impinged by coal dust particle and exposed to heat continuously, then the erosion takes place. The failure of the pulverizer pipe due to coal dust erosion demontrates in figure 1.

The erosion on mild steel is particularly induced by repeated plastic deformation of coal dust particle impact. Mbabazi et al [2] observed that the erosion rate of mild steel was affected by the impingement angle. The highest erosion rate of mild steel is 6.5 $mg.Kg^{-1}$ the impingement angle reached an inclination between 25-30 deg. Ceramic coatings are widely used for the protection of base metal or cement components in chemical, power, and refractory industries against high temperature corrosion and oxidation and for the minimization of wear or erosion [3].

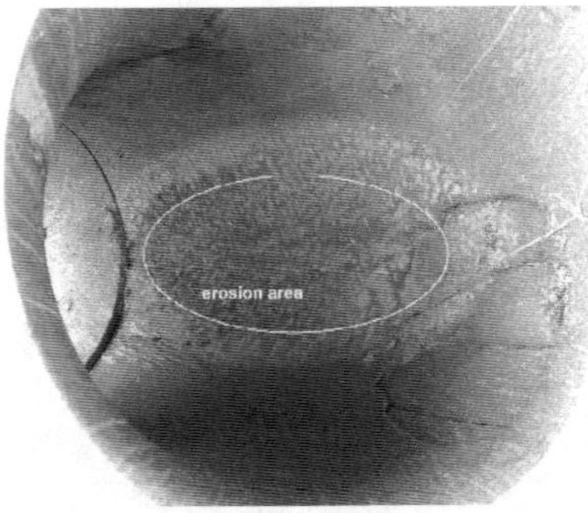

Figure 1. Erosion failure of pulverizer pipe.

Alumina ceramic coated on steel is used to provide better performance and help in decreasing erosion that is caused by pressurized air containing entrained abrasive solid coal particles such as oxides and minerals. When the angle of impingement is less than 30 deg, wear produced is closely similar to abrasion but if the angle of impingement is greater than 30 deg to surface, wear produced is released by impact erosion [4]. Ścieszka et al [5] observed the erosion resistance of mild steel coated-alumina compared to high alumina ceramic by solid particle erosion test using pulverized coal dust from a power plant. The result explained that the erosion resistance of alumina-coated mild steel is four times higher than the high alumina ceramic its self. Troczynski et al [6] have already investigated the behavior of alumina ceramic coating using phosphate binder for the variety of application, including high-temperature corrosion protection, wear resistance, dielectric properties, non-sticky surface, bioactive ceramic, thermal barrier ceramics, and others. Chen et al [7] expressed in their paper, phosphate inorganic binder plays a role in many applications i.e. repairing and joining of material.

This research focused on the use of alumina ceramic coating with phosphate binder on mild steel due to their superior properties and their suitability to the working conditions of pulverizer pipe. It is preferred over the sintering method which requires a process at an elevated temperature. The alumina phosphate ceramic coating consists of alumina particle in which aluminum phosphate used as a binder [8]. Aluminum phosphate binder is formed by reacting inorganic mineral or oxide with an acid-phosphate solution. It is a sort of inorganic binders used in refractory ceramic coating systems, which has been investigated and applied in the thermal spray coating systems. It is reported that refractories bonded with aluminum phosphate exhibit high strength, high-temperature stability, and abrasion resistance. Wang et al [9] reviewed that phosphate adhesive could be made in low temperature curing with high shear strength and excellent electrical properties. They observed the performance of phosphate adhesive which contains $Al(HPO_4)_2$ as the major constituent, has a heat-resistance of 1500°C.

In this study, the variety of mole ratio Al/P as a phosphate binder for alumina ceramic that coated mild steel were observed. Furthermore, it was characterized using XRD, SEM-EDS and erosion testing then compared with mild steel without any protection. The results will be recommended for erosion protection of pulverizer pipe.

2. Experiment Work

2.1. Materials and specimen preparation

The materials used for ceramic coating were alumina (Al_2O_3) as a filler and $Al(OH)_3$ – H_3PO_4 as a raw material of binder. Alumina (Al_2O_3) that has a size of 50 – 60 μm was obtained from the local market. All materials that are used come up with a technical grade. It is consist of phosphoric acid (85%), aluminum hydroxide ($Al(OH)_3$), methanol and aquades without further purification.

The specimen used in the investigation was mild steel of 5 mm thickness. The mild steel sheet was cut to a size of 25 mm x 25 mm for ceramic coating. Furthermore, the mild steel specimen was cleaned by methanol to remove scale, corrosion, and other contaminants.

2.2. Synthesis of binder

The phosphoric acid and aluminum hydroxide were synthesized to achieve monoaluminum phosphate (MAP) binder as the equation (1).

$$3H_3PO_4 + Al(OH)_3 \rightarrow Al(H_2PO_4)_3 + 3H_2O \quad (1)$$

The mole ratio of Al/P was varied to find out the influence of the binder composition to the erosion behavior of the coating material as shown in table 1.

Table 1. The mole ratio of Al/P in the synthesis of monoaluminum phosphate binder.

Materials	Various moles of $Al(OH)_3$: H3PO4		
	Al/P (25 : 75)	Al/P (28 : 72)	Al/P (30 : 70)
$Al(OH)_3$	12 %	14 %	16 %
H_3PO_4	55 %	52 %	51 %
Aquades	33 %	34 %	33 %

This mixed solution was reacted at 100° - 120° Celcius to achieve the final product reaction of monoaluminum phosphate. Furthermore, the reaction was dried at the ambient temperature. Figure 2 showed the whole process of this reaction.

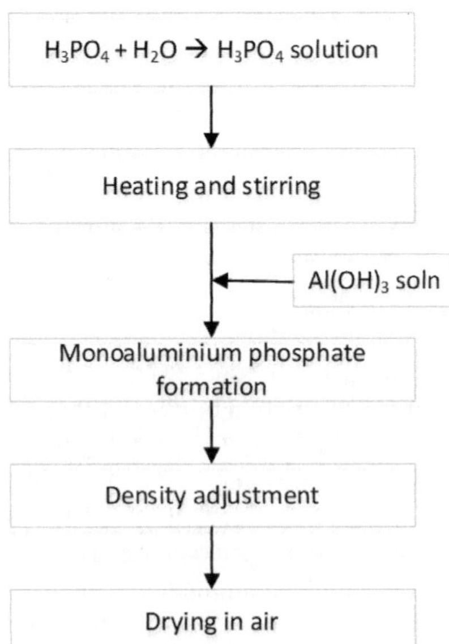

Figure 2. Synthesis of monoaluminum phosphate.

2.3. Synthesis of alumina phosphate ceramic coating

The alumina phosphate ceramic coating was synthesized from aluminum oxide particle and monoaluminum phosphate binder with 40:60 of the ratio [3] and applied to the mild steel which has already cleaned with methanol to remove the scale and grease, by dipping technique at the constant time. To gain the hardness of the aluminum coating, it was dried at the various temperature and illustrated in Figure 3.

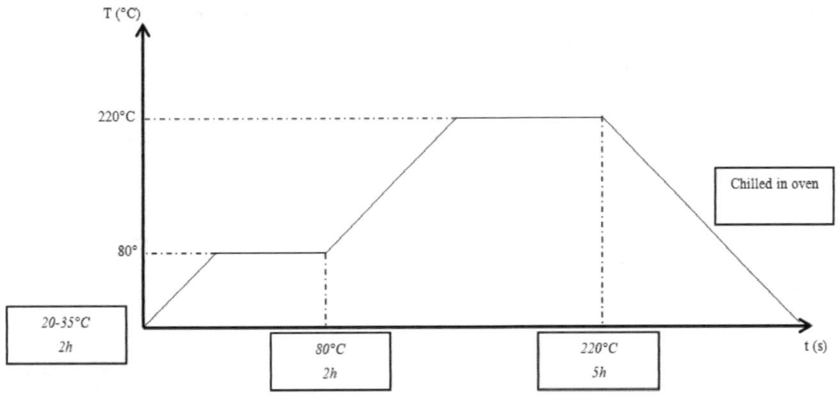

Figure 3. Heat treatment of the aluminum-phosphate ceramic coating.

2.4. Characterization

Several tools were used to characterize the coating. Optical microscope (OLYMPUS SZ61) was used to figure the coating surface, Scanning Electron Microscope (Hitachi SU3500 at 5kV and 40% spot intensity) to observe the microstructure of the coating and interface bonding between steel and coating. Energy Dispersive Spectrometer (EDS) was applied to check the alumina particle composition as depicted in figure 4. It showed the alumina was high purity.

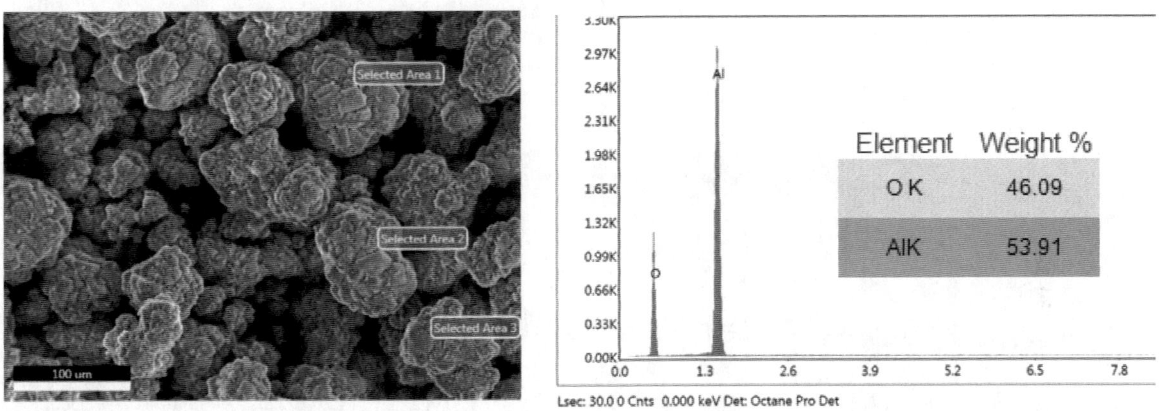

Figure 4. Elemental analysis of alumina by EDS.

The density of the coating was counted by Pycnometer (Pyrex, 5 mL). X-Ray Diffraction (SmartLab X-Ray Diffractometer) was used to characterize the binder and to operate at 30 kV and 30 mA scanning at 10° (2θ) to 60°. Erosion tester (TR-470) used to characterize the erosion properties of the coating. This test according to ASTM G76-02 (standard test method for conducting erosion test by solid particle impingement using gas jets). After the test piece being placed at a jig, it will undergo impact load of alumina particle from the nozzle with speed of 30 m/s until 70 m/s, 90° angle and particle feed rate 2 gr/min for 10 minutes.

3. Result and Analysis

3.1. Monoaluminum phosphate binder

He et al [3], Emmerson et al and Wagh [10], reviewed that aluminum hydrogen phosphate or $Al(H_2PO_4)_3$ is the phase being expected to be present on this binder which can be reacted with the aluminum oxide particle to form berlinite. The XRD result can be seen in figure 5.

The peak shows other compound aluminum hydrogen phosphate hydrate $Al(H_2PO_4)\ nH_2O$. This phase was detected because the binder aluminum still consists of water, due to the heating process before. The XRD result can be the basis of this research to continue the synthesized between binder and the aluminum oxide particles to achieve coating material. Aluminum hydrogen phosphate hydrate was formed as much as aluminum hydrogen phosphate based on their height of intensity peaks.

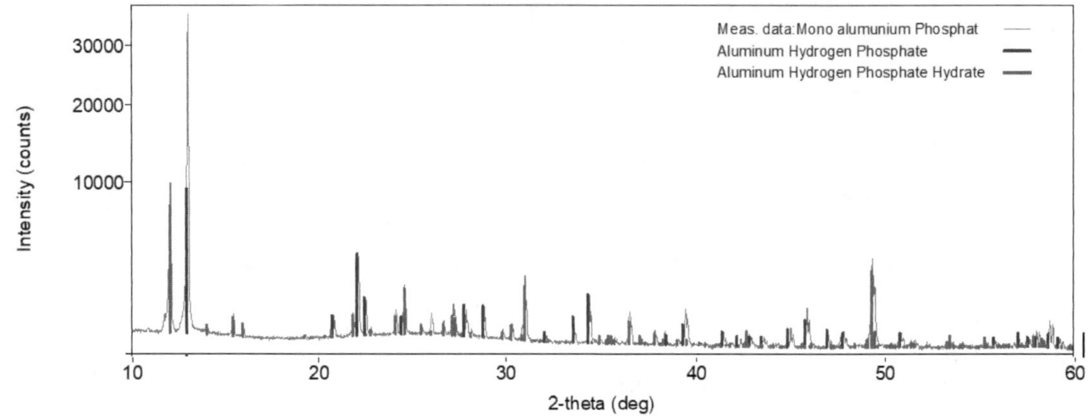

Figure 5. XRD result of synthetic binder.

3.2. Characterizations of alumina-phosphate ceramic coating
Observation of the coating surface was conducted to find the agglomeration of the particle, crack, and the impurity from other material. This observation is important explained by Colonetti research [11]. Colonetti reviewed that crack and porosity influence to the mechanical properties of the coating.

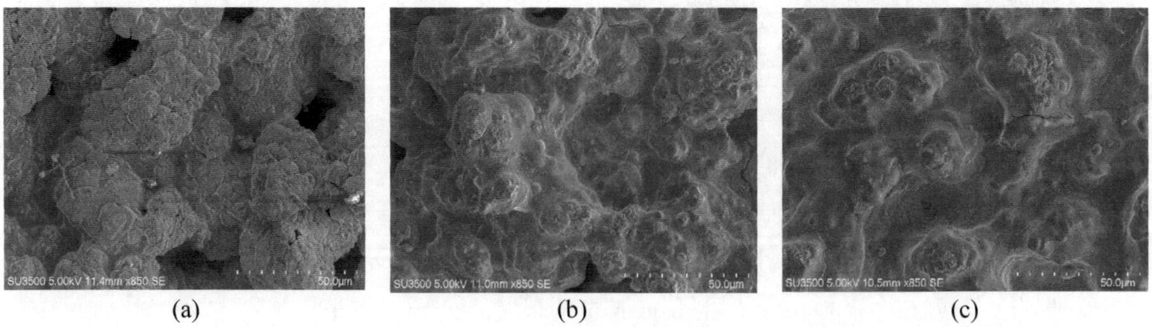

Figure 6. SEM micrograph of surface morphology of the coating affected by mole ratio of Al/P a. Al/P (25:75) b. Al/P (28 :72) c. Al/P (30:70).

As seen in figure 6, the surfaces of the coating has a uniform particle distribution, leak and fissures are observed in this coating. The mole ratio of Al/P leads the binder ability to form the compacted ceramic coating. The more the mole ratio of Al in Al/P the denser of the coating becomes.

Wagh et al [10] suggest that the bonding phase is a converted from a very small amount of alumina. Al_2O_3 particles disperse in the Al_2O_3-berlinite system and the product is a thin coating of berlinite on the alumina. As depicted in figure 6 (a-c), the increase of P ratio (as H_3PO_4) creates the amount of porosity due to gas formation when the dipping process. This phenomenon relates with the increase of MAP acidity which will react with mild steel surface and then releases hydrogen gas.

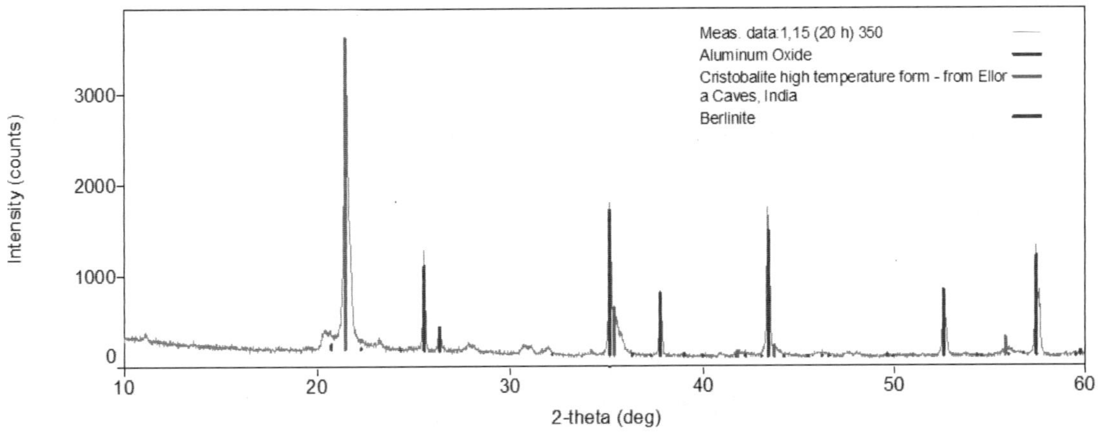

Figure 7. XRD result of alumina-phosphate ceramic coating contained Al/P binder (28:72).

XRD test was used to characterize the compound of the coating with binder ratio Al/P (28:72). Figure 7 shows the main compound of this coating are aluminum oxide, cristobalite and little amount of berlinite in 20°-30°. Berlinite is described as a small peak in a diagram, since it serves only as a binder in the material. He et al [3], and Colonetti et al [11] also detected berlinite at 20° - 30° with a small peak.

3.3. Effect of density to an alumina-phosphate ceramic coating

Characterization of coating morphology was conducted to prove the effectiveness of the binder to the alumina ceramic coating. Wagh et.al [10] observed the effect of binder density on the distance and compactness of the coating. Table 2 shows the various density of the binder with a different mole ratio of Al/P.

Table 2. The density of the alumina-phosphate ceramic coating.

	Density (g/cm^3)
Al/P (25:75)	1.6704
Al/P (28 :72)	1.7029
Al/P (30:70)	1.7158

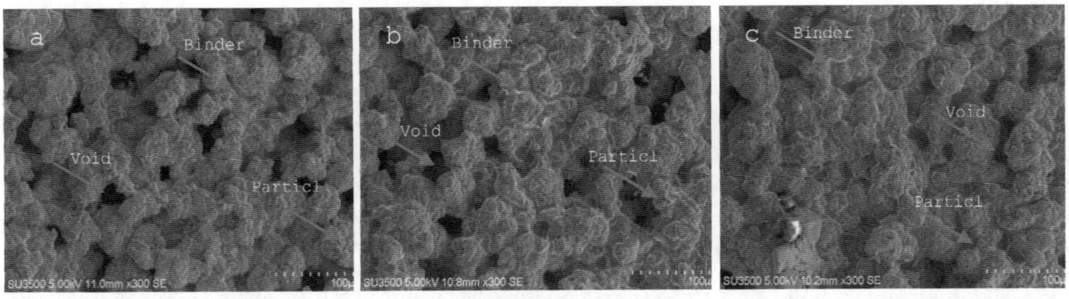

Figure 8. SEM micrograph a. Al/P (25:75) , b. Al/P (28:72) and c. Al/P (30:70).

Figure 8 shows in addition to the effect of the density, the void between the particle can also be caused by the distribution and size of the Al_2O_3 particle [10]. Void on aluminum ceramic coating was filled by the berlinite as synthesized between the binder and the Al_2O_3 particle. The reaction product will fill the space of the particle and make it compact. This condition occurrs when the product reaction has less density than the Al_2O_3 particle. Figure 8 (c) shows the product reaction of the binder with molar ratio Al/P (30:70) has less density than Al_2O_3 particle and make the coating more compact compared to other molar ratios. The compactness of the coating is important which related to the mechanical properties of the coating.

3.4. Erosion behavior of the alumina-phosphate ceramic coating

Figure 9 shows the result of erosion test, as described in the diagram, it can be seen that coating with the molar ratio of Al/P (30:70) has 5mg/Kg loss material, compare to the steel without any protection that has only 20 mg/Kg loss material.

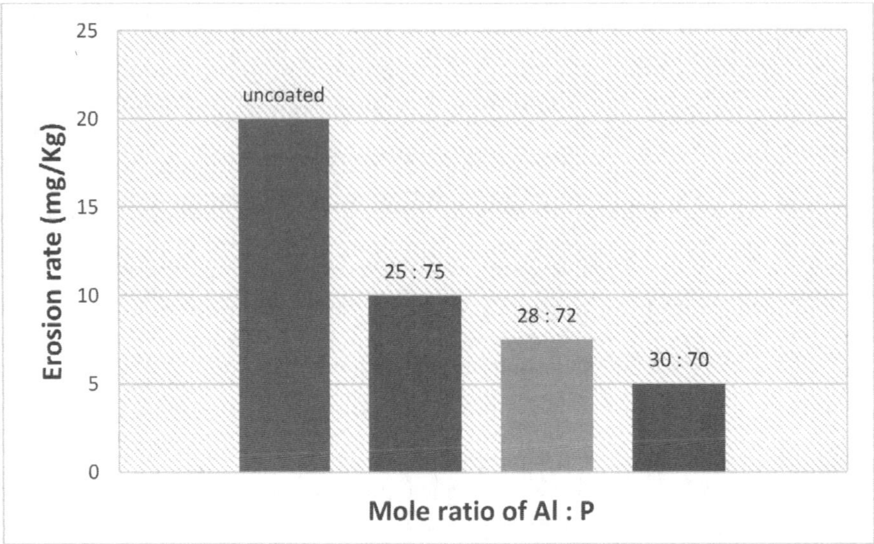

Figure 9. Erosion testing of alumina-phosphate ceramic coating and uncoated mild steel.

The result indicates that the erosion resistance of the coating steel is four-time than steel without any protection. This molar ratio also has a higher result than other ratios, it might be caused by the compactness of the coating and the void effect between the particle as the SEM test shows before. He et al [3] reviewed the abrasion resistance of the coating on their research which has two-time wear durability compared to steel without coating. The microstructure of the coating after erosion test can be seen in figure 9.

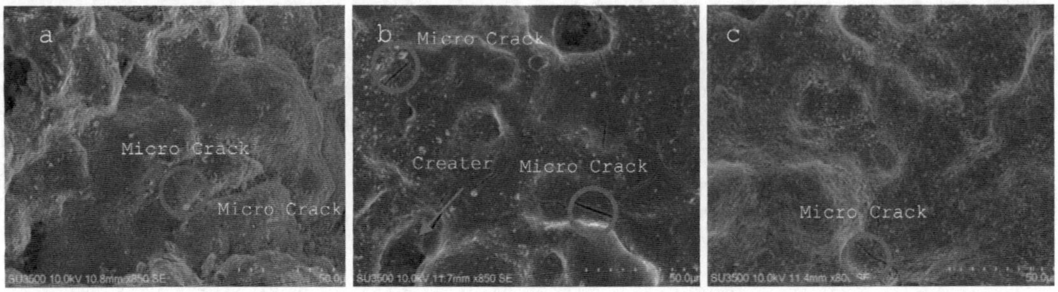

Figure 10. SEM micrograph on magnification x850 after erosion testing a. Al/P (25:75) , b. Al/P (28:72) dan c. Al/P (30:70).

The microcrack was found on the coating after the erosion test, in which the crack can develop due to the impact from erosion particle and also an indication of mechanical properties of the binder. Mol ratio Al/P (30:70) has a less amount of crack, and it is related to the erosion test result of this mol ratio which exibit less weight loss during the test compared to other ratios. A crater was observed in the mol ratio Al/P (28:72) this crater can happen through the test.

4. Conclusion
The mole ratio of Al/P in MAP binder influence the properties and erosion behavior of alumina ceramic coating. The aluminum phosphate ceramic coating was successfully coated to the steel and increase erosion resistance of the steel four times than steel without any protection. The best mole ratio of this aluminum phosphate ceramic coating was obtained in Al/P (30:70) and heated at 220°C to form aluminum phosphate or berlinite as the binding phase of this coating in this research.

References
[1] Bell T 2018 Steel Applications What Is Steel Used For?
[2] Mbabazi J G, Sheer T J and Shandu R 2004 *Wear* **257** 612–24
[3] He L, Chen D and Shang S 2004 *J. Mater. Sci.* **39** 4887–92
[4] BMW steel team Alumina Ceramic
[5] Ścieszka S, Grzegorzek W and Żołnierz M 2010 *Sci. Probl. Mach. Oper. Maint.* **4** 37–54
[6] Troczynski T and Yang Q 2001 *US Pat.* **US 6,284**
[7] Chen D, He L and Shang S 2003 *Mater. Sci. Eng. A* **348** 29–35
[8] Wagh A S and Dicosola G 2014 All-Inorganic Ceramic Performance Coatings *Ceram. Ind. Mag.*
[9] Chao W, Wen-bin L, Ji-jiang L and Tao S 2007 *Chem. Adhes.* **02**
[10] Wagh A S and Jeong S Y 2003 *J. Am. Ceram. Soc* **86** 1838–44
[11] Colonetti E, Hobold E and Noni A De 2014 *Ceram. Int.* **40** 14431–8

Acknowledgments
The authors acknowledge to RISTEKDIKTI which funded this research as a part of *HIBAH PUSN* research grant. Authors also would say thanks to POLMAN Bandung an all colleagues for the support in all sector.

IOP Conf. Series: Materials Science and Engineering **541** (2019) 012027 doi:10.1088/1757-899X/541/1/012027

Improvement of erosion resistance of alumina-phosphate ceramic coating on mild steel by SiC addition

D Idamayanti[1], I L Nurhakim[1], B Bandanadjaja[1], W Purwadi[1], N Lilansa[2]
[1] Foundry Engineering Department, Politeknik Manufaktur Bandung, Jl. Kanayakan no. 21 Bandung, Indonesia
[2] Automation Engineering Department, Politeknik Manufaktur Bandung, Jl. Kanayakan no. 21 Bandung, Indonesia

E-mail : idamayanti79@gmail.com

Abstract. This research is focused on the application of the Al_2O_3-phosphate ceramic coating on mild steel surface to protect mild steel from erosion in coal dust environment. Erosion resistance of mild steel could be improved by overlay it with SiC in the Al_2O_3-phosphate ceramic coating. As a filler, Al_2O_3 was mixed with 20%, 40%, and 60% SiC by using aluminium phosphate as a binder and heated at 220 °C for 5 hours. X-ray diffraction testing was conducted to observe the phase of Al_2O_3-SiC phosphate ceramic coating. Meanwhile, surface morphology and adhesion characteristic of Al_2O_3-SiC phosphate ceramic coating were analyzed by scanning electron microscope. To analyze the erosion resistance quantitatively solid particle impingement test by applying gas jets at the right angle (90°) against a sample surface has been conducted. The results showed that Al_2O_3-SiC phosphate ceramic coating is strongly bound to the mild steel surface without the presence of any void. The higher the SiC content can increase the ceramic coating density and its erosion resistance. The SiC 60% produces four times higher erosion resistance than uncoated mild steel. The material characterization of Al_2O_3-SiC phosphate ceramic coating proves that SiC gives a significant impact on the enhancement of erosion resistance of the Al_2O_3-SiC phosphate ceramic coating.

1. Introduction

The purpose of this research is to improve the resistance of the inner cone of a coal pulverizer against the erosion. The inner cone as shown in figure 1 is part of the coal pulverizer. When observed it has erosion failure due to coal dust scouring. The inner cone itself which generally made of mild steel serves to direct the coal dust to the classifier, therefore, it should show excellent resistance against erosion caused by the flow of the coal dust. The mild steel is widely used as a structural material in some of the engineering applications due to its good machinability, high thermal conductivity, and superior mechanical strength [1].

Nevertheless, this material is highly affected by corrosion and wear and last consequently shorter service life. Applying ceramic coating on the surface of the mild steel is one of the most reliable erosion protection methods. Ceramic coatings are widely applied to protect some metal or cement machine components in power and refractory industries against chemical corrosion, abrasive wear, and high-temperature oxidation environments [2,3].

The ceramic coating consists mainly of oxide particles such as Al_2O_3 (mostly used), CaO, MgO,Cr_2O_3, etc., and binder. Several types of binder can be applied for ceramic coatings such as polymers, inorganic materials, and chemically bonded ceramic. Polymer binder includes polyvinyl alcohol, Teflon, and acrylic [4] in the most cases cannot withstand against high temperature above 200 °C, while the inorganic binders namely sodium silicate, and hydraulic cement show high strength at the room temperature but less resistant at elevated temperature [5]. Type of binder used plays important roles in binding ceramic coating adhering to the substrate. The binder does not only adhere between the coating to the metal surface but also binds the ceramic particles together so that the ceramic particles is not peeled off the coating [3]. An additional challenge applies to binders for making alumina ceramic coating that operates at high temperatures. According to previous studies by the authors, aluminium phosphate is preferred as a refractory and alumina ceramic coating binder due to its high strength, high-temperature stability, abrasion resistance [1,6,7], good adhesion, and reduced risk of cracking under high heating rate after curing [8].

Figure 1. The failure of inner cone coal mill pulverizer.

Also, the aluminium phosphate binder also presents other interesting properties, such as excellent mechanical properties, low-cost preparation, short setting time and low-temperature ceramic "sol-gel like" coating which can be applied below 250°C [6,8]. The phosphate binder is firstly synthesized by reacting aluminium hydroxide and phosphoric acid to form monoaluminium phosphate (MAP). Subsequently, the MAP is mixed with ceramic particles such as alumina, magnesia, and zirconia to result in a ceramic slurry system which is known as chemically bonded phosphate ceramics (CBPC) [9]. Many researchers have observed CBPC and applied on metals to enhance corrosion and abrasion resistance of metals [1–3,6,10]. They modified a kind of ceramic materials to improve the mechanical properties of the phosphate bonded Al_2O_3 ceramic coating. He et al [5] has investigated the effect of SiC in the Al_2O_3-phosphate ceramic coating. It showed the improvement of wear rate of phosphate bonded Al_2O_3-SiC ceramic coating due to the increasing of SiC content. This phenomenon is very logical due to the high hardness of SiC particles compared to Al_2O_3 particles [11]. This research has been conducted to investigate the effect of SiC addition in phosphate bonded Al_2O_3 ceramic coatings as abrasive ceramic particles to enhance the erosion resistance of coated mild steel as representative materials for pulverizer inner cone. The SiC particles were mixed with Al_2O_3 to form ceramic slurry using monoaluminium phosphate as a binder. The performance of phosphate bonded Al_2O_3-SiC ceramic coating on the mild steel was tested by alumina particle impingement using gas jets at the right angle (90°) against a specimen surface and was then verified by surface morphology using a scanning electron microscope. The gas jet containing alumina particles is applied to simulate the real condition. The results can optionally be taken for increasing the life of the inner cone pulverizer.

2. Experimental Work

2.1. Raw materials and specimen

The materials which are used for phosphate bonded Al_2O_3-SiC ceramic coating were alumina (Al_2O_3) as a ceramic particle, SiC as a filler of ceramic system, and $Al(OH)_3$ – H_3PO_4 as raw materials of the monoaluminium phosphate (MAP) binder. Alumina (Al_2O_3) size was 50 – 60 µm with round shapes, obtained as technical grade material. The particle size of SiC was average of 126,36 µm with irregular shapes. Both Al_2O_3 and SiC have different particles morphology as shown in figure 2. Phosphoric acid (85%), aluminium hydroxide ($Al(OH)_3$), methanol and aquadest were used as technical grade material without further purification. The specimen used in the experiment was mild steel with 5 mm thickness. The chemical composition is listed in table 1. The specimen was cut to a size of 25 mm x 25 mm for implementing the ceramic coating. Furthermore, the specimen was cleaned by methanol to remove scale, corrosion, and other contaminants.

Table 1. Chemical composition of specimen.

C (%)	Si (%)	S (%)	P (%)	Mn (%)	Ni (%)	Cr (%)	Mo (%)	Cu (%)	Fe (%)
0,05	0,01	0,006	0,01	0,30	0,005	0,01	0,003	0,01	99,54

(a) (b)

Figure 2. The particles morphology of Al_2O_3 (a) and SiC (b).

2.2. Synthesis of monoaluminium phosphate binder (MAP)

Monoaluminium phosphate (MAP) as a binder was synthesized by reaction of phosphoric acid (H_3PO_4) and aluminium oxide [5]. The reaction between phosphoric acid and aluminium hydroxide was carried out in deionized water at an Al/P (1/3) molar ratio. The mixture was stirred for 15 minutes at 100 °C to form MAP. Subsequently, the reaction product was dried at ambient temperature and was characterized by X-Ray Diffraction (Rigaku SmartLab X-Ray Diffractometer). The design of the MAP molar ratio refers to the stoichiometry reaction that follows the equation (1).

$$3H_3PO_4 + Al(OH)_3 \rightarrow Al(H_2PO_4)_3 + 3H_2O \quad (1)$$

The density of the MAP was measured using pycnometer to control its viscosity for coating application.

2.3. The formation of phosphate bonded Al_2O_3-SiC ceramic coating on mild steel

Recently, phosphate bonded Al_2O_3-SiC ceramic coating is widely used because the process can be carried out at room temperature, easy to apply and low-cost process [10]. It was applied to the prepared mild steel using the brushing technique. To produce the Al_2O_3-SiC ceramic suspension, the MAP binder was slowly stirred with Al_2O_3-SiC particles at a ratio of 1:1. The various amount of SiC in the Al_2O_3-SiC system studied were 20%, 40%, and 60%, respectively. He et al [5] and Chen et al [3] used SiC as an abrasive material to improve the mechanical properties of the Al_2O_3 phosphate ceramic coating. Phosphate bonded Al_2O_3-SiC that coated on the mild steel was hardened by following heat treatment processes:

Phosphate bonded Al_2O_3-SiC ceramic coating on the mild steel was dried at ambient temperature to evaporate water content for two hours. The heat treatment of phosphate bonded Al_2O_3-SiC ceramic coating was continued at 80 °C for 2 hours, then was held at 220 °C for 5 hours to obtain the berlinite phase.

2.4. Characterization of phosphate bonded Al_2O_3-SiC ceramic coating

This research involved some instruments for the characterization process. Optical Emission Spectrometer (Spektris ARL 3460) was used to analyze the chemical composition of mild steel. Macroscopic observation of phosphate bonded Al_2O_3-SiC ceramic coating was done by using the optical microscope (Olympus). Scanning Electron Microscope (Hitachi SU3500) was used to observe the surface morphology of the ceramic coating and its binding strength on the mild steel and the measuring of alumina or SiC particles size. Erosion characteristic of the mild steel and ceramic coatings were observed using SEM magnification at 200x. X-Ray Diffraction (Rigaku SmartLab Diffractometer) were carried out to observed phase identification of the binder and ceramic coating. Diamond cutting saw was used to a cross-sectional cutting of the ceramic coating. Erosion tester (Ducom triboinovator) was used to examine the erosion resistance of the ceramic coating according to ASTM G76-02 (standard test method for conducting erosion test by solid particle impingement using gas jets). The velocity of the testing is 30 m/s up to 70 m/s for 10 minutes with an impact angle at 90°. The solid particles selected were alumina particles close to the real conditions in coal dust environment. The erosion resistance according to erosion rate (mg/Kg) and was measured as weight loss of the ceramic coating per weight of the erodent material (alumina).

3. Result and Analysis

3.1. Characterization of synthesized monoaluminium phosphate binder (MAP)

MAP binder has been successfully fabricated with its density of 1.71 g/cm^3 as much as resulted by He et al [5] for ceramic coating application. According to XRD patterns in figure 3, there is a mixture of several phases, namely aluminium hydrogen phosphate (known as a MAP) and aluminium hydrogen phosphate hydrate (known as MAP monohydrate). The intensity of the MAP as a function of its content is almost proportional to the intensity of MAP monohydrate. Chen et al observed initial dehydration of MAP trihydrate transforms to MAP as a dominant phase and MAP monohydrate as a secondary phase when the temperature is slightly elevated to 105 °C [3,5]. MAP binder plays an important role in binding between ceramic particles and also to bind the ceramic suspension to the mild steel surface. Chen et al [3] stated that MAP is the most effective bonding phase which fully formed when the temperature reaches 200 – 400 °C. A previous study concluded not only the concentration of MAP binder but also the reactivity and solubility of the selected oxide must be controlled during the processing steps to attain an improved bonding of the fine and coarse components [8].

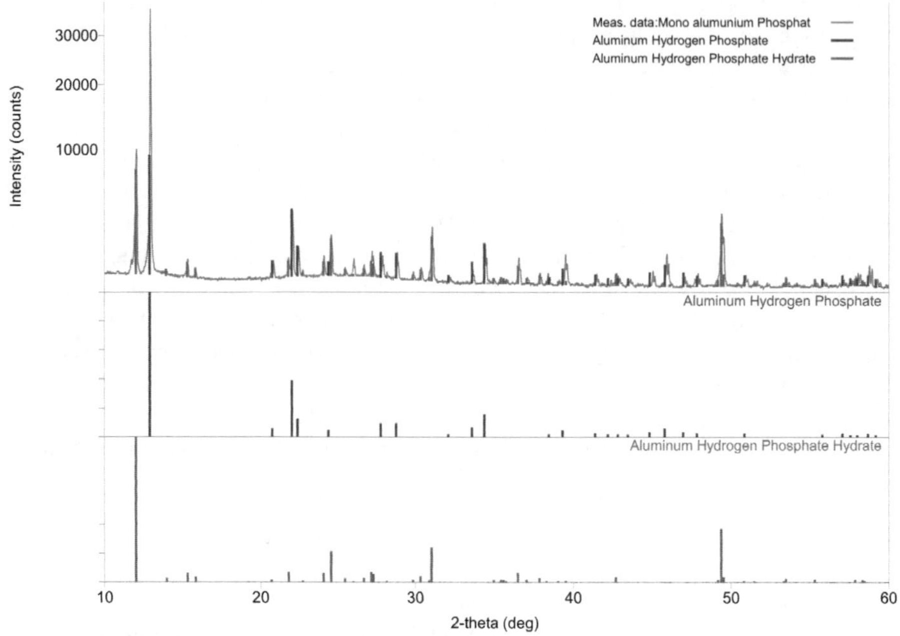

Figure 3. XRD pattern of the MAP binder consisting of monoaluminium phosphate (MAP) and monoaluminium phosphate hydrate (MAP hydrate).

3.2 Characterization of phosphate bonded Al_2O_3-SiC ceramic coating

Phosphate bonded alumina ceramic coatings are formed by slowly stirring alumina into a MAP binder solution. Alumina is a kind of amphoteric oxide that may slowly dissolve in aqueous MAP binder being weakly basic cation with small radii for the formation of cold-setting ceramics known as a kind of chemically bonded phosphate ceramics (CBPCs). The kinetics of formation CBPCs are the dissolution of alumina, the formation of aquosols by hydrolyzes, acid-base reaction and condensation into a CBPC, percolation and gel formation, saturation then the formation of crystallization into a monolithic solid of CBPC. Thus, unreacted alumina is available to form a good stable ceramic [12]. In this work, the addition of SiC in the Al_2O_3 ceramic system acts as a filler which enhances the mechanical properties of CBPC such as wear resistance, erosion resistance, and corrosion resistance [3,5]. The XRD patterns in Figure 4 show the composition of phosphate bonded Al_2O_3-SiC ceramic coating. The XRD peaks reveal the phases of residual aluminium oxide (Al_2O_3), berlinite, and cristobalite. The reaction of alumina particles and MAP binder produced aluminium phosphate ($AlPO_4$). As explained by Morris et al [13], $AlPO_4$ comes up as berlinite, tridymite cristobalite forms. Fully formed berlinite is obtained by heating precipitated, amorphous, aluminium phosphate to 450 °C with lithium fluoride, but berlinite inversion, take place at lower temperatures (150 °C, [9]). Meanwhile, the cristobalite form can be obtained in the presence of sodium chloride as promoters of crystallization or temperature stability relationship of berlinite phase [13]. As shown in figure 4, $AlPO_4$ as cristobalite form is a dominant phase. Bandyopadhyay [14] stated that $AlPO_4$ as cristobalite might be formed when heated to 150 – 200°C.

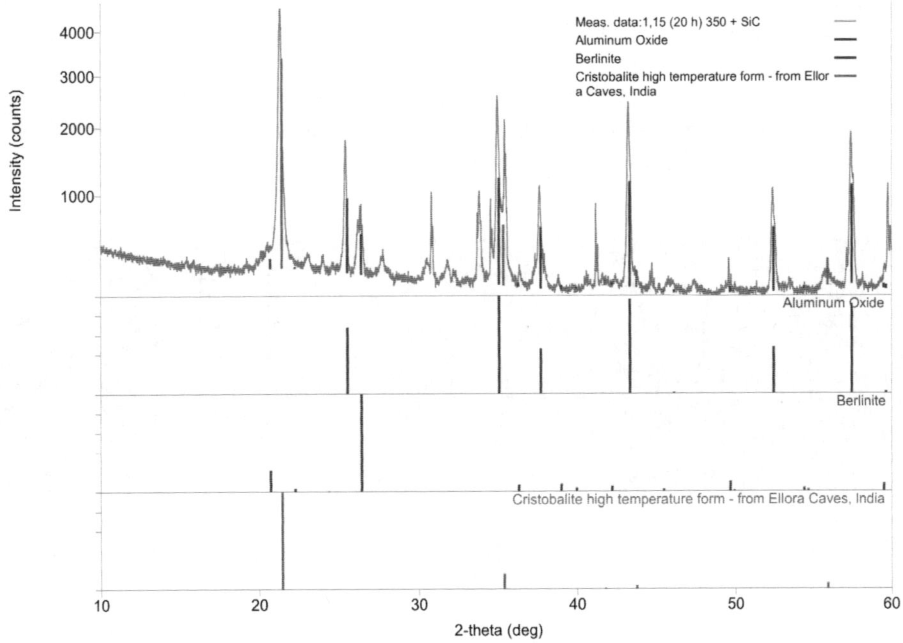

Figure 4. XRD pattern of phosphate bonded Al_2O_3-SiC ceramic coating on the mild steel.

Berlinite ($AlPO_4$) is the target phase compound of phosphate bonded Al_2O_3-SiC on the mild steel which the kinetic transformation of berlinite is similar to the one in the conventional sol-gel process [12]. According to wagh et al [7] once aluminium oxide dissolves, it will form intermediate phase of $AlH_3(PO4)_2.H_2O$ which will react with remaining oxide (Al_2O_3) on the surface of their particle to form berlinite and bind them. The reaction may be written as the equation (2).

$$Al_2O_3 + 2AlH_3(PO_4)_2.H_2O \rightarrow 4AlPO_4 + 4H_2O \quad (2)$$

Wagh et al [15] predict that a very small amount of alumina must be converted to form the bonding phase. In the Al_2O_3-berlinite system, Al_2O_3 particles disperse due to alumina sol-cluster and the Al_2O_3 surface being highly electrostatically repulsive [16] and the product is mostly alumina with a thin coating of berlinite on the alumina [15]. In the case of the SiC-Al_2O_3 system, SiC particles absorb alumina sol-gel, resulting in a steric repulsive force between particles [16]. This phenomenon is essential to get complete dispersion of ceramic particle and preserve the final quality of the products. The solubility of alumina is very low (ΔH° -837.8 kJ/mol) so that dissolution of alumina and its hydrate phases may be enhanced by elevating the temperature of the ceramic slurry. The heat treatment is also required for both drying and transforming crystalline and amorphous phases of aluminium phosphate [15].

Table 2. Parameters related to phosphate ceramic formulation for most common oxides [12].

Oxides	Reactions	ΔH (kJ/mol)
Al_2O_3	$2Al_2O_3 + H_3PO_4 \rightarrow 2AlPO_4 + 3H_2O$	-45.75
FeO	$FeO + H_3PO_4 \rightarrow FeHPO_43H_2O$	-136.7
Fe_2O_3	$2Fe_2O_3 + 3H_3PO_4 \rightarrow 2Fe(H_2PO_4)_3 + 3H_2O$	-95.1

The ability to chemically bond Al_2O_3 and other ceramic particles like SiC on mild steel without complicated mild steel preparation are technologically important. Table 2 shows FeO or Fe_2O_3 as

IOP Conf. Series: Materials Science and Engineering **541** (2019) 012027 doi:10.1088/1757-899X/541/1/012027

corrosion products which commonly found on mild steel may easily react with MAP binder due to highly exothermic enthalpy value. Mechanical adhesion due to surface roughness is also possible [13], and this can be increased by elevated temperature.

Figure 5 illustrates that the Al_2O_3-SiC phosphate ceramic coating strongly adheres to the mild steel. Thus, a good interface bonding between the mild steel and ceramic coating is created. Along in the interface line, there is no void or other discontinuities. The thickness of the ceramic coating results in 352-460 μm.

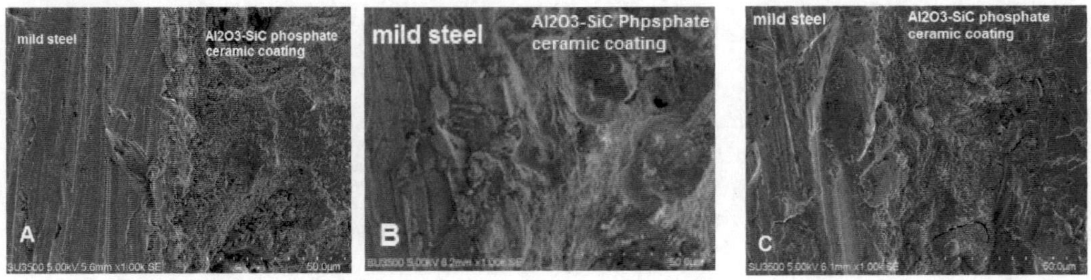

Figure 5. SEM Micrograph of the cross-section view of Al_2O_3-SiC phosphate ceramic coating on steel. (a) 20% SiC, (b) 40% SiC, (c) 60% SiC.

3.3 Erosion resistance of phosphate bonded Al_2O_3-SiC ceramic coating

3.3.1 Effect ratio of SiC in the phosphate bonded Al_2O_3-SiC ceramic coating

The filler and its proportion in the compact affect both strength and thermal expansion [17]. Idamayanti et al [18] applied phosphate bonded Al_2O_3 ceramic coatings (without SiC) on mild steel, then the ceramic coating shows the erosion rate of 5mg/Kg. Unfortunately, the Al_2O_3 phosphate ceramic coatings absorb the humidity in ambient temperature and also disintegrated when placed in water [7,17,18]. Chen et al [3] fabricated Al_2O_3-SiC ceramic coating on A_3 steel which its wear durability is about two times that of uncoated A_3 steel. As mentioned above, SiC particle act both filler and ceramic slurry stabilizer due its steric repulsive force. Considerable porosity was found in the phosphate bonded Al_2O_3-20%SiC ceramic coating (as seen in Figure 7A) caused by the releasing of hydrates when berlinite formed [7] and also by rapid evaporation of excess water in the slurry [15]. 20 % SiC in the ceramic system does not affect significantly to make dense ceramic coating. Also, the increase of SiC content 40 % to 60% generates compact and homogenous ceramic coating (figure 8A and 9A). In this work, an erosion testing was conducted to measure the erosion resistance phosphate bonded Al_2O_3-SiC ceramic coating influenced by SiC content of 20%, 40%, and 40%, respectively, when compared to that of uncoated mild steel. Figure 6 shows that the erosion rate of mild steel is four times higher (20 mg/Kg) than the Al_2O_3-60% SiC phosphate ceramic coating (5 mg/kg).

These results show a relation between the hardness of material and erosion behavior (brittle or ductile) [19,20]. Mohs hardness scale of each material used is alumina of 9, SiC of 9-10, and berlinite of 6-7 [21]. According to the hardness levels, berlinite will erode first due to its the lowest hardness. When erodent materials impact both berlinite and Al_2O_3-SiC ceramic coatings, their resistance depends on the most prominent phase in the impacted area.

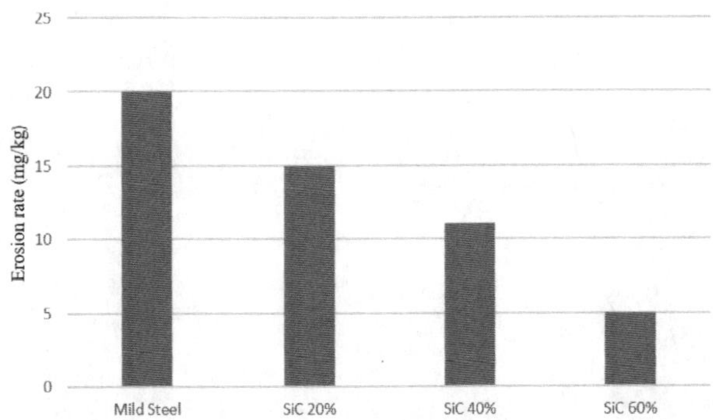

Figure 6. The effect of SiC addition on the erosion rate of phosphate bonded Al_2O_3-SiC ceramic coating.

3.3.2 Erosion behavior of the Al_2O_3-SiC phosphate ceramic coatings

The erosion behavior of materials was affected by its bonding strength to face the solid particle impingement then accelerated by the impact angle. According to Hutchings et al [22], the erosion rate of ductile materials will be high at low impact angle (25-30°), while the erosion rate of brittle material will be high at high impact angle (90°). The phosphate bonded Al_2O_3-SiC ceramic coating as a kind of brittle material was tested at the high impact angle 90° to ensure its performance in severe erosion [19].

Figure 7-9 (B) shows the morphology after erosion testing. The surface morphology of phosphate bonded Al_2O_3-20%SiC ceramic coating releases a lot of debris that comes from berlinite peel off, then a lot of cavities are formed. The result is comparable to the morphology of phosphate bonded Al_2O_3-SiC ceramic coating developed by Chen et al [3]. Almost certainly, well-covered berlinite of Al_2O_3-60%SiC ceramic coating (figure 9B) is the most resistant toward particles impingement.

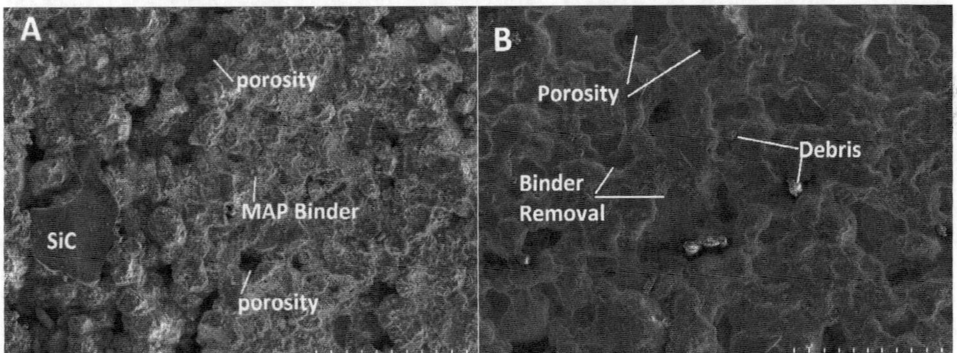

Figure 7. SEM micrograph of the morphology of phosphate bonded Al_2O_3-20%SiC ceramic coating, before erosion testing (A), after erosion testing (B).

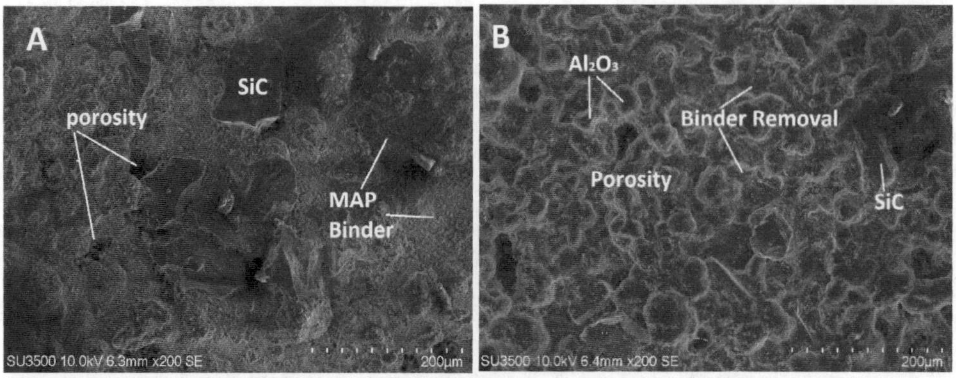

Figure 8. SEM micrograph of the morphology of phosphate bonded Al_2O_3-SiC ceramic coating before erosion testing (A), after erosion testing (B).

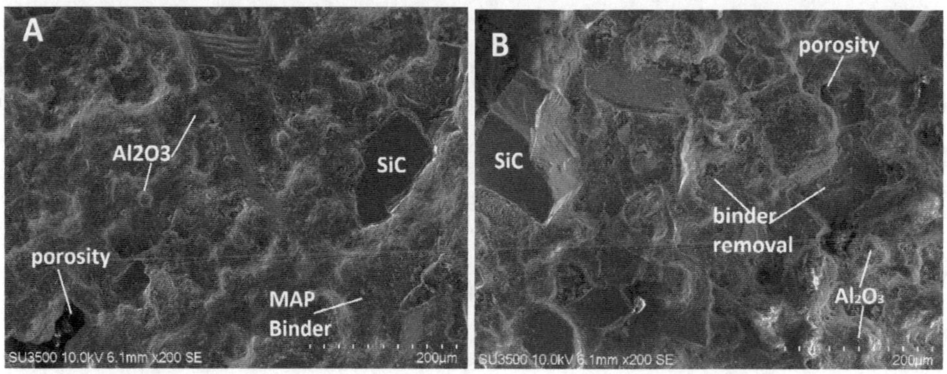

Figure 9. SEM micrograph of the morphology of phosphate bonded Al_2O_3-60%Si ceramic coating, before erosion testing (A), after erosion testing (B).

The higher the SiC content, the higher the particle density, the lower the porosity, and the lower debris releasing after erosion testing. The addition of 60% SiC in the Al_2O_3-SiC mixture showed the best performance against erosion.

4. Conclusion

A study of the SiC addition concerning the phosphate bonded Al_2O_3 ceramics system was carried out. It was found that SiC improves the erosion rate of phosphate bonded Al_2O_3-SiC ceramic coating. As high SiC was added in the ceramic coating, the erosion resistance of phosphate bonded Al_2O_3-SiC ceramic coatings increase. The addition of 60% SiC in the Al_2O_3-SiC ceramic coating produces four times erosion resistance if it compares respectively with uncoated mild steel. The Al_2O_3-SiC ceramic coating on mild steel results in good adhesion and produces 352-460 μm of the thickness.

References

[1] Ruhi G, Modi O P, Sinha A S K and Singh I B 2008 *Corros. Sci.* **50** 639–49
[2] Lv F, Wang L and Wang E 2013 *Adv. Mater. Res.* **821–822** 1256–60
[3] Chen D, He L and Shang S 2003 *Mater. Sci. Eng. A* **348** 29–35
[4] Chung D D L 2003 *J. Mater. Sci.* **38** 2785–91
[5] He L, Chen D and Shang S 2004 *J. Mater. Sci.* **39** 4887–92
[6] Kamo L and Saad P 2005 *Proceedings of WTC2005 World Tribology Congress III* (ASME) 3–4
[7] Wagh A S 2004 *Chemically Bonded Phosphate Ceramics* (Elsevier Ltd)
[8] Luz A P, Oliveira G R, Gomes D T and Pandolfelli V C 2016 *Ceram. Int.* **42** 8331–7
[9] Wagh A S 2013 *ISRN Ceram.* **2013**

[10] Devapal D, Sebastain T V, Prabhakaran P V and Packirisamy S 2012 *Mater. Sci. Forum* **710** 786–91
[11] Anon Mohs' Hardness (Typical) of Abrasives
[12] Wagh A S and Jeong S Y 2003 *J. Am. Ceram. Soc* **86** 1838–44
[13] Morris J H, Perkins P G, Rose A E A and Smith W E 1977 *Chem.Soc.Rev.* **2** 173-195
[14] Bandyopadhyay K 1985 *Phosphate Bonded Silion Carbide Refractories* (India: Indian Institute of Technology Kanpur)
[15] Wagh A S, Grover S and Jeong S Y 2003 *J. Am. Ceram. Soc* **86** 1845–9
[16] Moorlag C 2000 *Chemically Bonded Sol-Gel Ceramic : A Study of Alumina-Phosphate Reaction Products* (Canada: University of British Columbia)
[17] Morris J H, Perkins P G, Rose A E A and Smith W E 1977 *Chem. Soc. Rev.* **6** 173
[18] Idamayanti D, Ginanjar D, Bandanadjaja B, Purwadi W and Lilansa N 2018 Erosion behaviour of alumina ceramic coating on mild steel by the modified composition of phosphate binder *International Seminar Material and Metallurgy* (Tangerang)
[19] Ruff A W and Wiederhorn S M 1979 *Treatise Mater. Sci. Technol.* **16** 69–126
[20] Levy A V and Chik P 1983 *Wear* **89** 151–62
[21] Byrappa K, Prasad J S and Srikantaswamy S 1986 *J. Cryst. Growth* **79** 232–5
[22] Hutchings I M 1987 Wear by particulates *Chem. Eng. Sci.* **42** 869–78

Acknowledgments
All authors acknowledge to RISTEKDIKTI which funded this research as a part of HIBAH PUSN research. We also acknowledge to POLMAN Bandung for providing the research facilities

IOP Conf. Series: Materials Science and Engineering **541** (2019) 012028 doi:10.1088/1757-899X/541/1/012028

Microstructure and oxidation resistance of spark plasma sintered Yttria-Zirconia ODS steels

N A Rabbani[1], E A Basuki[1], T Sudiro[2], A Afandi[2]
[1]Metallurgical Engineering Department, Faculty of Mining and Petroleum Engineering-Bandung Institute of Technology, Jl. Ganesa 10, Bandung 40132, Indonesia
[2]Research Center for Physics, Indonesian Institute of Sciences, PUSPIPTEK Area, South Tangerang, Banten 15310, Indonesia

E-mail: rabbaninaufalazmi2@students.itb.ac.id

Abstract. In this experiment, the effect of Y_2O_3 and ZrO_2 additions on the microstructure and cyclic oxidation resistance of 17-Cr ODS steel containing Al consolidated using spark plasma sintering method were studied. In addition, density, porosity and hardness were also observed. Four compositions of ferritic steels as Fe-17Cr-3.5Al-Ti-Y ('FP'), Fe-17Cr-3.5Al-Ti-Y-0.4Y_2O_3 ('FY'), Fe-17Cr-3.5Al-Ti-Y-0.4ZrO_2 ('FZ'), and Fe-17Cr-3.5Al-Ti-Y-0.2Y_2O_3-0.2ZrO_2 ('FYZ') were prepared by mechanical alloying and followed by sintering at 1000 °C. Microstructure characteristics were observed for different aspect, viz. elemental distribution, composition, and phase identification. Cyclic oxidation tests were carried out at 800 °C for 8 cycles. The results show that ZrO_2 addition suppresses the formation of Y-Al-O clustering and promotes the formation of Y-Zr-O clustering and the enhancement of oxidation resistance of the alloy. However, in some condition it also has a deleterious effect on crack resistance. In contrast, the addition of Y_2O_3 decreases the oxidation resistance but increases the hardness.

1. Introduction

Ferritic/Martensitic Oxide Dispersion Strengthened (ODS) steel as material components of nuclear cladding has been developed for application at high corrosion and neutron irradiation environments of advanced nuclear reactor [1-4]. High temperature strength and resistance to swelling, thermal embrittlement, fatigue crack growth, and creep of ferritic/martensitic steel is superior compared to austenitic steel [5-8]. Moreover, the addition of oxide particle as dispersed materials (ferritic ODS steel) leads to increase in strength and resistance to thermally-induced creep [2]. Accordingly, the operation temperature of ferritic steel can be increased higher than 600°C [2]. However, there are several aspects affecting the mechanical properties, physical properties, and corrosion-oxidation resistance of ferritic ODS steel, such as material composition, consolidation method and processing parameter

The most common method used for consolidation of ODS alloy is extrusion and hot isostatic pressure (HIP). Comparatively, spark plasma sintering (SPS) is a relatively new consolidation method which has been developed. It utilizes pulsed direct electrical current to perform high-speed consolidation of powders. The properties of consolidated samples are improved because of intrinsic properties of nano-powders maintained in their fully dense products by a very high heating and cooling rates during SPS [9]. Although SPS is widely studied, the effects of Zr (and its oxide form) on ODS steel consolidation are still rare.

A sufficient amount of aluminium addition into ferritic ODS steel is known to give advantage in increasing corrosion resistance. It was found that the addition of 4 wt.% Al into 16 wt.% Cr ODS steel increases corrosion resistance through the formation of thin alumina film on the surface of the steel. The exact effect does not show up in 19 wt. % Cr ODS steel with the addition of 4 wt.% Al because of a rather dense chromia film formed on the surface [10-12].

In contrast, the addition of aluminium is also capable of giving some deleterious effects into ODS steel. Those are oxide particle coarsening by the formation of Y-Al-O phase and the reduction of high temperature strength [12-13]. Zr addition into the alloy is one of the methods that can be used to prevent aforesaid deleterious effects. The first principle calculation results indicate that the binding energy of Y-Zr-O cluster is higher than Y-Al-O cluster in Fe matrix. Consequently, the formation of finer Y-Zr-O nanoparticle is more favourable and more stable than coarse Y-Al-O, resulting the increase of the strength of the steel [14]. However, there is still a lack of research concerning the effect of Zr addition in its oxide form (ZrO_2) or in its stable form (Yttria-stabilized Zirconia) into ODS steel as dispersant on the oxidation resistance of the alloy.

In order to clarify the effects of Al-containing ferritic ODS steel with Y_2O_3 and/or ZrO_2 additions and SPS consolidation method, several experiments were conducted in this study. This report presents and discusses the effects of Y_2O_3 and/or ZrO_2 additions into 17Cr-3.5Al-ODS steel on microstructure, physical properties (density and porosity), mechanical property (hardness), and cyclic oxidation resistance.

2. Experimental Procedures

The alloys with nominal composition as 17Cr-3.5Al-Ti-Y (denoted as FP), 17Cr-3.5Al-Ti-Y-0.4Y_2O_3 (denoted as FY), 17Cr-3.5Al-Ti-Y-0.4ZrO_2 (denoted as FZ), and 17Cr-3.5Al-Ti-Y-0.2Y_2O_3-0.4ZrO_2 (denoted as FYZ) were prepared from high purity powders of Fe17Cr3.5AlTiY, Y_2O_3, and ZrO_2. Each composition was mechanical alloyed (MA) using a high-energy planetary ball mill (SFM-1 Desk-Top). MA process was performed for 6 h with a rotation speed of 280 rpm, a ball to mass ratio of 10:1, and hexane as MA solution.

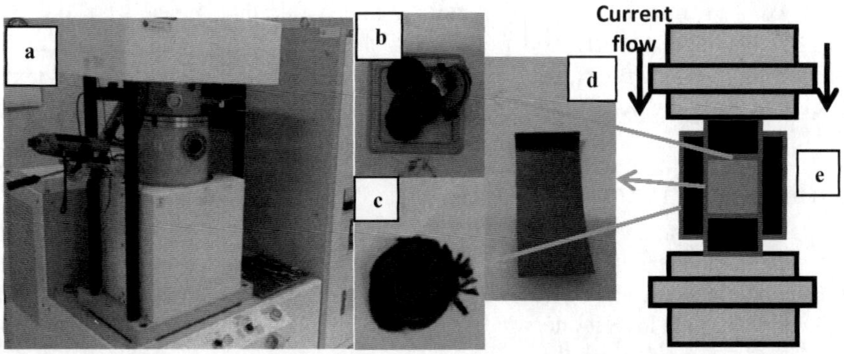

Figure 1. Sintering configuration: (a) Dr. Sinter 515S, (b) carbon paper for top and bottom part, (c) graphite dies and punches, (d) carbon paper between powder and dies, (e) carbon cover and container configuration.

The milled powders were consolidated into samples with a diameter of 15 mm using Dr. Sinter 515S Spark Plasma Sintering System at a constant axial pressure of 40 MPa and vacuum of 80 Pa. Carbon covers were used to increase the conductivity of samples (see Figure 1). Temperature was increased from a room temperature to 1000°C with a constant rate of 100°C/minute.

Density of sample was measured using Archimedes method. Porosity was calculated from optical microscopy data by ImageJ software analysis [15]. Micro-hardness (HV) was measured by a Vickers hardness tester Leco M-400-HI with 0.1 kgf load. High temperature cyclic oxidation test was performed at 800°C for 8 cycles. Microstructure observation was carried out using Hitachi SU3500 scanning electron microscope (SEM) equipped with energy dispersive X-ray spectrometer (EDX Horiba). Phase identification was determined using Rigaku SmartLab X-ray diffractometer. Etching was carried out only for samples before oxidation using etchant solution standard for ferritic grade steel (aqua regia + acetic acid) [16]. The etchant composition was 5 mL HNO_3, 5 mL $C_2H_4O_2$, and 15 mL HCl.

3. Results and Discussion

3.1. Density and Porosity

The results show that samples with higher ZrO_2 content tend to yield lower density. This can be observed in Figure 2 (a). The density of FZ and FYZ are lower, 6.447 and 6.513 g/cm^3, respectively compared to FP and FY, 6.619 and 6.637 g/cm^3, respectively. The density of sample from highest to lowest is FY, FP, FYZ, FZ.

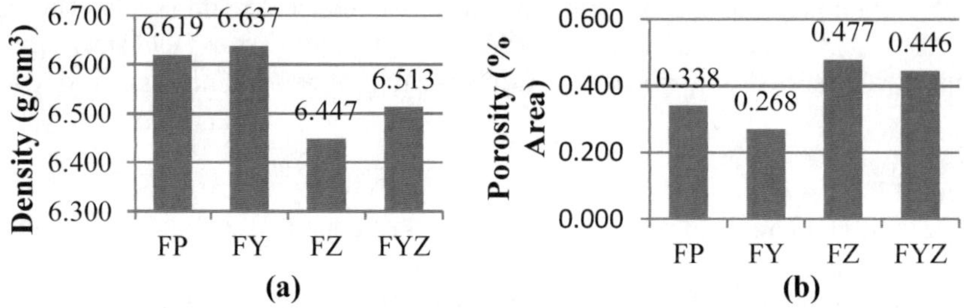

Figure 2. (a) Density and (b) porosity of FP, FY, FZ and FYZ.

It is well known that the greater porosity of the alloy is associated with lower density. This is in good agreement with the results of present study. This can be observed in Figure 2 (b). FZ and FYZ porosity are 0.864 and 0.684 %, respectively, which have greater value than FP and FY 0.476 and 0.400 %, respectively. Even though the density of ZrO_2 is higher than Y_2O_3 (5.68 compared to 5.01 g/cm^3), samples with the addition of ZrO_2 show a slightly lower value compared to samples with the addition of Y_2O_3. This suggest that the microstructure of sintered alloy seems to affect the density and porosity of the ODS steel.

3.2. Hardness

Hardness test results as presented in Figure 3 show a slightly different pattern compared to density. The highest hardness of the sample is FY (246.9 HV) followed by FYZ (206.9 HV), FZ (190.13 HV), and then FP (183.43 HV) as the lowest one.

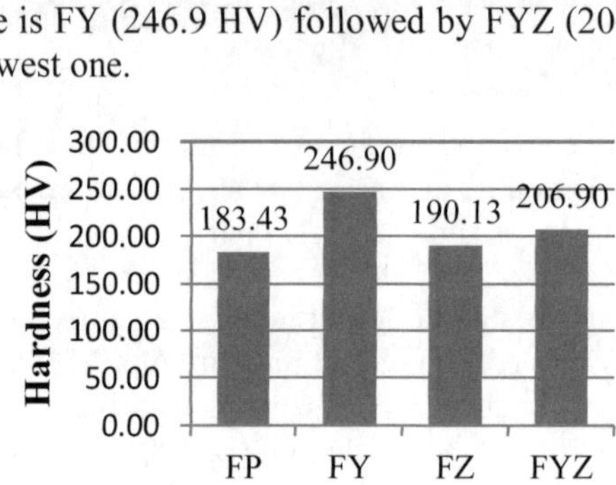

Figure 3. Hardness of FP, FY, FZ and FYZ sintered at 1000°C.

Although ZrO_2 hardness is higher compared to that of Y_2O_3, in accordance with the results as presented above, higher hardness of the samples seems to associate with its higher density and lower porosity. Since the size of starting material of ZrO_2 (10-15 μm) used in this study is higher than that of Y_2O_3 (5-10 μm) and also the effectiveness of particles size reduction may be lower due to high hardness of ZrO_2 (12.3 GPa) compared to Y_2O_3 (6 GPa) and steel milling ball (5 GPa), for ZrO_2 it can cause the distance between dispersed ZrO_2 particles becoming larger. Hence, it can reduce the strengthening effect.

3.3. *Before Oxidation*

3.3.1. *Phase identification*

The results of the XRD phase analysis after milling and after sintering (pre-oxidation) can be seen in Figure 4(a) and (b), respectively. According to the results, the phase detected for both conditions is α-Fe. This result is appropriate when compared with the Fe-Cr binary phase diagram. It indicates the homogeneity of Fe and Cr distribution. The similar results between these two conditions also indicate that a short sintering stage doesn't lead any other phase formation.

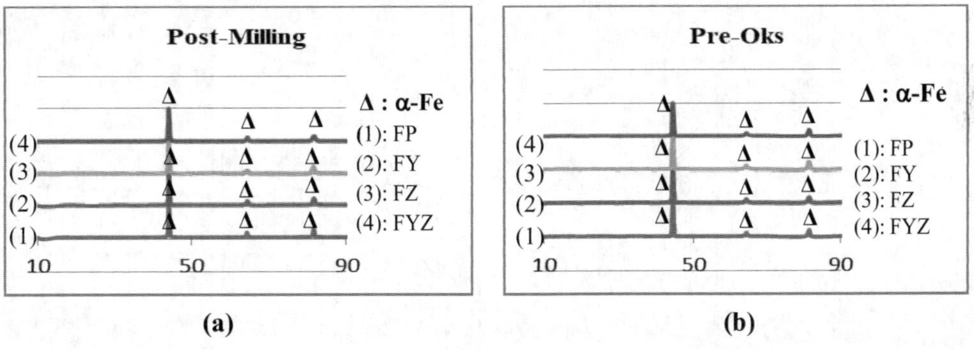

Figure 4. XRD patterns of FP, FY, FZ and FYZ: (a) after milling, and (b) after sintering.

3.3.2. *Microstructure*

Figure 5 shows the optical microstructure images of FP, FY, FZ and FYZ after sintering at 1000°C. It can be seen that intergranular crack are observed in alloys with higher ZrO_2 content. These intergranular cracks contribute significantly to the increase in porosity and decrease in density for FZ and FYZ. While there is no intergranular crack observed in samples without ZrO_2 addition, i.e. FP and FY.

The presence of intergranular crack in alloys with the addition of ZrO_2 can be attributed to the accumulation of stress due to the change of ZrO_2 crystalline structure from monoclinic to tetragonal during sintering. The change in crystal structure leads to an increase in crystal volume by 9%[17]. This crystal structure transformation occurs in ZrO_2 which has not been completely stabilized by Y_2O_3 at sintering stage (at 18 mol.% of $YO_{1.5}$ or 9.47 mol.% Y_2O_3[18]). The stabilization will be reached after subsequent heat treatment.

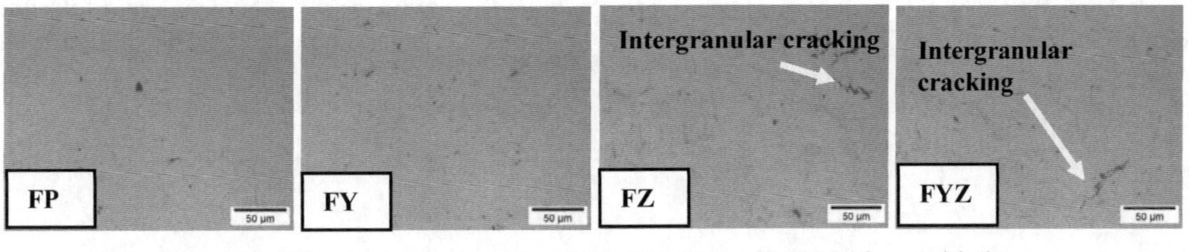

Figure 5. Microstructure images of FP, FY, FZ and FYZ before oxidation.

3.3.3. Elemental distribution

Figure 6 shows the distribution of Fe and Cr elements in the etched samples of FP, FY, FZ and FYZ after sintering. There is no concentration/cluster of Fe or Cr in all samples. Fe and Cr elements are uniformly distributed in the alloy matrix. This indicates that Fe maintains its ratio with Cr and vice versa. The homogeneous distribution of Fe and Cr also proves that 17 wt. % Cr can be maintained in all sample. Hence, α-Fe is the only possible phase formed in the samples as XRD phase analysis results tell.

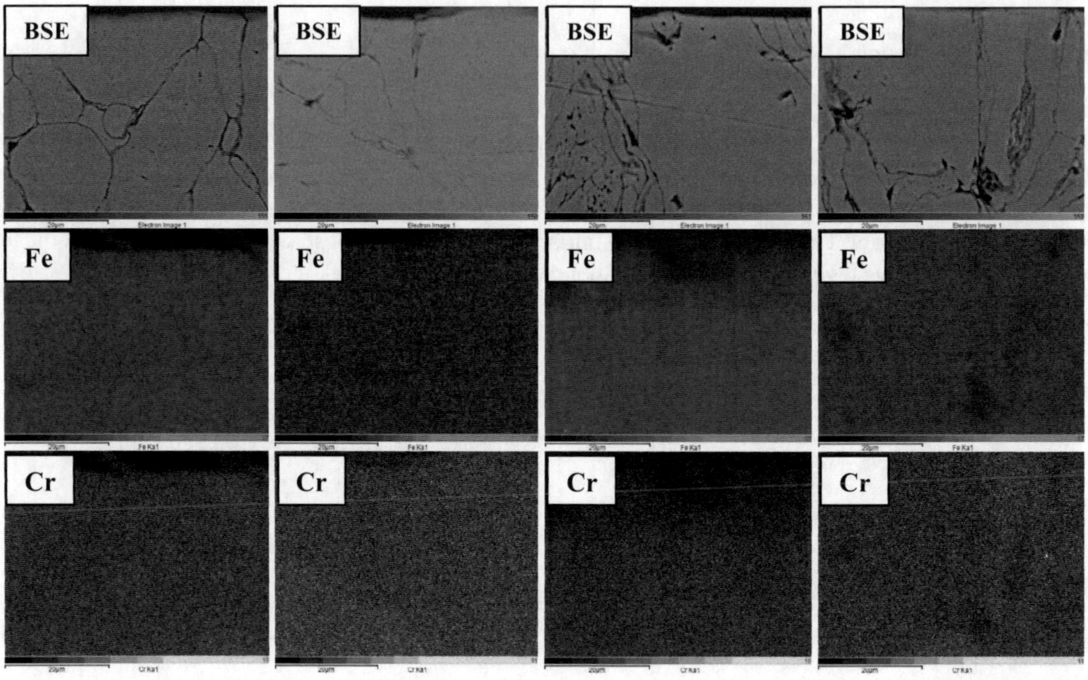

(a) (b) (c) (d)

Figure 6. Fe and Cr distribution in sample: (a) FP, (b) FY, (c) FZ, and (d) FYZ after sintering.

3.4. Oxidation kinetic

The results of cyclic oxidation test show that the initial stage of oxidation contributes significantly to the rate of weight gain. The rate of weight gain from the lowest to the highest is FZ, FYZ, FP, FY, respectively. By looking at Figure 7, it can be seen that the rate of weight gain follows a logarithmic pattern. The logarithmic pattern of weight gain rate can generally be caused by two things, viz. the presence of surface reactions and the presence of rapid ionic transport paths [19].

The surface reaction occurs at temperatures below 400 °C. In the case of oxidation at higher temperatures (such as 800°C), alloys usually undergo a rapid surface reaction while in the initial heating stage. The surface reaction occurs when a high electric field is formed due to changes in electron concentration. This results in an increase in the rate of ion migration resulting in the formation of the oxide layer [20]. The rate of ion migration decreases to a very low value with the decrease of electron movement from the metal through the oxide to the adsorbed oxygen atom due to the thickening of the oxide layer [21].

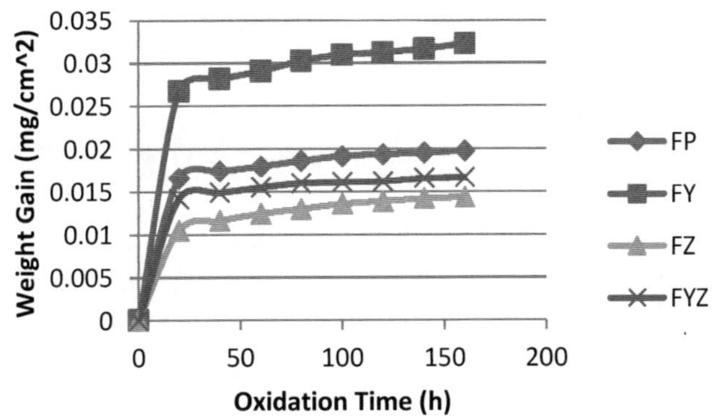

Figure 7. Cyclic oxidation kinetic curve of FP, FY, FZ and FYZ after exposure at 800°C for 8 cycles.

In addition to surface reactions, the presence of rapid ionic transport paths (along grain boundary, dislocation pipes, surface of the pores, residual stress) can also affect the difference in the rate of weight gain when the oxidation step is carried out. The presence of rapid ionic transport paths scattered in alloys can be the path of oxidation movement. These pathways increase ionic flux and prevent the uniform oxidation. This path will be closed along with recrystallization, residual stress removal, grain growth, or pore closure due to the growth stress by the oxide [22].

Even though FZ has the highest porosity in comparison with the other samples, surprisingly its weight gain rate is the lowest followed by FYZ, FP, and then FY. It is clearly in contrast with rapid ionic transport paths theory, hence indicating the existence of other variable affecting the result. The cause of this result can be acquired by means of microstructural analysis which will be covered in the next section.

3.5. After Oxidation

3.5.1. Phase identification

Figure 8 shows X-ray diffraction patterns of FP, FY, FZ and FYZ after oxidation at 800°C for 8 cycles. In all samples, α-Fe, $FeTiO_3$, and α-Al_2O_3 phases are observed. Based on the calculation of the enthalpy of oxide formation by Chinnappan (2014) [23], $FeTiO_3$ phase is a less stable phase compared to Y-Al-O, Y-Ti-O, TiO_3, etc. Thus, the formation of more stable phase is still possible.

In FP, FZ, and FYZ samples, Cr_2O_3 phase is not detected. This is because even though based on the kinetics Cr_2O_3 formation is preferred (due to the much higher concentration of Cr content), but thermodynamically, Al_2O_3 is more stable. Hence, Cr_2O_3 may be formed only at the beginning of oxidation, and will be replaced by Al_2O_3 subsequently. In this study, XRD tests were performed after oxidation for 160 hours. Therefore, re-oxidation may be the main cause of Cr_2O_3 formation in FY.

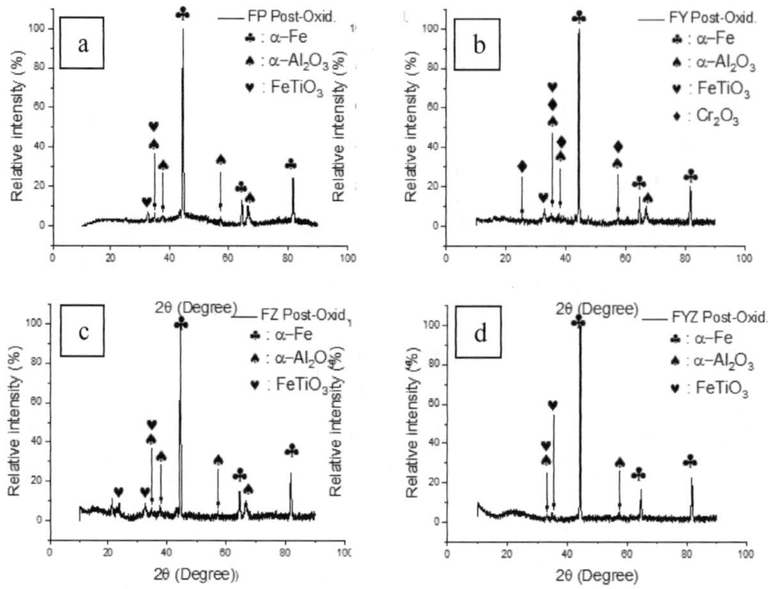

Figure 8. XRD patterns of (a) FP, (b) FY, (c) FZ and (d) FYZ after oxidation at 800°C for 8 cycles.

3.5.2. Microstructure

Figure 9 shows the cross-sectional SEM images of FP, FY, FZ and FYZ after oxidation at 800°C for 8 cycles. The presence of crack in the oxide layer initiated at the oxide/metal interface area in FY and FYZ is observed. The worst condition is shown in FY where the formed Al_2O_3 layer is discontinuous. This can be attributed to the difference in thermal expansion coefficients between Al_2O_3, ZrO_2, Y_2O_3, and ferritic steel.

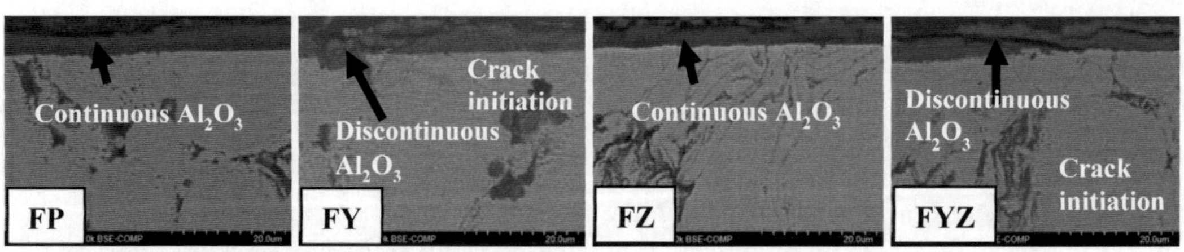

Figure 9. Microstructure images of FP, FY, FZ and FYZ after oxidation at the oxide/metal.

At the oxide/metal interface region, the high difference in thermal expansion coefficient between Al_2O_3 and substrate metal ($8x10^{-6}$: $13x10^{-6}$ /°C) may cause the spallation of the oxide layer when it was subjected to cyclic thermal loading. The same case is the most common cause of failure in coating applications in TGO interface (thermal-growth oxide) [24-26].

Moreover, the rise of Y_2O_3 content at the interface region of Al_2O_3 /substrate metal may induce crack initiation. This is because the coefficient of thermal expansion of Y_2O_3 which is almost similar to Al_2O_3 causes the stress to become stronger.

In contrast to Y_2O_3, the stabilized ZrO_2 (cubic crystal structure) has a thermal expansion coefficient closer to the substrate metal (ferritic steel), precisely between the substrate metal and the Al_2O_3 protective oxide layer, which is 10.5 x 10^{-6} /°C [27]. Therefore, the stress generation due to thermal expansion different becomes lower, hence lowering crack initiation and protective oxide layer discontinuity possibility. The difference in thermal expansion coefficient is what causes the highest Y_2O_3 content has the worst oxidation resistance. As for the FYZ sample (which has the second highest

Y_2O_3 content after FY), the initiated crack has not resulted in direct exposure of inner alloying elements yet. Accordingly, the oxidation rate is still well controlled.

3.5.3. Elemental distribution

Figure 10 shows the distribution of Cr and O elements near the surface of FY. The result indicates the formation of Cr_2O_3 cluster identified from XRD phase analysis. The formation of Cr_2O_3 may be caused by oxide layer discontinuity which enhances inward oxygen diffusion and promotes continuous oxidation of inner alloying elements.

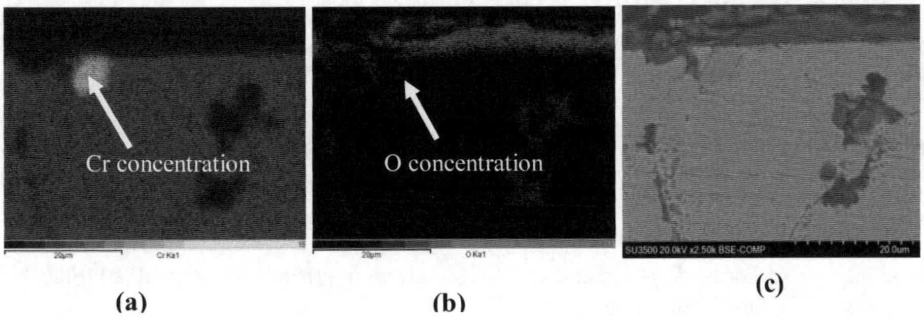

Figure 10. Clustering of: (a) Cr and (b) O in (c) FY sample after oxidation at 800°C for 8 cycles.

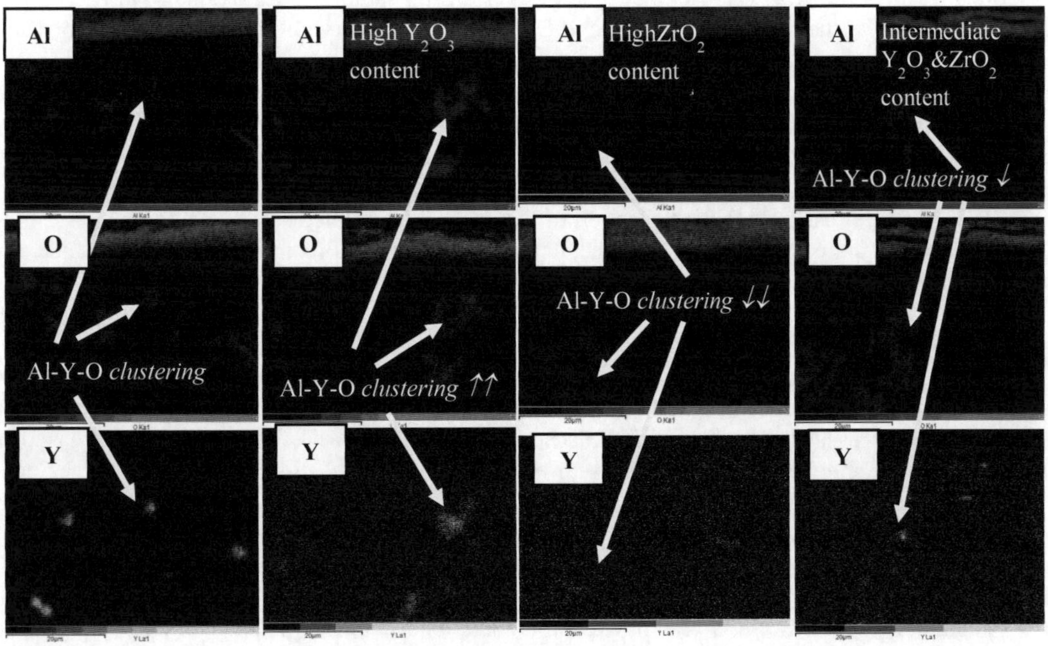

Figure 11. Distribution of Al, Y, and O alloying elements from samples: (a) FP, (b) FY, (c) FZ, and (d) FYZ after oxidation at 800°C for 8 cycles.

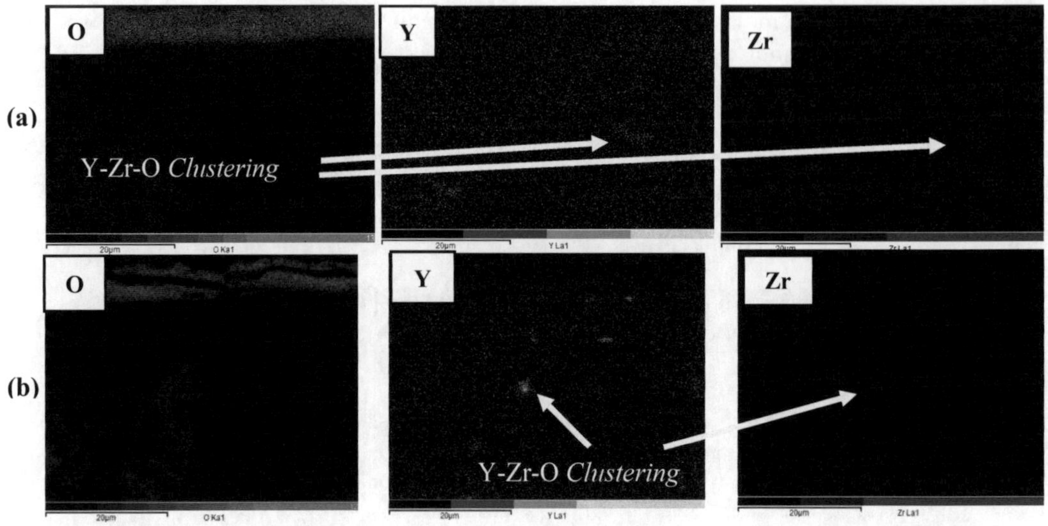

Figure 12. Distribution of Y, Zr, and O alloying elements from: (a) FZ, and (b) FYZ samples after oxidation at 800°Cfor 8 cycles.

Figure 11 shows the Al, Y, and O distribution over all the samples after oxidation at 800°C for 8 cycles. In FP and FY, the presence of Al, Y, and O clustering is observed. This indicates the formation of minor phase of Al-Y-O. Although Y_2O_3 is not added into FP sample, the clustering of these three elements could be derived from Y reactive element of base material composition in the FP sample.

In the sample with ZrO_2 addition, the formation of Al-Y-O cluster does not occur. This shows that the ZrO_2 addition can also suppress the formation of minor phase Al-Y-O. This result is supported by the identification of Y, Zr, and O clustering elements on FZ and FYZ (see Figure 12). The existence of a finer Y-Zr-O cluster can reduce the formation of the coarser Al-Y-O cluster due to the higher Y-Zr-O bonding energy compared to Al-Y-O[28]. Hence, it prevents the depletion of Al for protective oxide scale formation.

4. Conclusions

Overall research showed that the addition of un-stabilized ZrO_2 to the alloy might cause cracking when sintering was done, thus increased porosity and decreased density of the alloy. The addition of ZrO_2 had a capability to decrease the tendency of the secondary phase Al-Y-O formation and increased the tendency of the more refined Y-Zr-O phase formation. Excessive addition of Y_2O_3 might cause discontinuity in the protective oxide layer. The substitution of the oxide of Y_2O_3 using ZrO_2 at the right level would increase the cyclic oxidation resistance.

References

[1] Romanoski G R, Snead L L, Klueh R L and Hoelzer D T 2000 *J. Nucl. Mater.* **283** 642–646
[2] Ukai S and Fujiwara M 2002 *J. Nucl. Mater.* **307** 749–757
[3] Shimakawa Y, Kasai S, Konomura M and Toda M 2002 *Nucl. Technol.* **140(1)** 1–17
[4] De Carlan Y*et al.* 2009 *J. Nucl. Mater.* **386** 430–432
[5] Baluc N, Schäublin R, Spätig P and Victoria M 2003 *Nucl. Fus.* **44(1)** 56
[6] Klueh R L 2005 *Int. Mater. Reviews* **50(5)** 287–310
[7] Oka H, Watanabe M, Kinoshita H, Shibayama T, Hashimoto N, Ohnuki S, Yamashita S and Ohtsuka S 2011 *J. Nucl. Mater.* **417(1)** 279–282
[8] Wang M, Zhou Z, Yan Z, Yu P and Sun H 2013 *Acta Metall Sin.* **49(2)** 153–158
[9] Suárez M, Fernández A, Menéndez J L, Torrecillas R, Kessel H U, Hennicke J, Kirchner R and

Kessel T 2013 *Sintering Applications* InTech
[10] Cho H S, Kimura A, Ukai S and Fujiwara M 2004 *J. Nucl. Mater.* **329** 387–391
[11] Kimura A, Ukai S, Fujiwara M 2004 *Proc. Int. Cong. Adv. Nucl. Power Plants (13-17 June, Pittsburgh)* ISBN: 0-89448-680-2
[12] Kimura A *et al.* 2011 *J. Nucl. Mater.* **417(1)** 176–179
[13] Xia Y P, Wang X P, Zhuang Z, Sun Q X, Zhang T, Fang Q F, Hao T and Liu C S 2013 *J. Nucl. Mater.* **432(1)** 198–204
[14] Xu H, Lu Z, Wang D and Liu C 2017 *Fusion Engin. and Design* **114** 33–39
[15] Schneider C A, Rasband W S and Eliceiri K W 2012 *Nature methods* **9(7)** 671–675
[16] Vander Voort G F 1984 *Metallography, principles and practice* ASM international 649
[17] Yanagida H, Kōmoto K and Miyayama M 1996 *The Chem. Ceramics* John Wiley & Son Ltd
[18] Scott H G 1975 *J. Mater. Sci* **10(9)** 1527–35
[19] Birks N, Meier G H and Pettit F S 2006 *Introduction to the high temperature oxidation of metals* Cambridge University Press 69–71
[20] Cabrera N F M N and Mott N F 1949 *Reports on progress in physics* **12(1)** 163
[21] Hauffe K and Ilschner B 1954 *Berichte der Bunsengesellschaft für physikalische Chemie* **58(6)** 382–387
[22] Evans U R 1960 *The Corrosion and Oxydation of Metals 1st* ed. Edward Arnold (London) 868–871
[23] Chinnappan R 2014 *Calphad* **45** 188–193
[24] Karlsson A M, Levi C G and Evans A G 2002 *Acta Materialia* **50(6)** 1263–73
[25] Karlsson A M, Xu T and Evans A G 2002 *Acta Materialia* **50(5)** 1211–18
[26] He M Y, Hutchinson J W and Evans A G 2002 *Acta Materialia* **50(5)** 1063–73.
[27] Hund F 1951 *Zeitschrift für Elektrochemie und angewandte physikalische Chemie* **55(5)** 363–366
[28] Murali D, Panigrahi B K, Valsakumar M C, Chandra S, Sundar C S and Raj B 2010 *J. Nucl. Mater.* **403(1-3)** 113–116

Acknowledgment
The financial support from Indonesian Institute of Science is kindly acknowledged (Kegiatan Unggulan LIPI). The authors wish to express their gratitude to Research Center of Physics Indonesian Institute of Sciences (LIPI) for providing characterization facilities for this research. The authors would also like to thank Kemas Ahmad Zaini Thosin and Bambang Hermanto for technical support.

Modification of ledeburite microstructure on impeller blades by mean of heat treatment

W Purwadi[1] and D Idamayanti[1]
[1]Foundry Engineering Department, Politeknik Manufaktur Bandung, Indonesia

E-mail: wiwikpurwadi@yahoo.com

Abstract. Water pump machine impeller is made of gray cast iron with some variation of wall thickness. The thin part undergoes consequently a very rapid cooling and promotes the formation of hard ledeburite, This study removes the already formed ledeburite ((α + Fe_3C) + Fe_3C) by mean of heat treatment. By calculating the mass fraction of iron carbide at the level of 5%, heating the material at the austenite temperature, holding it for 2 hours and cooling it subsequently in the air, ledeburite changes to ferrite and graphite. Graphite is formed discrete along the grain boundaries of previous pearlite after the completion of heat treatment. The graphite resembles a randomly oriented eutectic undercooling type D and this is associated with a decrease of hardness to 115HV at the tip of the impeller blade.

1. Introduction

The impeller is a part of the water pump machine that serves to produce water pressure. Figure 1 illustrates a water pump image having thin involute blades.

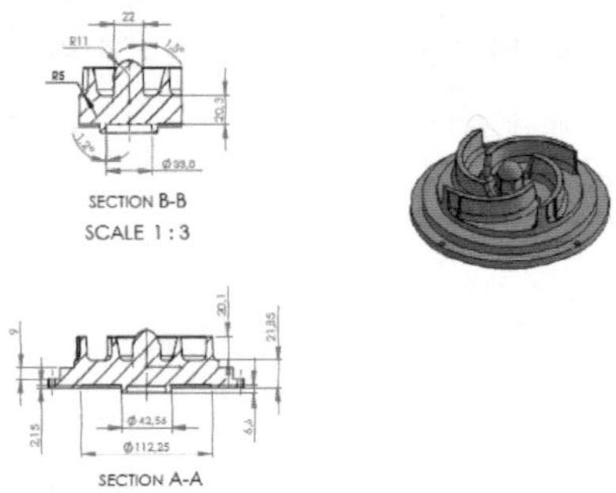

Figure 1. Water pump impeller.

The impeller is made of gray cast iron with a material composition based on Fe with carbon, silicon, manganese alloy and some accompanying elements such as phosphorus and sulphur (table 1), Grey Cast Iron belongs to the most cost effective material for the impeller and classified as is one of the stable types of cast iron, whose microstructure is characterized by the appearance of graphite, has a carbon content of above 2% and has good casting properties but low elongation value [1].

Appropriate manufacturing processes produce cast iron with microstructures consisting of eutectic lamellar graphite in the matrix of ferrite and pearlite (figure 2a). The most influencing parameters which affect the properties of the material and its microstructures are the elemental compositions and the cooling rate [2].

Since the impeller blade shows variation in its wall thickness, the edge encounters very thin part. In this section, the metastable solidification takes place and produces a microstructure of ledeburite [3] (figure 2b).

In addition, the occurrence of this ledeburite can also be initiated by several potential causes, such as high content of carbon, less silicon content (Si) and lack of adequate diffusion time [4].

Table 1. Chemical composition of grey cast iron for several wall thickness [5].

Grade	Wall Thickness (mm)	C	Si	Mn	P max	S max
FC100	-	3.4-3.9	2.1-2.6	0.5-0.8	0.3	0.15
FC150	<30	3.3-3.5	2.0-2.4	0.5-0.8	0.2	0.12
	30-50	3.2-3.5	1.9-2.3	0.5-0.8	0.2	0.12
	>50	3.2-3.5	1.8-2.2	0.6-0.9	0.2	0.12
FC200	<30	3.2-3.5	1.6-2.0	0.7-0.9	0.15	0.12
	30-50	3.1-3.4	1.5-1.8	0.8-1.0	0.15	0.12
	>50	3.0-3.3	1.4-1.6	0.8-1.0	0.15	0.12
FC250	<30	3.0-3.3	1.4-1.7	0.8-1.0	0.15	0.12
	30-50	2.9-3.2	1.3-1.6	0.9-1.1	0.15	0.12
	>50	2.8-3.1	1.2-1.5	1.0-1.2	0.15	0.12
FC300	<30	2.9-3.2	1.4-1.7	0.8-1.0	0.15	0.10
	30-50	2.9-3.2	1.2-1.5	0.9-1.1	0.15	0.10
	>50	2.8-3.1	1.1-1.4	1.0-1.2	0.15	0.10

Reducing the carbon content can considerably be done to avoid the formation of ledeburite, however, this will affect the thicker part of the object, whereby graphite will not be formed. Silicone content is additionally put into the material to prevent white solidification, whereas it causes a decrease of the hardness due to the formation of a larger number of ferrite and bigger size of graphite [6]. Meanwhile, the biggest problem in the manufacturing process of a blade is the ununiform thickness of the blade itself. The microstructure is consequently not uniform. The solidification of metal liquid in the area of smaller wall thickness will form a ledeburite microstructure. The area accordingly becomes very hard, which is also associated with poor elastic and plastic behavior. Due to the hardness and low impact value, this part is not to be machined [7].

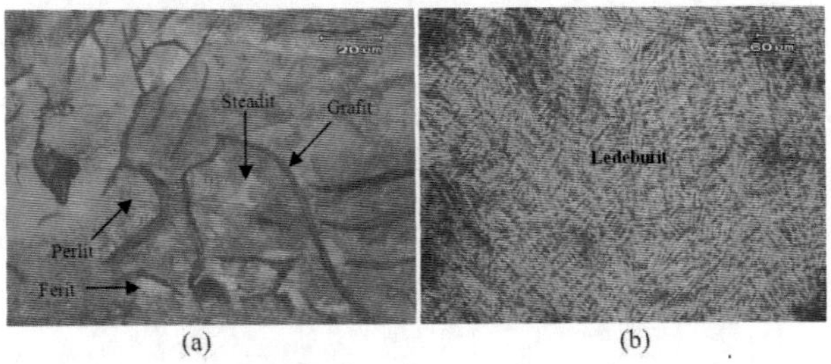

Figure 2. The microstructure of FC (A) and ledeburite (B), etched with Nital 3%.

Uniform wall thickness and appropriate chemical composition prevent a metastable solidification.

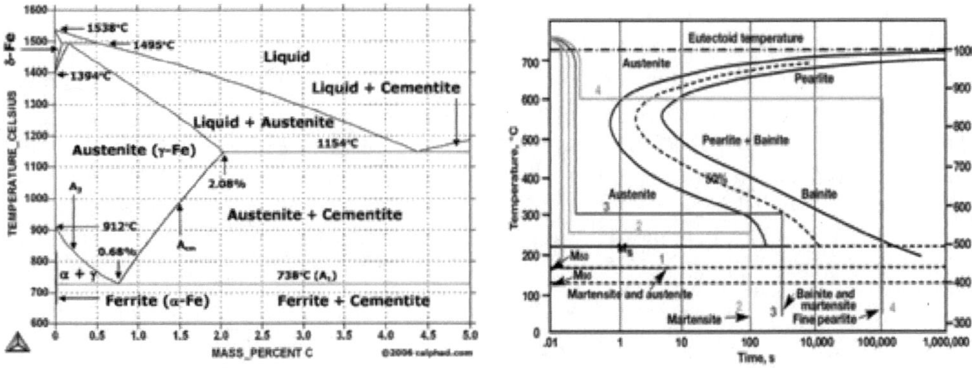

(a) (b)

Figure 3. Binary diagram Fe-C (a) and CCT diagram (b).

Occasionally heat treatment process can be done to eliminate the existing ledeburite. Heat-treatment to remove ledeburite is accomplished by heating the sample at the austenisation temperature, as shown in the binary diagram (figure 3a). Cooling rate should be set by using CCT diagram (figure 3b) [8].

By calculating the fraction of ledeburite, calculating the composition/carbon content in the matrix, determining the decomposition temperature, determining the decomposition time and cooling at the apropriate cooling rate. ledeburite can be transformed into Fe_3C which is dispersed at the grain boundary [9]. Ledeburite originally presented in the microstructure (figure 4a) can decompose to pearlite and ferrite (figure 4b). The heattreatment caused changes in the shape of pearlite and graphite size.

The decomposition of Fe_3C in the ledeburite is affected by several elements such as manganese (Mn). Manganese impedes the kinetics of graphitization of nodular graphite cast iron [10]. Manganeses stabilizes the cementite and slows down the decomposition kinetics (in iron and graphite).

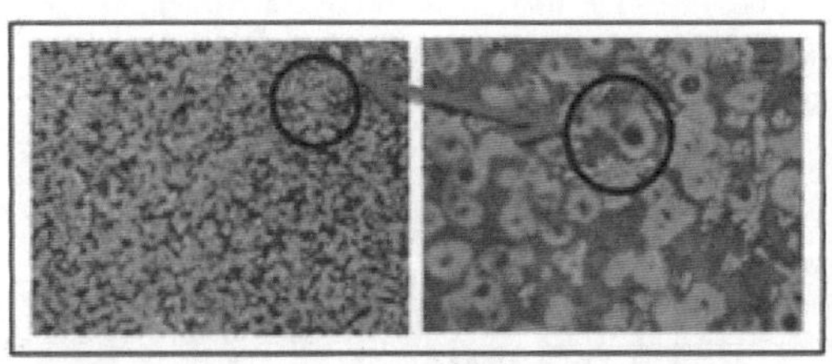

(a)

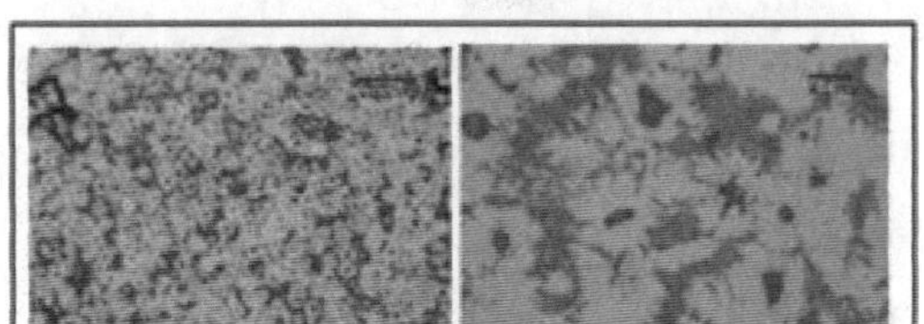

(b)

Figure 4. initial condition with ledeburite (A) and after heat treatment (B), etched with Nital 3%.

2. Design of experiment

2.1. Sampel (100% ledeburite)
The sample used is the tip of the impeller blade which has a 100% ledeburite micro structure. The calculation of the actual amount of C will be used for the formation of a new matrix after the decomposition of Fe_3C. The microstructure accordingly has a carbon content of 4.3%.

2.2 Solubility of Fe_3C in 5% Ledeburite
In this study, the maximum allowable level of ledeburite is set up at 5%. Ledeburite consists of a mixture of γ phase and cementite (Fe_3C). Calculation based on the Fe-Fe_3C diagram (figure 3) and lever rule can be applied for dissolving Fe_3C in the ledeburite.

Table 2. Percentage of ledeburite former

5% ledeburite in FCD		
Microstructure	Carbon content	Weight percent
Austenite (γ)	2.0 %	2.47 %
Fe_3C	6.67 %	2.53 %

Based on the data above it can be concluded that the austenite (γ) contained in the material of 5% ledeburite is 2.47%, while the cementite content (Fe_3C) is 2.53%. The removal of the ledeburite is done by decomposing Fe_3C into the matrix.

Table 3. Percentage of carbon in the decomposition of Fe_3C.

Microstructure	Max. Carbon content	Total carbon content
Fe_3C	6.67 %	0.977 %

The design of the heat treatment process is accordingly based on the carbon content of 0.977% C.

2.3. Determining the Heating Temperature
The carbon percentage is used as the initial reference of temperature determination for sample heating as it shown on the Fe-Fe_3C diagram. In detail it can be described as follows: From this Figure 5, the minimum temperature is at line A1 (727°C) and critical temperature at Acm line (± 830 ° C) and the maximum temperature is below solidus line (± 1360°C).

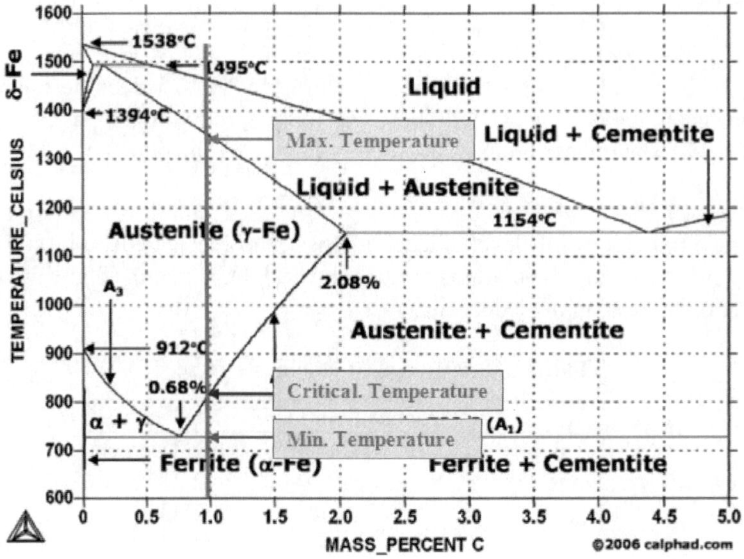

Figure 5. Diagram Fe-Fe_3C, decomposition of Fe_3C in ledeburite.

The heating is done exactly at the critical temperature of 830°C which is then followed by cooling in the open air.

2.4. Holding time
The heating time is set to 2 hours according to the minimum reaction time to disolve the ledeburite for a thickness of 5mm. The predetermined heating time can dissolve the ledeburite and the carbon from Fe_3C will subsequently diffuse into graphite or be present as Fe_3C at the grain boundary.

2.5. Testing
Tests were performed by applying metallographic analysis using an optical microscope and vickers hardness testing.

3. Result and Analysis

3.1. Initial microstructure
The impeller part having a thick wall (5 mm) shows a microstructure containing graphite with a dominant matrix of pearlite and a small amount of ferrite (figure 6a). Ferrite itself is formed around the graphite. The graphite that occurs has the shape type of A and random orientation [11]. At the edge of the impeller, the microstructure (figure 4b) shows a dominant fraction of the secondary ledeburite (pearlite + Fe_3C) [11]. In some other location assemblies of ferrite around the graphite are still to be found.

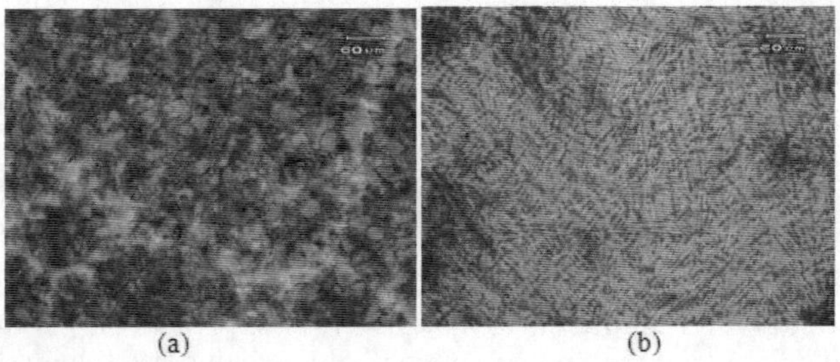

Figure 6. Micrograph at the thick wall (A) and at the tip (B), etched with Nital 3%.

3.2. Microstructure after heat treatment

Metallographic observation after the heat treatment process shows a complete change of the microstructure. Ledeburite decomposes completely into ferrite and graphite (figure 7a) [12]. The microstructure nevertheless shows nearly spherical primary phase, which is already transformed into ferrite. The cementite (Fe_3C) is not longer exist since all cementite is transformed into graphite and ferrite. Cementite (Fe3C), which was originally located at the grain boundary, has been transformed into a group of graphite with random orientation. Figure 7b the formation of D-type graphite which is commonly identified as the undercooling eutectic graphite [13]. Conditions without etching.

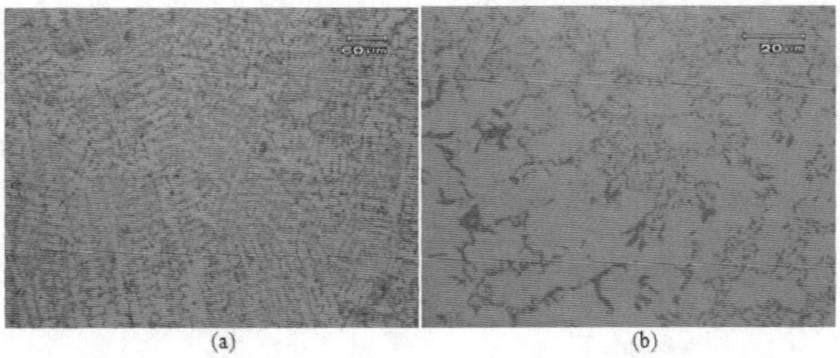

Figure 7. Micrograph after heat treatment process with graphite tipe D (a) and graphite at the grain boundary with random orientation. (b), without etching.

Figure 8 shows the undissolved (Fe3C) carbide in some particular location of the material, whereby the formation of graphite did not take place. The uneven distribution of manganeses and silicon is considered to be the cause of it [14].

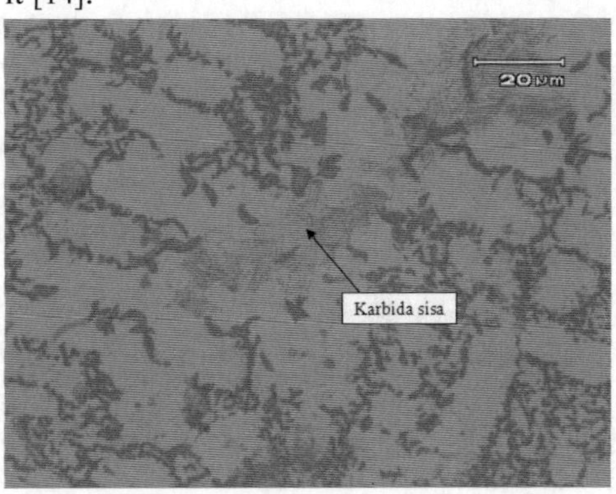

Figure 8. Retained iron carbide, etched with Nital 3%.

Related to the cange of its microstructure the heat treated material undergoes changes in mechanical properties in term of hardness. The initial hardness of the material on the thick part prior to the heat treatment was 165 HV by a test load of 1 kg. The thin impeller blade portion shows a hardness value of up to 705 HV, which decreases up to 115 HV after completing the heat treatment process.

3.3. Analysis

By heating at the upper temperature limit carbide in the ledeburite decomposes into graphite and ferrite. The total decomposition is achieved due to the sufficient 2 hour holding time, the small wall thickness and the availability of silicon in the matrix which is also supported by relatively low levels of Manganese. Longer holding time results in not only the breakdown of carbides but also the diffusion of carbon out of the pearlite matrix. The iron carbide in the matrix of pearlite also breaks down into ferrite and graphite. During the decomposition process, the basic form of pearlite in carbide that resembles a spherical shape does not change. Graphite that is formed from carbon diffusion has a random and discrete orientation. Finally, graphites resemble as D-type graphite which is commonly derived from a eutectic reaction and associated with undercooling. The decrease of hardness occurs at the tip of the impeller drastically. This pattern of graphite deployment will result in a decrease in the value of tensile strength and impact. The remaining carbides can still be found because the initial setting of temperature calculations based on the remaining carbide is 5%.

4. Conclusion

The fully ledeburitic microstructure can be transformed into stable cast iron by applying heat treatment process. To determined the heating temperature, a number of cementite in the matrix is put as the basic value for the calculation of carbon content while using the binary diagram of $Fe\text{-}Fe_3C$. Ledeburite decomposes into ferrite and graphite. However the decomposition of ledeburite does not cause any increase of carbon content in the matrix, so that ferrite is still dominant in the matrix. The process results in a structure of D-type graphite without any change in the shape and dimension of the previous primary phase. The hardness decrease up to 115HV. Furthermore, the presence of Si and the low level of manganese are the requisites for the effective decomposition of cementite in the ledeburite.

References

[1] Francis R 2003 Cost effective materials selection for pumps *PUMP engineer* 44–9
[2] Pluphrach G 2010 *Songklanakarin J. Sci. and Tech.* **32** 613–8
[3] Anton I V, Militaru C, Ştefan E M, Ivan N, Chişamera M and Ripoşan I 2009 *Scientific Bulletin, Series B: Chem and Mater. Sci.* **71** 115–30
[4] Sertucha J, Lacaze J, Serrallach J, Suárez R and Osuna F 2012 *Mater. Sci. and Tech.* **28** 184–91
[5] Allen C M and Boardman B 2005 ASM Handbook, Properties and Selection : Irons , Steels , and High Performance Alloys *Fonderie* **1** 1618
[6] Fras E, Górny M and López H F 2005. *Met. And Found. Eng.* **31** 37–52
[7] Ortega-Cubillos P, Nannetti-Bernardini P A, Celso-Fredel M and Campos R A 2015 *Revista Facultad de Ingenieria* **2015** 134–42
[8] Mrzygłód B, Kowalski A, Olejarczyk-Wozenska I, Adrian H, Głowacki M and Opaliński A 2015 *Arch. of Met. and Mater.* **60** 1941–9
[9] Purwadi W and Fazri 2013 Penghilangan Struktur Mikro Ledeburit pada FCD dengan Metoda Heat Treatment *RATMI* (Itenas)
[10] Briki J and Ben Slima S 2012 *J. Metallurgy* **2012** 1–4
[11] Stefanescu D M 2017 *ASM Handbook Cast Iron Science and Technology Prepared* vol 1A (Materials Park, Ohio: ASM International)
[12] Gebril M A, Aldlemey M S and Haider F I 2014 *Adv. Mater. Res.* **936** 1158–62
[13] Fraś E and Górny M 2012 *Arch. of Found. Eng.* **12** 39–46
[14] Ivanov V, Pirozhkova V and Lunev V 2017 *East.Eur J. Enterp. Technol.* **4** 26–30

Acknowledgments
This *research* was conducted in POLMAN Bandung. We want to thank for all the facilities and support

IOP Conf. Series: Materials Science and Engineering **541** (2019) 012030 doi:10.1088/1757-899X/541/1/012030

Optimization of ZnO-nanorods addition toward $Li_4Ti_5O_{12}$ (LTO) performance using sol-gel solid state method as half-cell lithium-ion battery anode

B Priyono[1], A A S Panjaitan[1] , A Z Syahrial[1], Faizah[1] and A Subhan[2]
[1]Departement of Metallurgical and Materials Engineering, Universitas Indonesia, Depok, 16424, Indonesia
[2]Research Center for Physics – LIPI, PUSPIPTEK, Tangerang, Banten, 15310, Indonesia

Email: bambang.priyono@ui.ac.id

Abstract. Performance optimization for the anode of lithium-ion batteries (LIBs) can be conducted by adding ZnO through sol-gel solid-state reaction. In this research, the $Li_4Ti_5O_{12}$ (LTO) used was synthesized through the sol-gel solid-state process and added with ZnO-nanorods obtained ZnO synthesis after LTO synthesis done. LTO-ZnO obtained was characterized to determine the main phase and chemical composition by XRD and SEM-EDS respectively. Electrochemical performance of LTO-ZnO was tested by EIS, CV, and CD. ZnO-nanorods characterization with SEM-EDS results shows that the ZnO inside the LTO dispersed homogeneously. Characterization using XRD revealed that the ZnO successfully enter the LTO with the variation of the amount of 4, 7, and 10 wt% of ZnO. Electric conductivity test shows improvement at an optimum addition amount of ZnO at 4 wt%, although BET result shows at the optimum amount of surface area with 75.545 m2/g. Electrochemical performance result shows optimum performance in ZnO at 4 wt% for its ability to withstand EIS test at 20C compared to 7 wt% and 10 wt%. Also, the capacity of 4 wt% added is 110.2 mAh/g compared to 7 wt% with 109.1 mAh/g and 10 wt% with 96.7 mAh/g.

1. Introduction

In this current era, energy is one principal concerning matters in current ages. With every appliance that helps human need energy. For this, lots of energy is necessary to maintain the equilibrium of supply and demand. Fossil fuel had been long occupying this section, with the line already oversaturated by this product, alternative energy came as the solution to this problem. Alternative energy came ranging from many natural inexhaustible forces, such as wind, solar, water, geothermal, and so on. For this to be used, there is a necessary vessel that able to store this energy so it could be harvested [1].

The battery is an electrochemical device, which converts electrochemical reaction into electrical energy by oxidation-reduction reaction, reversal process in recharging is used for the rechargeable battery system, involving reaction in the form of transfer of electrons, with an electrical circuit, the transfer of an electron from one material to another. The battery is a vital component in our life, our phone, laptops, and other devices mostly using the battery as their energy storage. Development of battery is very fast over the past decades; lithium-ion battery is one of the most promising prospects for the future [2].

In this research, the experiment was focused on synthesizing of $Li_4Ti_5O_{12}$ (LTO) with the addition of ZnO nanorods. Both of them is mixed before slurry made with a weight content of ZnO is 4, 7, and 10 wt%. Then, the slurry is coated to make an anode sheet and assembled into a Li-ion battery (LIB) half-cell with copper foil sheet and the Li chip is used as the counter electrode. The influence of ZnO nanorods content (in weight %) in the LTO slurry mixture was observed using Electrochemical Impedance Spectroscopy (EIS), Cyclic Voltammetry (CV) and Charge-Discharge (CD) testing. The EIS is conducted to measure the conductivity and CV is performed to measure the specific capacity and working voltage using $LiPF_6$ as an electrolyte. Then, the CD test is carried out to know the rate capability and coulombic efficiency, with 20 cycles to test the LTO/ZnO battery.

2. Experimental

2.1 Materials synthesis and electrode fabrication

Starting with TiO_2 synthesis from titanium (IV) butoxide reagent grade 97% from Sigma Aldrich using sol-gel method. Preparation of primary solution consisted of 40 ml ethanol pH 3 and 6.80 gr of titanium (IV) butoxide, followed by a secondary solution of made of 10 ml ethanol pH 3 and 1.26 gr of distilled water. Then secondary mixed with a primary solution while being stirred with a magnetic stirrer. The solution being stirrer until it becoming sol-gel for 60 minutes. Then, we hold sol-gel for 24 hours and open the beaker glass for a week. Next, we grind the powder until it finer in size. Next, we use tube furnace for calcination and put a sample inside alumina tube for 2 hours and 300°C with on hold time 15 minutes in the end with 300 °C and let the sample to cool to room temperature under slow cooling inside the furnace after turn the furnace off. Sample mixed with Lithium source LiOH along TiO_2 by ball mill both sample for 30 minutes and put back to a petri dish for next stage. In the sintering process, we put a sample that already mixed to alumina tube and heated inside a tube furnace with 750°C and 3 hours, along with on hold time in the end for 15 minutes. LTO spinel is obtained after finishing this synthesis [3].

Several methods prepared for ZnO synthesis, starting with Zn and Hexamethylene Tetramine (HMTA) with 0.05 M concentration consist of Zn mass 1.3072 gram and HTMA mass of 0.70095 gram. Measure both of them and pour 100 ml of distilled water to sample. Next, we hold sample inside icebox for 1 hour; then we put a sample inside the furnace for 3 hours with temperature 90°C. Then we put the sample to filter out the residue and ZnO nanorods. The results are ZnO nanorods [4].

After both materials prepared, we mixed both materials with agate mortar with the composition of ZnO used is 4wt% and LTO is 96% respectively. Further information on LTO and ZnO grinding composition presented in Table 1. The results are LTO/4% ZnO nanorods, LTO/7% ZnO nanorods, LTO/10% ZnO nanorods.

Table 1. Mass composition of each sample.

Material	Mass (gram)	
	ZnO nanorods	LTO
LTO-AC/4% ZnO nanorods	0.08	1.92
LTO-AC/7% ZnO nanorods	0.14	1.86
LTO-AC/10% ZnO nanorods	0.20	1.80

Coin cells fabrication started with preparing active material which is LTO/ZnO nanorods, binder materials PVDF or polyvinylidene fluoride, and acetylene black for conductive agents. The ratio of the material of active material, binder, and conductive agents respectively 8:1:1. All material mixed along with dimethyl acetamide (DMAC) 5 gram using a magnetic stirrer. The mixture was stirred for 3 hours with hot plate temperature 80°C. Slurry results were coated to Cu-foil and dried for 1 hour in 80°C with doctor blade as the media. The coated material then assembled inside a glove box with argon atmosphere [5].

2.2. Characterization and Testing
The LTO/ZnO characterized using Scanning Electron Microscopy (SEM) and Energy Dispersive Spectroscopy (EDS) mappings to analyse the topography, microstructure and elements distributions. X-ray Powder Diffraction (XRD) test was used to identify the phase and content of each composite.Further, Brunauer-Emmet-Teller (BET) is used to measure the surface area of the samples.

The Cyclic voltammetry (CV) of coin cell was carried out at a scan rate of 100 mVs^{-1}. Electrochemical impedance spectroscopy (EIS) tests were performed to show the impedance value. The charge and discharge (CD) testing were conducted to evaluate the cell electrochemical performance from low current (C/5) to high current (20C) discharge rate.

3. Results and discussion

3.1. The structure of LTO/ZnO
In this characterization, LTO/ZnO is observed to gain the microstructure and morphology of material by conducting SEM-EDS characterization. Figure 1 shows us all 3 parameters of LTO/ZnO of 4, 7, and 10 wt% with a magnification of 10.000 X.

Figure 1. SEM image of 10.000 X of LTO/ZnO with different variables (a)LTO/ZnO 4%, (b)LTO/ZnO 7%, (c)LTO/ZnO 10%.

We could see several rounded points of nanorods, showing that ZnO nanorods are successfully generated. Along with sample, there are several agglomerates that came from ununiform of the sample during the grinding process. Thus, the sample of LTO/ZnO 10% by increasing the ZnO content the agglomeration become massive and make the grain looks coarse and bigger than the grain of LTO/ZnO 7% and 4%.

3.2. The surface area LTO/ZnO
Brunauer–Emmett–Teller is conducted to know the surface area of LTO/ZnO nanorods. In this Table 2, we could see the effect of ZnO percentages on surface area with fluctuation between each part, with the highest surface area is LTO/ZnO 4%, followed by LTO/ZnO 10%, and LTO/ZnO 7%.

This BET results of each sample correlate with the SEM photomicrograph result. The addition of 7% ZnO leads to lowering the surface area compare to the 4% ZnO. It could be predicted that the some of ZnO particle may block the pore of LTO particle. Nevertheless, as a result of some agglomeration of ZnO particle at the composition of 10% ZnO, at the certain level the agglomeration of ZnO may reduce the particle that blocking the LTO pore. That would be the possible cause why the sample of 10% ZnO possesses the higher surface area than that of 7% ZnO.

Table 2. The surface area of LTO/ZnO nanorods sample.

Samples	Surface Area
LTO/ZnO with 4%	75.545 m^2/g
LTO/ZnO with 7%	40.103 m^2/g
LTO/ZnO with 10%	63.158 m^2/g

3.3. Phase analysis of the LTO/ZnO

The structure of the synthesized powder was characterized using X-ray diffraction (XRD). The data is processed using XRD results X'Pert High Score Plus and Based on the data we could see several parameters that are $Li_4Ti_5O_{12}$ (LTO), ZnO, and rutile phase. LTO/ZnO parameter consists of LTO/ZnO in 4, 7, and 10 wt%, with angle that been taken within 20° and 80° by conducting with Xpert High score analysis program, with LTO along with ZnO found in all parameter and rutile phase in this process occurs because it could not construct LTO/$Li_4Ti_5O_{12}$ and in this case it becoming impurities inside LTO/ZnO nanorods [6]. The peak at 2θ for LTO with JCPDS no. 00-049-0207 [7], with peak number 18.33°, 43.24°, 62.84°, 66.08°. Next is ZnO with JCPDS no. 00-036-1451 [8], with peak number 34.42°, 36.25°, 56.60°, 62.86°, 66.38°.

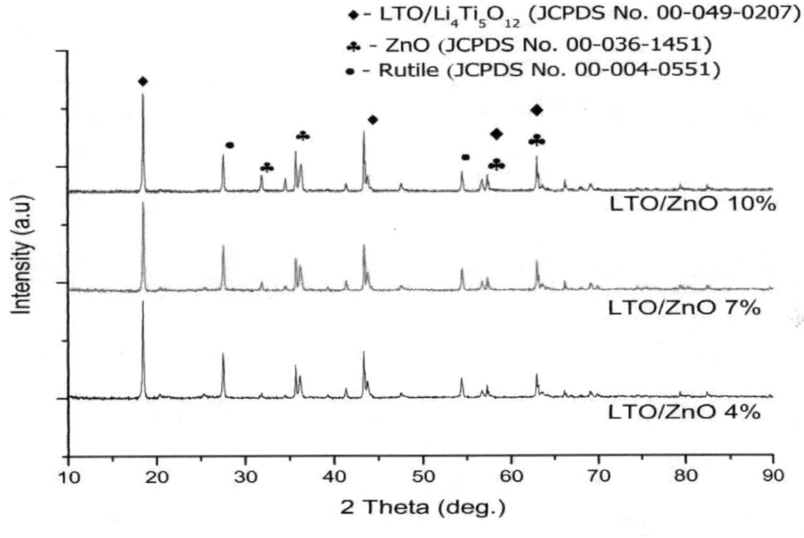

Figure 2. X-ray diffraction patterns of LTO/ZnO powder.

3.4. EIS characterization of LTO electrodes

Half-cell measurement of LTO and a lithium counter electrode using electrochemical impedance spectroscopy (EIS) using the frequency of 0.1 – 2.000 Hz. In this characterization, we conduct this method by taking semi-circle graph from sample LTO/ZnO with various compositions.

IOP Conf. Series: Materials Science and Engineering **541** (2019) 012030 doi:10.1088/1757-899X/541/1/012030

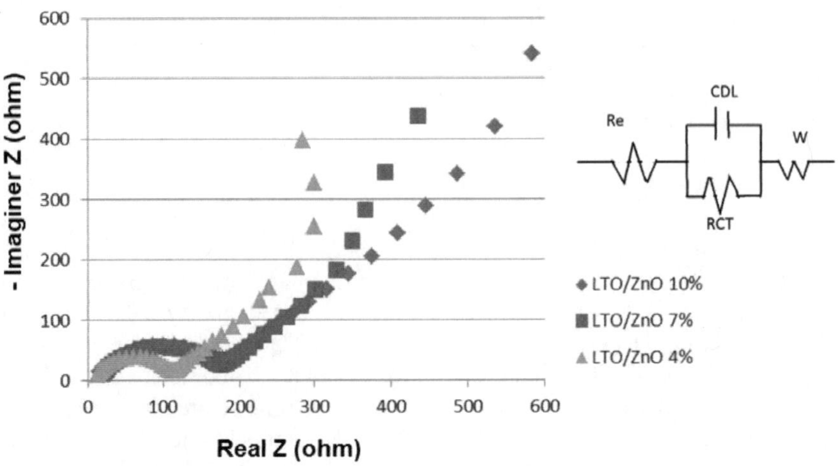

Figure 3. Complex impedance plots of the half-cell with LTO/ZnO anode with its equivalent circuit.

The electrode-polarization resistance increased of ZnO, the resistance also increasing along with it, that we can be seen in table 3 shows resistance to charge transfer (Rct) to determine the resistance or impedance at the half-cell batteries.

In this EIS data, there is Randles circuit that works as equivalent circuit. Within Randles circuit there is several parameter, such as Re as solution resistance, CDL as capacity double layer and RCT resistivity charge transfer is representing the impedance in the porous electrode, and then W for warburg is constant for semi-infinite diffusion condition. The fitting was conducted using EISSA software. The value of each element is shown in Table 3.

Table 3. Rs, Rct, Cdl and W impedance values of the experimental with variation ZnO content.

Sample	Rs (ohm)	Rct (ohm)	Cdl (F)	W
LTO/ZnO 4%	11.44	87.70	1.210 E-06	221.24
LTO/ZnO 7%	19.58	129.36	1.234 E-06	285.08
LTO/ZnO 10%	16.69	131.77	1.145 E-06	388.66

Based on Table 3 we could see with increasing ZnO content, it affects sample conductivity, with the highest conductivity is LTO/ZnO 4%, while the lowest conductivity is LTO/ZnO 10%.

3.5 Cyclic Voltammetry of LTO/ZnO electrodes

Cyclic Voltammetry is used to gain work potential and specific capacity of the sample, where every sample is tested, it shows us differences between each of them are insignificant, with sample LTO/ZnO 4, 7, and 10 wt% is 1.594 V, 1.589 V, and 1.593 V, respectively.

The specific capacity of sample LTO/ZnO could be obtained by cyclic voltammetry test. Sample capacities of sample LTO/ZnO 4% are 110.2 mAh/g, LTO/ZnO 7% are 109.1 mAh/g, and LTO/ZnO 10% are 96.7 mAh/g. Based on the data, there is decreasing specific capacity, and also ZnO content didn't improve specific capacity, that could be proof by the content, with LTO/ZnO 4% is the highest specific capacity with 110.2 mAh/g, while LTO/ZnO 7 wt% and 10 wt% is lower than that.

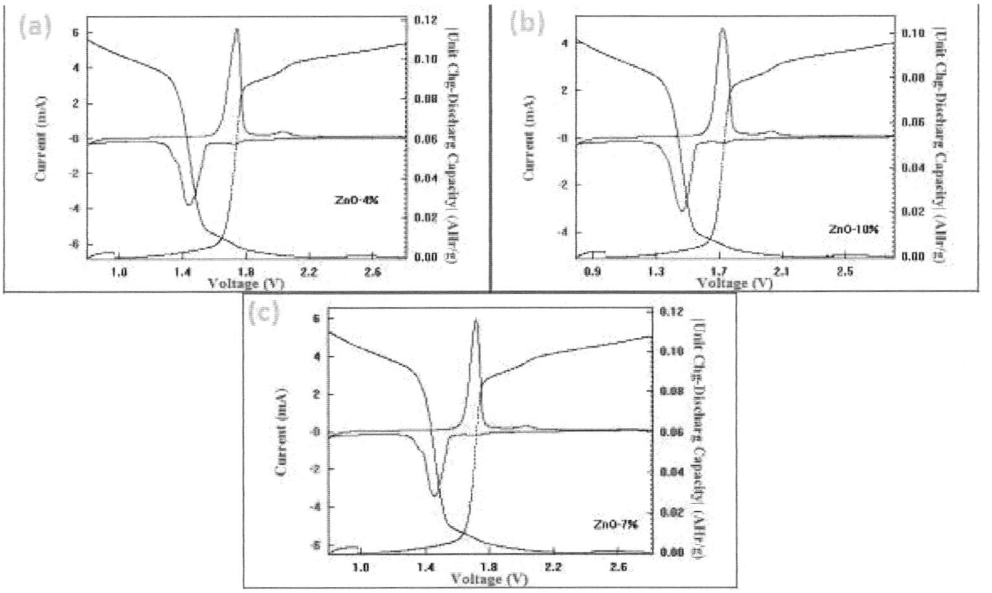

Figure 4. Cyclic Voltammetry curve for LTO/ZnO 4% (a), LTO/ZnO 7% (b), and LTO/ZnO 10% (c).

3.6. Charge-Discharge (CD) characterization of LTO/ZnO electrodes

Charge-Discharge (CD) is done to gain an understanding of battery performance by conducting charges discharges from 0.2 C to 20 C, with 1C means the charges and discharges current that occur for 1 hour for battery performance reaching top performance.

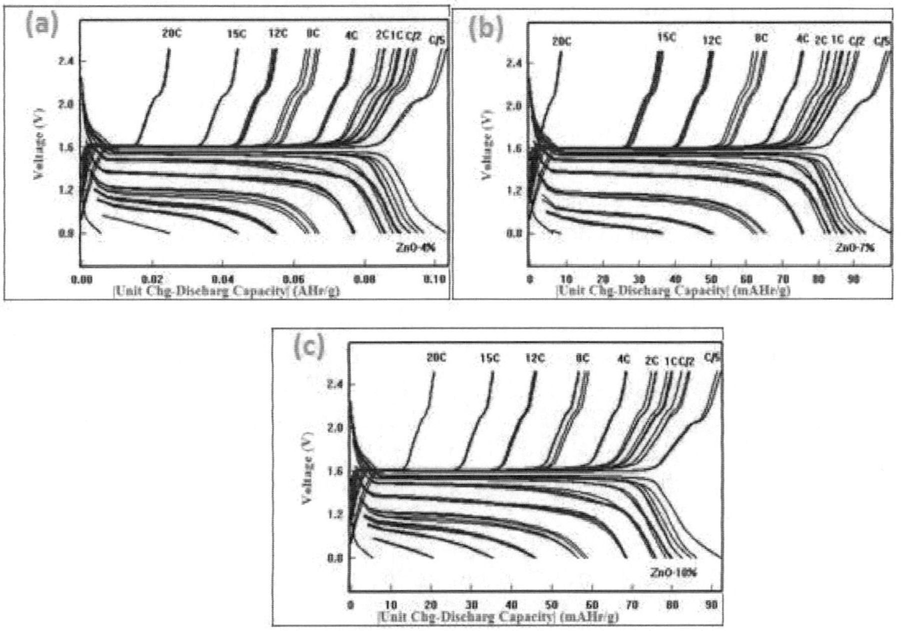

Figure 5. Charge Discharges for LTO/ZnO 4% (a), LTO/ZnO 7% (b), LTO/ZnO 10% (c) of half-cell with current rate 0.2C to 20C.

Based on Figure 5 it shows that with increasing C-rate, there is degradation of sample durability of LTO/ZnO with different results. For LTO/ZnO 4 % it started with 0.2 C and end with 20 C, charges from 103.1 mAh/g to 24.87 mAh/g and discharges from 102.7 mAh/g to 24.8 mAh/g. LTO/ZnO 7% started with 0.2C to 20C with charges from 95.6 mAh/g to 8.6 mAh/g and discharges from 99.5 mAh/g to 8.4 mAh/g. LTO/ZnO 10% started with 0.2C to 20C, with charges from 91.8 mAh/g to 20.76 mAh/g and discharges from 91.6 mAh/g to 20.6 mAh/g. In the previous experiment, specific capacity at a high C-rate of 10C [9], while with the addition of ZnO our C-rate achieves charge-discharge until 20C.

Based on Figure 6, we could see the highest capacity within 0.2 C is LTO/ZnO 4%, followed by LTO/ZnO 7% and LTO/ZnO 10%. While at this point, there is a declining rate with increasing current rate, that toward 20C it becoming lowered down, while it could also be seen in Figure 6.

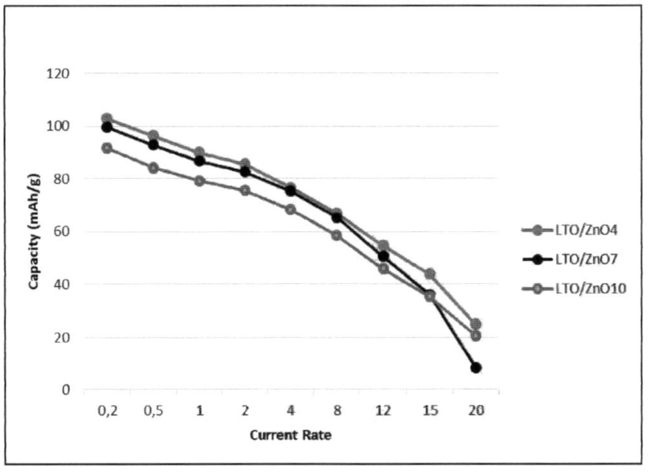

Figure 6. Capacity discharge vs current rate of LTO/ZnO.

Within the graph, above we could observe that the current rate within 20C LTO/ZnO 4% still the highest capacity followed by LTO/ZnO 10% and LTO/ZnO 7%, where 7 wt% had steep decline within 15C becoming the lowest capacity. From here we could see with lesser ZnO component in LTO/ZnO, the higher capacity of battery material that we could see at LTO/ZnO 4%. Although all the sample unable to reach the theoretical capacity of LTO itself at 175 mAh/g [10], but the addition of nanorods increasing the discharge capability of the LTO/ZnO until 20 C.

4. Conclusion

$Li_4Ti_5O_{12}$ (LTO)/ZnO synthesis occurs successfully, with mixing of xerogel TiO_2 and LiOH in milling process becoming LTO, and mixing with ZnO, becoming LTO/ZnO in slurry preparation. Based on battery fabrication of LTO/ZnO with each composition of them are 4, 7, 10 wt% of mass within battery, with the highest capacity is LTO/ZnO 4% with 110.2 mAh/g, followed by LTO/ZnO 7% with 109.1 mAh/g, and finally LTO/ZnO 10% with 96.7 mAh/g. Coin cell battery conductivity of LTO/ZnO has different value of resistivity, with LTO/ZnO 4% is the lowest of resistance value, while it implies that it has the highest conductivity value, as LTO/ZnO 4% is 103 Ω, while LTO/ZnO 7% and 10% is 160 Ω and 163 Ω respectively. It is shown that increasing ZnO content influence the conductivity of the material, with ZnO mass content affect the capability of battery conductivity.

References

[1] Su X, Wu Q, Wu J and Wei S 2012 *J Mater Sci* **47** 2519–2534
[2] Nitta N, Wu F, Lee J T and Yushin G 2015 *Materials Today* **18** 252–264
[3] Mahmoud A, Amarilla J M, Lasri K and Saadoune I 2013 *Electrochim. Acta* **93** 163–172

[4] Sholehah A and Yuwono A H 2015 *IJTech* **6** 565–572
[5] Syahrial A Z, Priyono B, Yuwono A H, Kartini E, Jodi H and Johansyah 2016 *IJTech* **7** 392–400
[6] Yang L, Li H, Liu J, Lu Y, Li S and Min J 2016 *J Alloys Compd* **689** 812–819
[7] Chen S, Xin Y, Zhou Y, Ma Y, Zhou H and Qi L 2014 *Energy Environ Sci* **7** 1924
[8] Ashok A and B Raj 2014 *RSC Adv.* **4** 58902-58909
[9] Abdurrahman N M, Priyono B, Syahrial A Z and Subhan A 2017 *Journal of Physics: Conference Series* **877** 012008
[10] Goriparti S, Miele E, Angelis F D, Di E, Proietti R, and Capiglia C 2014 *Journal of Power Sources* **257** 421–443

Acknowledgements
The authors would like to thank the Directorate Research and Public Services Universitas Indonesia for their financial support to do this research under HIBAH PITTA 2018 with contract PITTA/1062/FT/2018.

Corrosion of carbon steel after exposure in the river of Sukabumi, West Java

A Royani[1], S Prifiharni[1], L Nuraini[1], G Priyotomo[1], Sundjono[1], I Purawiardi[2] and H Gunawan[3]
[1]Research Center for Metallurgy and Materials - Indonesian Institute of Sciences, Kawasan Puspiptek 470 Tangerang Selatan Banten, 15314, Indonesia
[2]Research Center of Physics, Indonesia Institute of Science, Kawasan Puspiptek, Tangerang Selatan, Banten, 15314, Indonesia
[3]Research Center and Development of Roads and Bridge, Indonesia's Ministry of Public Works and Housing Jl. A.H. Nasution No.264, Bandung, West Java, 40294, Indonesia

E-mail : ahmad.royani@lipi.go.id

Abstract. In this study, the corrosion behavior of carbon steel had been investigated in the river of Sukabumi Region, West Java. The corrosion rates of those steels were measured by weight-loss method after exposure for certain periods of time in various depth of water (0 meters and 1 meter). The surface morphology and the composition of their corrosion product were characterized using scanning electron microscopy (SEM), XRD (X-ray diffractometer) and energy dispersive spectroscopy (EDS). After exposure, the entire surface of carbon steel was covered by corrosion product in all depth of water. On the basis of results, the corrosion rate of those steels was 19.86 mpy for 0 meters of depth and 7.01 mpy for 1 meter of depth after exposure for 76 days, respectively, which means the no protective oxides on the surface of carbon steel at 0 meters.

1. Introduction

Carbon steel has been widely applied in different applications and the industry significantly increased its production in recent years. The mechanical properties of those alloys are excellent [1]. Recently, carbon steel is used in different engineering applications for the production of some automobile components, structural shapes and sheets that are used in pipelines, buildings, plants and bridges [2]. So, the evaluation of corrosion on carbon steel in natural water environments is very important for engineering constructions in bridges, as this phenomenon is responsible for costly social, economic and sometimes even human losses.

Corrosion is a serious problem to the long-term function and integrity of a steel bridge. Structural steel will corrode if left unprotected or inadequately protected from the natural environment. This corrosion can take the form of general uniform thickness loss or concentrated pitting depending on exposure to the environment. The engineer should view corrosion as a long term threat to the integrity of the bridge structure – a critical design consideration that must be addressed in a rational manner during the design process.

Several studies have been performed to determine the corrosion rate of carbon steel and those alloys [3-8]. The rate of corrosion of plain carbon steels is dependent on both the environmental conditions, structural and compositional properties of the steel [6]. Corrosion behavior of mild steel in seawater was

IOP Conf. Series: Materials Science and Engineering **541** (2019) 012031 doi:10.1088/1757-899X/541/1/012031

studied by Sundjono et al [7] who found that uniform corrosion occurred on the mild steel during the immersion test. The corrosion resistance of the steel decreased with increasing exposure time. Prifiharni et al [8] studied the corrosion performance of steel in seawater at Karangsong and Limbangan region and found that the surface of the steel was covered by oxide film and the main corrosion product of steel was Fe_3O_4 and FeOOH. In another study, the Corrosion behavior of carbon steel in river water was investigated by mass- loss method [4]. Corrosion rates of commercial mild steel have been determined in underground water samples using conventional gravimetric measurements [9]. The corrosion behavior of steel has a correlation with several factors such as the concentration of free chlorides, electrolytic conductivity, and total dissolved solids. The experimental of the corrosion behavior for mild steel in five different media show that corrosion of mild steel decreasing intensity in the order of 0.1M of hydrochloric acid, underground (soil), atmosphere, salt water and fresh water [10]. The corrosion rate of carbon steel in the natural water is generally lower than that in seawater. Although corrosion of carbon steel was studied in a variety of aqueous media ranging from natural water to artificially water, the result of their corrosion rates always different.

In this work, the primary objective of the present study is to determine corrosion rates of carbon steel in the river water with different of depth at Sukabumi and to study correlate the water quality parameters with corrosion rate of those steel samples. The corrosion rate of these samples was investigated by mass-loss methods. The differences between corrosion products formed of these samples after exposure are described.

2. Materials and Methods

2.1. Sample preparation

Commercial carbon steel was used in the present study. Specimens (dimensions 150 x 80 x 2 mm) of carbon steels was used. The chemical composition of the samples is presented in Table 1.

Table 1. Chemical composition (wt%) of mild steel used.

Elements	Carbon	Manganese	Silicon	Sulphur	Phosporus	Iron
Percentages	0.051	0.734	0.126	0.004	0.016	Balancing

The specimens were prepared, before and after the test, using procedures as described in ASTM G-1 and the procedures are briefly described here. The specimens were polished with fine quality silicon carbide (SiC) emery sheet (400 grits). After polishing, specimens were washed with distilled water, degreased by using acetone, ultrasonically cleaned for 2 min, dried and stored in desiccators before use.

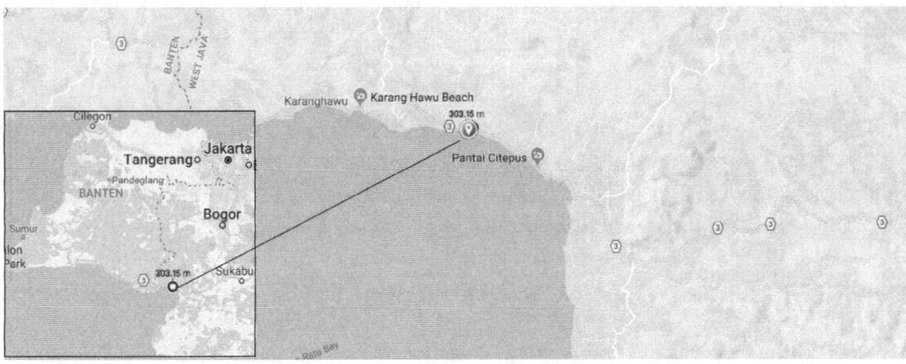

Figure 1. Map of site location for specimen testing.

The specimen testing locations in the river of Sukabumi (6°58'13.5"S; 106°31'28.6"E & -6.970427; 106.524611), 1.2 km from the shore south marine region of the district of Sukabumi, West Java -

Indonesia. The geographic site location of the experiment is shown in Figure 1. The specimens were installed exposed to the racks on PVC panels for a period of certain as shown in Figure 2.

Figure 2. The rack of specimen testing.

2.2. Weight loss analysis

Before the specimens were cleaned with the pickling solution, which contained hydrochloride acid and hexamethylenetetramine [11], they were weighed for initial weight by using an analytical balance (accuracy of 0.0001 g). After that, the specimens were immersed in the river at Sukabumi. After exposure, each specimen was taken out at a certain interval (27 and 76 days). Before weighing, the specimens were cleaned with distilled water and immersed in the pickling solution to remove the corrosion product. The specimens were washed with distilled water, dried and weighed in order to get the final weight. The method for the calculation of the corrosion rate is based on weight loss, as expressed in Eq. (1) as follows [7]:

$$CR\ (mm/y) = 87.6 \times (W / DAT) \quad (1)$$

Where:
W = weight loss in milligrams, D = metal density in g/cm^3, A = area of the sample in cm^2, T = time of exposure of the metal sample in hours.

2.3. Water parameter analysis

The natural water quality parameters were measured using an HQ40d Advanced Portable Meter. This instrument is a handheld system for field measurement of dissolved oxygen (DO), salinity, conductivity, temperature, total dissolved solids (TDS), and pH. At the certain interval period immersion experiment, the parameters were measured.

2.4. Corrosion products analysis

The morphologies of corrosion products were observed by using a scanning electron microscope (SEM JEOL JSM-6390A) equipped with Energy Dispersive Spectrometer (EDS). The working distance is 10 mm and the accelerating voltage is 20 kV. The characterization of corrosion products was analyzed using X-ray diffractometer (Shimadzu XRD 7000).

3. Result and Discussion

3.1. Visual Inspection

The surface visual appearances of carbon steel specimens exposed in the river natural environment of Cidaho for 27 and 76 days are shown in Figure 3. Products of carbon steel in 1 meter of depth have more outer rust. In the surface zone (0 meters), the products are not protective and porous.

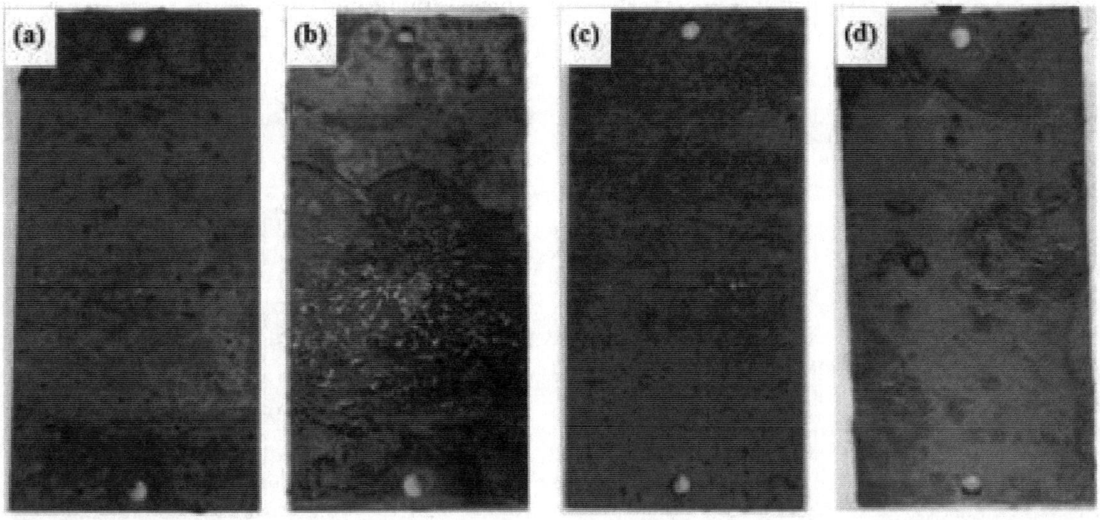

Figure 3. Visual images of specimens in river water for: (a) 0 meter, 27 days, (b) 1 meter, 27 days, (c) 0 meter, 76 days, (d) 1 meter, 76 days.

3.2. Weight Loss

Figure 4 shows the corrosion rates of the carbon steel used at site location exposure and under different exposure time. The results show that the carbon steel at the surface water level has the maximum corrosion rate whilst the carbon steel in 1-meter depth shows the lowest corrosion rate. The corrosion rates of the specimens at surface water are higher than the specimens at full immersion. This attack could be due to the formation of differential aeration of water (wet and dry phenomena). Due to low oxygen solubility in water, the oxygen concentration will be higher above the water surface [12].

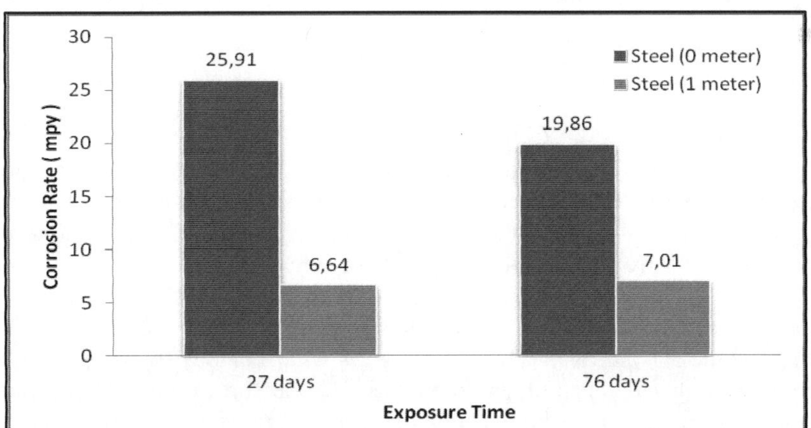

Figure 4. Corelation of corrosion rate to exposure time.

The corrosion rate from weight loss after 27 days corresponds with 25.91 mpy and 6.64 mpy for the 0 meters and 1 meter of depth, respectively. It is evident that the corrosion of specimens located in surface water zone is more serious than that of full immersion specimens, which can be explained by that the different content of oxygen[13]. Surface water zone (0 meters) also suffer dry-wet alternate and the oxygen is sufficient, which make it corrode more severe than immersion zone.

The wet-dry cycle accelerates these rusting process, especially precipitation and transformation with deprotonation and dehydration. The mechanism of the rust product in the wet-dry cycle is explained by Liu et al [12] and Stratmann et al [14] through 3 stages:
Stage 1: wetting of the dry surface
a corrosion cell starts where the anodic dissolution of iron is balanced by the cathodic reduction of Fe(III) in the rust layer:

$$Fe \rightarrow Fe^{2+} + 2^{e-} \quad (2)$$

$$2FeOOH + 2H^{+} + 2^{e-} \rightarrow 2FeO.H_2O \quad (3)$$

During this stage, the cathodic O_2 reduction reaction is very slow compared to anodic iron dissolution. The metal dissolution rate is high, but the amount of dissolved iron is restricted to the amount of reducible FeOOH in the rust layer [14].
Stage 2: wet surface
Once the reducible FeOOH has been used up, the O_2 reduction reaction becomes the cathodic reaction:

$$Fe \rightarrow Fe^{2+} + 2^{e-} \quad (4)$$

$$\tfrac{1}{2}O_2 + H_2O + 2^{e-} \rightarrow 2OH^{-} \quad (5)$$

The metal dissolution rate is determined by the diffusion limited current density of the O_2 reduction reaction on the pore surfaces. Because the pores in the rust layer are filled with electrolyte, the corrosion rate is quite slow during stage 2, as the diffusion rate is lower in the electrolyte than in the gas phase.
Stage 3: drying-out of the surface
During drying out, the rate of the diffusion limited O_2 reduction reaction is extremely fast due to thinning of the electrolyte film on the inner surface of the rust layer. Accordingly, the corrosion rate is very high, O2 reduction again being the cathodic reaction.

$$Fe \rightarrow Fe^{2+} + 2^{e-} \quad (6)$$

$$\tfrac{1}{2}O_2 + H_2O + 2^{e-} \rightarrow 2OH^{-} \quad (7)$$

$$2FeO.H_2O + 2O_2 \rightarrow 2FeOOH + H_2O \quad (8)$$

In addition to this, O_2 can reoxidize the reduced Fe^{2+} formed in stage 1. As a consequence of the high corrosion rate, stage 3 seems to dominate the metal loss during the whole wet-dry cycle. That is, after each wet-dry cycle, so the FeOOH (rust) is increased. The rust itself took part in the corrosion processes and hence increased the corrosion rate of the wet-dry sample [12]

Also, from Figure 4. It is obvious that corrosion rates decrease with exposure time at 0 meters of depth. The decrease in corrosion rate of carbon steel with time could be attributed to the thickening of the passive film on the alloy surface. This result agrees with previous studies [15,16] on corrosion of steel after exposure in different natural environments. The reason for this phenomenon can be due to the corrosion product (rust product) builds upon the surface, therefore the diffusion rate of oxygen and other agents decreasing to the metal surface. This results in the slowing down of the kinetics of the cathodic reactions. Therefore, the increase in corrosion product thickness decreases the corrosion rate [8].

3.3. Effect of Water Parameter on Corrosion Rate
The resulting test of water quality parameters on the river – Sukabumi is clearly shown as summarized in Table 2. The results of water quality in Table 2, show that the water in this study is included in freshwater because content chlorides are less than 1000 ppm. the classification of water based on chloride content can be divided into freshwater containing less than 1000 ppm chlorides, brackish has

1000 to 25,000 ppm chlorides, sea water has between 2.5 and 3.5 percent sodium chloride, and salt water with a higher chloride concentration [17]. From a corrosion standpoint, a significant water component is dissolved oxygen (DO). Oxygen acts both as a cathodic depolarizer and as an oxidizer. As a cathodic depolarizer, DO can remove hydrogen from the cathode during electrochemical corrosion and accelerate the corrosion attack. As an oxidizer, DO can be reduced on the metallic surface and participate directly to the electrochemical processes [18].

The effect of DO on the corrosion of carbon steel at full immersion conditions is illustrated in Figure 5. The corrosion rate of carbon steel for 27 days was calculated to be 6.64 mpy with 4.95 mg/L of dissolved oxygen. It was significantly increased to 7.01 mpy with increasing 8.3 mg/L of dissolved oxygen at 76 days of exposure time. Similar results were observed for different carbon steel [10]. Corrosion rates for mild steel were increased to 72% by increasing DO concentration from 9±0.5 mg/L to 2±0.5 mg/L [13]. This result clearly expresses that DO significantly influence the corrosion properties of water rivers.

Table 2. Result of the water quality parameters.

Exposure Time	Temp. (°C)	pH	Salinity (ppt)	Conductivity (μs/cm)	DO (mg/L)	TDS (mg/L)
0	27.1	6.92	0.33	667	6.23	332
27	27.2	6.86	0.5	1018	4.95	500
76	25.5	7.61	0.03	61.5	8.3	28.8

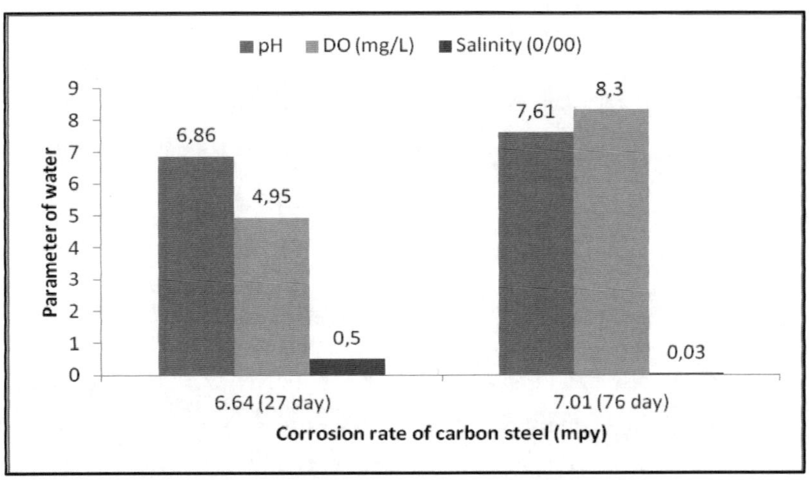

Figure 5. Corrosion rate as a function of the quality water and exposure time of carbon steel under the immersion test in 1 meter.

The temperature greatly influences the corrosion behavior of steels in water as has been indicated by the values of corrosion parameters from electrochemical studies [15]. In general, with the same exposure time, the severity of corrosion attack increases with increases in temperature. The value of temperature was not significant in this experiment as shown in Table 2. This result seems that the stability of the specimen is situated in the same corrosion area in the room temperature (25.5 – 27.2 °C). In this experiment, the pH level obtained was also almost the same as the temperature was not significant. The water conditions in the river of Sukabumi are categorized as fresh water because the pH levels are in the range of 5.8 and 8.5 [9].

Table 2 shows the value of chloride ion is very small. Salinity is evaluated by determination of the [Cl−] ion concentration in the water, which empirically states that salinity equals 1.80655 x [Cl−] [19].

It is known that an increase of salinity can increase corrosion risk. However, on the basis of the present results, the magnitude of salinity in the freshwater is very low at various exposure times. Therefore, the corrosion risk of the carbon steel in all the test solutions did not depend on the salinity.

3.4. Corrosion Products

SEM-EDS of corrosion products for carbon steel samples is shown in Figure 6 and Figure 7. The samples were exposed for 76 days in river water at Sukabumi. Differences in surface elevation are visible as the sample surface is not polished. The sample in this study shows some dark-brown corrosion product, or nodules, which are randomly arranged over the corroded surface, and also shows a separate layer on the surface (Figure 6). A sample in 0 meters of depth from surface water reveals a few of rust attached to the sample surface.

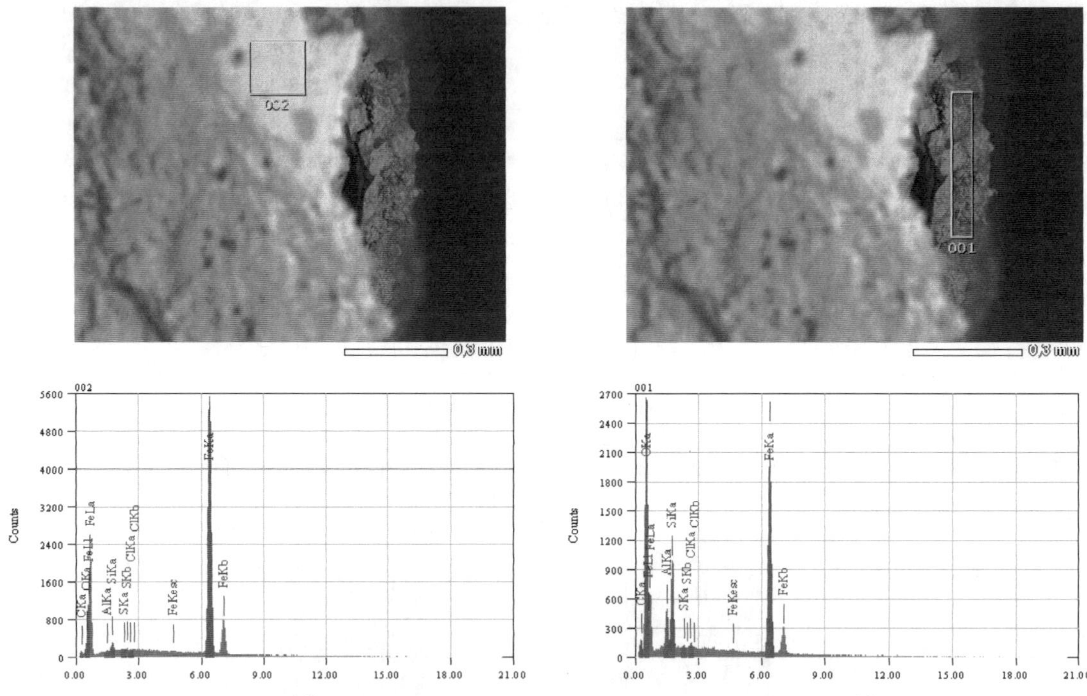

ZAF Method Standardless Quantitative Analysis
Fitting Coefficient : 0.2913

Element	(keV)	Mass%	Error%	Atom%
C K*	0.277	1.72	0.28	7.35
O K*	0.525	0.77	0.21	2.46
Al K*	1.486	0.06	0.32	0.11
Si K	1.739	0.76	0.27	1.39
S K*	2.307	0.03	0.18	0.05
Cl K*	2.621	0.02	0.19	0.03
Fe K	6.398	96.63	0.55	88.61
Total		100.00		100.00

ZAF Method Standardless Quantitative Analysis
Fitting Coefficient : 0.3519

Element	(keV)	Mass%	Error%	Atom%
C K*	0.277	9.17	0.28	19.99
O K	0.525	28.50	0.21	46.62
Al K	1.486	2.82	0.24	2.73
Si K	1.739	5.69	0.21	5.30
S K*	2.307	0.13	0.15	0.11
Cl K*	2.621	0.35	0.16	0.26
Fe K	6.398	53.34	0.50	24.99
Total		100.00		100.00

(a) Base metals (b) Corrosion products

Figure 6. SEM-EDS for the morphology of carbon steel after exposure time 76 days in surface water (0 meters).

The SEM image of the carbon steel samples corroded in 1 meter of depth immersion shows corrosion product embedding on the surface which inhibits further corrosion of those sample in this environment. The sample in 1 meter of depth immersion shows the generation of large corrosion products which forced the layer apart sometime after the corrosion commenced (Figure 7). The corrosion products might be basic iron oxides and iron carbonates under these conditions [4] because, in addition to oxides, river water contains carbonate salts [9]. Identification of the corrosion products would be confirmed by Energy Dispersive Spectrometer (EDS) analyses exactly over the products.

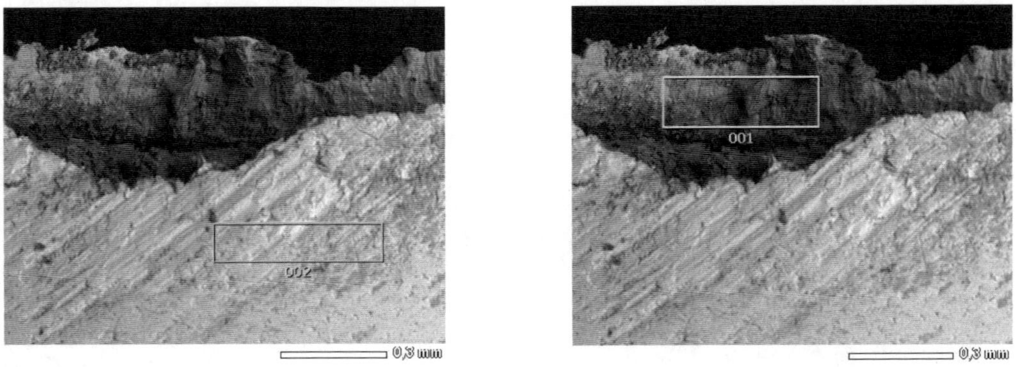

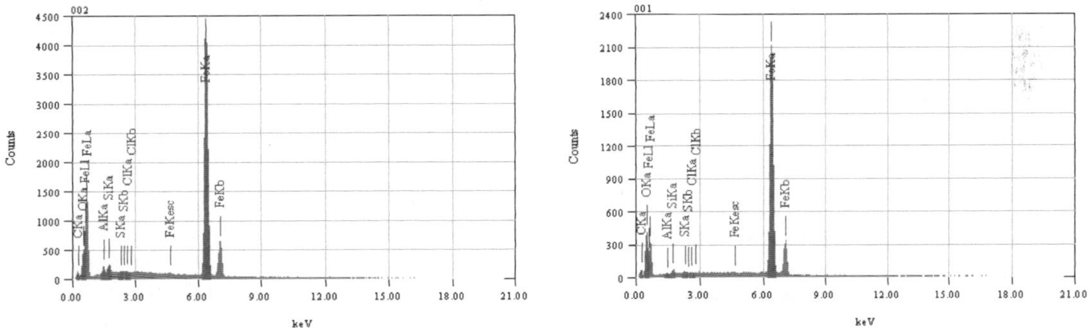

ZAF Method Standardless Quantitative Analysis
Fitting Coefficient : 0.2870

Element	(keV)	Mass%	Error%	Atom%
C K*	0.277	2.65	0.26	10.57
O K*	0.525	2.11	0.19	6.32
Al K*	1.486	0.70	0.29	1.24
Si K	1.739	0.79	0.25	1.35
S K*				
Cl K*	2.621	0.00	0.17	0.00
Fe K	6.398	93.76	0.51	80.52
Total		100.00		100.00

ZAF Method Standardless Quantitative Analysis
Fitting Coefficient : 0.3074

Element	(keV)	Mass%	Error%	Atom%
C K*	0.277	4.51	0.25	15.87
O K	0.525	6.07	0.19	16.04
Al K*	1.486	0.12	0.28	0.19
Si K*	1.739	0.39	0.24	0.59
S K*	2.307	0.06	0.16	0.09
Cl K*				
Fe K	6.398	88.85	0.49	67.23
Total		100.00		100.00

(a) Base metals (b) Corrosion products

Figure 7. SEM-EDS for morphology of carbon steel after exposure time 76 days in depth water (1 meter).

Analysis of corrosion products on carbon steel specimens exposed to surface water (0 meters) from figure 6, revealed common features for the iron, carbon, oxygen, and chloride-containing corrosion products. whereas the results of EDS on corrosion products at a depth of 1 meter was not detected the chloride content as shown in Figure 7. It implies that the presence of chloride ions contributes to increasing the corrosion rate of carbon steel at 0 meters so that the corrosion rate at a depth of 0 meters is higher than 1 meter.

The analysis of corrosion products on carbon steel samples exposed to fresh water in the river shown in Figure 8. The result of the analysis by X-Ray diffractometer at Figure 8 shows that the phases formed after certain exposure were iron hydroxide (FeOOH) and black iron oxide or magnetite (Fe_3O_4).

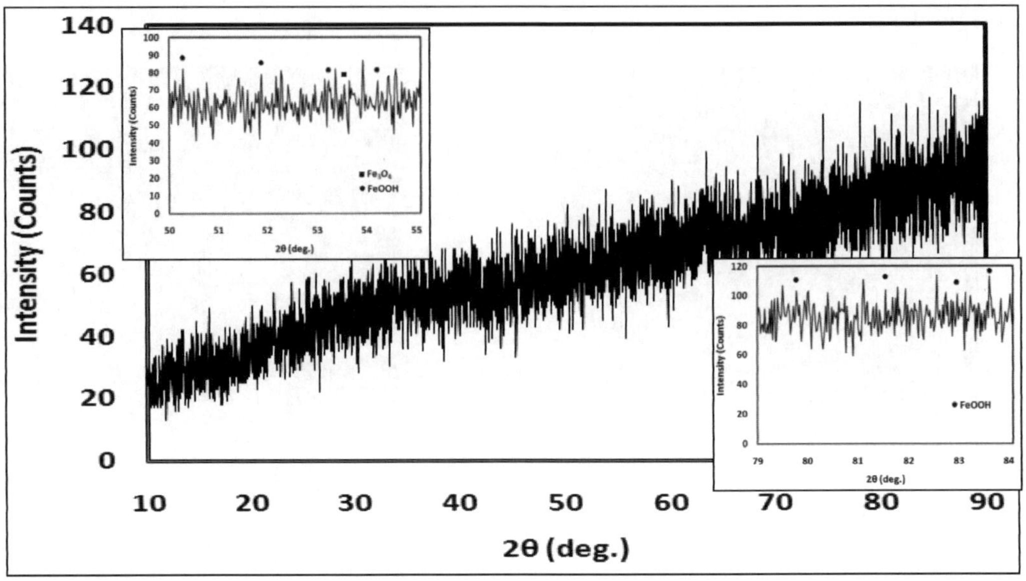

Figure 8. The result of XRD pattern of carbon steel after 76 days immersion in the river.

4. Conclusion

The corrosion of carbon steel was investigated in the river of Cidaho at Sukabumi – West Java. Weight loss analysis was conducted and the result shows that the corrosion rate increases with respect to the immersion period. On the basis of results, the corrosion rates of those steels were 19.86 mpy for 0 meters of depth and 7.01 mpy for 1 meter of depth after exposure for 76 days, respectively. These results indicate that corrosion rate of carbon steel at the wet-dry level more higher than that full-immersion. This results, of course, it still needs research and development or simulated under wet and dry cycles.

References

[1] Dobatkin S, Zrnik J, and Mamuzic I 2009 *Metalurgija.* ***48*** 157-60
[2] Gandy D 2007 *Carbon Steel Handbook. Final Report* (California, USA: Electric Power Research Institute)
[3] Rajesh V, Monica C L, Sarada Kalyani D, and Srinivasa Rao S 2014 *Int J Adv Res Sci Eng Technol* **12** 287-93
[4] Rose A I, Rani F R S, Regis A P P and Rose C M 2012 *Int.J. ChemTech Research.* **4** 157-64
[5] Royani A, Nuraini L, Prifiharni S, Priyotomo G, and Sundjono 2018 *IJETT* **59** 51-58
[6] Uzorh A C 2013 *Int. J of Eng. & Sci.* **11** 18-24
[7] Sundjono, Priyotomo G, Nuraini L and Prifiharni S 2017 *J. Eng. Technol. Sci.* **6** 770-84
[8] Prifiharni S, Nuraini L, Priyotomo G, Sundjono, Gunawan H and Ibrahim P 2017 *Proc. of the Int. Seminar on Metallurgy & Materials ISMM2017* (Jakarta) (USA:AIP Conf. Proc. 1964) pp 020038-1-6
[9] Rajesh V, Monica C L, Sarada Kalyani D and Srinivasa Rao S *Int J Adv Res Sci Eng Technol* **12** 287-93
[10] Chinwko E C, Odio B O, Chukwuneke J L, and Sinebe J E 2014 *Int. J. of Sci. & Tech. Research.* **7** 306-10
[11] ASTM Committee G1 1999 *Standard Practice for Preparing, Cleaning, and Evaluating Corrosion Test Specimens, Book of ASTM Standards* (USA: ASTM Committee) pp 15-21

[12] J G Liu, Z L Li, Y T Li, and B R Hou 2014 *Int J Electrochem Sci* **9** 8175 – 81
[13] Ismail A and Adan N H 2014 *AJER* **3** 64-67
[14] M Stratmann, K Bohnenkamp, and T Ramchandran 1987 *Corros Sci* **27** 905-26
[15] Asadi V, Danaee I and Eskandari H 2015 *Mat Res* **18** 706-13
[16] Wan Nik W B, F Zulkifli F, Rahman M M, and Rosliza R 2011 *Int J of Basic & App Sci* **11** 75-80
[17] Roberge, PR 2008 Corrosion *Engineering: Principles and Practice* (New York: McGraw-Hill)
[18] Revie, R W and Uhlig, H.H 2008 *Corrosion and Corrosion Control: An Introduction to Corrosion Science and Engineering, Fourth Edition* (Canada: John Wiley & Sons)
[19] Zakowski K, Narozny M, Szocinski M and Darowicki K 2014 *Environ Monit Assess* **186** 4871-79

Acknowledgments
Grateful acknowledgment is made for the technician staff and chemical analysis support for these experiment by Joko Triwardono, Heri Nugraha, and Sugiarti, respectively.

IOP Conf. Series: Materials Science and Engineering **541** (2019) 012032 doi:10.1088/1757-899X/541/1/012032

Synthesis of graphene oxide from used electrode graphite with controlled oxidation process

M Handayani[1], M Ganta[2], D N A Susilo[2], M S Yahya[2], G. K. Sunnardianto[2], N Darsono[1], E Sulistiyono, I Setiawan, F. P. Lestari[1], A. Erryani[1]

[1] Research Centre for Metallurgy and Materials, Indonesian Institute of Sciences, Indonesia, Building 470, Puspiptek Area, Tangerang Selatan

[2] Faculty of Engineering and Technology, Sampoerna University, South Jakarta, Indonesia

E-mail: murni.handayani@lipi.go.id

Abstract. Graphene oxide and reduced graphene oxide attract a lot of attention due to graphene provides some properties that suitable for wider application. Properties of graphene are unique since it can be tuned regarding to its controlled synthesis method and several treatments. This work, graphene oxide was synthesized from used graphite electrode by modified Hummer's method in which oxidation process controlled. Zinc was used to produce reduced graphene oxide. Graphene oxide and reduced graphene oxide were characterized by SEM, EDS, FT-IR, UV-vis, and XRD. Effective strategy was studied to produce graphene oxide and reduced graphene oxide by controlling its oxidation process. These results prove that chemical and structural properties of graphene oxide and reduced graphene oxide depend on oxidation duration. Furthermore, carbon/oxygen ratio was studied to evaluate effective oxidation process.

1. Introduction

Carbon got tremendous attention since last decade. Its nanostructure and carbon allotropes properties overcome other potential atoms. Among all various classes of carbon allotropes, graphene provides adorable properties for numerous applications. Those adorable properties of graphene showed in superior electronic, optical, mechanical and other properties. Graphene, is a two dimensional material made up of densely packed carbon atoms through sp2 hybridization [1].

Graphene based material have numerous applications such as light weight, flexible, thin, durable display screens and electrical circuits, super capacitor [2], transistor [3], transparent conducting electrodes for the replacement of ITO, conductive polymer composite [4], energy storage [5] and solar cells [6]. Lots of efforts have been intended to produce graphene in recent year by several ways. However, effective graphene's synthesis method still remains challenging task.

Various graphene oxide synthesis methods are learned for last decade to provide most effective way. Graphene provides remarkable properties since its properties can be tuned and modified for certain application. Chemical method is the prominent method of producing graphene based material in mass production [7]. However, conventional chemical methods were dangerous and toxic. Modified Hummer's method appears to be one of safe method and one of the oldest technique yet it is one of the most suitable methods to produce graphene oxide [8].

Graphene oxide (GO) comes out with advantages which are easily to produce. Furthermore, it consumes low cost and abilities to convert to graphene. The most common ways synthesizing graphene oxide is from bulk of graphite. Flake graphite is naturally occurring mineral that is purified by removing the heteroatomic contamination [8]. However, graphene exfoliation from used battery illustrates a possible way of recycling used batteries [9].

Graphene oxide was prepared by treating used electrode with $NaNO_3$, H_2SO_4, H_2O_2, and $KMnO_4$. However, effective oxidation duration of graphene oxide still remains unclear. Among all the protonic acid used to produce graphene from electrochemical exfoliation method, Sulfuric Acid is the most commonly used one because its interlayer spacing between graphite layers (0.335 nm) is quite comparable to SO_4 ion size (0.46 nm) in which facilitates easy intercalation of sulphate anion within the carbon layers in graphite [10,11]

2. Method

Chemicals used in this synthesis were Potassium Permanganate ($KMnO_4$), Sulfuric Acid (H_2SO_4), Sodium Nitrate ($NaNO_3$), Hydrogen Peroxide (H_2O_2) and used electrode graphite.

In order to synthesize graphite oxide, it was done by mixing 0.5 gr of Sodium Nitrate, 2.5 gr Potassium Permanganate, 0.5 gr used electrode and last but not least 23.3 mL of H_2SO_4 as oxidant. Dissolved and mix all those chemicals under 20°C, and stir for about 2 hours.

After being stirred about 2 hours, mixture's temperature should be lifted up to 35° C-40° C controlled by water bath. This oxidation process will be controlled by varying stirring time 3h, 5h, 7h and 14h. Texture of the mixture and the colour changed into brown paste starts from 3h of oxidation which indicated oxidation process was occurred.

Distilled water then came into the place in order to stop the reaction. 41.67 mL of distilled water pour into the beaker then followed by 3.3 mL of hydrogen peroxide (H_2O_2) to indicate the level of oxidation. When fully oxidation (high level of oxidation) occurs, the addition of H_2O_2 resulted in yellow colour and mixture stirred for 30 minutes to speed up reaction. 133.3 mL of distilled water added up into mixture and 5% of HCl was added into the mixture. Washed the sample with aqueous until the pH neutral and furnace the sample at 70°C until dried and brown black sample was obtained. 0.1 g of graphite oxide added up into 50 mL of distilled water and sonicated for 30 minutes. Then sample was done by centrifugation and after that dried samples using oven at 60°C.

Reduced graphene oxide was prepared by green reductant. 0.2 gr of graphene oxide dissolved in 200 mL of distilled water, sonicated mixture for 2 hours. 9.5 mL of HCl 37% and 1 gr of zinc powder added up into solution and sonicated until mixture become transparent. 5 mL of HCl 37%was added up to solution and washed by ethanol and aqueous to remove metal ions from the solution. Then the result was dried at 60°C for several times until it dried.

3. Results and Discussion

Graphite oxide provides phenomenon where colour and texture changed. A mixture of sulfuric acid and used electrode graphite added by potassium permanganate and sodium nitrate resulted dark green colour. It stirred for 2 hours under 20°C to keep mixture away from explosion. Texture and colour changed after oxidation occur with slight exothermic it changed into brown paste sample. After adding distilled water into the mixture, no purple colour shown which indicated $KMnO_4$ reacted or not fully reacted. Hydrogen peroxide added into the mixture to show the level of oxidation. H_2O_2 would consume the residue of oxidant agent besides showing the level of oxidation. The obtained graphite oxide was dispersed in H_2O upon the mechanical stirring, this might be through the introduction of functional groups which are responsible for hydrophilic nature of graphite oxide [12].

To obtain graphene oxide powder without metal ions, the mixture filtrated and washed by HCl to remove the rest of metal ions. Graphite oxide powder resulted in brown paste. Reduced graphene oxide by zinc as reductant was observed by visual observation. Colour of resulted sample changed, black colour of rGO can be assigned to the partial restoration of the π-network between the sheets due to removal of oxygen functional groups resulted in electronic conjugation with in reduced sheets [13].

IOP Conf. Series: Materials Science and Engineering **541** (2019) 012032 doi:10.1088/1757-899X/541/1/012032

Table 1 Synthesis Parameter and designations of prepared sample

Chemical used	GO AA	GO AB	GO AC	GO D
Used Electrode [gr]	0.5	0.5	0.5	1
KMnO4 [gr]	2.5	2.5	2.5	5.0
$NaNO_3$ [gr]	0.5	0.5	0.5	1
H_2SO_4 [mL]	23.3	23.3	23.3	46.6
H_2O_2 [mL]	3.3	3.3	3.3	6.6
Stirring Duration [h]	3	5	7	14

Table 1 showed the designations of the prepared sample. GO AA, AB, AC, and D were experimental parameter to synthesis graphene oxide and reduced graphene oxide. Those parameters were synthesized from used electrode graphite with the 1:2 of used electrode graphite and $KMnO_4$, 1:1 with sodium nitrate. Same amount of reaction temperature was used but the reaction time was varied as shown in table 1.

The amount of oxidant, reaction temperature, and reaction time were the main things to determine resulted graphene oxide and reduced graphene oxide. The effect of the parameter on oxidation level of graphite oxide was studied by visual observation. Yellow colour indicates the reaction fully happened while a bit of green colour means the reaction is not fully occurred. For further analysis, the degree of oxidation can be found and studied by using UV-Visible spectra. The effect of reaction duration of oxidant potassium permanganate, the equal concentration of graphite powder of GO AA, AB, and AC are collected. Resulted graphite oxide were grounded and dispersed into same amount of distilled water.

Dispersed graphene oxide of sample was characterized by UV Vis and the spectra for each sample were recorded as shown in figure 1. While each sample characterized, the sample of 14 hours reaction time also observed to compare with less than 10 hours reaction time. The adsorption peak was at 240 nm for all graphite oxide samples with different intensity level. The peaks are related to C and C bonding which similar to reported value in theoretical literature [23-24].

Enough time is required for higher degree of oxidation as observed by UV-Vis spectra. It could be over oxidation whenever the reaction time overcame the limits, and 14 hours is the limits for oxidation reaction time.

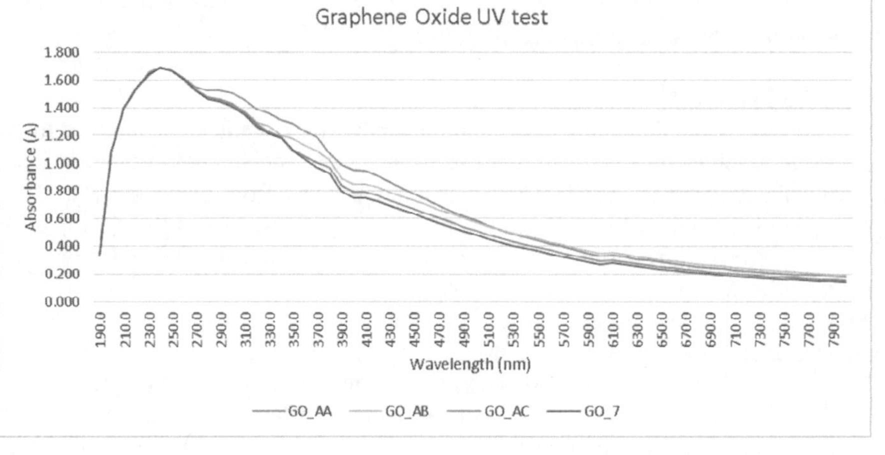

Figure 1. UV-Visible spectra recorded in distilled water at different reaction time

IOP Conf. Series: Materials Science and Engineering **541** (2019) 012032 doi:10.1088/1757-899X/541/1/012032

While the adsorption peak same for each sample at 240 nm, the intensity level different. GO AA provides 1.694, GO AB provides 1.692, GO AC provide 1.691 and GO D provide 1.687. The successful of reduced graphene oxide was also observed by UV Vis spectroscopy as shown in figure 2. GO D which resulted better than other sample in visual observation will be reduced by zinc as reductant. After reduction by zinc, the peak of reduced graphene oxide shifted to 220 nm which indicates decreased electron concentration after reduction.

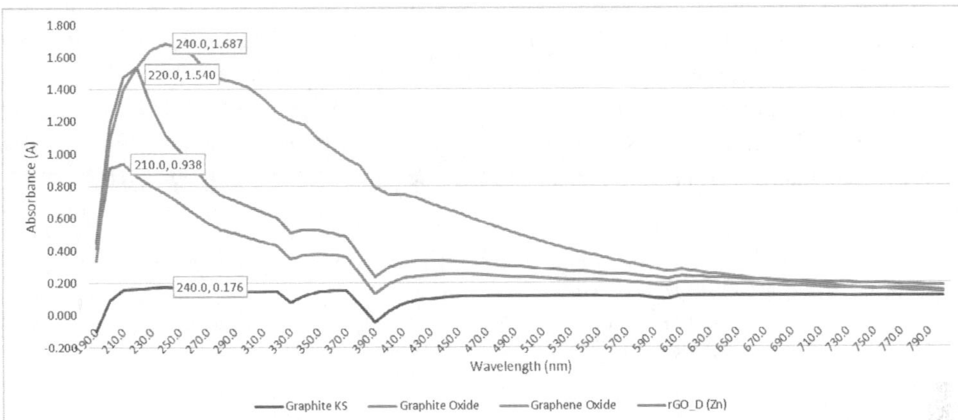

Figure 2. GO to rGO comparison UV Vis spectra of GO D

Three different SEM magnification of graphene oxide sample was collected. GO AC was used for Scanning Electron Microscopy (SEM) analysis. Based on SEM image, it was indicated the typical surface of graphene oxide. At the edge of morphological characterization, it is a little bit thick rather than in the middle as shown in figure 3. It is due to oxygen functional groups contained in resulted sample. As reaction time increase to enough time, amount of oxygen functional groups will be decreased. SEM image would be supported by Energy Dispersive X-Ray Spectroscopy (EDS) data analysis to investigate its composition. It is proven that surface of graphene oxide consists mostly by carbon and oxygen.

(a) (b) (c)

Figure 3 SEM analysis of graphite oxide sample GO AC (a) 10K magnification. (b) 25K magnification. (c) 40K magnification.

Furthermore, the purities of GO AC were observed by EDS analysis. Material contains such carbon, oxygen, sulphur, potassium and manganese were detected. The detailed energy and mass atom stated in figure 4. Element composition of graphene oxide was obtained by Energy Dispersive Spectroscopy (EDS) analysis. Graphene oxide was made up of 69.09 wt% C, 23.94 wt% O, 4.09 wt% S, 0.73 wt% K, 2.15 wt% Mn. In terms atomic weight, it was composed of 77.38 at% C, 20.13 at% O as main composition of graphene oxide.

IOP Conf. Series: Materials Science and Engineering **541** (2019) 012032 doi:10.1088/1757-899X/541/1/012032

According to EDS analysis, maximum peak of C element occurred at approximately 0.277 keV. Furthermore, O element peak were obtained at 0.525 keV and 2.307 keV as Sulphur element's peak. Mn and K element peak also obtained as 5.894 keV and 3.312 keV. Graphene oxide should only contain carbon and oxygen for its composition. Meanwhile, we found such as sulphur, potassium and manganese. This defect is due to synthesize process while oxidation happened. Excess sulphur from sulphuric acid, potassium and manganese from potassium permanganate are shown in the table 2.

Sulphuric acid and potassium permanganate as oxidant agent were left on the mixture due to stirring duration at 40° C. It is due to time limitation for oxidation process and full oxidation was not happened. For better oxidation process, stirring duration should be increase more than 12 h. Another factor that affects the impurities is raw material. Due to electrode was used for this experiment rather than pure graphite, we can assume that electrode is not pure carbon. It contains impurity inside it so will effect in last analysis on result. Thus, electrode itself may contain sulphur even before oxidation process.

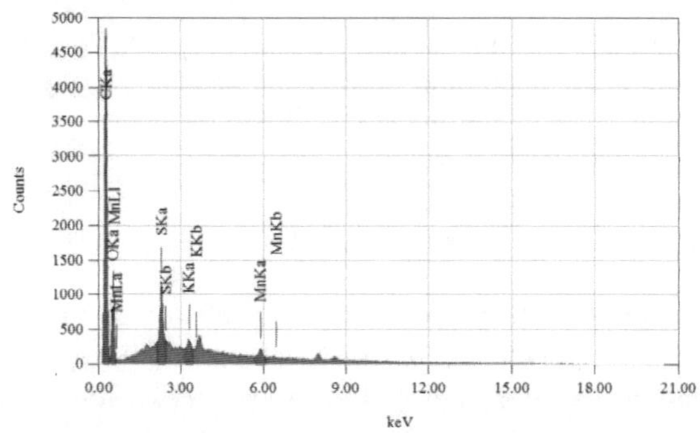

Figure 4. EDS analysis of GO AC

Table 2 Material composition

Element	keV	Mass %	Error %	Atom %
C	0.277	69.09	0.27	77.38
O	0.525	23.94	1.49	20.13
S	2.307	4.09	0.27	1.72
K	3.312	0.73	0.43	0.25
Mn	5.894	2.15	1.14	0.53
Total		100.00		100.00

The presence of oxygen functional group that detected in SEM analysis, was observed by FT-IR characterization. GO D spectrum was recorded and a broad peak between 3500 cm^{-1} to 2900 cm^{-1} indicates the presence of O-H carboxyl or oxygen functional groups as shown in figure 5. The peaks between 2924 cm^{-1} and 2854 cm^{-1} occurred due to asymmetric and symmetric CH_2 stretching of GO respectively while the peak around 1620 cm^{-1} is attributed to C=C stretches [14]. 1087 cm^{-1} corresponds to C-O stretching vibration of C-O-C [12].

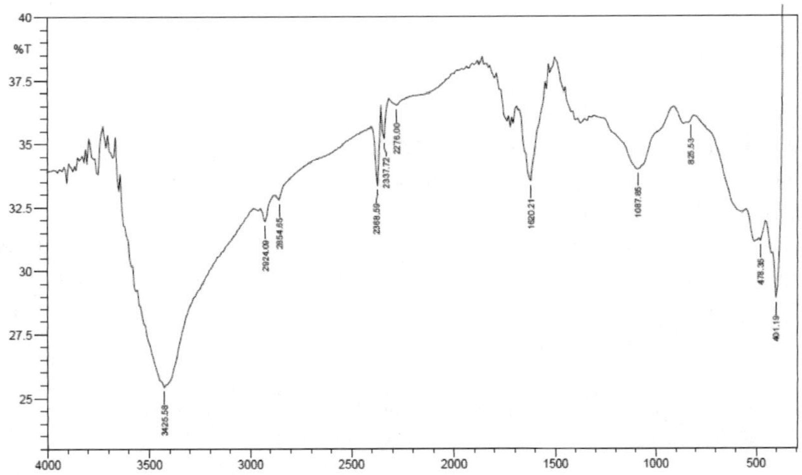

Figure 5. FTIR spectra of graphene oxide GO D

X-ray diffraction pattern of graphene oxide was collected from GO AC sample as shown in figure 6. The diffraction peak of the sample at around 2θ=9 till 12 degree corresponding to a layer to layer distance of 0.88 nm more or less which is close to the literature [15]. The patterns showed a larger interlayer spacing of graphene oxide due to oxygen functional groups between the layers [16], and exhibits turbostatic disorder [17]. The peak at 26 was not detected which confirm that the used electrode graphite is successfully oxidized after treatment of the oxidation and exfoliation resulted the formation of oxygen-containing functional groups of the graphene oxide.

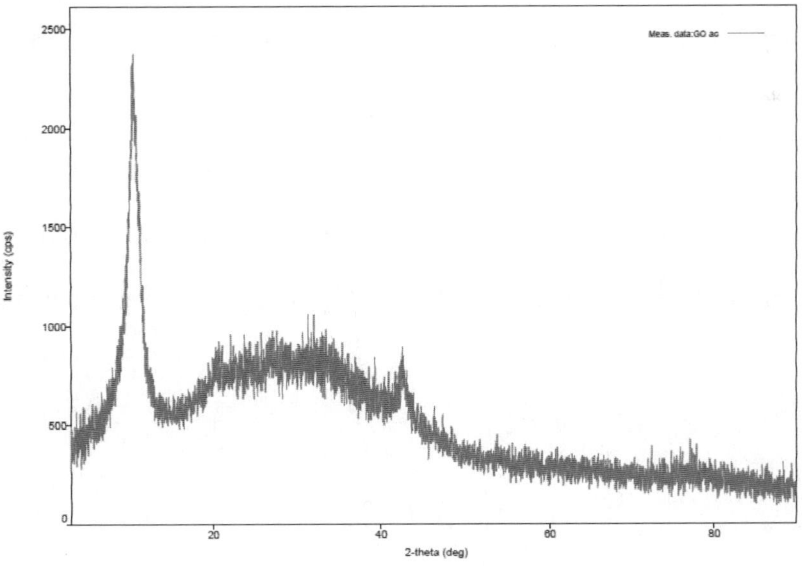

Figure 6. X-ray diffraction analysis of graphite oxide from GO AC sample

Reduction of graphene oxide is mainly to remove oxygen functional groups. The interlayer distance of reduced graphene oxide (rGO) by zinc should be decreased due to removal of oxygen functional groups which resulted in restacking of rGO sheets. Further analysis should be characterized rGO by zinc to compare graphene oxide and rGO.

IOP Conf. Series: Materials Science and Engineering **541** (2019) 012032 doi:10.1088/1757-899X/541/1/012032

4. Conclusion

In this study of graphene oxide and reduced graphene oxide were successfully synthesized. The synthesis was done by convenient method. Graphene oxide obtained by oxidation treatment with potassium permanganate and reduced graphene oxide by zinc as green reductant. Both results characterized by UV vis, FT-IR, SEM and EDS, XRD.

This work showed that synthesizing graphene oxide and reduced graphene oxide could possible from used electrode and resulted a good result for numerous applications. Reaction duration observed showed that enough time is better to obtained higher oxidation level when using certain reaction temperature which is 40 C. 12 hours to 14 hours still remains the best reaction duration, but used electrode could be an alternative way to produce graphene oxide and reduced graphene oxide.

References

[1] Geim A . and Novoselov K . 2004 *Nature Materials* **6** 183–191
[2] Jiang X, Setodoi S, Fukumoto S, Imae I and Komaguchi K 2013 *Carbon* **67** 662–672
[3] Lemme M C, Member S, Echtermeyer T J, Baus M and Kurz H 2007 **28** 282–4
[4] Stankovich S, Dikin D A, Dommett G H B, Kohlhaas K M, Zimney E J, Stach E A, Piner R D, Nguyen S T and Ruoff R S 2006 *Nature* **442** 282–286
[5] Stoller M D, Park S, Zhu Y, An J and Ruoff R S 2008 *Nano Lett.* **8** 3498–3502
[6] Zhu B Y, Murali S, Cai W, Li X, Suk J W, Potts J R and Ruoff R S 2010 *Adv Mater.* **22**3906-24 3906–3924
[7] Ruoff R 2008 *Nature Nanotechnology* **3** 10–11
[8] Alam S N, Sharma N and Kumar L 2017 *Graphene* **06** 1–18
[9] Kumar M K P and Srivastava C 2016 *Appl. Mater. Sci. Eng* **68** 374–383
[10] Liu J, Kok C, Zhan D, Lai L, Hua S, Wang L, Liu X, Gopal N, Li C, Shen Z and Lin J 2013 *Nano Energy* **2** 377–386
[11] Parvez K, Li R, Puniredd S R, Hernandez Y, Hinkel F, Wang S, Feng X, Mu K, 2013 *ACS Nan.,* 2013 **7** 3598–3606
[12] Emiru T F and Ayele D W 2017 *Egypt. J. Basic Appl. Sci.* **4** 74–79
[13] Gilje S, Kaner R B, Wallace G G, Li D A N and Mu M B 2008 *Nature Nanotechnology* **3** 101–105
[14] Guo H L, Wang X, Qian Q, Wang F and Xia X 2009 *ACS Nano*. **3** 2653–2659
[15] Allahbaksh A, Sharif F, Mazinani S and Kalae M 2014 *International Journal of Nano Dimension* **5** 11–20
[16] Lerf A, Buchsteiner A, Pieper, Schottl S, dekany, Szabo T, Boehm H P 2006 *J. Phys. Chem. Solid.* **67** 1106–10
[17] Nethravathi C, Rajamathi J T, Ravishankar N, Shivakumara C and Rajamathi M 2008 *Langmuir* **24** 8240–8244

Fabrication, microstructure and corrosion study of porous Mg-Ca-Zn using $CaCO_3$ pre-treatment with sodium trisilicate

A Erryani[1], I Al-aziz[2], F P Lestari[1], I Kartika[1]
[1]Research Center for Metallurgy and Material, Indonesian Institute of Science (LIPI) Kawasan Puspiptek Gedung 470 Serpong Tangerang Selatan, 15314, Indonesia
[2]Department of Metallurgy, Faculty of Engineering, University of Sultan Ageng Tirtayasa, Cilegon, Banten 42435.

E-mail: apri011@lipi.go.id

Abstract. Magnesium-based alloys have the good potential as a biodegradable implant that can be degraded in biological tissue. In this research, the fabrication of porous implant materials based on Mg-Ca-Zn alloy develop by powder metallurgy. $CaCO_3$ as a foaming agent which previously coated by using sodium trisilicate has been used to obtain open celullar pores. The alloy compositions used in this research are Mg-Ca-3Zn-5$CaCO_3$ and Mg-Ca-3Zn-10$CaCO_3$. Sintering process is carried out at temperature of 600, 650 and 700 °C with holding time of 5h. The alloy is characterized by using XRD (microstructure analysis), SEM (surface and pore analysis) and corrosion test of alloys (electrochemical analysis using tafel graph). The microstructures of the alloy show MgO and Mg_2Ca phase SEM results show the pores in the alloy with the size approximately arround 194.843-266,938μm. The corrosion rate in alloys is obtained in the range of 0.4-3.2 mmpy. The morphology of the alloy surface after corrosion process shows the volcano type. The corrosion process occurs localized where the electrochemical reaction occurs, so as to form a distinctive structure with the shape of the circle and the hole in the middle.

1. Introduction

Metal foams or porous metal are engineering material that have been unique combination of mechanical, physical and thermal properties [1]. These pure or alloy metals have shown a significant role in many parts of daily life, especially in the medical field such as material implant [2]. The chemical composition of implant determines the nature of the changes in the living body [3]. Metal alloys for the biomaterial should also have good corrosion resistance in an environment of tissues and body fluids and mechanical properties, which are required for load transfers [4][5].

One of the most widely used metal and alloy as an implant material is magnesium and magnesium alloys [6][7]. Magnesium has low corrosion resistant therefore its used in biomedical applications. Magnesium alloys have been regarded as materials for biodegradable orthopedic implants. High reactivity of magnesium in corrosive media can be used as an advantage [8]. Magnesium alloys have been regarded as materials for biodegradable orthopedic implants [8].

Biodegradable implant materials which were identical to porous materials have similar properties with human bones. It is expected that porous materials are also capable of fossilization in the process of growth of new bone tissue [9]. Cellular metals and metal foams are metals with pores that are intentionally connected in their structure. The terms cellular metals or porous metals are general expressions referring to metals having large volume of porosities, while the terms foamed metal or metallic foams applies to porous metals produced with processes where foaming take place. Besides, the term metal sponge refers to highly porous materials with complex and interconnected porosity where the porosity cannot be subdivided into well-defined cells [10][11].

In this research, Mg Alloy foam was prepared by powder metallurgy method. Porous structure in Mg-Zn-Ca alloy can be obtained by using a foaming agent. Foaming process is performed with the gas-releasing agent and other additives that produce a foamed alloy, which differs greatly from the matrix alloy [12]. The foaming agent used in this study is $CaCO_3$ which was previously coated with a passive layer of SiO_2. SiO_2 passive layer was coated on the surface of $CaCO_3$ to decrease its foaming speed and conditioning the foaming process more simultaneously resulting in a more uniform pore size [13].

2. Material and Method

Mg, Zn and Ca as main alloys are mixed with $CaCO_3$ which serves as a foaming agent. The composition of alloys (wt.%) are Mg-Ca-3Zn-5$CaCO_3$, Mg-Ca-3Zn-10$CaCO_3$. Previously, $CaCO_3$ is coated using sodium trisilicate with sol-gel method. The alloying process is carried out by using powder metallurgy. Mixing process is done using a shaker mill for approximately 2 hours in order to obtain a homogeneous powder size in the alloy. Then, the alloys are compacted with a pressure of 100 MPa for 2 minutes and 200 MPa for 3 minutes. The green compact is heated in a tube furnace with the Argon atmosphere. Sintering process is carried out in the temperature of 600, 650 and 700 °C and holding time for 5 h. Rate of temperature rise in the furnace is 5°/ min.Phase analysis was performed using X-Ray diffraction technique (Shimadzu XRD-7000), while microstructure of alloys are characterized by using scaning electron microscopy (JEOL, JSM-6390A Japan).

Corrosion testing is performed using Gamry Instrument G750 according to ASTM G5-94 standard. Three electrode cells are used for corrosion measuring i.e; calomel electrode (SCE), graphite electrode (opposing electrode) and sample as awith a pH of 7.4. The SBF solution temperature is 37 °C which is prepared using laboratory chemicals with the following composition:

Table 1. Chemical Composition of Hank's Solution.

Component	Composition (g/ml)
$CaCl_2$	0.14
KCl	0.4
KH_2PO_4	0.06
$MgCl_2$.6H2O	0.1
$MgSO_4$.7H2O	0.1
NaCl	8.0
$NaHCO_3$	0.35
Na_2HPO_4	0.048
Glucose	1.0

Microstructures of the alloy after corrosion testing were observed by using scanning electron microscopy (JEOL, JSM-6390A Japan).

3. Results and Discussion

3.1 X-Ray Diffraction Analysis

XRD analysis is used to determine the phase or crystals formed after various temperatures sintering process of 600, 650 and 700°C with a 5-hour holding time in Mg-Zn-Ca alloys with $CaCO_3$ coated by natrium trisilicate (Figure 1- 3).

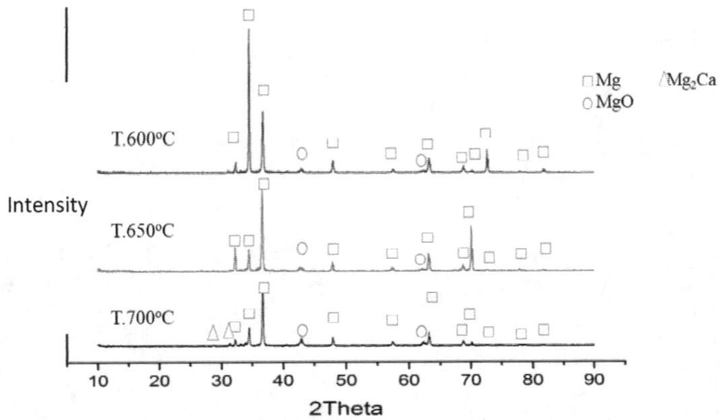

Figure 1. XRD Pattern of Mg-3Zn-1Ca-5$CaCO_3$ with various temperatures sintering process.

In general as shown in Figure 1, Mg and MgO phases was obtained in the XRD pattern, while the Mg_2Ca was revealed in the alloys at sintering temperature of 700°C. The predominantly formed Mg phase in the alloy is due to Mg element as base metal, whereas Zn and Ca are merely small metal compositions.

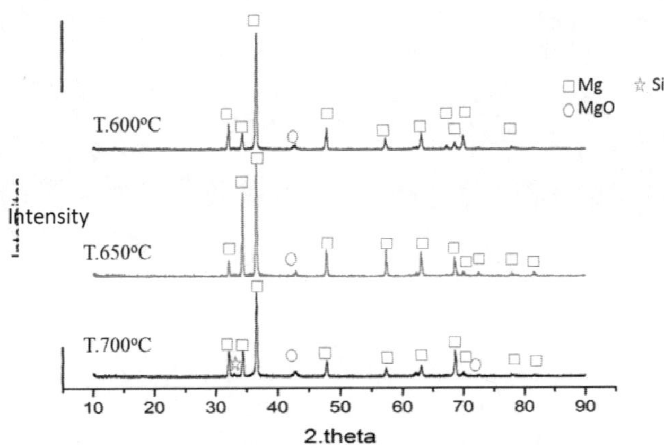

Figure 2. XRD Pattern of Mg-3Zn-1Ca-10$CaCO_3$ with various temperatures sintering process.

The lattice formed of Mg crystallographic system is Hexagonal, with an average distance between lattices of a = 3.2093 Å and MgO is cubic-with the distance between atoms a = 4.2112 Å. The MgO indicates that oxygen is still presented during sintering process. MgO in its manufacture is avoided because it can decrease the mechanical properties of Mg-based materials. The non-toxic MgO oxide in the body only affects the mechanical properties of the implant [14].

3.2 Scanning Electron Microscopy Analysis

Figure 3 shows SEM results of Mg-3Zn-1Ca-5$CaCO_3$-coated with natrium trisilicate at various sintering temperatures with 5 h holding time.

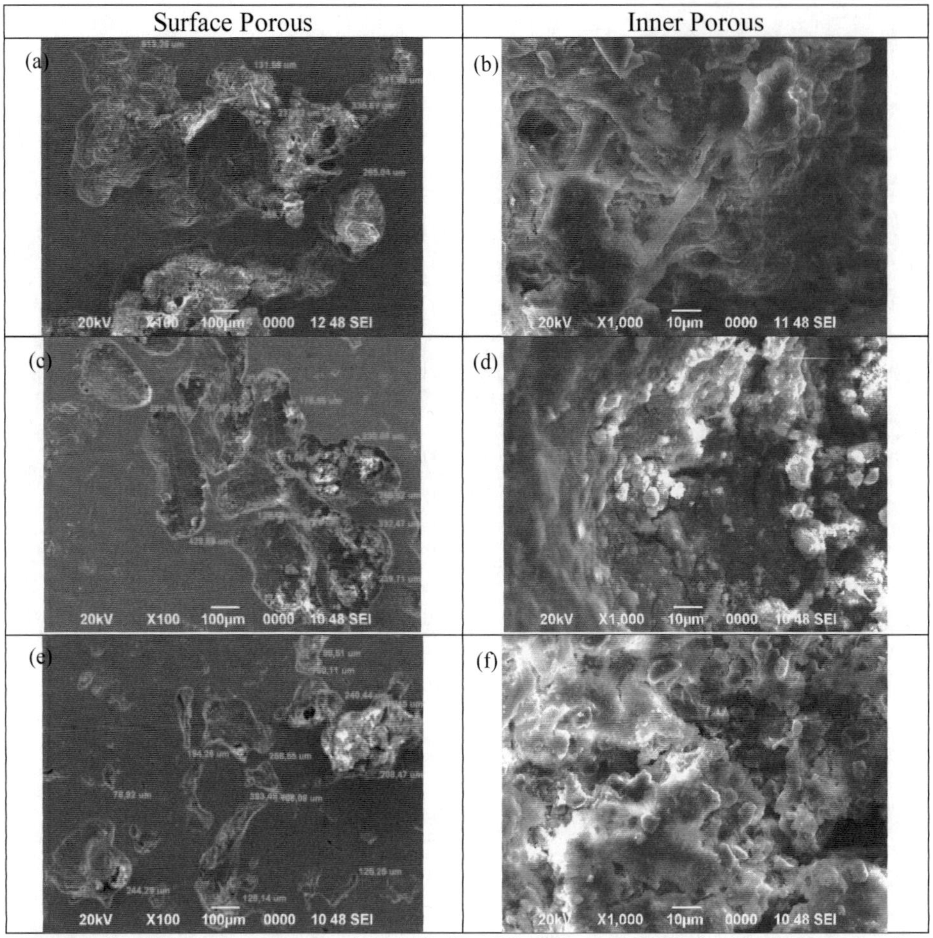

Figure 3. SEM Micrograph of Mg-Ca-Zn-5$CaCO_3$ at 600°C (a, b), 650°C (c, d), 700°C (e, f).

The analysis performed by SEM shows that the alloy composition has a large pore cavity of 254μm at 600°C (figure. 3a), slightly different from the pore size that is in figure 3 (c) with average size of pores approximately around 266 μm at 650°C. As shown in figure 3 (e), the alloy at temperature sintering of 700°C has pores with a fine structure and uniform compared to the structures in figure 3 (a) and 3 (c). These result approved that pretreatment in foaming agent can obtain homogeneous of fine pore strucure. Meanwhile, the pore size obtained in the previous study showed that the average pore size without pre-treatment SiO_2-Coated $CaCO_3$ at 600°C and 650°C was 10-158 μm and 15-168 μm [15].

This difference in number and distribution is influenced by the difference in sintering temperature, along with the temperature sintering allowing the number and distribution of pores to be formed to be larger. Temperature of 700°C the pore number decreases this is due to the sintering critical temperature in the use of $CaCO_3$ foaming agent. Critical temperature is evaluated from the comparison of base metal melting temperature of Mg of 650°C and the sintering temperature variation of $CaCO_3$ which is used is 700°C with a holding time of 5 hours so that the formed pore is shrinkage because Mg having melting temperature below the decomposition temperature of $CaCO_3$.

IOP Conf. Series: Materials Science and Engineering **541** (2019) 012033 doi:10.1088/1757-899X/541/1/012033

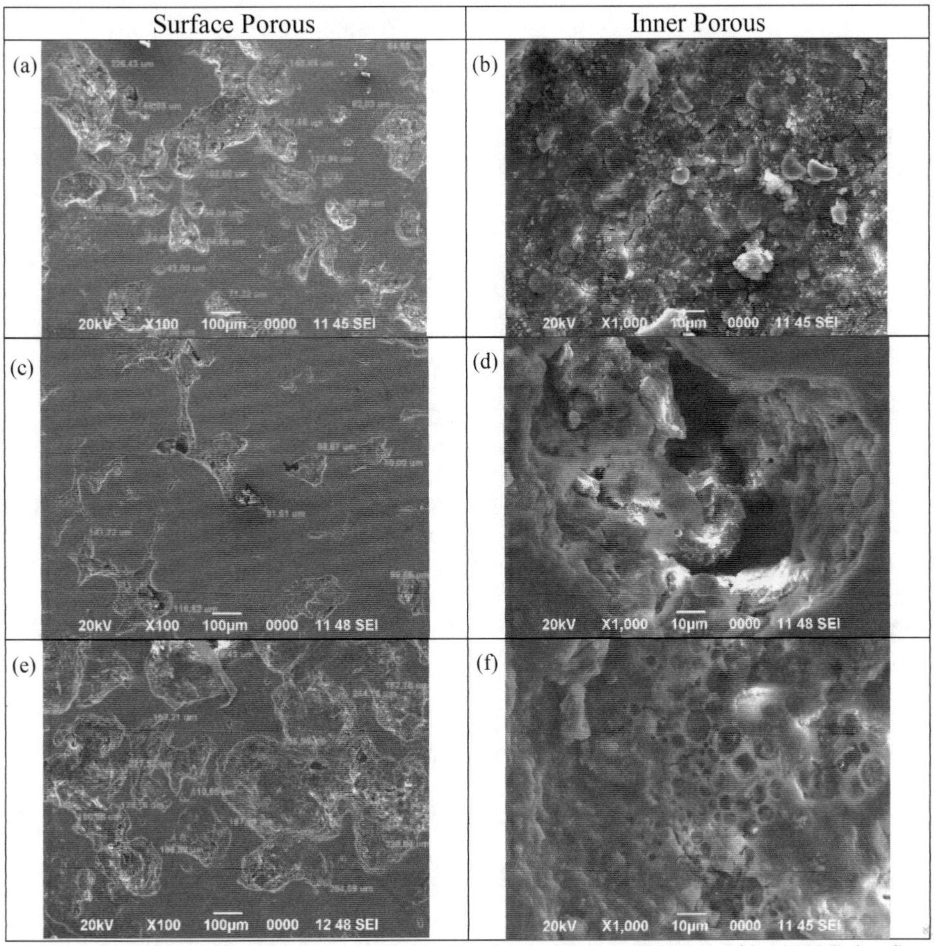

Figure 4. SEM Micrograph of Mg-Ca-Zn-10$CaCO_3$ at 600°C (a, b), 650°C (c, d), 700°C (e, f).

SEM analysis with 100x magnification seems to be seen that the pores are very diverse. The number and distribution of pores formed in this study have characteristics as shown in figure 4 (a) more evenly and larger than Figure 3.4 (c) while figure 3.4 (e) shows a larger pore size than the size and distribution of pores 3.4 (a). The average size of the pores formed by the addition of 10% wt of $CaCO_3$ coated natriun trisilicate at sintering temperature is 83.1147µm at 600°C, 103.08µm at 650°C and 195.335µm at 700°C. Table 3.1 shows the comparison of average pore size that can be with the composition of foaming agent 5 and 10%.

Table 2. Porosity of Mg-Ca-Zn-$CaCO_3$ alloys.

Temperature (°C)	Mg-Ca-Zn-$CaCO_3$			
	5 wt. % (μm)	*Range* (μm)	10 wt. % (μm)	*Range* (μm)
600	254.942	111.8 – 513.26	83.115	30.07 – 140.01
650	266.938	107.81 – 429.9	103.080	70 – 141.72
700	194.843	79.92 – 393.49	195.335	110.65 – 255.56

Table 2 show the pore size comparison between 5 wt. % and 10 wt. % of coated $CaCO_3$ in these alloys after sintering. Mg can still be stable with the growth of pores throughout the base metal, so that the pores formed do not experience of shrinkage. The influence of temperature may also lead to the possibility of smaller pore size. As shown on that table that in the temperature of 700°C at a composition of 5wt% coated $CaCO_3$ smaller pore size was formed. In the addition of foaming agent up to 10wt.% coated $CaCO_3$ at 700°C is related to the decomposition of $CaCO_3$, it takes time and sufficient heat energy to decompose $CaCO_3$ into gas and form pores.

3.3 Electrochemical Analysis
Electrochemical test is performed to determine the rate of corrosion of implanted material to the simulated body fluid. Mg is one of the most reactive metal elements, meaning that this metal is easily oxidized and very easily degraded in the body due to direct contact with human biological system.

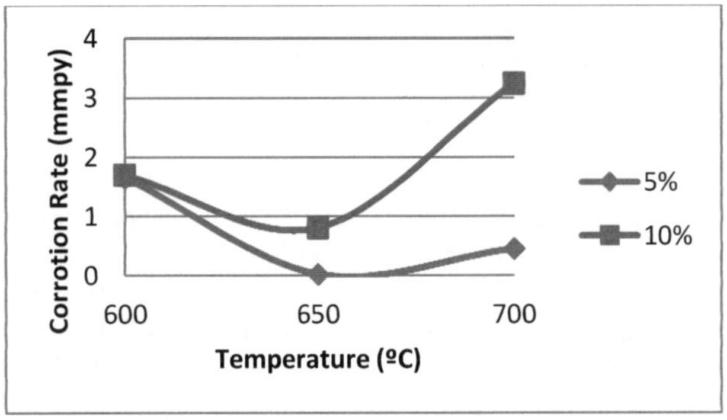

Figure 5. Corrosion rate of Mg-Ca-Zn-$CaCO_3$ alloys.

Figure 5 shows an increase in corrosion rate at a temperature of 700°C. This is thought to be due to the high shaft count and MgO and Mg_2Ca phase formation more than the other two temperatures there by impacting low corrosion resistance compared to 600°C and 650°C temperatures. The lowest corrosion rate is found in the composition of 5% foaming agent with 650°C sintering temperature. The lower corrosion rate is due to the more uniform pore size and the influence of the less intermetallic phase generated (MgO and Mg_2Ca). Therefore, it can be said that the alloy composition with 5% foaming agent and sintering temperature of 650°C is the optimum corrosion rate generated.

3.4 Corrosion Morphology

After exposure in Hank's solution, the material surface looks rough and corrosion occurs. The pore structures that form on the surface of the alloy after corrosion like volcano. The results show that the corrosion is localized, as shown in figure 6 where Mg concurrent dissolution occurs with the evolution of H_2. The evolution of H_2 occurs when bubbles form. The bubbles of H_2 gas are formed on a specific surface or localized where electrochemical reactions occur, thus forming a distinctive structure with a circular shape and a hole in the middle [16].

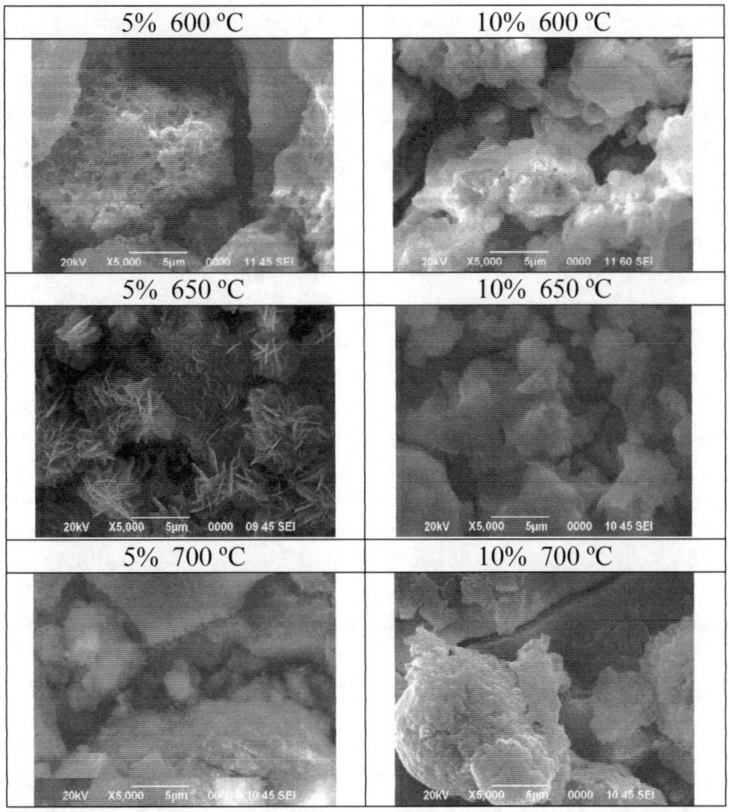

Figure 6. SEM micrograph of Mg-Ca-Zn-$CaCO_3$ alloys after exposure in Hank's solution.

Evolution of H_2 is always from cathodic, then with a second phase with a spherical structure is the water reduction area acting as a cathode.

$$\text{Anodic: } Mg \rightarrow Mg^{2+} \text{ dan } 2^{e-} \quad (1)$$

$$\text{cathodic: } 2H_2O + 2^{e-} \rightarrow H_2 + 2OH^- \quad (2)$$

$$\text{Product: } Mg^{2+} + 2OH^- \rightarrow Mg(OH)_2 \quad (3)$$

According to the cathodic reaction in equation 2, generally local pH increases because of the OH^- ions. As a result, corrosion products are easily formed in the second phase area. The $Mg(OH)_2$ formation is stimulated by cathodic reduction of water liberating OH^- and by Mg^{2+} free anodic corrosion.

In addition to the $Mg(OH)_2$ formation, other corrosion products formed during the metals exposed in Hank's solution are O, Ca, P and Mg. The presence of OH^- on the Mg surface causes the transformation of $HPO4^{2-}$ into $PO4^{3-}$. It is desirable that Ca^{2+} and $PO4^{3-}$ ions diffuse from Hank's solution to the Mg surface and coalesce in the $Mg(OH)_2$ layer which is a corrosion product. The solubility of

$Ca_5(PO_4)_3(OH)$ and $Mg_3(PO4)_2$ products is less than $Mg(OH)_2$ [17]. Thermodynamically, $Ca_5(PO_4)_3(OH)$ and $Mg_3(PO_4)_2$ are more stable than $Mg(OH)_2$. So it can be said that the final product of corrosion is $Mg(OH)_2$, $Ca_5(PO_4)_3(OH)$ and $Mg_3(PO_4)_2$[18].

4. Conclusion

Mg-Ca-Zn alloy using $CaCO_3$ coated natrium trisilicate as foaming agent are successfully fabricated by using powder metallurgy method. The phase analysis results using XRD show MgO and Mg_2Ca phases on the alloy. SEM results show the formation of pores from the decomposition of $CaCO_3$ as a foaming agent with the size approximately around 194.843-266.938μm. The corrosion rates of this alloy obtained in the range of 0.4-3.2 mmpy. The morphology of the alloy surface after corrosion process shows a volcano type. The corrosion process occurs localized where an electrochemical reaction occurs, so as to form a distinctive structure with the shape of the circle and the hole in the middle.

References

[1] Yang D, Yang C and Hur B 2008 *Mater. Sci. Techno.* **24** 302–304
[2] Orinak A, Orinakova R, Kralova Z O, Turonova A M, Kupkova M, Hrubovkakova M, Radonak J, and Dzuunda R 2014 *J. Porous Mater* **21** 131–140
[3] Gonzlez S, Pellicer E, Suriach S, Bar M D and Sort J 2013 *Bio. and Mech. Int. of Mag. and* **12**
[4] Nowosielski R, Cesarz R, Babilas K 2013 *J. Achiev. Mater. Manuf. Eng.* **58** 7–15
[5] Zander D and Zumdick N A, 2015 *Corros. Sci.* **93** 222–233
[6] Sahinoja M T 2013 Tampere University Of Technology
[7] Zivić F, Grujović N, Manivasagam G, Richard C, Landoulsi J and Petrović V 2014 *Tribol. Ind.* **36 (1)** 67–73
[8] Wang H and Shi Z M 2008 *Mater. Forum* **34** 100–104
[9] Erryani A, Pramuji F, Annur D, Amal M I, and Kartika I 2017 *IOP Conf. Ser. Mater. Sci. Eng.* **202** 012028
[10] Lefebvre J, Banhart L P and Dunand C David 2008 *Adv. Eng. Mater.* **10** 775–787
[11] Zhao X, Shi L L and Xu J 2013 *J. Mech. Behav. Biomed. Mater.* **18** 181–190
[12] Seyedraoufi Z S and Mirdamadi S 2013 *J. Mech. Behav. Biomed. Mater.* **21** 1–8
[13] Lu G Q, Hao H, Wang F Y and Zhang X G 2013 *Trans Nonferrous Met. Soc. China* **23** 1832–1837
[14] Niederlaender J, Walter M, Krajewski S, Schweizer E, Post M, Schille C, Geis-Gerstorfer J and Wendel H P 2014 *J. Mater. Sci. Mater. Med.* **25** 835–843
[15] Erryani A Lestari F P, Annur D and Kartika I, Sriyono 2015 *Proceeding 2nd Int. Conf. Mater. Metall. Technol.* 423–434
[16] Noviana D, Paramitha D, Ulum M F and Hermawan H 2016 *J. Orthop. Transl.* **5** 9
[17] Fajardo S and Frankel G S 2015 *Electrochim. Acta* **166** 403
[18] Hou L, Li Z, Pan Y, Du L, Li X, Zheng Y and Li L 2014 *Prog. Nat. Sci. Mater. Int.* **24** 466–471

Acknowledgment

The author would like to thank for the financial support by Excellent Research Program of Indonesian Institutes of Sciences (UNGGULAN LIPI) 2017. This paper was supported by Research Center for Metallurgy and Materials – LIPI, especially biocompatible implant research group.

Optical properties of ZnO nanorods derived from chemical bath deposition process with different seeds solution concentration

L Suhaimi and A H Yuwono
Department of Metallurgical and Materials Engineering, Faculty of Engineering Universitas Indonesia, Depok-West Java, 16424, Indonesia

E-mail: ahyuwono@eng.ui.ac.id

Abstract. Owing to their high surface to volume ratio and fast electron transfer, zinc oxide (ZnO) nanorods have been well-known as potential nanostructured material for various applications including sensors, dye sensitized solar cells, optoelectronic, transparent heater and biomedical devices. Among other synthesizing techniques for obtaining ZnO nanorods, chemical bath deposition (CBD) has been thought as a simple and low-cost method. However, there are several processing parameters that need to be investigated for the above-mentioned applications where the highly optical transparency of thin film ZnO nanorods grown on glass substrates is one of important targets to be achieved. In this work, ZnO nanorods were synthesized through CBD process at low temperature (0°C) by using seed solution prepared by dissolving 1: 1 equimolar zinc nitrate tetrahydrate and hexamethylene tetraamine. For investigation purposes, three different concentration of seed solutions i.e. 0.005, 0.025 and 0.05 M were used. Thin films containing ZnO nanoseeds were formed by spin coating the precursors on the glass substrates, followed with annealing at 200°C for 5 minutes. Finally, the ZnO nanorods were further grown at 90°C for 3 hours in the beaker glass using the same solution. X-ray diffraction (XRD) analysis showed that all ZnO nanorods demonstrated a strong (002) peak belong to wurtzite phase. It was found that the estimated crystallite size and band gap energy (E_g) for ZnO nanorods derived from the seed solutions of 0.005, 0.025 and 0.05 M were 21.42, 137.11, 171.39 nm, and 3.60, 3.20, 3.18 eV, respectively. However, the optical transparency was adversely lowered from about 75 to 40 % as a result of increased coverage of ZnO nanorods on the glass substrate. For transparent heater application where, a desired combination of high optical transparency and suitable electronic properties is needed, the current results were considered to be promising.

1. Introduction

The development of technology in various area today cannot be separated from the use of semiconductor materials. One of the most widely used semiconductor materials is zinc oxide (ZnO). ZnO has unique properties such as direct band gap energy of 3.37 eV, high-electrochemical stability, high thermal and mechanical stability at room, nontoxic, excellent optical and electrical properties, and easily formed into nanostructured materials [1]. These unique properties make the material useful for optoelectronic devices such as sensors, solar cell energy, laser technology, optoelectronics, and transparent conducting oxide [2]. ZnO nanorods is one dimensional (1D) nanostructured material that has received considerable attention for photovoltaic devices due to its unique optical and electrical properties including much higher surface to volume ratio and faster electron transfer.

ZnO nanorods have been synthesized by using different physical and chemical methods such as chemical bath deposition (CBD) [3,4], RF sputtering [5], spray pyrolysis [6], metal chemical vapor deposition [7], vapor phase transport [8] and pulsed laser deposition [9]. Among these fabrication techniques, chemical bath deposition (CBD) has been considered as versatile one because it can offer many advantages such as easily available equipment, affordable cost, simple process, and harmless. In addition, the synthesis process takes place at low temperatures which make it suitable for producing large areas of thin film.

There are several important CBD parameters that significantly influences the formation of ZnO nanorods. These include precursor concentration, pH of the solution, bath temperature, thickness of the seed layer, type of substrates and time. It has been reported that the precursor concentration has affected the size and morphology of the formed ZnO nanorods [10]. For various optical applications, the size, shape and morphology of the ZnO nanorods is of importance and they need to be optimized. In this paper, therefore, we report the influence of seed solution concentration on the nanostructural characteristics and optical property of ZnO nanorods.

2. Methodology

2.1. Preparation of Seed Solutions and Layer Deposition

Prior to seed solutions preparation and seed layer deposition, the glass substrates as a growth medium for the ZnO nanorods formation were ultrasonically cleaned in beakers containing deionized water, acetone and ethanol, respectively for 8 minutes each and dried in water. The seed solution concentration starting from 0.005, 0.025 and 0.05 M were prepared by dissolving 1: 1 equimolar zinc nitrate tetrahydrate ($Zn (NO_3)_2.4H_2O$, Merck) and hexamethylene tetraamine ($C_6H_{12}N_4$ / HMTA, Merck) in water at 0°C for 1 hour. Thin films containing ZnO nanoseeds were obtained through spin coating the precursors on the glass substrates with 2000 rpm for 20 seconds, followed with annealing at 200°C for 5 minutes.

2.2. Growing of ZnO Nanorods with Chemical Bath Deposition Method

Further growth of ZnO nanorods was carried out by soaking the glass substrate in each seeding solution for 3 hours at 90°C. Furthermore, the glass substrates were cleaned using distilled water and dried in air. The morphology of ZnO nanorods derived from different seed solution concentration was characterized by using field emission scanning electron microscopy (FE-SEM JEOL JIB tipe 4610F), while the crystal structure was investigated using X-ray diffraction (XRD, Pan Analytical X-Pert Pro). For analysis purposes, the crystallite size was calculated based on Debye-Scherrer method [11]. The optical absorbance-and transmittance of the ZnO nanorods for different seed solution concentration was obtained by using spectrometer UV-Vis (Shimadzu UV, 2450), while the resulting band gap energy, E_g of nanorods was calculated by using Tauc equation [12].

3. Results and Discussion

3.1. XRD Pattern of ZnO Nanorods with Different Seed Solution Concentration

Figure 1 shows XRD patterns of ZnO nanorods thin film with different seed solution concentration. The films are polycrystalline with hexagonal wurtzite ZnO according to JCPDS No. 36-1451. The (002) peak intensity of ZnO nanorods thin film increased significantly with the increase of seeding solution concentration. These results indicate that different seed solution concentration has affected the crystallinity of ZnO nanorods formed in this work. This can be attributed to the nucleation velocity of ZnO nuclei in the higher concentration which was faster than that of in the early stages with lower one, which resulted in bigger formation of ZnO crystallites. To obtain a more quantitative data on the crystallites growth, the diffraction data was processed further by using Debye-Scherrer method and the calculation results shows that crystallite size of nanorods increased significantly from 21.42 to 137.11 and finally 171.39 nm when the seed solutions concentration was added from 0.005 to 0.050 M were, respectively.

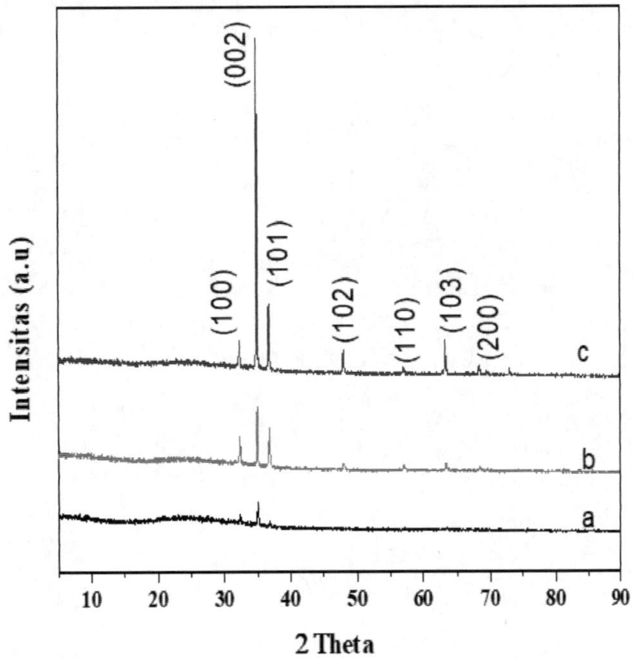

Figure 1. X-ray diffraction patterns for ZnO nanorods thin films with variation in seed solution concentration of: (a) 0.005, (b) 0.025, and (c) 0.050 M.

3.2. The Morphology Properties of ZnO nanorods with Different Seed Solutions Concentration

The FE-SEM results of ZnO nanorods morphology for each seed solution concentration is shown in Figure 2. It shows that at concentration of 0.005 M the ZnO nanorods have not resulted in a good coverage on glass substrate (Figure 2a and b). It can be understood since at low concentration the supersaturation level of ZnO nuclei was still weak. The growth of nanorod was thus retained due to slow nucleation and growth rate. With an increase in the precursor concentration up to 0.025 M, the morphology of sample was improved where a quite uniform distribution of nanorods with hexagonal shape was obtained (Figure 2c and d). Furthermore, with a higher concentration up to 0.050 M the coverage of ZnO nanorods appeared much more uniform where the nanorods grew in a hexagonal vertical direction. It was found that the average diameter for ZnO nanorods was increased from 161 to 205 and finally 427 nm when the above seeding solution concentration was increased. This is in agreement with the previous results of XRD analysis on the crystallite size growth of the nanorods.

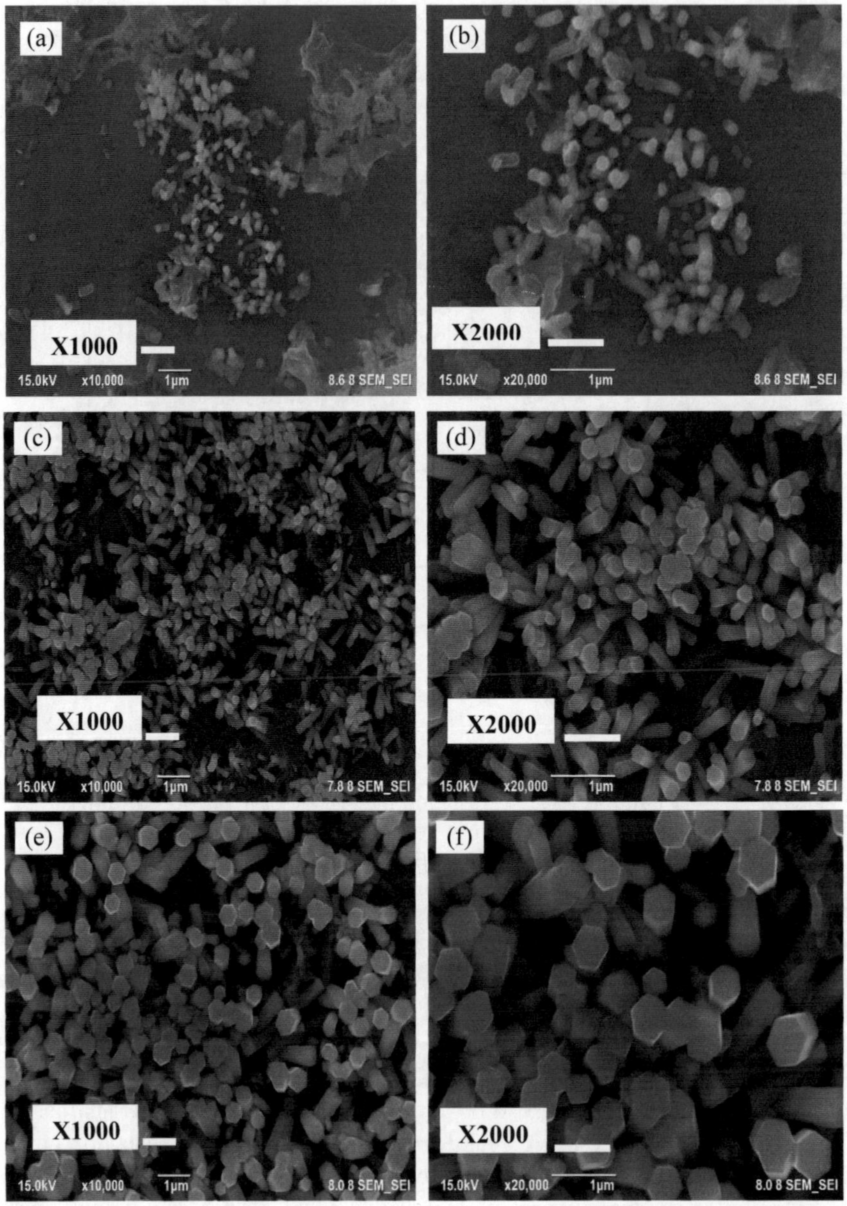

Figure 2. FE-SEM images of ZnO nanorods grown on the glass substrate with different seed solution concentration of: (a-b) 0.005, (c-d) 0.025 and (e-f) 0.050 M. The left figure is provided with magnification of 1000 X, while the right one with magnification of 2000 X.

3.3. The Optical Properties of ZnO Nanorods with Different Seed Solutions Concentration

Figure 3 shows the optical absorbance spectra of ZnO nanorods derived from different seed solution concentration. All the samples provided quite high transparency in the visible range (800-400 nm) and started to demonstrate absorption edge at around 380 nm. Sample with a seed solution concentration of 0.050 M has the highest absorbance in comparison to the other two samples, this result can be related to the higher crystallinity of nanorods with bigger crystallite size as has been studied by previous XRD analysis.

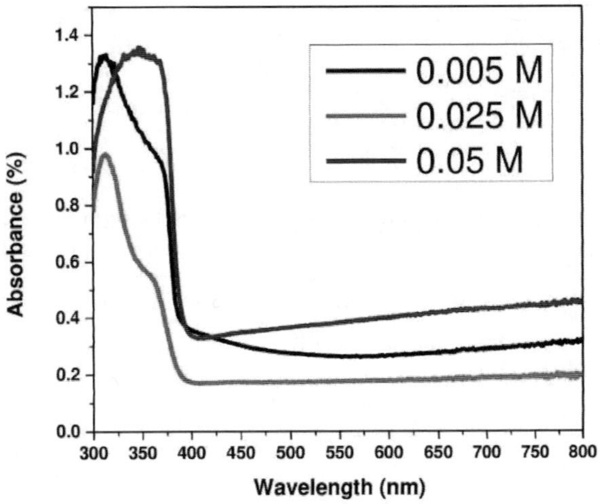

Figure 3. Optical absorption spectrum of ZnO nanorods for difference seed solution concentration (a) 0.005, (b) 0.025 and (c) 0.050 M.

The results of plotting $(\alpha h v)^2$ versus photon energy $(h v)$ of ZnO nanorods with three different seed solution concentrations by using Tauc equation is presented in Figure 4. It was found that with increasing seed solution concentration from 0.005 to 0.050 M the band gap energy, E_g decreased from 3.60 to 3.18 eV. The decrease in the band gap energy value can be correlated to the increase in the carrier concentration and the declining in the strain value [13,14]. Again, this complies with the XRD analysis previously showing the trend of increase in crystallite size. In nanometer regime, the smaller E_g is obtained with bigger nanocrystallite size of semiconductor materials, which relates to the electron excitation from valence band to conduction band. We already have observed this phenomena in our previous studies on conventionally annealed and post-hydrothermally treated TiO_2 nanoparticles derived from the sol-gel process [15], TiO_2 nanotubes [16,17], ZnO nanoparticles [18], and ZnO nanorods [19].

Figures 5a - c shows optical transmittance spectrum of ZnO nanorods thin films by using different seed solution concentration. It can be clearly seen that the transmittance value of the samples has decreased from 75 to 40% with the respective increase in the seed solution concentration. The drop of transmittance at higher seed solution concentration was due to the increase in photon scattering by increase of the roughness of the surface morphology [13]. Furthermore, the decrease in transmittance with increasing concentrations of seed solutions can also be associated to the increasing number of molecules involved in absorption of light energy [14]. Smaller band gap energy own by ZnO nanorods derived from higher concentration of seeding solution, as shown in previous analysis by Tauc equation, also confirmed the easiness of electron excitation from valence band to conduction band, resulting more absorbance and thus lower transmittance.

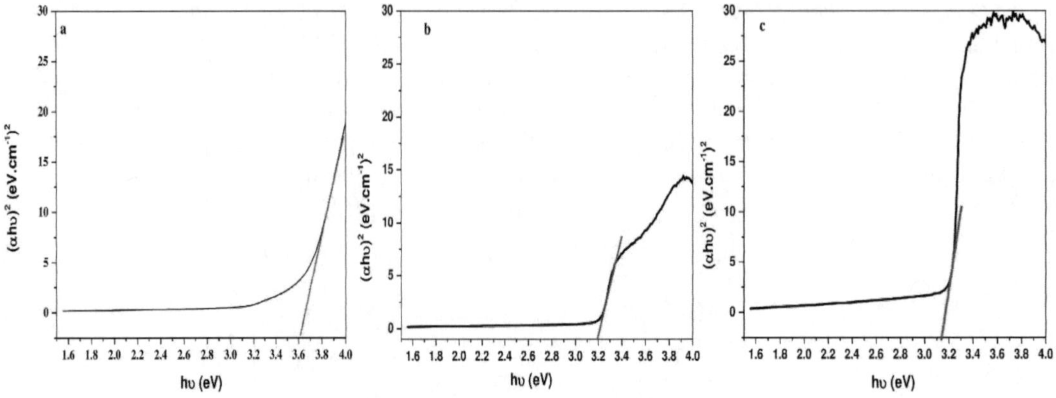

Figure 4. The plot of $(\alpha hv)^2$ versus phonon energy (hv) of ZnO nanorods thin film with different seed solution concentration of (a) 0.005, (b) 0.025 and (c) 0.050 M.

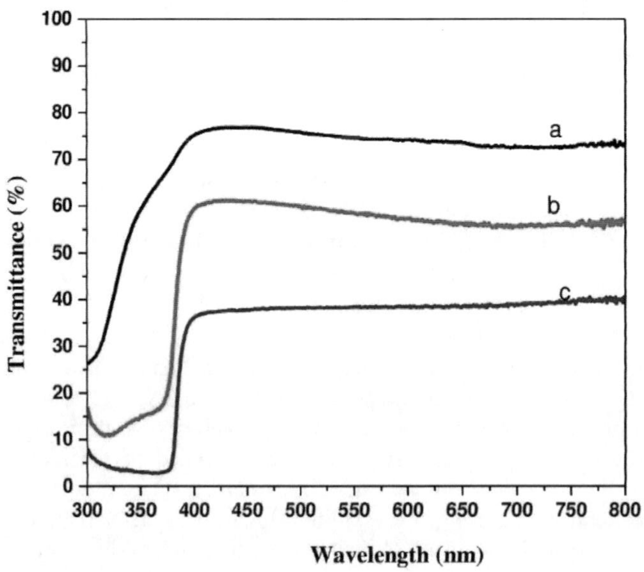

Figure 5. Optical transmittance spectra of ZnO nanorods thin film with different seed solution concentration of: (a) 0.005, (b) 0.025 and (c) 0.05 M.

4. Conclusion

The effect of seed solution concentration by CBD process on optical properties, morphological properties, and crystal structure of ZnO nanorods thin film have been studied in this work. The band gap energy from the thin film of ZnO nanorods decreased from 3.60 to 3.18 eV with increasing concentration of seed solution from 0.005 to 0.05 M. Meanwhile, the optical transmittance spectrum of ZnO nanorods was reduced to 40% as a result of bigger crystallite size causing more photon energy absorption. The average diameter of nanorods was found to increase from 161 to 427 nm as a consequence of crystallite growth within the nanorods.

References

[1] Yi G C, Wang C and Park W I 2005 *Semiconductor Science and Technology* **20** 22–34
[2] Godlewski M, Guziewicz E, Łuka G, Krajewski T, Łukasiewicz M, Wacnicka A, Kopalko K, Sarem A and Dalati B 2009 *Thin Solid Films* **518** 1145–148
[3] Lang J, Yang J, Li C, Yang L, Han Q, Zhang Y, Wang D, Gao M and Liu X 2008 *Crystal Research and Technology* **43** 1314–1317
[4] Gurav K, Patil U, Pawar S, Kim J and Lokhande C 2011 *Journal of Alloys and Compounds* **509** 7723–7728
[5] Tang N J, Wang H, Xu H, Peng and Fan C 2009 *Science in China Series E: Technological Sciences* **52** 2200–2203
[6] Shinde S, Patil G, Kajale D, Gaikwad V and Jain G 2012 *Journal of Alloys and Compounds* **528** 109–114
[7] Kim K S and Kim H W 2003 *Physica B: Condensed Matter* **328** 368–371
[8] Mofor A C, Bakin A, Fuhrmann D, Bertram F, Hangleiter A, Christen J and Waag A 2006 *Physica Status Solidi (c)* **3** 1046–1050
[9] Sun Y, Fuge G M and Ashfold M N 2004 *Chemical Physics Letters* **396** 21–26
[10] Abdulrahman A, Ahmed S and Almessiere M 2017 *Digest Journal of Nanomaterials & Biostructures (DJNB)* **12** 1001–1009
[11] Cullity B D 1978 *Elements of X-ray Diffraction* (USA: Addison-Wesley Publishing Company Reading-Massachusetts)
[12] Tauc J, Grigorovici R and Vancu A 1966 *Phys. Status Solidi* **15** 627–637
[13] Alami Z Y, Salem M, Gaidi M and Elkhamkhami J 2015 *Advanced Energy: An International Journal (AEIJ)* **2** 11–24
[14] Timuda G E and Maddu A 2010 *Journal of Study Science and Technology* **28** 1–4
[15] Yuwono A H, Liu B, Xue J, Wang J, Elim HI, Ji W, Li Y and White TJ 2004 *Journal of Materials Chemistry* **14** 2978–2987
[16] Yuwono A H, Ferdiansyah A, Sofyan N, Kartini I and Pujianto T H 2011 *AIP Conference Proceedings* **1415** 159–162
[17] Yuwono A H, Sofyan N, Kartini I, Ferdiansyah A and Pujianto T H 2011 *Advanced Materials Research* **277** 90–99
[18] Yuwono A H, Ramahdita G and Sofyan N 2012 *Advanced Materials Research* **557-559** 468–471
[19] Sholehah A, Yuwono A H and Rimbani C R 2013 *Materials Science Forum* **737** 28–32

Acknowledgement

This project was financially supported by the Directorate of Research and Community Services of Universitas Indonesia through the PITTA Research Grant of Universitas Indonesia, Year 2018, contract number 2370/UN2.R3.1/HKP.05.00/2018.

Ductility improvement of high density polyethylene (HDPE)-modified bitumen through adding modified lignin

E A Septimorien[1], L M D Friskila[1], B Natasya[1] and M Chalid[1]
[1]Department of Metallurgical and Materials Engineering, Universitas Indonesia, Depok 16424, Indonesia

E-mail : chalid@metal.ui.ac.id

Abstract. High Density Polyethylene (HDPE) is added to bitumen to improve the mechanical properties as Polymer Modified Bitumen (PMB), but both of them have different chemical properties. To achieve best mechanical properties of the mixture coupling agent is needed such as lignin that has bipolar characteristic. In the previous research, unmodified lignin tendency of polarity is bigger thus modified with polyurethanee PEG to increase its non-polar parts. In order to confirm the potential of modified lignin as a coupling agent for HDPE-Modified Bitumen. Several stages of this experiment consists of sample preparation, characterization of raw materials, mixing, and characterization of the PE-Modified Bitumen. This experiment used hot melt mixing to mix modified lignin, HDPE, and bitumen. The result of this experimental was analysed by using Ductility test, FTIR and FESEM. The addition of modified lignin to HDPE-Modified Bitumen increase distribution and dispersion of HDPE particulates inside bitumen matrix, increasing PMB HDPE mechanical properties. Furthermore, FESEM characterization indicated that the addition of modified lignin gave better mixing of PE-Bitumen. FTIR showed an increase of intensities at non-polar peaks, and down at polar peaks compared with unmodified lignin. Ductility test showed that modified lignin has better compatibility as coupling agent than unmodified lignin. From this experiment, modified lignin can improve mixing between HDPE and Bitumen better than modified lignin. So, we can use modified lignin as the new coupling agent.

1. Introduction

The applications of bitumen as matrix material have a weakness such as plastic deformation when receiving pressure such as a vehicle. For countering that weakness, reinforcement should be added as filler. In order to obtain bitumen with better quality, an increasing number of research began to focus on bitumen modification. Bitumen modification commonly used polymer materials. Polymer for bitumen modification include polyethylene waste (WPE), polypropylene, and tire crumb rubber [1-5]. From the research result, it shows a positive effect, which solves the waste problem in the beginning of this part. But when it is viewed chemically, bitumen has a hydrophilic characteristic that is polar and polymer is the opposite, hydrophobic and is non-polar, which make the two the exact opposite making them hard to interact with one another. What needed for the two of them to become one is the presence of coupling agent that acts as a bridge between these two.

2. Experimental

2.1. Materials

The materials used in this research are HDPE Plastic bags from commercial market. Bitumen come from Pertamina with pen 60/70 grade. While the modified lignin made in accordance to our laboratory[6]. With the materials to produce modified tignin are lignin from Technical Chemicals Indonesiaku, 4,4-methylenebis (cyclohexyl isocyanate) (HMDI), polyethylene glycol (PEG) with MW 6000 g/mol, and dibutyltin dilaurate (DD) as catalyst purchased from Sigma Aldrich.

2.2. Preparation of *HDPE-modified bitumens*

New HDPE plastic bag (purpose to minimize impurities) was crushed in to small part and prepared 5% w/w of bitumen. Secondly, lignin was prepared 0.5% w/w of bitumen, respectively and then bitumen 60/70 was modified with 5% HDPE and modified lignin with those variables. Bitumen modification was performed at 160°C in 30 min using Hot Melt Mixing apparatus shown in Figure 1.

2.3. Characterizations

HDPE-modified bitumen will go through for characterization, FE-SEM images were taken at an

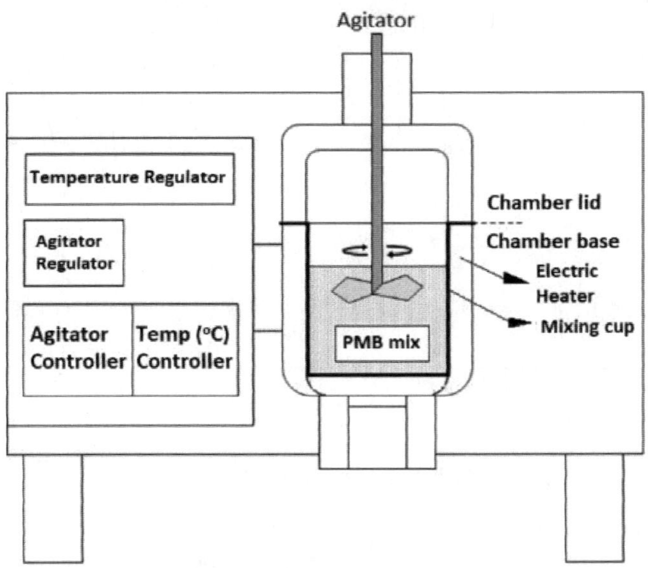

Figure 1. Hot Melt Mixing.

accelerated voltage of 20 kV, FTIR (Fourier Transform Infrared) were recorded using a Perkin-Elmer FTIR spectrometer, and each sample recording consisted of 30 scans recorded from 500 to 4000 cm-1. The ductility test of high density polyethylene (HDPE)-modified bitumen was tested based on ASTM D113.

3. Result and Discussion

3.1. Base Material Characteristics

HDPE, bitumen, and modified lignin consist of certain compounds which give their own identity. Compounds are pure and homogeneous substances which consist of 2 or more different elements with certain comparisons that produce different properties from the constituent elements. The compound has a specific functional group which signifies its type. The functional group contained in a compound can be known by FTIR test. Modified lignin characteristics could be found in our laboratory previous research [7].

Bitumen is a basic ingredient of a mixture which is a hydrocarbon compound which has a functional group -CH3 which is an aromatic lime subtituent on bitumen, C-H which is the deformation of -CH2 and -CH3, double chains C = O and C = C. From the graph in figure 2, there are several peaks that are produced from pure bitumen FTIR testing that is 876 cm-1, 1037 cm-1, 1118 cm-1, 1262 cm-1, 1436 cm-1, 1620 cm-1, 2851 cm-1, 2921 cm-1, and 3405 cm-1 where each peak shows a functional group. Peak widths at wavelengths between 3050-3700 indicate that the O-H bond is in the asphaltenes bitumen. Similar results were obtained by Brown [8]. The peak at 2850-3000cm-1 shows the presence of C-H alkane bonds in the main chain of bitumen. Furthermore, the peak which has a wavelength of 1620 cm-1 indicates the existence of a C = C bond of an aromatic group. At peaks between 1350-1400 cm-1 shows the presence of aromatic C-H bonds. Similar findings were reported by Xiangdao *et al.* [9].

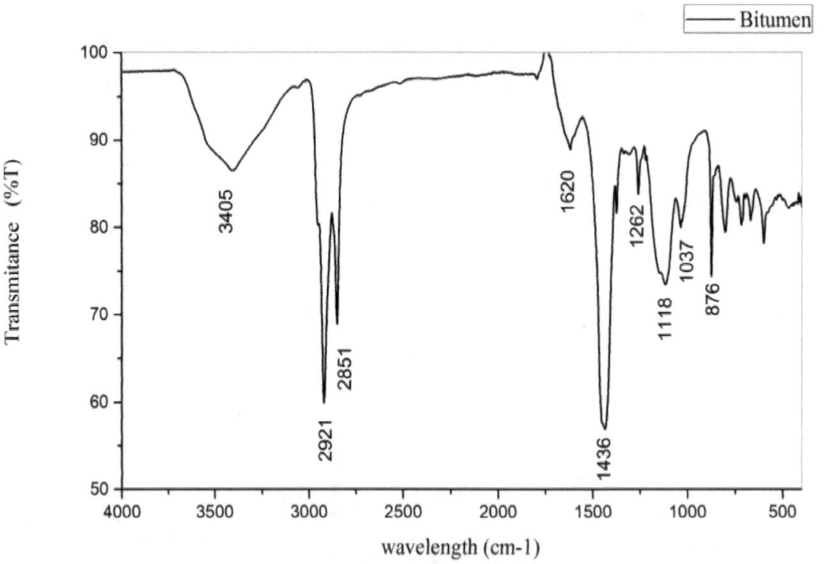

Figure 2. FTIR of Bitumen.

Addition of HDPE into the bitumen was aimed to contribute to the mechanical properties of the mixture to be better than the mechanical properties of pure bitumen so that the properties of the mixture have better mechanical strength than pure bitumen. This is because, the nature of HDPE which is harder than bitumen makes the results of the mixture tend to be more rigid than before it was given HDPE filler. Figure 3 have five peaks, namely 2916 cm-1, 2848 cm-1 1424 cm-1, 874 cm-1, and 720 cm-1. This shows that there is a C-H functional group which is an alkane bond with different deformations. Especially at the peak of 2916 cm-1 and 2848 cm-1 are asymmetric C-H and symmetric C-H which show HDPE identity. Similar tendency also was reported by Gulmine *et al.* [10].

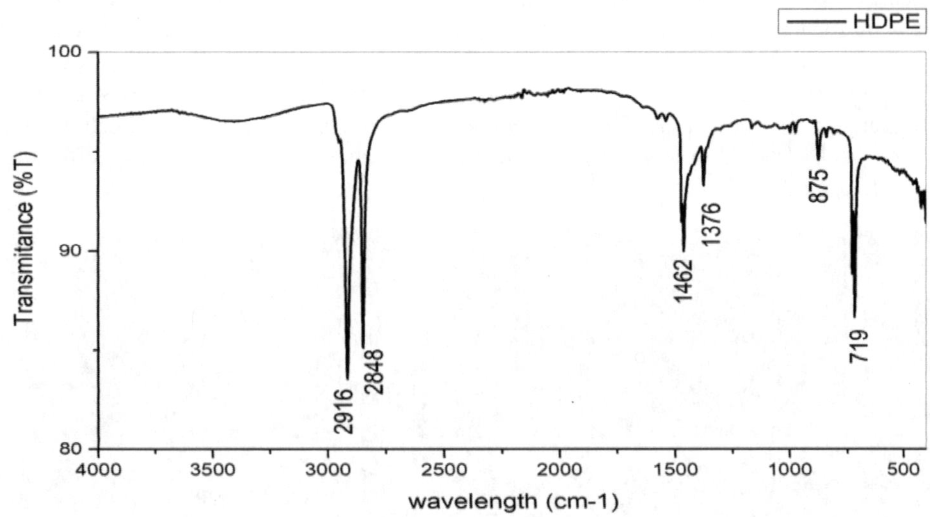

Figure 3. FTIR of HDPE.

3.2. Compatibility of Modified Lignin

The compatibility discussion of modified bitumen, HDPE, and lignin mixture aims to see the effect of modified lignin concentration as a coupling agent so that we can find out the optimal modification of modified lignin. The role of coupling agent is a bridge that will form a chemical bond between bitumen and HDPE so that bitumen and HDPE plastic have good compatibility. FT-IR testing on the mixture aims to see whether a new functional group is formed between bitumen, HDPE, and modified lignin so that the mixture can be said to have good compatibility or not. From Figure 4 it can be compared to the functional groups owned by each mixture with a concentration of 5% wt lignin without modification, modified lignin 0% wt, and 5% wt. The visible peaks of PMB use lignin, without modified lignin, and using modified lignin is almost the same but the intensity is different.

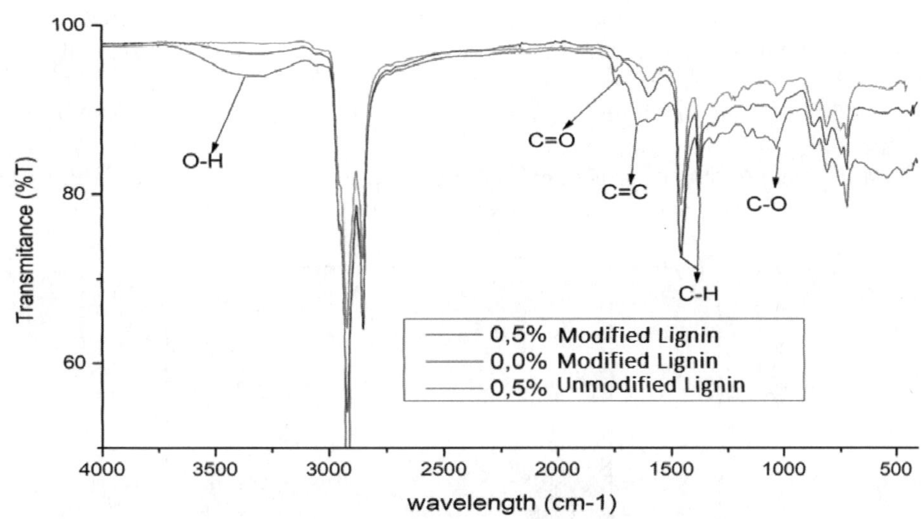

Figure 4. FTIR of PMB HDPE without lignin, with lignin and modified lignin.

Figure 4 indicated PMB HDPE with modified lignin compared to 0% wt lignin modification and 0.5 wt lignin at a peak of 2921 cm-1 intensity that increased dramatically indicates the presence of CH2 bond. In the fingerprint area and 3200-3500 cm-1 shows the hydroxyl group (-OH) which is hydrophilic 0.5% wt modified lignin is between 0.5 wt lignin and 0% wt lignin modification. This shows that modified lignin has higher hydrophilic properties than 0% wt lignin and is lower than 0.5% lignin without modification and its hydrophobic properties are increased compared to 0% wt lignin modification and 0.5% lignin without modification.

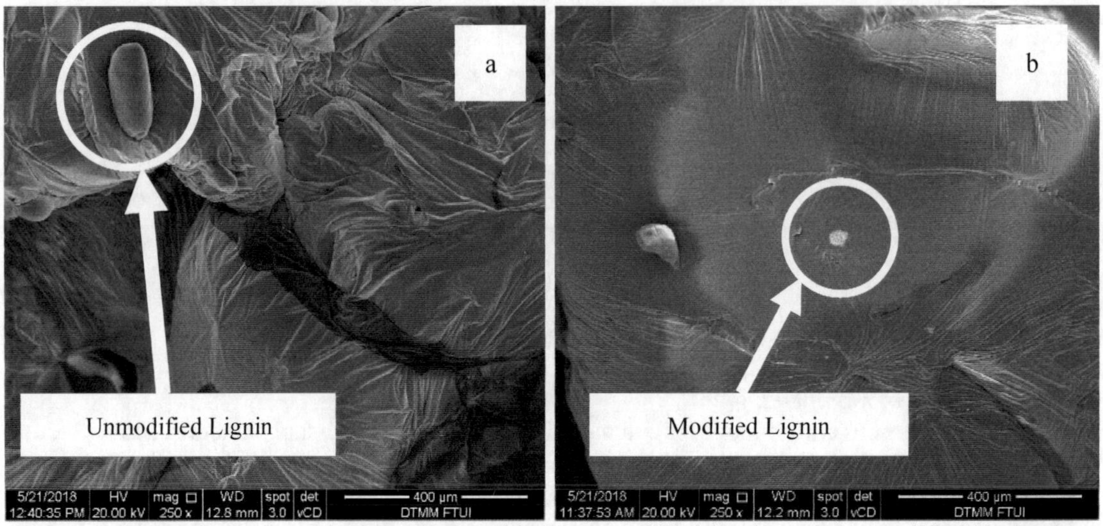

Figure 5. Fracture Surface of PMB HDPE with (a) unmodified lignin and (b) modified lignin as a coupling agent.

Figure 5 showed that PMB HDPE between unmodified lignin and modified lignin shows a drastic difference in fracture surface in the same magnification. In the modified PMB HDPE lignin, the flat fracture surface indicates a strong bond on the bitumen interface and the modified lignin is perfectly attached. Whereas, PMB HDPE with unmodified lignin shows uneven fracture surface shape. The surface shape shows that the unmodified lignin does not adhere perfectly to the PMB HDPE, indicated weaker interface strength between bitumen and unmodified lignin. This shows that modified lignin has compatibility with bitumen and HDPE which is better than lignin without modification. The result of ductility test PMB HDPE using modified lignin is 25,8 cm while using unmodified lignin is 14,3 cm and the penetration test PMB HDPE using modified lignin is 12,4 mm while using unmodified lignin is 31,3 mm showed in figure 6, thus it furthermore indicated the compatibility is better in PMB HDPE using modified lignin than PMB HDPE using unmodified lignin.

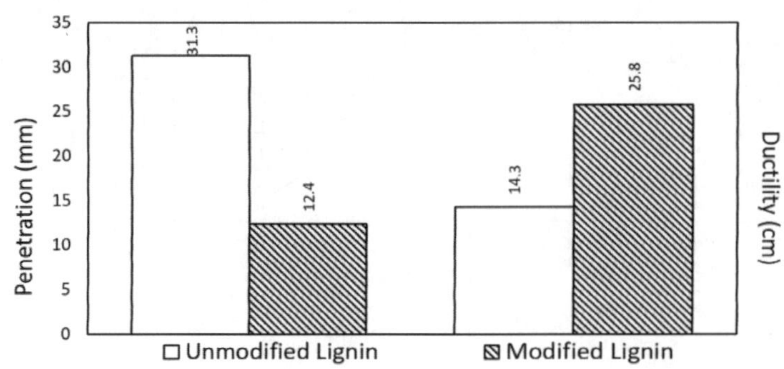

Figure 6. Comparation of PMB HDPE with modified lignin and unmodified lignin.

4. Conclusion

Modified lignin can be used as a coupling agent between the HDPE with bitumen which was indicated by the formation of ether peak and the loss of hydroxide peak better than unmodified lignin. FE-SEM result indicated that addition of modified lignin compared to unmodified lignin produce a better mixture between HDPE and bitumen. In addition, penetration value is lower and ductility value is higher in PMB HDPE with modified showed that the compatibility of the bitumen and HDPE with modified lignin was better than unmodified lignin.

References

[1] Hınıslıoğlu S and Ağar E 2004 *Materials letters* **58**(3) 267- 271
[2] Kumar U A and Satyanarayana P V V 2015 *International Journal of Engineering Trends and Technology* **22**(7) 325–330
[3] Zhu J, Birgisson B and Kringos N 2014 *J. Eur. Polym.* **54** 18-38
[4] Tapkın S 2008 *Building and Environment* **43**(6) 1065-71
[5] González V, Martínez-Boza F, Gallegos C, Pérez-Lepe A and Páez 2012 *Fuel Processing Technol.* **95** 137-43
[6] Ilmiati S, Mustafa J H, Yaumal A, Hanum F and Chalid M 2017 *IOP Conf. Ser.: Mater. Sci. Eng.* **223** 012033
[7] Yuanita E, Hendrasetyawan B E, Firdaus D F and Chalid M 2017 *IOP Conf. Ser.: Mater. Sci. Eng.* **223**(1) 012028
[8] Brown W H and Poon T 2004 *Introduction to Organic Chemistry* (3rd ed.) (Kanada: John Wiley & Sons) pp 313
[9] Chabannes M, Ruel K, Chabbert B, Arata Y, Jauneau A, Joseleau J P and Boudet A M 2001 *The Plant Journal* **82** 271
[10] Rojas O J, Bullón J, Ysambertt, Forgiarini A, Argyropoulos D S and Salager J 2006 *American Chemical Society Symposium Series* **954** 182-199

Acknowledgement

We appreciate and thank for the support from DRPM UI through *The Research Grant Scheme Program of PIT 9-2019* which makes this research possible.

IOP Conf. Series: Materials Science and Engineering **541** (2019) 012036 doi:10.1088/1757-899X/541/1/012036

The Influence of Sn addition on the microstructure and mechanical properties of the new β-type Ti-Mo-Nb based alloys for implant material

C Sutowo[1], G Senopati[1], S Supriadi[2], A W Pramono[3], B Suharno[1]
[1]Metallurgy and Material Engineering Department, Faculty of Engineering Universitas Indonesia, Depok 16424, Indonesia
[2]Mechanical Engineering Department, Faculty of Engineering Universitas Indonesia, Depok 16424, Indonesia
[3]Research Center of Metallurgy and Material, Indonesia Institute of Sciences, Kawasan Puspiptek Serpong, Tangerang Selatan 15314, Indonesia

Email :csutowo@yahoo.com, suharno@metalurgi.ui.ac.id

Abstract. Titanium alloys are very interesting for biomedical applications due to excellent biocompatibility, corrosion resistance, lower density, and lower young modulus compared to cobalt and stainless steel alloys. However, compared to bone, young modulus of pure titanium and Ti-6Al-4V are still relatively high and the mechanical properties are still insufficient to meet the needs of biomaterials replacing the hard tissues. In this paper, a newly Ti-Mo-Nb based alloys were designed and the effect of Sn content in Ti-6Mo-6Nb-xSn alloys (x=0,4,8 and 12 wt%) after homogenized on microstructure, hardness, and young modulus were investigated. The alloys were produced by electric vacuum arc furnace with non-consumable tungsten electrode then homogenized at 1100 °C for 7 under controlled argon atmosphere. Optical microscope, scanning electron microscopy, x-ray diffraction hardness test and ultrasonic test were used for alloys characterization. The results showed that Ti-6Mo-6Nb-xSn has equiaxed structure and Sn addition could promote the formation of β phase. The elasticity modulus of Ti-6Mo-6Nb-xSn alloy with the addition of 12% Sn was 88 GPa, this is better since it is below the elastic modulus of Ti6Al4V implant material.

1. Introduction

The biomaterial is a synthetic biocompatibel material to replace or improve the function of body tissues, one of which is a bone replacement [1]. The basic properties that must be possessed by orthopedic implants, namely biomechanical properties and biomedical properties. The biomechanical properties of orthopedic implants must match the body's tissues without the rejection reaction of body tissues when applied [2-4].

Selection of titanium as an implant material has been done since it has several advantages such as light-weight nature, high strength, good corrosion resistance and lower modulus of elasticity compared to other biomaterials such as stainless steel and cobalt alloys [5]. In addition to its superior mechanical properties, another reason is that stainless steel and cobalt alloys contain harmful elements such as Ni, Cr, Co. However, commercial pure Titanium (cp Ti) has limitations and is considered insufficient to meet the needs of biomaterials that require high strength, for example to replace hard tissue so that it is developed in the form of alloys [3].

The titanium alloy currently popular on the market is the Ti-6Al-4V with the α-β phase, but the fact that the release of small amounts of vanadium and aluminum ions from this aalloy into the human body is detrimental. V and l can induce possible cytotoxic and neurological effects causing serious health problems if they enter the human body [6,7] having problems with allergies [3]. The development of titanium alloys has been carried out using molybdenum or niobium as a vanadium substitute which functions as a beta phase stabilizer [8].

Besides that, the presence of aluminum can increase the potential for Alzheimer's disease development[9-10].Several studies have been carried out by substituting vanadium element using molybdenum and niobium but alpha-beta type titanium alloys still have the relatively high modulus of elasticity [11,12].

The development of the latest titanium alloy is done by replacing V and Al as well as adding another β (beta) phase stabilizing element so that it can function to increase strength and decrease the modulus of elasticity [13,14].

Beta type single phase titanium alloys are still in the research and development stage to reduce the modulus of elasticity so that it can avoid the effect of shield states on bone repair. Stress phenomenon shielding effect take place because of the inequality between the nature of the elastic modulus of the implant material and that of the bone so that the load is seemingly received solely by implant material. Consequently the terosteoporosis bone decreases and the bone that is not cracked will experience re-fracture due to the stress concentration around the implant. Some elements that have good biocompatibility and are not toxic to beta-type titanium alloys include Ta, Nb, Zr, Mo, W, Sn [4]. The purpose of this study was to design a new titanium alloy type β (beta) through the addition of beta phase stabilizer elements of niobium and molybdenum and to investigate the effect of adding low-cost stannum elements to reduce or substitute the use of niobium and molybdenum.

2. Material and Method

Design of of Ti-6Nb-6Mo-(x)Sn alloy with variations in Sn composition (x=0, 4, 8, and 12 wt%) in this study pure titanium (99, 9%), pure molybdenum (99.9%), pure niobium (99.9%) and pure stannum (99.9%). Melting process was performed using electric arc vacuum furnace in argon environment. The melting process was carried out five times to get homogeneity of the solid solution in the resulting ingot. The specimen produced in the remelting process was a button ingot weighing about 20 grams with 20 mm in diameter and 5 mm in thickness. The chemical composition of titanium alloy produced was tested using standard Q4 Mobile Broker testing.

The ingot was further homogenized by heating in a tube furnace with argon at 1100°C for 7 hours followed by furnace cooling. The purpose of homogenization is to increase uniform distribution of the element in the titanium alloy.

The sample ws then mounted to facilitate grinding and polishing. The grinding of sample was carried out using wet emery papers no 80, 120, 400, 800 and 1200, while the polishing used alumina micropolish up to 0.3μ followed by etching with Kroll's reagent solution (5 ml HF, 10 ml HNO_3, and 85 ml H_2O). Microstructure examination was conducted using optical microscope Meiji Techno Japan and scanning electron microscope (SEM/EDS) by JEOL JSM5390A. X-ray diffraction (XRD) measurements for phase analysis was conducted using Rigaku smart-lab diffractometer at room temperature using Cu Kα radiation and operating at 30kV/15mA. The hardness value was measured using the Mitutoyo Hardness Testing Machine HM-200 with a load of 0,3N for 12s. The elastic modulus was examined using the DM4 DL Krautkramer ultrasonic test and DA312B type probe with the ASTM-E494-95 standard [15].

Ultrasonic test will be obtained in the form of ultrasonic longitudinal velocities (V_L), the calculation of elastic modulus using equation 1, while the relationship between ultrasonic longitudinal velocities and shear wave velocities (V_T) with poisson ratio is shown in equation 2 [16].

$$E = \frac{\rho V_T^2 (3V_L^2 - 4V_T^2)}{V_L^2 - V_T^2} \quad (1)$$

$$\frac{V_T}{V_L} = \sqrt{\frac{1-2\sigma}{2(1-2\sigma)}} \tag{2}$$

3. Results & Discussion

The material used in this study was as-cast Ti-6Mo-6Nb-(x)Sn alloy, with the chemical composition shown in Table 1.

Table 1. Chemical composition of Ti-6Mo-6Nb-(x)Sn

Unsurewt%	Ti	Mo	Nb	Sn	Cr	Fe	V	Cu	Si
Ti-6Mo-6Nb	93.2	3.15	2.79	-	0.02	0.05	0.45	0.09	0.04
Ti-6Mo-6Nb-4Sn	84.8	5.75	4.55	3.93	0.03	0.05	0.59	0.11	0.04
Ti-6Mo-6Nb-8Sn	81.5	5.06	3.92	8.06	0.04	0.04	0.50	<0.01	0.04
Ti-6Mo-6Nb-12n	76.9	5.25	4.92	12.20	0.03	0.05	0.42	0.06	0.04

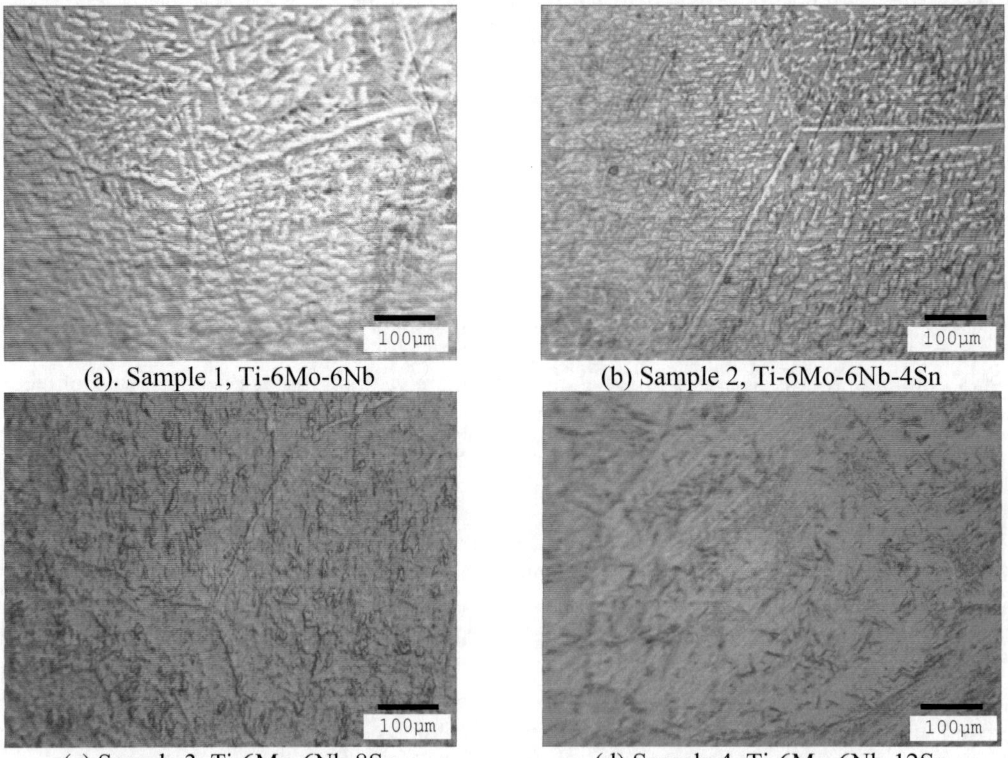

(a). Sample 1, Ti-6Mo-6Nb (b) Sample 2, Ti-6Mo-6Nb-4Sn
(c) Sample 3, Ti-6Mo-6Nb-8Sn (d) Sample 4, Ti-6Mo-6Nb-12Sn

Figure 1. The microstructure of Ti-6Mo-6Nb-(x) Sn as-cast homogenized, etching using Kroll's reagent solution (5 ml HF, 10 ml HNO_3, and 85 ml H_2O).

The microstructure of Ti-6Mo-6Nb-(x)Sn alloy in Figure 1 shows as-cast microstructure after homogenized at 1100°C for 7 hours with slow cooling in the furnace, which aims to eliminate segregation and to achieve uniform microstructure. The microstructure of the globular alloy is the result of static recrystallization due to the cooling process after homogenized which occurs along with the decrease in temperature of the β-transus. The results showed that Ti-6Mo-6Nb-xSn have equiaxed structure and Sn element addition could promote the formation of β phase, the higher the Sn content the less the alpha phase contained in the beta matrix.

The influence of the homogenization results in a large amount of grains evenly distributed in various compositions, but the material of Ti-6Mo-6Nb-12Snwith a 12% Sn have smaller grain size average about 500 microns causing the increase in hardness.

Figure 2 shows diffraction patterns of as-cast Ti-6Mo-6Nb after homogenization consisting of β phase. The XRD results showed that in the Ti-6Mo-6Nb alloy without Sn, beta stabilizer was still not sufficient to form the beta phase as seen from the intensity peak that is still more dominant in the phase α than the β phase. The addition of Sn has an effect on increasing the phase volume fraction of β, as evidenced by the increasing peak/intensity of the β phase in the alloy containing Sn.

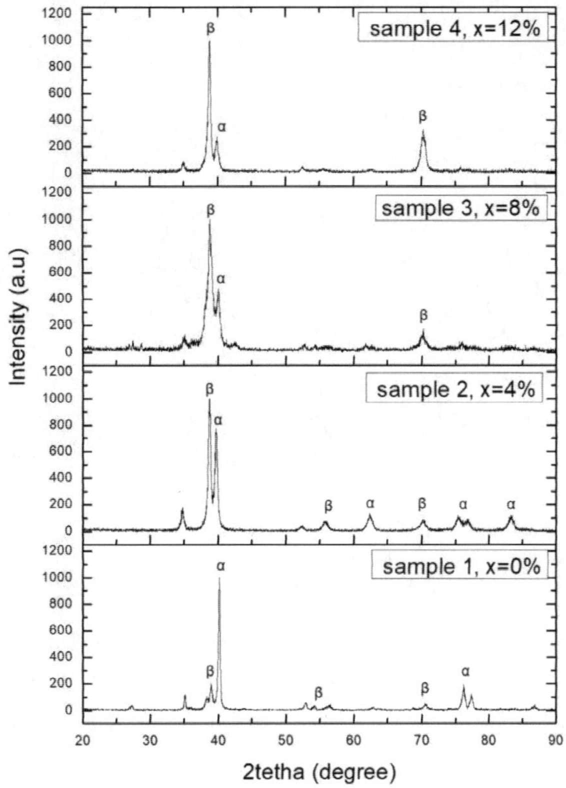

Figure 2. Diffraction patterns of Ti-6Mo-6Nb-(x)Sn alloys.

Further observations on the microstructure of the alloy were carried out using SEM / EDS. The sample grain size is almost the same as sharing the variation of Sn elemental content. In the Ti-6Mo-6Nb-4Snalloy with the addition of Sn = 4% and Sn = 8%, the appearance of α precipitates in the β matrix is still obvious; whereas in the sample with Sn = 12% does not indicate the presence of precipitates/α phases. In EDS observation the presence of precipitate alpha is marked by the higher dominant content of the alpha phase stabilizer.

Yang et al (Huet al., 2008) have investigated the effects of alloying elements of Nb, Zr, and Sn on the stability of Ti alloy and the results show that Sn is a beta phase stabilizer stronger than Nb, and an additional amount of Sn can reduce the elasticity modulus of alloy by suppressing the formation phase ω in the Ti-Nb-Zr-Sn alloy.

IOP Conf. Series: Materials Science and Engineering **541** (2019) 012036 doi:10.1088/1757-899X/541/1/012036

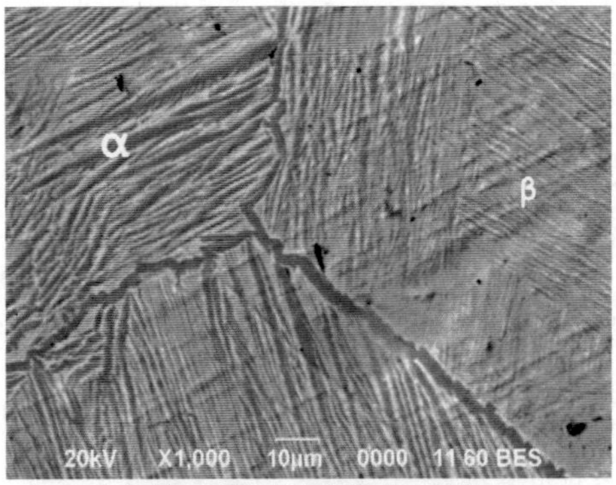

(a). Sample 1, Ti-6Mo-6Nb

(b) Sample 2,Ti-6Mo-6Nb-4Sn

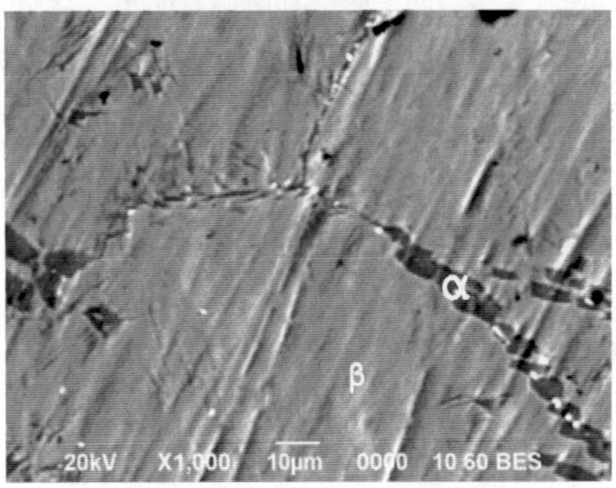

(c).Sample 3, Ti-6Mo-6Nb-8Sn

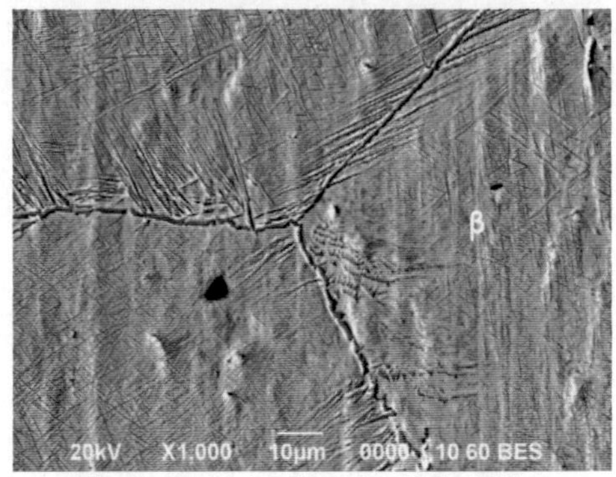

(d). Sample 4, Ti-6Mo-6Nb-12Sn

Figure 3. SEM photograph of the Ti-6Mo-6Nb alloy, etching using Kroll's reagent solution (5 ml HF, 10 ml HNO_3, and 85 ml H_2O).

Figure 4 shows the average Vickers hardness value of as cast Ti-6Mo-6Nb before and after homogenization condition. The effect of homogenization on Ti-6Mo-6Nb and Ti-6Mo-6Nb-4Sn alloy causes a decrease in the value of Vickers' hardness whereas in the Ti-6Mo-6Nb-8Sn and Ti-6Mo-6Nb-12Sn alloy with higher Sn content(x=8% and x=12%) the hardness increased.

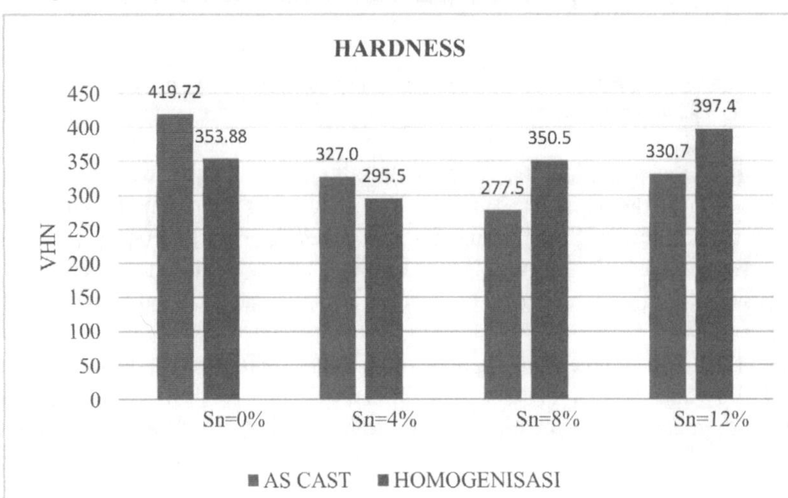

Figure 4. Vickers hardness value of Ti-6Mo-6Nb-(x)Sn alloy.

Figure 5 shows the modulus of elasticity before and after the homogenization process. In the study of P. Majumdar et al, it is said that the β phase in titanium can reduce the elastic modulus compared to the α phase [16].

Elasticity modulus of titanium alloy is related to alloy composition and phase constitution. In the Ti-6Mol-6Nb-based alloy, the elastic modulus is determined by the phase and volume fraction. The modulus of elasticity of titanium alloy is lower with increasing Sn content. This proves that the addition of Sn element can promote the formation of phase β, as shown in the figure 2 which shows the increase in beta phase intensity as the Sn content increases. In the Ti-6Mo-6Nb-12Sn alloy with 12% Sn, the lowest achieved modulus of elasticity is 88 GPa and increases after homogenization to 92 GPa.

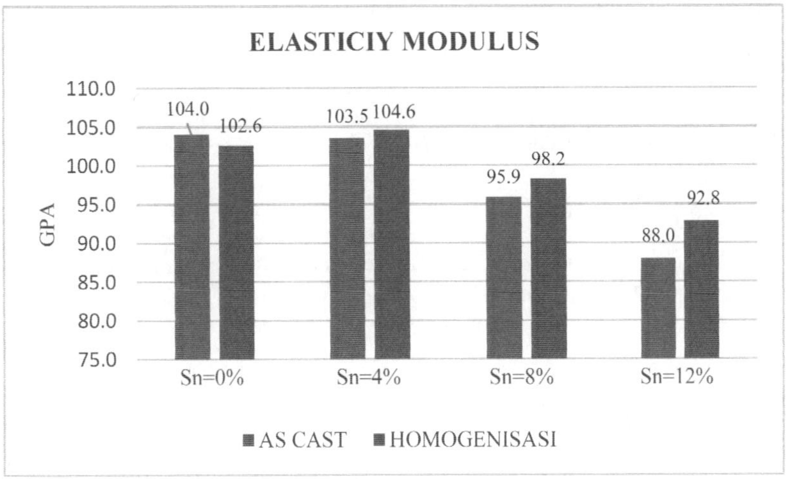

Figure 5. Elasticity modulus of Ti-6Mo-6Nb-(x)Sn alloy.

In the homogenized samples, the modulus of elasticity is relatively higher compared to that o the as-cast samples. This indicates the occurrence of phase changes due to the cooling effect in the furnace. So that the alloy samples with the addition of Sn homogenization experience an increase in the value of the modulus of elasticity compared to the as cast sample as shown in figure 5.

4. Conclusion

The results of the x-ray diffraction analysis of Ti-6Mo-6Nb alloy after homogenization showed the increase of peak intensity of phase β in the alloy with the addition of Sn. Phase intensity β increases with increasing percentage of Sn. The hardness value off the alloy has increased with the addition of 12% Sn and has increased further with the homogenization. The elasticity modulus of Ti-6Mo-6Nb alloy with the addition of 12% Sn was 88 GPa, this value is better than other implant material even below the elastic modulus of Ti6Al4V.

References

[1]. Davis J R 2003 *ASM International Handbook of Materials for Medical Devices editor* (United States: ASM Internasional) 1-11
[2]. Mitsuo N 2002 *Metallurgical and materials transactions A, 33*(3), 477-486 *SM International, OH*
[3]. Yuhua L, Chao Y, Haidong Z, Shengguan Q, Xiaoqiang L, and Yuanyuan L 2014 *Materials, 7*(3), 1709-1800
[4]. Abdel-Hady *Gepreel* M, Niinomi M 2013 Journal of the Mechanical Behavior of Biomedical Materials **20** 407-415
[5]. Niinomi M, Nakai M, and Hieda J 2012 *ActaBiomaterialia* **8** (11) 3888–3903

[6]. Okazaki Y, Ito Y, Kyo K, Tateisi T 1996 Mater. Sci. Eng. **A213** 138–147
[7]. Okazaki Y, Gotoh E 2005 Biomaterials **26**, (1) 11-21
[8]. Marsumi, Yuswono, and Pramono A W 2014 *Advanced Materials Research* **900** 53–63
[9]. Ikeda M, Komatsu S Y, Sowa I, et al. 2002 *Metall. Mater.Trans. A.* **33** 487-493
[10]. Perl D P and Brody A R 1980 *Science* **208** 297-08
[11]. Sutowo C, Rokhmanto F, Senopati G 2017 Widyariset **3** (1) 47-54
[12]. Senopati G, Sutowo C, Amal M I 2017 Journal of Physics **817** (1) 012012
[13]. Takahashi K, Shiraishi N, Ishiko-Uzuka R, Anada T, Suzuki O, Masumoto H and Sasaki K 2015 *Int. J. Mol. Sci.* ***16***, 5779-5788
[14]. Abd-elrhman Y, Gepreel M A -H, Abdel-Moniem A, Kobayashi S 2016 Materials and Design **97** 445 –453
[15]. ASTM-E494-95 Standard Practice for Measuring Ultrasonic Velocity in Materials
[16]. Majumdar P 2008 Material Science and Engineering A 419-425

Acknowledgment
The authors would like to thank Direktorat Riset dan Pengabdian Masyarakat Universitas Indonesia (DRPM UI) for their financial support through PITTA grant funds with contract number of 2394/UN2.R3.1/HKP.05.00/2018 for performing this work and Technical Support from Research Center for Metallurgy and Material Indonesia Institute of Sciences.

Thermal characteristics of ferronickel slag on roasting process with addition of sodium carbonate (Na_2CO_3)

A B Prasetyo[1,3], **A Maksum**[1,2], **J W Soedarsono**[1], **and F Firdiyono**[3]
[1]Center of Minerals Processing and Corrosion Research, Department of Metallurgical and Materials Engineering, Universitas Indonesia, Depok 16424, Indonesia
[2]Department of Mechanical Engineering, Politeknik Negeri Jakarta, Depok 16425, Indonesia
[3]Research Center For Metallurgy and Materials, Indonesian Institute of Sciences, Serpong, Tangsel, Indonesia

Email: chencen_abp@yahoo.com

Abstract. Thermal characterization of ferronickel slag waste has been studied using TG / DTA, XRD, and SEM-EDX. The characterization of the initial samples of ferronickel slag was carried out TG / DTA to 1200°C. The result obtained is at a temperature of 800°C, there is an increasing mass and up to 4.68% at a temperature of 1200°C. At temperature of 807.4°C to 845.8°C, an exothermic reaction occurs. The increasing mass is due to the ferronickel slag which is originally in the form of metal, and then it was roasted to undergo an oxidation reaction so that the metal that has been formed, it returned into oxides. So that the weight of the sample mass increase. Samples of ferronickel slag added with sodium carbonate were also analyzed using TG / DTA. The results obtained are 2 endothermic peaks at temperatures of 90.6°C and 858.9°C with a total mass reduction of 49.3%. At a temperature of 90.6°C, there is a heavy loss caused by 2.38% loss of surface water. The XRD result of ferronickel slag is composed of enstatite ($MgSiO_3$), forsterite (Mg_2SiO_4), fayalite (Fe_2SiO_4), and quartz (SiO_2) structures. From the XRD analysis, the composition of silica oxide associates with magnesium and iron in the form of enstatite, forsterite, and fayalites is a very dominant composition. The roasting process of a mixture of ferronickel slag with sodium carbonate was carried out by heating at a temperature of 800-1000°C for 1 hour, and the sample result of roasting were analyzed using XRD. The result of roasting shows that the roasting process takes place more perfectly at higher temperature; it is indicated by the increasing phase intensity of SiO_2 and the formation of sodium silicate (Na_2SiO_3). The result of SEM shows that the higher the temperature, the distribution of Na, Si, and O elements tend to cluster in the same place or spot, while the elements of Mg, Si, and O are less bonded.

1. Introduction

The depletion of world mineral reserves has led researchers to try to reuse waste from the mining industry, such as tin slag [1] and ferronickel slag, as alternative sources of high value metals. Ferronickel slag is a by-product of the metals separation from ore in nickel smelting with pyrometallurgical processes. During this time, ferronickel slag is only used as a material layer for the manufacture of mine road access and as a material to increase soil carrying capacity [2, 3]. Nearly 38% of ferronickel slag is used for road construction, 48% is used for the cement industry mixture and the remainder is used for fertilizer, geopolymer, hydraulic technique, and so on [4-7]. There is not much further utilization of ferronickel slag, so if the technology is not found to process it, the slag will accumulate in the waste storage pond which will eventually disrupt the surrounding environment. Ferronickel slag waste is B3 waste (Hazardous and Toxic Material). So, it requires special handling [8].

As an effort to support the increase of minerals value and the implementation of sustainable mining, it is by utilizing the slag. Therefore, it becomes more valuable. The use of ferronickel slag can be done by extracting it in order to extract valuable elements from the slag. The biggest elements contained in ferronickel slag are 30% silica, 20% magnesium, 12% iron, 1-2% aluminum, and so on [9]. In addition, ferronickel slag is also indicated that it contains of rare earth metal elements. The rare earth elements contained in ferronickel slag include lanthanum, cerium, neodymium, yttrium, and so on. Soil metal elements rarely have special characteristic, namely being able to react with other elements to produce something new. Rare earth metals are able to increase the ability of materials in the form of strength, hardness, and resistance to heat; so that these elements are added to the manufacture of high-strength steel. Almost all today's high-tech products, from televisions, cell phones, hybrid cars, and nuclear missile guidance devices require rare earth metals [10, 11]. Extractions that have been carried out by several researchers from slag include magnesium, nickel, cobalt, chromium, zinc, cadmium, copper, and silica [12-14]. Indications of rare earth metals contained include lanthanum, cerium, neodymium, yttrium, and so on. Therefore, the study of rare earth extraction from ferronickel slag will be very interesting. With the presence of valuable elements in ferronickel slag, a preliminary study of the characteristics and roasting experiments of ferronickel slag are needed to determine the suitable process or technology for the extraction of precious metals in ferronickel slag.

The purpose of the study is to observe the changes in phase and mass; morphology and distribution of elements contained and the structure of ferronickel (FeNi) slag minerals by adding sodium carbonate (Na_2CO_3) additives then roasted at temperature variations of 800°C, 900°C, and 1000°C for 1 hour. With the addition of sodium carbonate, it is expected a decomposition reaction between magnesium silicates as the main compound in slag with sodium carbonate to form sodium silicate compound.

2. Materials and Experimental

The raw material used in this study is ferronickel slag from smelted nickel at PT Morowali Indutrial Park. The preparation stage was done by smoothing the ferronickel slag to get the size of -100 meshes. The sample was crushed using a crusher then mashed again using a disk mill to get the size of -100 meshes. The characteristics of ferronickel slag were analyzed using XRF, TG / DTA, SEM EDX, and XRD analysis. The chemical composition (wt. %) of ferronickel slag which was analyzed using XRF is listed in Table 1. TG/DTA was used to analyze the phase changes and mass of ferronickel slag at a certain temperatures. The analysis of TG / DTA was carried out using the TG/DTA with brand name of SETARAM ex France type Setsys in the temperature range of 100°C -1200°C with a heating rate of 10°/minute and air atmosphere. The analysis of SEM EDX was used to determine the morphology and distribution of the elements contained in ferronickel slag. SEM used is the SEM with JSM 6390 A. Jeol brand. The structures of ferronickel slag minerals were analyzed using XRD of the Schimadzu MAXima XRD-7000 brand with Cu Kα radiation ($\lambda = 1.54$). Operation of the generator was done at 40 kV and 30 mA, with a measurement speed of 2 degrees / minute, 0.020 degree pitch sampling.

Ferronickel slag roasting experiments were carried out by adding sodium carbonate (Na_2CO_3) additive. Raw materials of ferronickel slag which have been crushed to grain the size of 100 meshes were then mixed with additives in the mixer. The composition of the mixture between ferronickel slag and sodium carbonate is 50:50. Roasting was done by heating the sample in crucible graphite in a carbolite type of muffle furnace at temperatures of 800°C, 900°C, and 1000°C for 1 hour. Roasting with the addition of sodium carbonate is expected to be able to bind silicates which are the most elements that are contained in ferronickel slag.

The reactions that occur in the binding of SiO_2 with the addition of Na_2CO_3 (reaction 1, 2, and 3) are as follows[15] :

$$Na_2CO_3 + SiO_2 \rightarrow Na_2SiO_3 + CO_2 \quad (1)$$

The reaction is preceded by the process of decomposition of Na_2CO_3 to Na_2O and CO_2. When the temperature is increased there will be a reaction between Na_2O and SiO_2:

$$Na_2CO_3 \rightarrow Na_2O + CO_2 \quad (2)$$

The products between Na_2O and SiO_2 depend on the temperature and composition of the Na_2O and SiO_2 mixtures. The reactions that occur are as follows:

$$Na_2O + SiO_2 \rightarrow Na_2SiO_3 \quad (3)$$

3. Result and discussion

3.1. XRF Analysis

The analysis results of chemical composition with XRF are presented in table 1. The results of the analysis show that in the ferronickel slag based on table 1, the most abundant element contained is silica at 25.02%. Other contained elements are iron and aluminum as much as 7.20% and 3.61% respectively. While the Ni element is detected very small at ppm, namely 430.6 ppm, this indicates that the Ni element has been extracted all into ferronickel metal as a product in smelting. Rare earth metals are also detected in this XRF analysis. The detected elements of rare earth metals are lanthanum, cerium, yttrium, neodymium, samarium, and so on. With the potential for rare earth metals, separate research can be done for rare earth metal extraction which has very high economic value.

Table 1. XRF results in elemental content contained in ferronickel slag

Element	Wt(%)	Element	Wt(ppm)	Element	Wt(ppm)
Si	25.02	S	402	Pr	70.9
Al	3.61	Cl	267.1	Zn	142.1
Na	0.22	Sc	40.6	Y	102.1
P	0.09	V	218.5	La	346.7
K	0.10	Zr	211	Ce	1109
Ca	0.65	Co	64	Nd	303.3
Ti	0.09	Mn	5511	Nb	26.1
Fe	7.20	Ni	430.6	Sm	116.7
Cr	0.95	Ba	62.7	Sr	39

3.2. TG/DTA Analysis

TGA analysis (Thermo Gravimetric Analysis) / DTA (Differential Thermal Analysis) was carried out to see the thermal kinetics of ferronickel slag when heated, so that the temperature can be determined effectively in the roasting process and to determine the phase changes that occur at a certain temperatures. TGA/DTA was carried out on ferronickel slag samples without the addition of other substances, and the ferronickel slag added with sodium carbonate.

The results of TGA / DTA Analysis can be seen in Figures 1 and 2. The TG / DTA analysis of ferronickel slag samples without the addition of other substances can be seen in Figure 1. The results of TG / DTA analysis in Figure 1 show that heating to ferronickel slag samples shows exothermic peaks at temperature of 807.4°C to temperature of 845.8°C. The exothermic reaction occurs at the maximum peak at a temperature of 824.7°C. The occurrence of exothermic reactions is also followed by an increasing sample mass in the TG / DTA analysis of ferronickel slag without the addition of another substance of 4.68%. This increasing mass is probably due to the reaction between the slag and oxygen from the outside during roasting. This is because the TG / DTA analysis was carried out on a small number of samples with an atmosphere that did not use inert gas or with an open system, so the metal on the surface might react with oxygen.

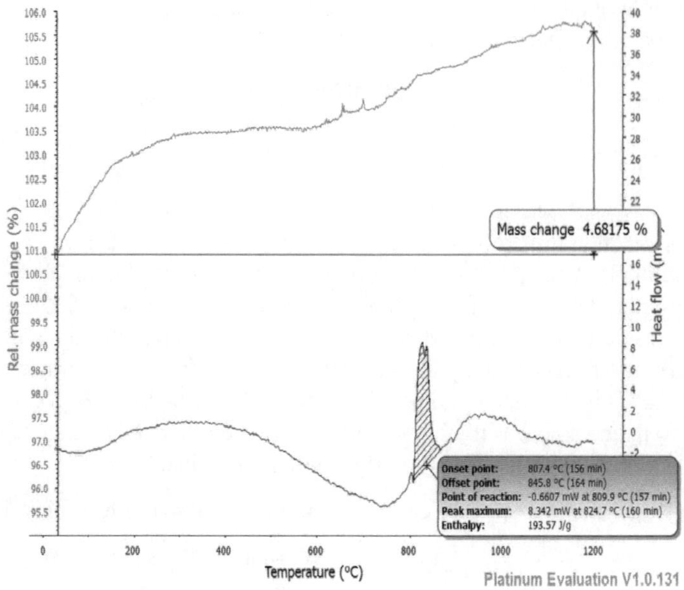

Figure 1. TG / DTA curve in ferronickel slag.

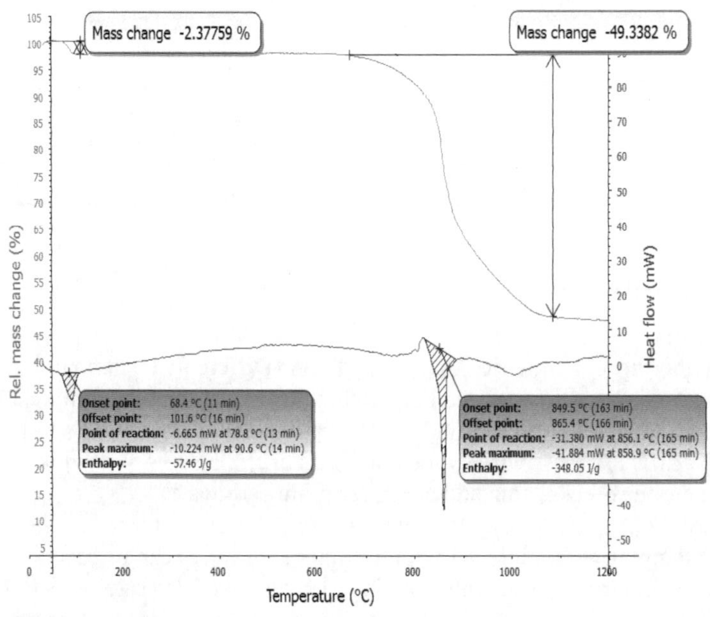

Figure 2. TG / DTA curve of a mixture of ferronickel slag and Na_2CO_3.

The TG / DTA curve of the mixture of ferronickel slag and sodium carbonate can be seen in Figure 2. In Figure 2 shows 2 reaction peaks that are quite clear. Those are at temperatures of 90.6°C and 849.5°C. The first endothermic peak on the curve of figure 2 begins at temperatures of 68.4°C to a temperature of 101.6°C and the peak of endothermic reaction 1 occurs at a temperature of 90.6°C. At a temperature of 90.6°C, there is a heavy loss of 2.38% because the surface water in the ferronickel slag begins to vaporize. At a temperature of 849.5°C, the formation of an endothermic reaction begins and ends at a temperature of 865.4°C. The endothermic peak reaction occurs at a temperature of 858.9°C accompanied by a very drastic loss of mass of 49.34%. This is because at the temperature interval there is a released CO_2 like reaction 2 above, namely the decomposition reaction of Na_2CO_3 to $Na_2O + CO_2$. By increasing the temperature of process, an endothermic reaction occurs which does not cause a decrease in mass. At this temperature, the possibility of Na_2O to SiO_2 binding becomes Na_2SiO_3 (sodium / sodium silicate) as seen in reaction 3 above. This is in accordance with the phase diagram of the formation of sodium silicate from Na_2O and SiO_2 that can occur at temperatures of 800°C to 900°C, with 1: 2 to 1: 2.3 mole ratios of Na_2O and SiO_2 [16].

3.3. XRD Analysis

XRD analysis was conducted to determine the composition of the dominant chemical compounds and other compounds. The XRD analysis of the initial sample of ferronickel slag and the roasting process with temperature variables of 800°C, 900°C, and 1000°C is presented in Figure 3. The result of XRD analysis of ferronickel initial slag is identified that ferronickel slag is composed of forsterite (Mg_2SiO4), fayalite (Fe_2SiO_4), quartz (SiO_2), and enstatite ($MgSiO_3$). From this XRD analysis, there is showed that the composition of silicate oxide which associates with magnesium and iron in the form of enstatite, forsterite, and fayalite is a very dominant composition. It can be seen in the graph of the highest peak diffraction, which is the composition of enstatite, fosterite, and fayalite. Based on the analysis with GSAS software, the composition of this ferronickel slag compound can be estimated to be 24.5% fayalite, 31.5% forsterite, 20.2% enstatite, and 25.5% quartz (in% mass). This condition is in accordance with the results of XRF analysis that the element with the largest content is silica which associates with magnesium and iron. This result is in accordance with the research that has been done before, namely the characterization of XRD on ferronickel slag shows that the dominant compositions are enstatite and forsterite [3, 9, 14].

The results of the roasting process with temperature variations of 800°C, 900°C, and 1000°C will be more clearly understood by comparing the XRD diffraction graph from roasting with the addition of sodium carbonate at various temperatures as in Figure 3. In figure 3 it is showed the XRD results of the initial samples of ferronickel slag contain of compounds with crystal structures consisting forsterite, enstatite, quartz, and fayalite as the dominant minerals. In figure 3, roasting at a temperature of 800°C with the addition of sodium carbonate as an additive causes the silica bound in enstatite, forsterite, and fayalite to start decomposing. Although the forsterite phase is still detected at a temperature of 800°C, but at this temperature, there are forsterite and enstatite which start to transform into MgO, and fayalite transforms into ferrosilite ($FeSiO_3$).

The result of roasting at a temperature of 900°C shows that sodium silicate is formed even though it does not show a high peak. This result is in accordance with the TG / DTA analysis shown in Figure 2 and in accordance with the phase diagram of Na_2O and SiO_2 that sodium silicate is formed at temperatures between 800°C-900°C as seen in Figure 3. At this temperature the enstatite phase ($MgSiO_3$) starts to separate into MgO and SiO_2, so that it dominates high peaks. The SiO_2 phase experiences an increase in intensity; this intensity change indicates that the roasting reaction is more perfect due to rising temperatures.

At the temperature of 1000°C, sodium silicate (Na_2SiO_3) has been formed and the SiO_2 phase has a very significant increase. It indicates that the addition of sodium carbonate is very influential on the process of sodium silicate formation and SiO_2 decomposition. With the very high intensity of SiO_2, it is expected that after the roasting process, a further process can be carried out, namely leaching with hot water to separate SiO_2 from ferronickel slag.

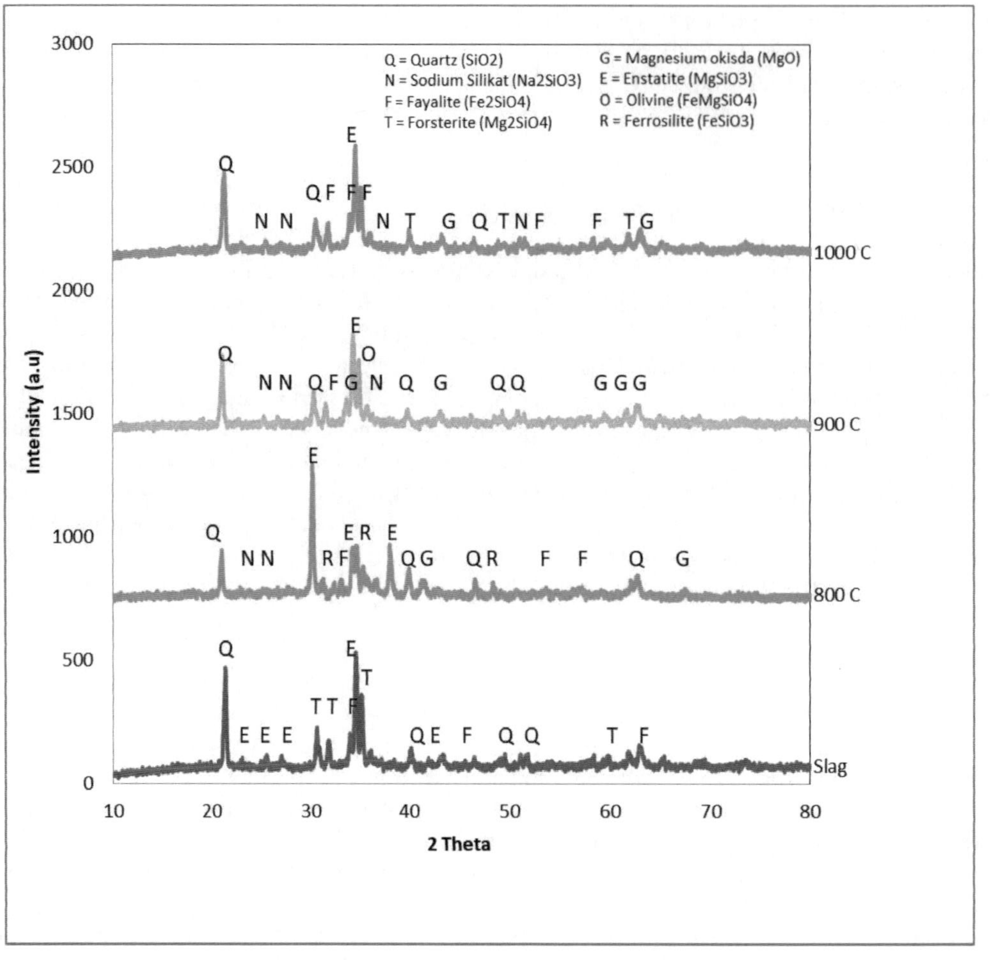

Figure 3. Curve of XRD Diffraction from the Initial slag and Roasting with the addition of Na_2CO_3 at various temperatures of 800°C, 900°C, and 1000°C.

3.4. SEM EDX Analysis

To find out the mineralogical structure and distribution of elements in ferronickel slag and the results of roasting, a SEM EDX analysis was done. The results of observations and mapping of elements using SEM EDX are shown in Figure 4. From figure 4 (a) it can be observed that silica and magnesium cover almost all parts of the ferronickel slag along with oxygen. This shows that indeed most silica is present in the form of silica oxide compounds and associates with magnesium. This corresponds to the XRD analysis that the dominant composition is silica that associates with magnesium and iron, namely enstatite, fosterite, and fayalite as seen in the XRD analysis of ferronickel slag. Calcium, aluminum, iron, and chromium elements are very small and evenly distributed in almost all parts. These results are in line with the research conducted by Mubarok et al. And Tangahu et al. which explains that in SEM analysis of ferronickel slag, the dominant distribution are magnesium, silica, and oxygen [3, 14].

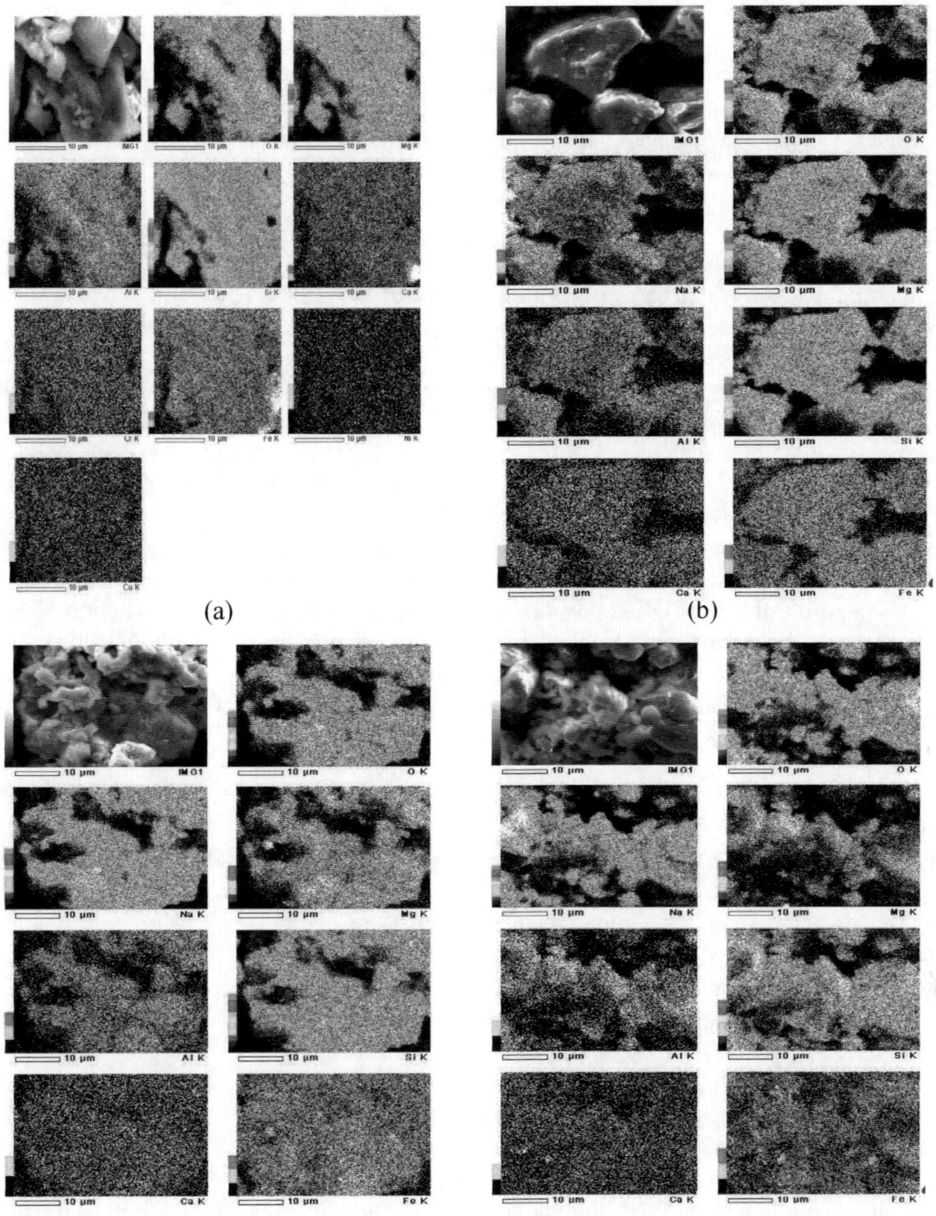

(a) (b) (c) (d)

Figure 4. SEM EDX Mapping Analysis of (a) ferronickel slag, Roasting with the addition of Na_2CO_3 at temperatures of (b) 800°C, (c) 900°C and (d) 1000°C.

In figure 4 (b) the results of 800°C roasting show that sodium is visible even though it is still in certain spots and the association between magnesium, silica, and oxide is still very strong which is shown by the red distribution of the three elements. In figure 4 (c), the results of 900°C roasting show the distribution of sodium is very high. It indicates that the red color of the distribution is greater and it blends with silica and oxide. Magnesium in its distribution has begun to decrease in association with silica. So it can be assumed that at this temperature sodium silicate (Na_2SiO_3) has been formed completely. It is indicated from the red color between the associations of sodium, silica, and oxide which are increasingly red, while magnesium has begun to appear, and its distribution is not too dominant. In figure 4 (d) shows the distribution of elements that are higher for the distribution of sodium, silica, and oxide in one distribution / spot, whereas the distribution of magnesium begins to thin or decrease. This is consistent with the analysis of TG / DTA and XRD where the higher the temperature, the formation of sodium silicate will be stronger and can be separated from magnesium.

With the separation between silica and magnesium, it is expected to facilitate the process of dissolving silica by using hot water, so that silica can be separated from the valuable elements contained in the ferronickel slag. Through these steps of process, it is hoped that it would facilitate the next process, namely the extraction of precious metals from ferronickel slag.

4. Conclusion

The result of TG / DTA ferronickel slag analysis shows that at temperatures of 807.4°C to 845.8°C, there is an exothermic reaction which causes the mass to increase by 4.68%. The result of TG / DTA analysis on ferronickel slag with sodium carbonate adds 2 endothermic peaks at a temperature of 90.6°C and 858.9°C with a total mass reduction of 49.3%. At a temperature of 90.6°C, there is a heavy decrease caused by 2.38% of surface water loss. In endothermic reaction the temperature of 858.9°C shows the release of CO_2 which results in significant mass loss. With increasing temperature, the formation of sodium silicate will be more perfect.

The result of XRD shows ferronickel slag is composed of enstatite, forsterite (Mg_2SiO_4), fayalite (Fe_2SiO_4), and quartz (SiO_2) structures. The result of XRD analysis on roasting of ferronickel slag added with sodium carbonate shows that the roasting process takes place more perfectly at higher temperature. This is indicated by the increasing of phase intensity of SiO_2 and the formation of sodium silicate (Na_2SiO_3). The result of SEM EDX shows that with higher temperatures, the distribution of Na, Si, and O elements tends to be in the same place or spot, while the elements Mg, Si, and O decrease.

The formation of sodium silicate is expected to facilitate the extraction process of precious metals contained in ferronickel slag. The next process is leaching with hot water to separate SiO_2 from ferronickel slag. So that by separating SiO_2 as the dominant element, it is expected that other valuable elements will be easy to extract.

References

[1] Permana S, et al. 2016 *IOP Conference Series: Materials Science and Engineering*. IOP Publishing
[2] Kang S S, K Park and D Kim 2014 *Materials* **7(10)** 7157-7172
[3] Tangahu B V, et al. 2015 *Adv. Chem. Engineer. Sci.* **5(03)** 408
[4] Maragkos I, Giannopoulou I P, and Panias D, 2009 *Miner. Eng.* **22(2)** 196-203
[5] Choi, Y.C. and S. Choi, 2015 *Constr. Build. Mater.* **99** 279-287
[6] Rahman M A, et al 2017 *Constr. Build. Mater.* **140** 194-202
[7] Sugiri S and D T S 2005 *Jurnal Infrastruktur dan Lingkungan Binaan* **1(1)**
[8] Shen H and Forssberg E 2003 *Waste Manage.* **23(10)** 933-949
[9] Demotica J S, et al. 2012 *Int. J. Environ. Sci. Develop.* **3(5)** 470
[10] Weber R J and Reisman D J 2012 *US EPA Region*
[11] Tsamis A and Coyne M 2015 Directorate General For Internal Policies Policy Department A: Economic And Scientific Policy
[12] Chu Y S, et al. 2010 *J. Korean Ceram. Soc.* **47(6)** 613-617
[13] Park J H, et al. 2016 *Wa.t Sci. Tech.* **73(5)** 993-999
[14] Mubarok M and Yudiarto A 2017 *Energy Technol.* 247-258
[15] Firdiyono F, et al. 2016 *Maj. Metalurgi*, **27(1)** 15-26
[16] Holand W and Beall G H 2012 *Glass ceramic technology* (United States : John Wiley & Sons)

Acknowledgment

The author would like to thank the Ministry of Research and Higher Education as a provider of scholarships and financial support through the Postgraduate Team Grant with contract number 553 / UN2.R3.1 / HKP05.00 / 2018. In addition, this study is also supported by the Research Center for Metallurgy and Materials, the Indonesian Institute of Sciences with its laboratory facilities.

IOP Conf. Series: Materials Science and Engineering **541** (2019) 012038 doi:10.1088/1757-899X/541/1/012038

Investigations on mineralogical characteristics of Indonesian nickel laterite ores during the roasting process

I Setiawan[1], E Febrina[1], R Subagja[1], S Harjanto[2] and F Firdiyono[1]
[1]Research Center For Metallurgy and Materials, Indonesian Institute of Sciences, Serpong, Tangsel, Indonesia enter of Minerals Processing and Corrosion Research
[2]Department of Metallurgical and Materials Engineering, Universitas Indonesia, Depok 16424, Indonesia

Email: iwan028@gmail.com

Abstract. Nickel is obtained from laterite ores by a metallurgical process to produces ferronickel. Nickel metal is present in a carrier mineral, a nickel magnesium silicate hydrate compound called garnierite. Garnierite is the common name of magnesium silicate hydrate minerals including chlorite, clay and serpentine. This study to investigate the phase formed when the nickel ore is roasted from a temperature of 600°C to a temperature of 1000°C under atmospheric enviromental. Microscopic analysis of roasted samples was analyzed by SEM-EDX and the dehydroxylation and recrystallization was studied using DTA-TGA. The endothermic peak was appear at 78.8°C, 293.7°C and 638.0°C, this corresponds to the release of free water and the release of crystalline water. While the exothermic peak occurs at 828.0°C. The exothermic temperatures associated with structural changes due to dihydroxylation of serpentine into a forsterite phase. Based on the XRD graph it is known that the sample of the roasting will produce the olivine and forsterite phases.

1. Introduction

The most important mineral carrying nickel from a laterite nickel ore is Garnierite. This mineral is a common name for a group of minerals which include nickel magnesium silicate minerals such as chlorite, clay and serpentine silicate. Garnierite has a fine grain structure, has a low crystalinity and consists of a mixture of several minerals. The constituent components of garnierite are usually Lizardite, Antigorite, Chrysotile and which have a nickel-rich structure namely serpentine [1], [2]. In the type of nickel sulfide ore, concentration can be done in traditional ways such as flotation and milling. Laterite ores can only be treated with minor upgrading by sieving. Apart from a lot of development process, laterite nickel ore can be processed well and economically by smelting while other techniques that are desired do not provide good results. Another problem is that nickel carrier minerals are finely dispersed in silicates and oxide ores. Although various investigations for the use of laterite ore in ferronickel and stainless steel smelting processes have been carried out, there are still many problems that need to be studied. The knowledge of nickel carrier minerals in laterite ores and their behavior at high temperatures is important for developing a pre-reduction process [3]. The nickel recovery process from laterite ores by pyro metallurgy has been intensively studied [4]. Rhamdhani [5] study on microstructure, phase characteristics and thermodynamic analysis and transformation during roasting of laterite nickel ore. Kim performs an increase in levels by looking at the effect of calcination from Indonesian laterite nickel ore before magnetic separation on various calcination and pulp density temperature variations. And the optimum conditions for calcination temperature at 500°C for 1 hour at 0.5T magnetic field intensity will increase nickel grade from 1.5% to 2.9% with 48% recovery. Zhu et al. [6] performed elevated levels by selective reduction followed by magnetic separation. Reduction is carried out using coal (coal-base reduction) with calcium sulfate additive was carried on by Iwan [7] and with sodium sulphate by Li [8]. Bunjaku [9] studied the phenomenon that occurs when laterite nickel ore from Colombia is heated to 1300°C, so that it can provide information for reduction in order to increase levels. Nickel lateritic ore from Indonesia has been succesfully reduced to produce ferronickel at low temperature reduction by Iwan [10]. However, the nature and minerology of the ore is very typical for each place, so to achieve optimum results a reduction process needs to study the character of the ore roasting. This study will study the behavior of laterite due to heating treatment and phase changes and morphology of laterite nickel ore when heated to 1000°C as a basis for reduction.

2. Methode and Materials

Nickel ore used in this experiment is laterite nickel ore from Sulawesi, Indonesia. Laterite or nickel oxide nickel ores are classified in the saprolitic category because they contain Ni and Fe in the range of 1.7% and 18.93%. One of the main characteristics of saprolitic ore is that it has a high content of MgO and SiO_2. The ore used for this study contained high levels of MgO and SiO_2, each at 13.4% and 24.64%, so that the ratio of MgO/SiO_2 was 0.54. Samples are made in the form of dry pellets in a pelletizer to produce pellet diameters ranging from 12 mm to 15 mm.

The characteristics of the ore used were also further explored by conducting XRD testing. This test aims to find out the minerals contained in the ore. Based on the results of the X-ray Diffractometer (XRD) obtained, it is known that the ore is composed of group garnierite minerals, namely clinochrysotile [$(Mg,Ni)_3Si_2O_5(OH)_4$)], magnesium silicate hydroxide/ lizardite [$Mg_3Si_2O_5$ $(OH)_4$], haematite [Fe_2O_3] mineral goethite [FeO(OH)], with hexagonal closed-packed oxygen crystal structure with Fe atoms in the octahedral interstition, and quartz mineral (SiO_2). XRD results show that the ore is dominated by magnesium silicate minerals. The results of the mineral mineralogy XRD testing are shown in Figure 1.

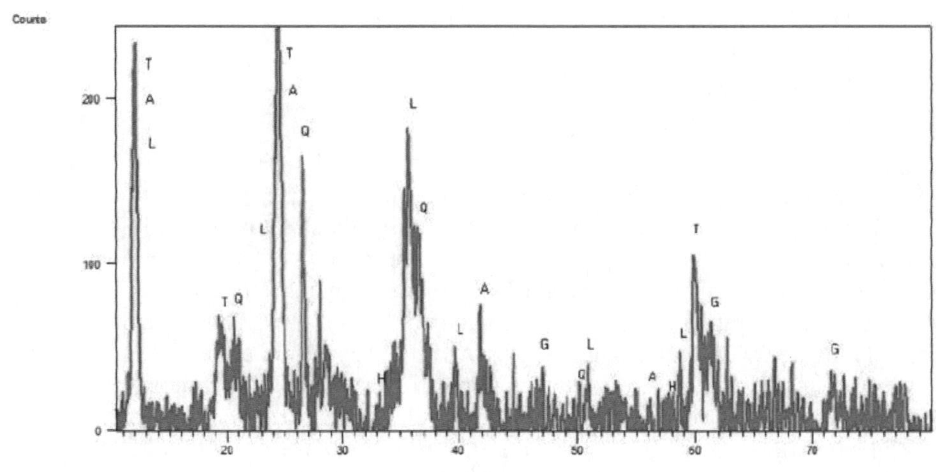

Figure 1. Diffraction peak of nickel ore (T: Clinochrysotile, A:Antigorite, L:Lizardite, Q:Quartz, G:Goethite, H:Haematite).

IOP Conf. Series: Materials Science and Engineering **541** (2019) 012038 doi:10.1088/1757-899X/541/1/012038

Table 1. Minerals contained in laterite ores.

Phase	Formula
Clinochrysotile	$Mg_3Si_2O_5(OH)_4$
Antigorite	$(Mg,Al)_3(Si,Al)_2O_5(OH)_4$
Lizardite	$(Mg,Al)_6(Si,Al)_4O_{10}(OH)_8$
Hematite	Fe_2O_3
Goethite	FeOOH
Quartz	SiO_2

The composition of ore material based on analysis using Shimadzu VF320 X-ray Fluorescence (XRF) is listed in Table 2.

Table 2. Chemical composition of ore (%).

NiO	Ni total	Fe_2O_3	Co_2O_3	MgO	Cr_2O_3	MnO_2	SiO_2	Al_2O_3	LOI	others
2.17	1.71	27.04	0.08	13.40	0.19	0.68	24.64	1.93	17.21	12.65

In this experiment the heating treatment was carried out on the initial ore by roasting about 10-20 grams of pellet-shaped ore samples in a muffle furnace with a sample container that is crucible alumina, with a roasting range of 600, 700, 800, 900 and 1000°C with open atmosphere. The roasting detention was carried out for 1 hour at these temperatures and the samples were analyzed using XRD and SEM-EDX (Scanning Electron Dispersive X-ray Electron Microscopy). Thermal characterization of the ore using Shimadzu DTG-60H Thermo Gravimetry Analyzer (DTA/TGA) Differential Thermal Analyzer aims to determine the reactions that occur during roasting. Scanning Electron Microscopy (SEM) tools used are Jeol 6390A.

3. Result and Discussion

3.1. DTA-TG Analysis

Figure 2 is DTA / TG data from laterite nickel ore. In the picture there are 4 reactions that occur during the roasting process. There are endothermic reaction temperatures at 78.8°C, 293.7°C and 638.0°C, and exothermic temperatures at 828.7°C.

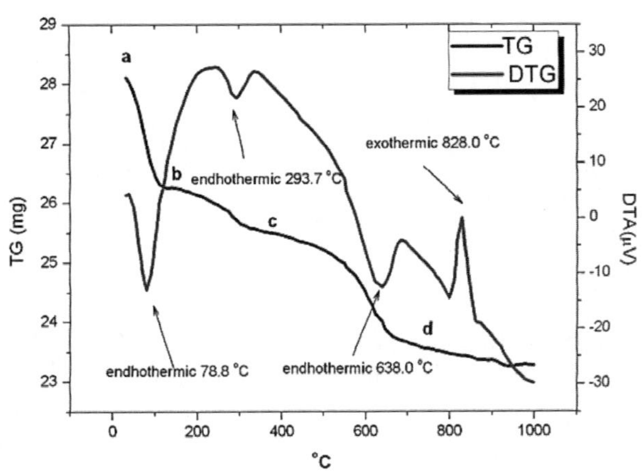

Figure 2. DTA/TGA graphic of nickel ore.

The endothermic reaction at 78.8°C is the beginning of free water evaporation. In this roasting, the temperature is held at a constant temperature so that the water evaporates all. Then the endothermic reaction then occurs at 293.7°C. At this stage there is a goethite dissociation reaction, namely goethite transformation into hematite with the following equation Error! Reference source not found.:

$$2\alpha\text{- FeOOH} \rightarrow \alpha\text{-}Fe_2O_3 + H_2O \quad (1)$$

The water formed evaporates, characterized by a decrease in the sample mass so that a sample weight loss occurs on the TGA curve. Endothermic reaction at 638.0°C is a dehydroxylation reaction of hydroxy groups from serpentine minerals followed by exothermic reactions at 828.0°C which is a forsterite phase recrystallization. Based on the analysis using XRD it is known that the main minerals contained in the ore are serpentinic categories with the main lizardite minerals and contain little goethite. To obtain beneficiation from nickel, the structure of the lizardite and goethite must be damaged. Seen at a temperature of 659.4°C there is dehydroxylation of the lizardite according to the following reaction Error! Reference source not found.:

$$Mg_3Si_2O_5(OH)_4 + MgSiO_3 \rightarrow Mg_2SiO_4 + H_2O \quad (2)$$

Based on Figure 2 it can be estimated that the weight loss due to water vapor evaporation from A to B is 15.4%, goethite crystal water loss is from B to C of 2.3% and the loss of crystal water to forsterite from C to D is 6.0%.

3.2. Structure analysis by XRD

Diffraction analysis using X-rays is done to identify the phases formed before and after roasting treatment. The main mineral contained in laterite is garnierite and is a mixture of lizardite [$(Mg,Al,Ni)_6(Si,Al)_4O_{10}(OH)_8$)], clinochrysotile [$(Mg_3Si_2O_5(OH)_4$] and antigorite [$(Mg,Al,Ni)_3(Si,Al)_2O_5(OH)_4$], after heating are forsterite [Mg_2SiO_4], Quartz [SiO_2], Olivine [$(Fe_{0.24}Mg_{1.76}O_4Si)$].

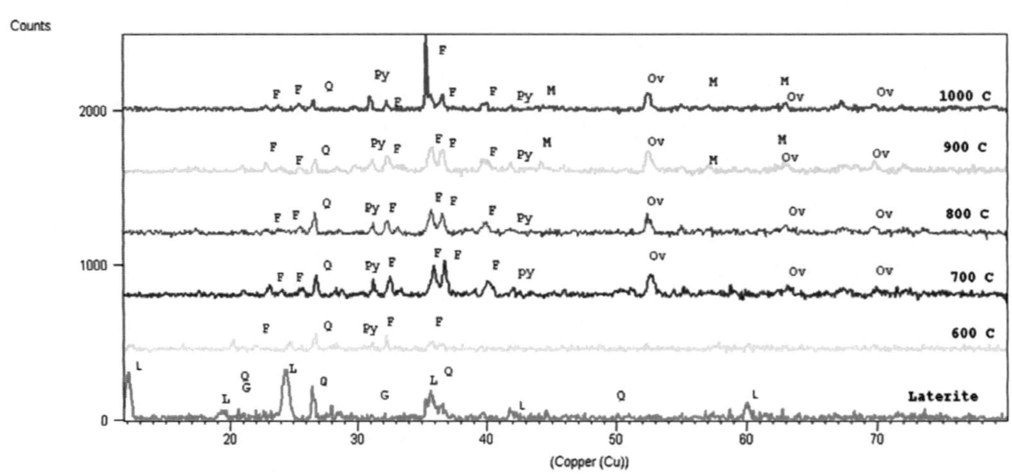

Figure 3. Diffraction patterns from ore samples to roasting 1000°C (F:Forsterite, Q:Quartz, Ov:Olivine, L: Lizardite, G:Goethite, Py:Pyroxene, M:Magnetite).

Table 3. The main minerals formed at some roasting temperature.

Raw	600°C	700°C	800°C	900°C	1000°C
lizardite, quartz goethite	lizardite, quartz, forsterite	quartz, forsterite, olivine, pyroxene	quartz, forsterite, olivine, pyroxene	quartz, forsterite, olivine, pyroxene, magnetite	quartz, forsterite, olivine, pyroxene magnetite

Based on XRD analysis in Figure 3, the minerals formed after roasting are forsterite, olivine and trevorite which are oxidation products from minerals. Garnierite group (clinochrysotile, antigorite, lizardite) transforms into forsterite and olivine.

$$\text{Garnierite, Goethite} \rightarrow \text{Forsterite} + \text{Olivine} + \text{Magnetite} + H_2O \qquad (3)$$

Based on Figure 3 and Table 3, it can be seen that the formation of forsterite has occurred at a temperature of 700°C, and according to the DTA-TG around a temperature of 659.4°C , which at this temperature occurs dehydroxylation of crystalline water from the lizardite. At that time the structure of the lizardite became open so that the nickel and iron metals in the ore body lizardite were easy to reduce. But if heated further, the fast exothermic reaction will occur at the crystallization temperature of forsterite. At this stage nickel and iron metal will be difficult to reduce further because the phase formed will trap iron and nickel metal into it so that the reduction process will be hampered [3,9].

3.3. Analysis by SEM-EDX

Analysis of element distribution in the surface of the sample before and after roasting was investigated using an elemental mapping analysis using the SEM-EDX tool where the gradation of colors from points from black to red indicates the possibility of finding elements in red indicating more elements in the place. Figure 4 is a microstructure image of the initial laterite before roasting. From the picture, the main elements such as Mg, Si, O, Fe and Ni are spread evenly. Fe and Ni metals do not appear to be grouped so that they indicate Ni and Fe metals are present in the mineral structure of garnierite. Chemical composition analysis, with the same test conditions being made, was carried out on the surface of the initial sample found that the peak intensity of Fe and Ni metal was higher than the results of chemical composition analysis on the roasting samples as shown in Figure 5, this is likely because after partially roasting nickel and iron trapped in forsterite crystals so that the chemical composition of nickel and iron on the surface of the sample will decrease.

IOP Conf. Series: Materials Science and Engineering **541** (2019) 012038 doi:10.1088/1757-899X/541/1/012038

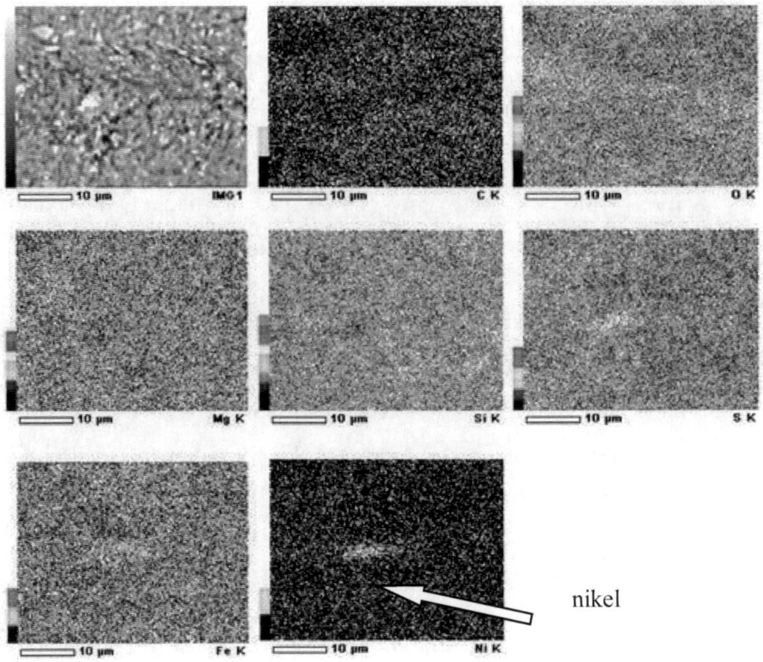

Figure 4. Element mapping analysis from SEM-EDX on the surface of raw laterite samples.

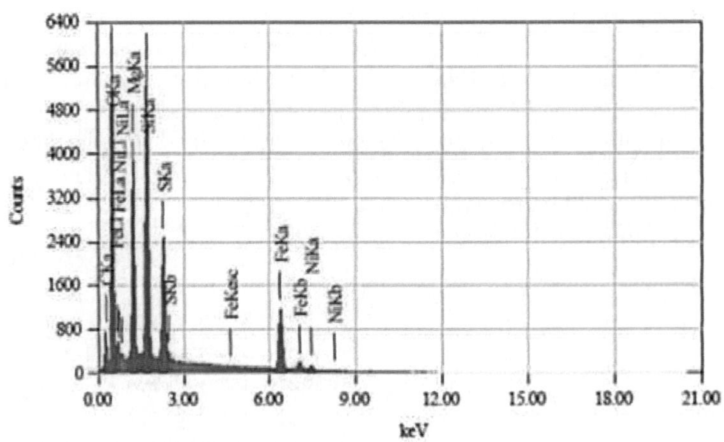

ZAF Method Standardless Quantitative Analysis
Fitting Coefficient : 0.3252

Element	(keV)	Mass%	Error%	Atom%	Compound	Mass%	Cation	K
C K	0.277	19.62	0.30	30.24				4.4184
O K	0.525	41.18	0.24	47.65				48.4788
Mg K	1.253	9.77	0.16	7.44				8.1505
Si K	1.739	11.01	0.19	7.26				12.4466
S K	2.307	5.39	0.12	3.11				7.9014
Fe K	6.398	11.81	0.42	3.92				16.8813
Ni K	7.471	1.22	0.66	0.38				1.7231
Total		100.00		100.00				

Figure 5. Chemical composition of the sample surface in the sample before roasting.

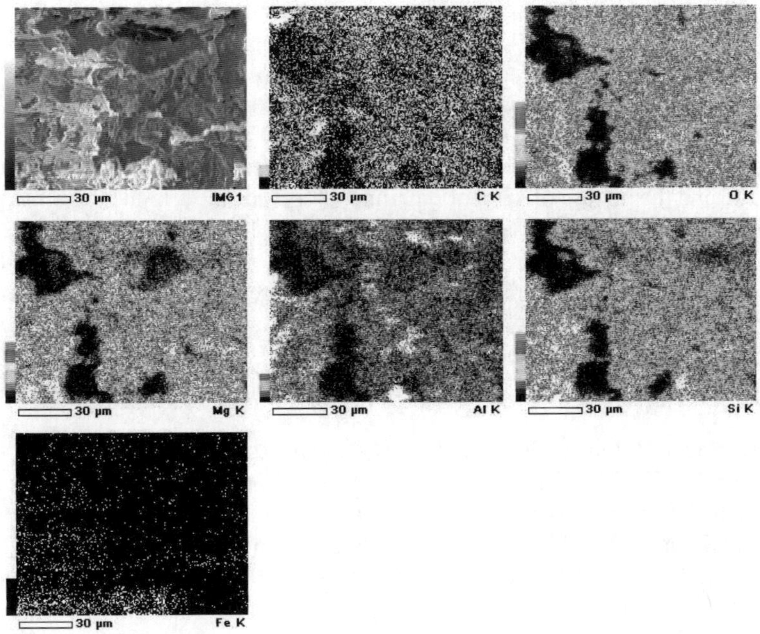

Figure 6. Analysis of elemental mapping from SEM-EDX on the surface of laterite samples after roasting 1000°C.

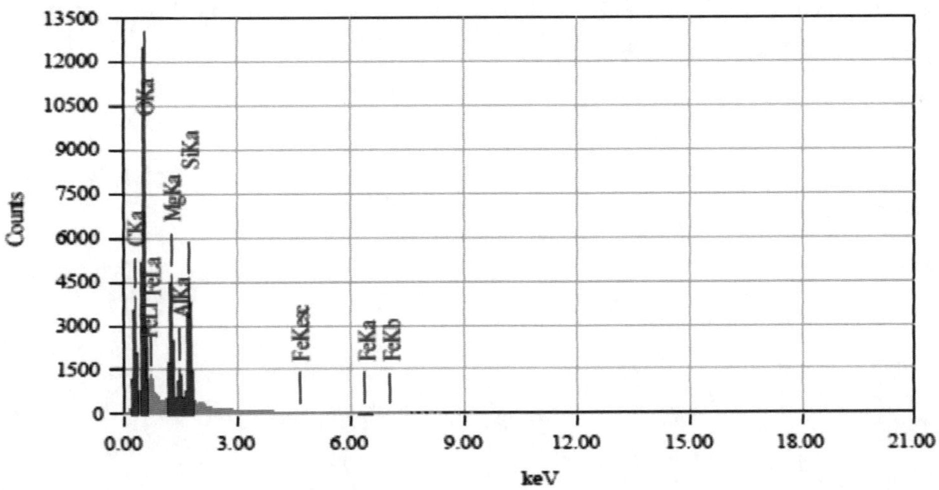

ZAF Method Standardless Quantitative Analysis
Fitting Coefficient : 0.2854

Element	(keV)	Mass%	Error%	Atom%	Compound	Mass%	Cation	K
C K	0.277	30.58	0.14	38.85				14.6378
O K	0.525	55.62	0.22	53.06				69.9082
Mg K	1.253	6.68	0.13	4.19				6.5906
Al K	1.486	1.81	0.13	1.02				1.9452
Si K	1.739	5.23	0.12	2.84				6.7913
Fe K	6.398	0.08	0.38	0.02				0.1268
Total		100.00		100.00				

Figure 7. Chemical composition of the sample surface in the sample after roasting 1000°C.

Even in Figures 6 and 7 the Ni element is not detected in the surface of the sample being tested which indicates that at that temperature the nickel and iron metals are inside the forsterite.

4. Conclusion

The main mineral type of nickel metal carrier is the serpentine group with the Lizardite mineral type. Based on the analysis using the DTA-TG that the endothermic peak occurs at a temperature of 78.8°C, 293.7°C and 638.0°C and this is related to the evaporation of water vapor and the release of crystalline water from goethite and the release of serpentinic mineral crystals to form the forsterite phase. Forsterite crystallization temperature occurs at a temperature of 828.0°C.

At a roasting temperature of 700.0°C, there has been a reaction of dehydroxylation of laterite nickel to forsterite, at which temperature the mineral undergoes a structural change and at that time the time is right to reduce nickel and iron in it before the forsterite crystallization reaction will cause nickel and iron to be inert against outside influences.

References

[1] Dalvi A D, Bacon W G and Osborne R C 2004 *Int. Conv. Trade Show Investors Exch.* 1–27
[2] Brindley G W and Kao C C 1980 *Clays Clay Miner.* **28** 435–443
[3] M Kawahara, J M Toguri, and R A Bergman 1988 *Met. Trans. B* **19** 181–186
[4] R R Moskalyk and A M Alfantazi 2002 *Miner. Eng.* **15**(5) 593–605
[5] Rhamdhani M A, Hayes P C and E Jak 2009 *Miner. Process. Extr. Metall. (Trans. Inst. Min. Met. C)* **118(3)** 146–155
[6] Zhu D Q *et al.* 2012 *Int. J. Miner. Process.* **106** 1–7
[7] Setiawan I, Harjanto S and Subagja R 2014 *Int. J. Eng. Technol.* **14** 56–66
[8] Li G, Shi T, Rao M, Jiang Tand Zhang Y 2012 *Miner. Eng.* **32** 19–26
[9] Bunjaku A, Kekkonen M, Pietila K and Taskinen P 2012 *Miner. Process. Extr. Metall. (Trans. Inst. Min. Met. C)* **121(4)** 156–165
[10] Setiawan I, Harjanto S and Subagja R 2017 *IOP Conf. Ser. Mater. Sci. Eng.* **202 (012019)**

IOP Conf. Series: Materials Science and Engineering **541** (2019) 012039 doi:10.1088/1757-899X/541/1/012039

The effect of iron content on the reduction of nickel laterite by direct reduction method

I Setiawan[1], E Febrina[1], A B Prasetyo[1], R Subagja[1], and F Firdiyono[1]
[1]Research Center For Metallurgy and Materials, Indonesian Institute of Sciences, Serpong, Tangsel, Indonesia

E-mail: iwan028@gmail.com

Abstract. Reduction and the compounds formed when the reduced laterite nickel has been investigated. Nickel laterite with average grain size below 0.149 mm mixed with 13% (weight percent) of coal as reductant then reduced in muffle furnace with temperature from 700 °C to 1000 °C. Type of laterite is used which has high iron (saprolite) and low iron content (limonite). Reaction time varied 30, 60 and 90 minutes. The reduction of nickel during reduction is determined by an analysis of metallization values. Metallization analysis used bromine methanol method, while the compound analysis was formed using x-ray Diffractometer. Based on the metallization test, it is known that the laterite nickel ore containing high iron will produce a higher nickel reduction than the laterite nickel ore containing low iron. This corresponds to the oxide phases formed. The higher iron contain on the ore tend to produce the fayalite mixture and slightly forsterite, while at low iron content will form a forsterite. Forsterite tends to inhibit reduction of nickel.

1. Introduction

Nickel consists of two types of common ores, sulfides and laterite ores. Laterite nickel ore, which comprises 73% of the world's proven nickel reserves is the dominant source of nickel. However, only 42% of the world's production of nickel comes from laterite nickel ore [1], [2]. Ferronickel processing from laterite ores requires high energy costs, especially when laterite ores with low nickel content are processed. Because energy is mostly used to melt ore and produces only a little nickel metal and a large amount of slag. In addition, areas where large deposits of laterite ore are located do not have access to cheap electricity sources. Unlike sulfide ores, nickel in laterite is finely dispersed in ore and cannot be increased by fine grinding or other physical methods [3].

There are three main processes of commercial processes: rotary kiln electric furnaces (RKEF) and rotary kiln magnetic separators, Caron process, and high pressure acid leaching (HPAL). In general, laterite ore types are suitable for certain process, such as crystals for leaching high pressure acids or saprolite for RKEF. A significant problem is the presence of MgO and SiO_2 contents which will affect reductibility level on temperature used in the process. Reduction results from laterite will produce forsterite (Mg_2SiO_4), fayalite ($FeSiO_3$), and enstatite ($MgSiO_3$) simultaneously. This intermediate phase is very irregular and thus has high chemical reactivity. Kawahara [4] and Hallet [5] have found that nickel oxides will be bound in pyroxene and are particularly strong in the olivine phase, and consequently the reduction of nickel oxide becomes much more difficult. Stevens, et. al. [6] and Hayashi [7,16] concluded the role of forsterite structure formation in saprolite at high temperatures, inhibiting the reduction of nickel and cobalt. Several investigations of the reduction of nickel oxide or laterite nickel ore have been carried out and are related to the effects of phases on the reduction and formation of nickel metal particles. For example, Krasuk et al. [9] studied the reduction of nickel oxide with CO,

Utigard et al. [10] studied the nickel oxide reduction kinetics using hydrogen, and James et al. [11] using X-ray diffraction to study the reduction of nickel oxide by hydrogen. Reduction of laterite nickel ore by gases containing H_2 and CO in horizontal tube furnaces and in laboratory-scale reactors was studied by Utigard [10]. Valix et al. [12] study the effect of sulfur on laterite ores during temperature reduction. Li et al. [13] completed a simulation study of pre-reduction and calcination of nickel laterite ore with coke to determi.lkjuyne the effect of temperature, time, atmosphere and CaO content on nickel reduction rates. Enriching metallic nickel from laterite nickel ore using the process of separating thermal carbon magnetic reduction with activated carbon powder as a reductant was investigated by Xu et al [14]. Other additives have been added in order to study the effect on the reduction of nickel metal which in principle suppresses the growth of the inert phase forsterite by adding chemicals such as $CaSO_4$ [14,15], Sulfur [12], silica [4] and CaO [13]. There several main factors influenced the reductibility of nickel from ore. These parameter are minerology of ore, composition of reductor, temperature, reduction time, aditif and the composition of slag. There are two general type slag as by product forsterite and fayalite. The type of slag produced will affected the success of reduction. This study has investigated the effect of iron content due to type of slag produced. The reduction level of ore with the difference iron contained was determined by reducing it with coal followed by analysis using bromine-methanol for determination of the degree of metallization. So that the reduction properties of laterite types can be known.

2. Methodology

The research was carried out at first by grinding the saprolite and limonite ore from Sulawesi Island into powder form. After that, sieving with a size of > 100 mesh, to get a uniform powder size. Then the powder is dried at a temperature of 100 °C. After drying the laterite powder, initial characterization to obtain the chemical content of laterite by XRF and thermal characterization using TG/DTA. There are three composition of samples: limonite, saprolite, and mixed of them. The mixed samples are made by mixing saprolite with limonite with a ratio of 1: 1 (% by weight). The chemical composition listed in table 1.

Table 1. Chemical composition of sampel limonite, saprolite and mix saprolite-limonite 1:1 (%w/w).

Ore	Ni Total	Fe Total	MgO	SiO_2	Al_2O_3	LOI
Limonite	1.31	47.5	2.35	8.27	5.32	10.4
Saprolite	1.71	18.9	13.0	15.7	0.53	17.2
Mix 1 : 1	1.51	31.2	4.32	11.3	4.20	14.2

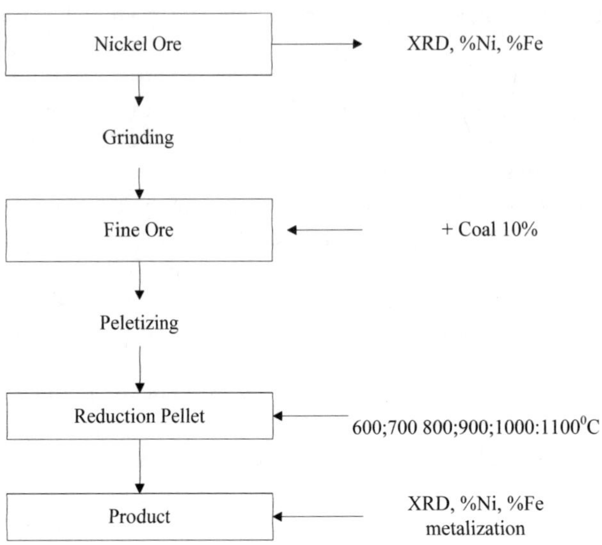

Figure 1. Flowchart of Reduction.

The weight of metal nickel is obtained by dipping the reduced sample in bromine-methanol solution to calculate "% reduced metal" or can also be called "metallization degree" using the equation [8].

$$\% \text{ metallization} = \frac{weight\ of\ reduced\ nickel}{weight\ of\ nickel\ total}\ x\ 100\% \quad (1)$$

3. Result and Discussion

3.1. Thermal Characteristic

Figure 2 is DTA/TG data from laterite nickel ore. The picture shows the endothermic reaction temperature at 75, 290, and 638 °C, and exothermic temperature at 830 °C.

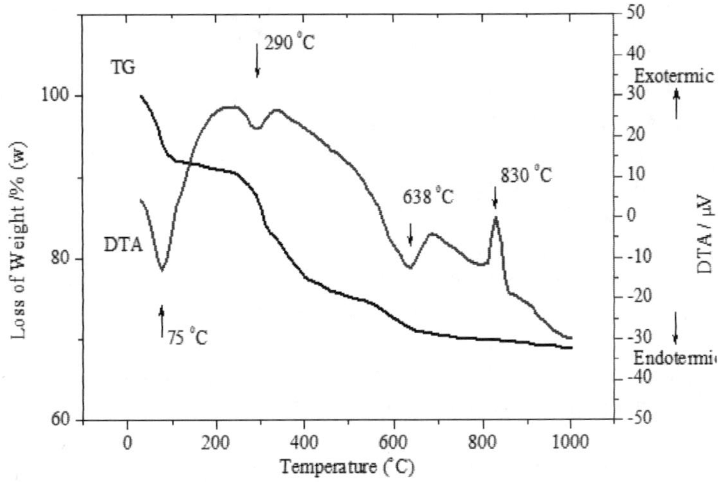

Figure 2. Graph of DTA / TGA of laterite ore (heating conditions : 20 °C /minute, CO2 50mL/s).

The endothermic reaction at 75°C is the beginning of free water evaporation. In this heating, the temperature is held at a fixed temperature so that the all water evaporates. Then the endothermic reaction then occurs at 290°C. At this temperature there will be free water evaporation and from goethite with the following equation.

$$2\alpha\text{-}FeOOH \rightarrow \alpha\text{-}Fe_2O_3 + H_2O \quad (2)$$

The water formed evaporates, marked by a decrease in the sample mass so that there is a decrease in the curve on the graph. The endothermic reaction at 638°C is a dehydroxylation reaction of hydroxy groups from serpentine minerals followed by an exothermic reaction at 820°C which is a forsterite phase recrystallization. According to Brindley [18], the water loss reaction in this dehydroxylation event is because the migration of protons from H + ions joins OH-ions to form H_2O molecules, and magnesium and silicon ion ions migrate in opposite directions to keep the charge neutral. At a temperature of 638°C, the structure of the lizardite and goethite is damaged so that at this temperature it is a good time to process nickel because the structure of the lizardite is open to outside influences. If the final product is forsterite and silica, the overall chemical equation can be written as follows:

$$\underset{\text{Lizardite}}{2Mg_3Si_2O_5(OH)_4} \rightarrow \underset{\text{Forsterite}}{3Mg_2SiO_4}+SiO_2+4H_2O \quad (3)$$

At a temperature of 820 °C there is a crystallization reaction from the formed forsterite. This is according to what has been explained in detail by Ball that serpentine will undergo several stages of reaction when heated (evaporation of free water), dehydroxylation (evaporation of crystal water) and crystallization of forsterite at temperature about 800°C.

3.2. Effect of the amount of coal on metallization

The degree of reduction of laterite samples at various temperatures and the time of the reduction process of saprolite ore and limonite samples shown on figure 3. Figure 3 is the relationship between the reduction temperature of saprolite samples at various variations of reduction time There is the influence of the amount of reducing agent on metallization of nickel at a reduction temperature at fixed temperature 900 °C. Based on figure 3 known that the optimum amount of coal is 13%(w/w). Metallization, sometimes referred as a "% reduction level" or "% reducibility", is defined as the weight of a the total weight metal as element in the sample, metal element in a sample which is in a state of element or oxidation number "zero", compared to the total of element weight.

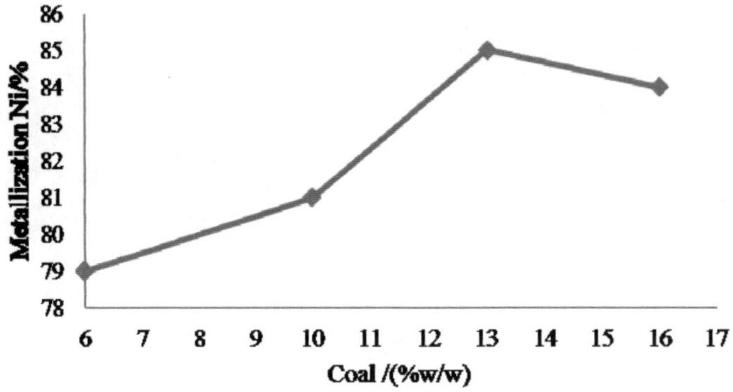

Figure 3. Relationship between the amount of coal and nickel metallization (%) at a reduction temperature of 900°C.

3.3. Comparison of metallization of limonite, saprolite and mix

For saprolite ore (ore mixed with coal 13%) the amount of reduced nickel metal increases after reduction is carried out for more than 30 minutes. From the temperature of 700°C, the nickel reduced was significant increase. But after 700°C, the degree of reduction will rise slowly. For limonite ores with a rise in temperature will increase the degree of reduction but slightly different from saprolite, the increase in the degree of reduction of each temperature change is higher. The highest degree of metallization is 85% from 900°C to 1000°C for 60 minutes reduction time. The degree of metallization does not seem to increase significantly with the increase in temperature even though it is higher than 1000°C. The metallization degree of saprolite is lower than limonite at the same temperature, this is related to the ratio of SiO_2 to MgO.

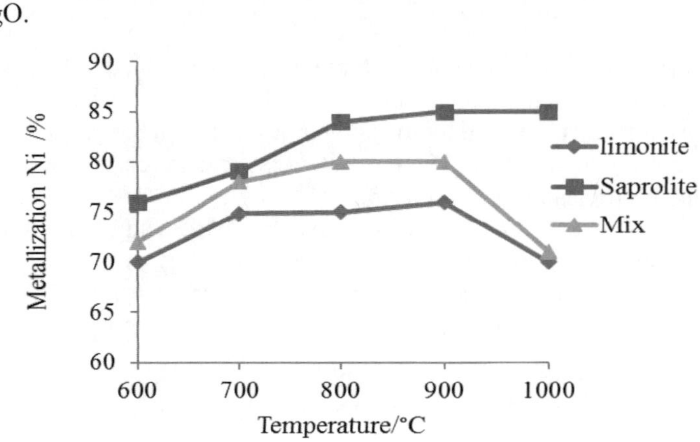

Figure 4 Relationship between the reduction temperature of saprolite samples (13%) and limonite with% reduced nickel (metallization) at various reduction times.

Olivine which has a chemical structure of $2MgO.SiO_2$. Based on the above results it is very clear that nickel reduction is very dependent on the type of minerals contained in the ore. Based on its iron content, limonitic ores have a much higher iron content. It is known that nickel oxide ore contains relatively high magnesium and silica. This will inhibit the reduction of nickel at high temperatures. This phenomenon is related to the formation of olivine, which is an iron-magnesium-silicate. The atomic radius of the Ni^{2+} atom is very close to the radius of Mg^{2+} (0.068 and 0.065 nm) so that the Ni atom can exchange places with the Mg. Limonite ores have relatively low levels of MgO and SiO_2 so formation of olivine is less significant. When reduced at high temperatures the interference of the formation of olivine can be ignored so that the amount of reduced nickel will be relatively proportional to the increase in temperature.

Phase analysis and diffraction on figure 5 is an XRD graph of the saprolite sample as initial ore and after reduction at several temperatures. All reductions were carried out within 60 minutes using coal in a closed muffle furnace.

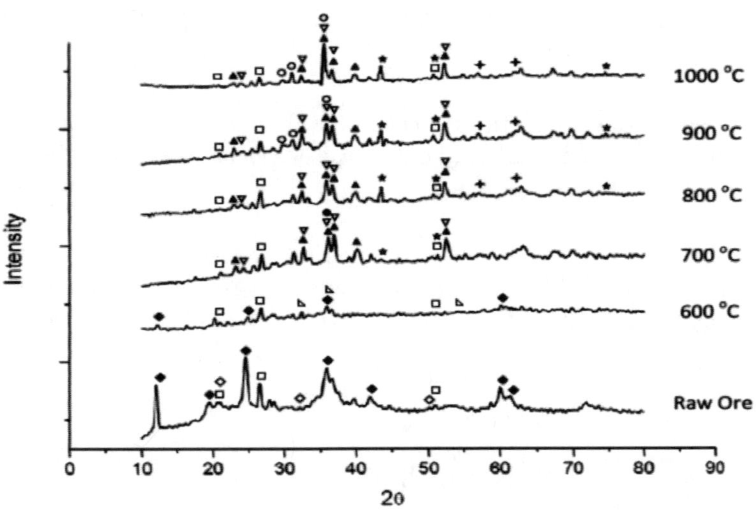

Figure 5 Diffraction pattern of *Saprolite* before and after reduced (□ silika,♦ Lizardite, ◇ Goethite, ▲ *Olivine*, ★ Taenite, ∇ *Forsterite*, ✚ Magnetite, O *Pyroxene*, ◣ Hematite.

The main mineral contained of saprolite mineral there are lizardite, quartz and goethite. Lizardite contains 3 main peak there are 2θ 12.04; 24.30 and 35.87 (ICDD No. 09-0444). The silica peak appear on 2θ: 26.64; 20.86 and 50.13. (ICDD No. 79-1910). Peaks of *goethite* appears with the low intensity on 2θ: 21.24; 33.28 and 33.62 (ICDD No. 17-539) from goethite to hematite.

$$FeO.OH \rightarrow \alpha\text{-}Fe_2O_3 + H2O \qquad (4)$$

Further reduction of the hydroxyl group Garnierite group serpentine (clinochrysotile, antigorite, lizardite) transforms into forsterite and olivine $(Mg,Fe)_2SiO_4$.

$$\text{Garnierite, Goethite} \rightarrow \text{Forsterite + Olivine + Magnetite} + H_2O \qquad (5)$$

Ferronickel metal in the form of taenite (ICDD no. 47-1417) with 3 strongest peaks, that 2θ are 43.5; 50.7 dan 74.5 appears on 700°C The main mineral that appears at high temperatures is magnesium silicate in the form of forsterite and olivine. Figure 6 is an XRD graph of the limonite sample as initial ore and after reduction at several temperatures. All reductions were carried out within 60 minutes using coal in a closed muffle furnace.

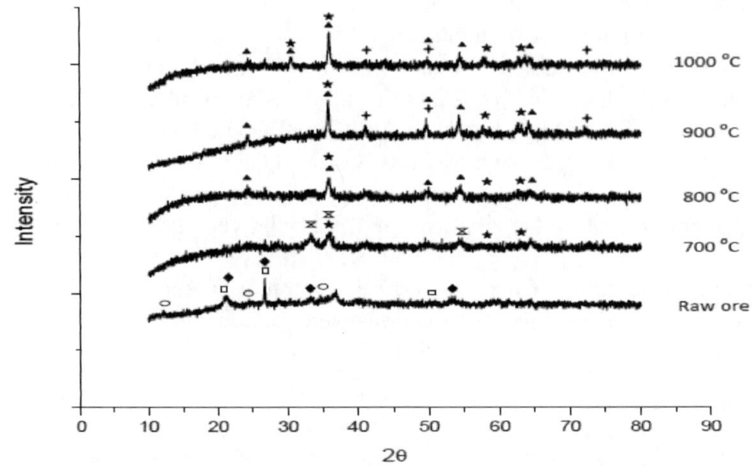

Figure 6. Diffraction pattern of *limonite* ore from 700 °C to 1000 °C before and after reduced (▲Fayalite, ■ Silica, ◆ Goethite, ● Antigorite, ★ Magnetite, ✦ Taenite, ⧗ Hematite).

The main mineral that appears after reduction at a temperature of 1000 °C is magnetite (ICDD No. 84-308) dan *fayalite* (ICDD No. 76-0852) marked of appearence of 2θ on 21.2; 33.3; 33.6 for *magnetite*, and 35.9; 31.7; 51.5 for *fayalite*. FeNi substance as *taenite* (ICDD No. 47-1417) first appears on 900 °C.

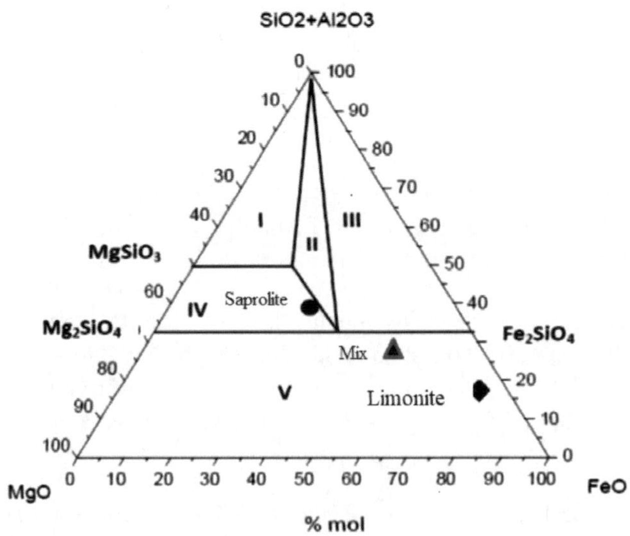

Figure 7. Composition of *saprolite* and *limonite* and mix, on SiO2-MgO-FeO system phase diagram.

Figure 6 is the ternary diagram of silica, magnesia dan iron oxide. These three types of ore have a significant difference from the chemical composition of the main constituent elements, there are composition main element such as iron, magnesium, silicon and aluminum oxide. To further explore the reductibility of these ores, the composition as % mol of each of these oxides was calculated and plotted on a ternary diagram of FeO-SiO_2-MgO from Kawahara and this diagram is shown in Figure 7. Based on Figure 7 from Kawahara, There are 5 parts in the graph that can be grouped as below;

- Part I, each containing pyroxene and SiO2,

IOP Conf. Series: Materials Science and Engineering **541** (2019) 012039 doi:10.1088/1757-899X/541/1/012039

- Part II, *pyroxene*, SiO_2 and *olivine*
- Part III, *olivine*, SiO_2
- Part IV, *olivine*, *pyroxene*
- Part V, magnesio *Wustite* and *olivine*.

Saprolite samples used in this study will be in part IV which in this composition contain *olivine* and *pyroxene*. NiO will be difficult to reduce in these phases, but if NiO is present in magnesioferrite or fayalite phases then NiO is easily reduced as well as on part V. The chemical composition of *limonite* oxides is in part V which contains a little magnesium silicate so that nickel in the *limonite* sample is easier to reduce.

4. Conclusion

Increased iron content in laterite nickel ore samples will increase the reducibility of nickel metal. The goethite mineral in the ore functions as a fayalite or magnesioferite forming which does not trap nickel when the reduction is carried out as well as *forsterite*.

References

[1] Dalvi A D, Bacon W G, and Osborne R C *2004 Int. Conv. Trade Show Investors Exch.* 1–274
[2] Mudd G M 2010 *Ore Geol. Rev.* **38** 9–26
[3] Norgate T and Jahanshahi S 2010 *Miner. Eng.* **23** 65–73
[4] Kawahara M, Toguri J.M and Bergman R A 1988 *Met. Trans. B.* **19** 181–186
[5] Hallett C J 1997 *Proc. Nickel-Cobalt* 299–312
[6] Stevens L G, Goeller L A and M Miller M 1975 *14th Annual Conference of Metallurgists, the Canadian Institute of Mining and Metallurgy*
[7] Hayashi M 1971 Effect of phase transition of reductive roasting of nickel bearing serpentine (United States: *University of Utah)*.
[8] Setiawan I, Harjanto S and Subagja R 2017 *IOP Conference Series: Materials Science and Engineering* **202**
[9] Krasuk J H and Smith J M 1972 *AlChE J.* **18** 506–512
[10] Utigard T A, Wu M, Plascencia G and Marin T 2005 *Chem. Eng. Sci.* **15** 169–179
[11] James T R, Robert S and Martyn V T 2003 *Appl. Catal. A Gen.* **246** 137–149
[12] Valix M and Cheung W H 2002 *Miner. Eng.* **15** 523–530
[13] Li G, Shi T, Rao M, Jiang T and Zhang Y 2012 *Miner. Eng.* **32** 19–26
[14] Xu D, Liu Y, Li J and Zhai Y C 2010 *J. Northeast. Univ. Natural Sci.* **31** 559–563
[15] Zhu D Q, Cui Y, Hapugoda S, Vining K and Pan J2012 *Trans. Nonferrous Met. Soc. China* **22** (**4**)907–916
[16] Setiawan I, Harjanto S and Subagja R 2014 *Int. J. Eng. Technol.* **14** 56–66
[17] Brindley G W and Kao C C 1980 *Clays Clay Miner.* **28** 435–44

IOP Conf. Series: Materials Science and Engineering **541** (2019) 012040 doi:10.1088/1757-899X/541/1/012040

The Effect of Slow Evaporation on Ca, Na, K, and Mg Concentration in Natural Brine from Ciseeng, West Java

A R Rhamdani[1, a)], **L H Lalasari**[1], **E Sulistyono**[1], **I Setiawan**[1], **F Firdiyono**[1]
[1]Research Center for Metallurgy and Material, Indonesian Institute of Sciences, PUSPITEK area building 470, South Tangerang, Banten, Indonesia

E-mail: [a)]rizkyrhamdani@gmail.com; ahma075@lipi.go.id

Abstract. Indonesia's location in the ring of fire gives Indonesia a lot of natural resources such as natural brine water from Ciseeng, West Java. Natural brine contains a lot of minerals such as Na, K, Ca, Mg. and could potentially be the source of those elements. One of the common process for recovering minerals in brine water is through solar evaporation process. This process is ineffective because the evaporation stops in the night time. The purposes of this study are to find another method for concentrating brine, characterize natural brine from Ciseeng, West Java, and investigate the effect of new evaporation method to the brine. In this work, the author conducted the evaporation using electric heater to keep the solution in the desired temperature coupled with blowers to blow air to the surface of the solution for 10 days. The volume of the system was maintained by making up the evaporated water with fresh brine. The results indicate that the evaporation can increase the Mg, Na, and K concentration up to 282%, 283%, and 207%, respectively, while the Ca concentration stays relatively constant compared to the fresh brine. The brine's °Bé increase from 2.5 to 6. In the process, about 158% of Ca and 51% of Na were recovered as precipitate.

1. Introduction

Natural brine is a source of a lot of minerals such as sodium, potassium, calcium, and magnesium. Those minerals could be used for a lot of application. Sodium compound can be used as feedstock for chemical industry, potassium compound as fertilizer, calcium compound as building stones and cement, and magnesium as alloying agent. The concentration of those elements in the brine could vary depending on its source. Indonesia's location in the ring of fire gives a lot of natural brine sources. One of unexploited natural brine in Indonesia is located in Ciseeng, West Java. The brine now used as public bathing area. People believes that the mineral content and its hot temperature could cure some of skin disease.

Figure 1. Natural brine pond in Ciseeng.

To harness the potential of minerals contained in the brine, some study has been conducted. The usual approach is using evaporation method to concentrate the brine up until certain degree of salinity. In the process of concentrating the brine, calcium, sodium and some magnesium are recovered as precipitate. Turk[1], Balarew[2], and Nie[3] study the evaporation of `brine using solar evaporation, but using solar energy to evaporate the brine is ineffective because the evaporation is stopped during the night and also during rainy season, thus an alternative for this method need to be studied. The concentrated brine is then processed by precipitation[4–7], ion-exchange[8], or solvent extraction[9] to recover other precious minerals such as magnesium and lithium.

This study will conduct preliminary research of brine evaporation using electric heater in a semi-pilot plant scale. Brine will be evaporated for 10 days and concentrated brine and precipitate will be analyzed. The purpose of this study is to characterize the effect of evaporation to the mineral's concentration and to the precipitated mineral. The concentrated brine from this study will be used as raw material to extract the remaining minerals using another method in the next work.

2. Materials and Methods

Preliminary research was carried out to see the effect of evaporation to the concentration of minerals in the natural brine. The evaporation apparatus consists of electric heater to keep the temperature of brine in the desired temperature and blowers to blow air to the surface of brine. The composition of brine from Ciseeng was analyzed using ISP-OES to determine the chemical composition. The results are listed in Table 1.

Table 1. Chemical composition of brine from Ciseeng.

Component	Concentration (ppm)
Sodium	6194.99
Potassium	483.54
Calcium	763.33
Magnesium	427.23

The experiment consists of a loop which was circulated in 3 open vessels. 1st vessel equipped with air blower and rack of thin sheet of plastic on which the brine will flow through, 2nd vessel contain electric heater to keep the temperature of brine around 40°C, and 3rd vessel contain electric heater also set up at 40°C and air blower. Fresh brine was introduced to the system until all 3 vessels are filled.

After the vessels are filled, a pump was used to circulate the brine from 1st, 2nd, 3rd, and back to 1st vessel again. Fresh brine was added for making up the evaporated water every day. Brine was sampled each day from 1st and 2nd vessel and filtered using Whatman paper no. 42 and then analyzed using ICP-OES to determine its chemical composition. Salt precipitated in the vessels were then characterized using SEM, XRF, and XRD to determine its morphology, dry basis chemical composition, and phase. The flowsheet of experiment is shown in Figure 2.

Baume degree (°Bé) was used to see the density of the processed brine. °Bé have been used by traditional salt farmer to know when they should harvest their salt. °Bé correlation to spesific gravity (s.g.) at temperature of 60°F can be seen in equation (1)

$$s.g. = \frac{145}{(145 - {}^{o}Bé)} \quad (1)$$

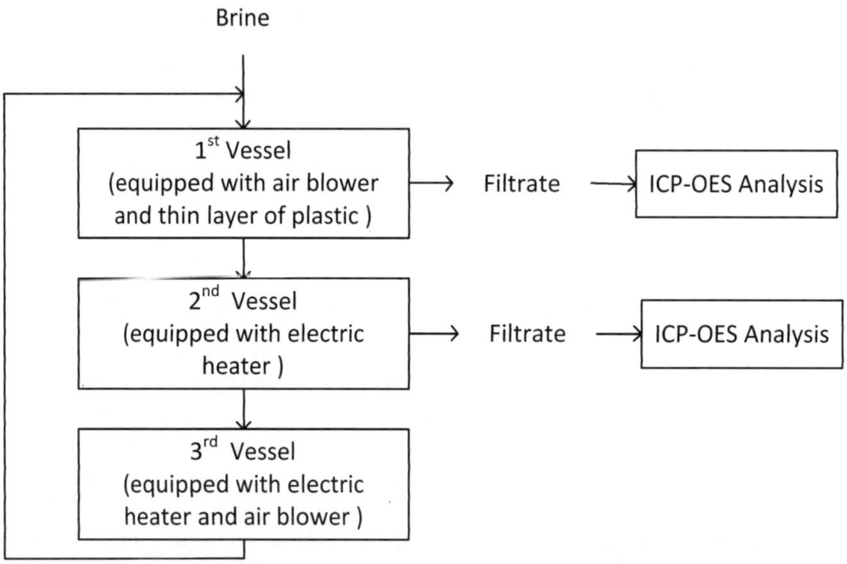

Figure 2. Schematic diagram of experiment.

3. Result and Discussion

The experiment was conducted for 10 days straight. The average temperature of the system is 40.4°C and 61.33% of brine is evaporated. Figure 3 shows the chemical composition of Ca, K, Mg, and Na. It can be seen that the concentration of Ca is relative constant while the concentration of K, Mg, and Na is increasing in the same pattern throughout the experiment. The concentration of Mg, Na, and K increase for 282%, 283%, and 207%, respectively. This result is in line with previous study for seawater evaporation. Study by Abdeel-Aal, et al [10] stated that Ca will precipitate first in the form of $CaCO_3$, followed by $CaSO_4$, NaCl and $MgSO_4$, respectively[10].

But the result is slightly different from that previous study. Ca already precipitate at °Bé 2.5 (can be seen from the relatively constant value of Ca concentration from day 1 to day 3 of experiment), while those study stated that Ca will start to precipitate at °Bé > 3. Figure 4 shows the °Bé value throughout the experiment. This is possible because the brine from Ciseeng contain a lot more Ca (763.33 ppm) compared to the Ca concentration in sea water (416 ppm) [11]. This high concentration of Ca makes brine already saturated with Ca and start precipitating earlier. 61.18% of Ca was precipitated, calculated using mass balance.

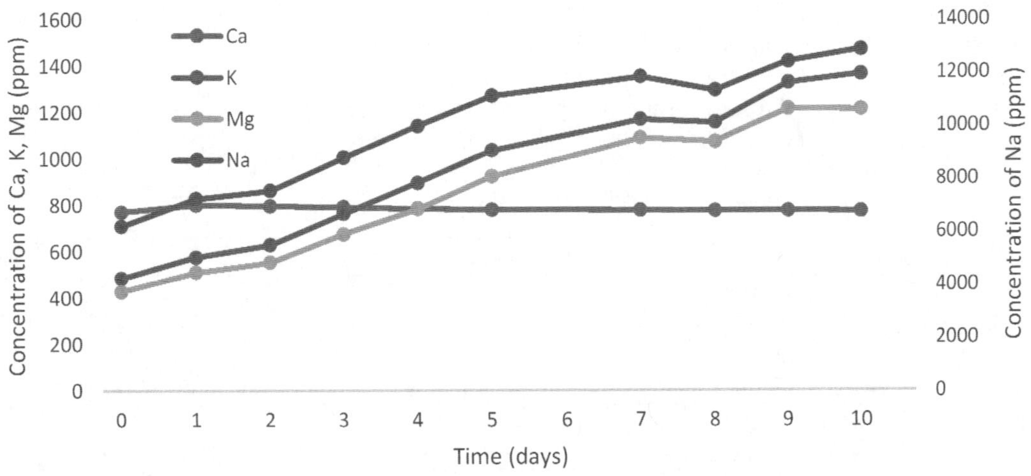

Figure 3. Concentration of Ca, K, Mg, and Na in the system for 10 days.

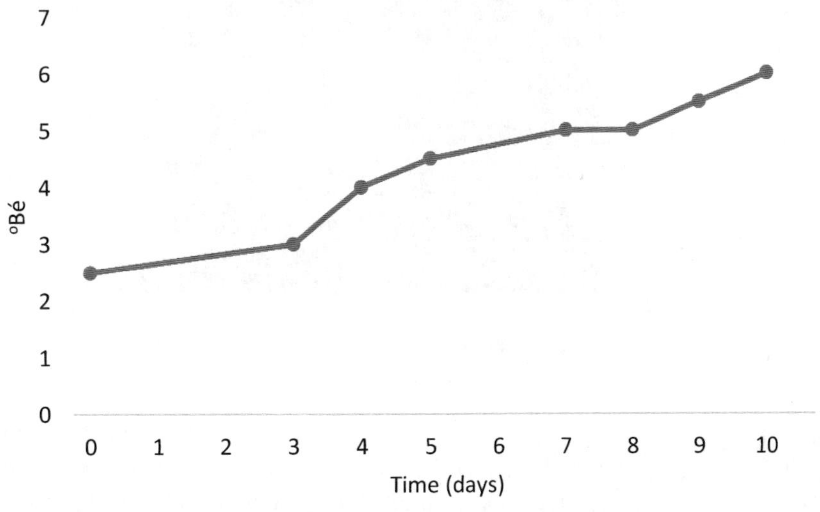

Figure 4. °Bé changes during experiment.

The results of this research compared to other research can be seen in table 2. Natasha et. al.[12], and Lalasari et. al[13] did the evaporation on brine from ciseeng site B while this experiment used brine from ciseeng site A, the maps of ciseeng site A and site B can be seen in Figure 5. The distance from those two brine sources were only around 300 m but the composition is different. Brine from site B has higher Ca, Na, and Mg content compare to brine from site A. Those two other experiments were conducted in a lab scale apparatus while this experiment was conducted in a pilot plant scale. The results from this study are different when compared to the other study. Ca concentration seems to stayed constant and Na and Mg concentration were increased while study by Natasha et. al.[12] saw a decrease in Ca and Na concentration. These could happen because of the differences in the starting concentration of Ca and Na and also the differences in the evaporation treatment. This experiment was conducted semi-batch system while the other experiments were conducted in the batch system. Fresh brine was introduced every day and the evaporation happened in lower temperature (average 40.4°C). These lower temperature means that the solubility of $CaCO_3$ were higher[14] compared to the higher temperature evaporation in those other experiment. The rise of Na and Mg concentration is still in line with theory that Na and Mg was not supposed to precipitate under °Bé 25 and 30, respectively[2,10].

Table 2. Results comparison with other researches.

	Raw Material	Water evaporated	Concentration (ppm)					
			Before Evaporation			After Evaporation		
			Ca	Na	Mg	Ca	Na	Mg
This research	Brine from Ciseeng Site A	61%	763	6159	472.23	772.24	12851	1211.1
Natasha et. al. [12]	Brine from Ciseeng Site B	50%	2010	27050	-	170.06	4188.85	-
Lalasari et. al.[13]	Brine from Ciseeng Site B	80%	2940	28870	9170	53020	9500	2100

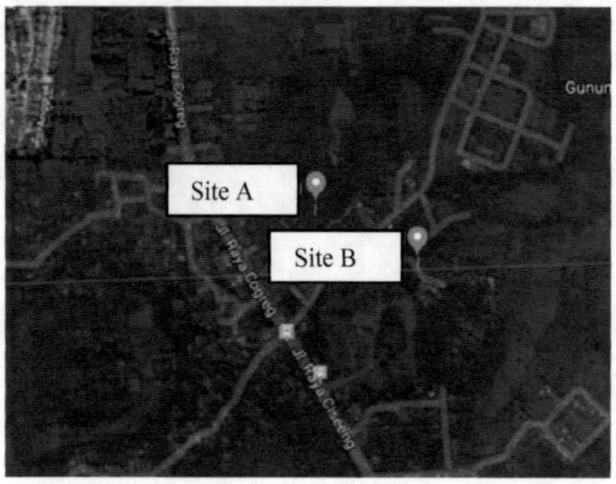

Figure 5. Map of brine source location Site A and Site B.

3.1. SEM Analysis

SEM Analysis was performed to see the morphology of the precipitate from the evaporation process, the results can be seen in figure 6. It can be seen that the precipitation product has various particle size and crystal. This could happen because small amount of sodium chloride was co-precipitated alongside calcium carbonate as the main product and supported by EDX quantitative analysis results of the sample (table 3) which shows that sodium was also co-precipitated.

IOP Conf. Series: Materials Science and Engineering **541** (2019) 012040 doi:10.1088/1757-899X/541/1/012040

Table 3. EDX quantitative analysis of SEM sample

Component	Mass (%)
Ca	38.14
O	38.02
C	13.14
Cl	3.89
Na	3.02
Si	2.01
Al	0.69
K	0.61
Mg	0.49

(a) **(b)**

Figure 6. Morphology of precipitated salts at (a) 1000x and (b) 2000x magnification

3.2. XRF and XRD Analysis

Table 4 show XRF analysis result of precipitated salts. XRF analysis was conducted using Bruker S2 Puma and was able to see the concentration of element starting from sodium. It can be seen that calcium are the main constituent of the salts followed by strontium, sodium. This co-precipitation of sodium and magnesium was not expected since based on previous study [2,10] sodium and magnesium would not precipitate until °Be value reached around 25 and 30, respectively. This co-precipitation could happen because some of the brine might be evaporated on the wall of the evaporation apparatus, Figure 7, and reached high °Be value, thus sodium and magnesium were precipitated. Extremely similar chemical properties between strontium and calcium [15] made the co-precipitation of strontium is possible. XRD pattern was analyzed using High Score Plus Software. Figure 8 show XRD analysis result of precipitated salts. Phases of calcium carbonate ($CaCO_3$) and sodium chloride (NaCl) are present in the result. This result confirms SEM and XRF analysis that calcium carbonate was the main phase while sodium chloride was the minor.

Table 4. Chemical composition of precipitated salts.

Component	Mass (%)
Ca	85.94
Sr	3.63
Cl	2.93
Na	2.12
Si	1.59
Fe	1.17
Mg	1.12
Others	1.5

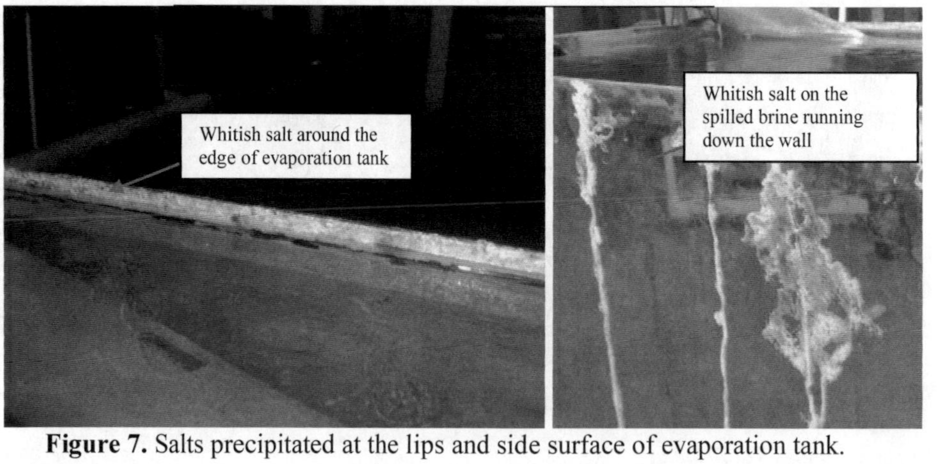

Figure 7. Salts precipitated at the lips and side surface of evaporation tank.

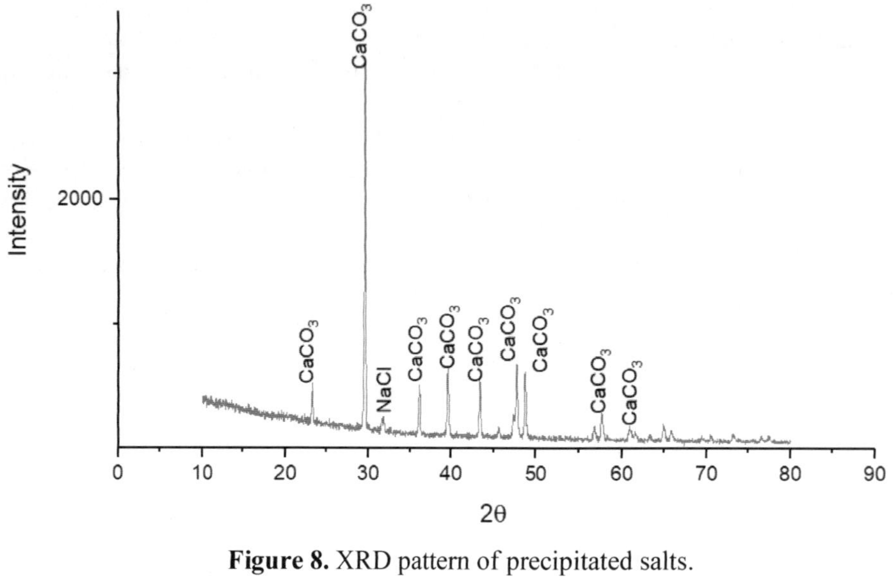

Figure 8. XRD pattern of precipitated salts.

4. Conclusion

Based on preliminary research of evaporation-precipitation of brine from Ciseeng, heating the brine at average temperature of 40.4°C could increase the °Be values up until 6. The concentration of Mg, Na, and K increase for 282%, 283%, and 207%, respectively, while the concentration of Ca stays relatively the same. In the process, 61.18% of Ca was precipitated and only small amount of Mg and Na were precipitated. Mg and Na precipitation was possible due to the evaporation on the lips of evaporation tank. The concentration of Mg, Na, and K increase for 282%, 283%, and 207%, respectively. The main constituent of the salt precipitated are Ca (60.31%) and Sr (2.33%). Sr was co-precipitated because of extremely similar chemical properties with calcium.

References

[1] Turk L J 1970 **6** 1209–15
[2] Balarew C 1993 *Pure Appl. Chem.* **65** 213–8
[3] Nie Z, Bu L, Zheng M and Huang W 2011 *Sol. Energy* **85** 1537–42
[4] Dong H, Unluer C, Yang E H and Al-Tabbaa A 2018 *Desalination* **429** 88–95
[5] Dong H, Unluer C, Yang E H and Al-Tabbaa A 2017 *Hydrometallurgy* **169** 165–72
[6] Sulistiyono E, Lalasari L H, Mayangsari W and Prasetyo A B 2018 **020007** 020007
[7] Thi K, Luong T Van, An J, Kang D, Kim M and Tran T 2013 *Hydrometallurgy* **138** 93–9
[8] Muraviev D, Noguerol J and Valiente M 1996 *React. Funct. Polym.* **28** 111–26
[9] Lee J M and Jackson L 1970 3–5
[10] Hussein A A, Zohdy K and Abdelkreem M 2017 *Int. J. Waste Resour.* **07** 1–6
[11] Waters T H E and Ocean O F T H E 2003 Chemical and Physical Features of Seawater and the World Ocean
[12] Natasha N C and Lalasari L H 2017*AIP Conf. Proc.* **1805** 70002
[13] Lalasari L H, Widowati M K, Natasha N C, Sulistiyono E and Prasetyo A B 2017 *IOP Conf. Ser. Mater. Sci. Eng.* **176** 012040
[14] Brečević L and Nielsen A E 1989 *J. Cryst. Growth* **98** 504–10
[15] Wang X, Zhao K, Li L, Guo Y, Meng L and Deng T 2017 *IOP Conf. Ser. Mater. Sci. Eng.* **274**

Acknowledgment

This research was carried out in Research Center for Metallurgy and Material, Indonesian Institute of Sciences and funded by INSINAS budget 2018

IOP Conf. Series: Materials Science and Engineering **541** (2019) 012041 doi:10.1088/1757-899X/541/1/012041

Titanium dissolution from decomposed Ilmenite using NaOH into the aqueous sulphuric acid solutions

R Subagja[1] and A Royani[1]
[1]Research Centre for Metallurgy and Material, Indonesian Institute of Sciences, Gedung 470, Kawasan Puspiptek, Tangerang Selatan, Indonesia

E-mail: rudi.subagja@lipi.go.id

Abstract. At present work, the experiment to dissolve titanium from the Indonesian Ilmenite into the aqueous sulphuric acid solutions has been done through the following process: Ilmenite decomposition using NaOH at temperature 800°C for 5 hours to produced decomposed Ilmenite, followed with titanium dissolution from the decomposed Ilmenite into the aqueous sulphuric acid solutions in the glass reactor. The variables used for dissolution process were covering: dissolution temperature from 30°C to 70°C, sulphuric acid concentrations from 5% to 15%, and dissolution time from 30 minutes up to 180 minutes. The influence of those all variables were observed on the titanium dissolution from the decomposed Ilmenite. Result of the dissolution experiment show that the dissolved titanium in the aqueous sulphuric acid solutions increased when the dissolution temperature was increased from 30°C to 60°C, sulphuric acid concentration were increased from 5% to 15% and dissolution time were extended from 30 minutes to 180 minutes.

1. Introduction

Ilmenite is a mineral with chemical formula $FeTiO_3$. It is one of the primary raw materials for titanium containing material and iron productions. In the world, almost 95% of titanium containing in titanium feedstock was for TiO_2 production and only 5% in feedstock were converted to titanium metal [1-2]. TiO_2 is widely used in many applications such as a pigment for paint production, additive for paper manufacturing, ceramic industries, pharmaceutical industries, environment purification, gas sensor, and in photovoltaic for photo catalyst application due to its unique characteristics [3-7].

Indonesia has a potential resource of Ilmenite scattered at some areas, however those Ilmenite are not well utilized due the appropriate processing technology for Indonesian Ilmenite is un available. The problem face to utilize the Indonesian Ilmenite is the mineral structure of Indonesian Ilmenite is sometimes very complex that lead to the processing of Indonesian Ilmenite becomes difficult. Therefore, in order to developed an appropriate processing technology for Indonesian Ilmenite, the Research centre for Metallurgy and Material – the Indonesian Institute of sciences (RCMM-LIPI), conducts some experiment on the processing of the Indonesian Ilmenite [8-10] and at present research activity, the titanium dissolution into aqueous sulphuric acid solutions was carried out through the following steps of processes : the decomposition of Ilmenite at elevated temperature with NaOH followed with the titanium dissolution from the decomposed Ilmenite into the aqueous sulphuric acid solutions. The experiment was carried out at laboratory scale experiments. The purpose of experiments is to investigate the influence of some parameters such as sulphuric acid concentration, dissolution temperatures and dissolution time on the titanium dissolution from the decomposed Ilmenite.

2. Experiment

2.1. The raw materials for experiment

The raw material used for experiment were Ilmenite, analytical grade sodium hydroxide (NaOH), and aqua demineral water. Table 1 shows the chemical composition of Ilmenite obtained by X-Ray Fluorescence and figure 1 shows the X-Ray Diffraction pattern of Ilmenite used in the experiment. The chemical composition data at table 1 shows the raw material Ilmenite has 49,44% Fe_2O_3 and 38,30% TiO_2. Other compounds exist at the Ilmenite sample are SiO_2, Al_2O_3, MnO, MgO, CaO, K_2O, P_2O_5, Cr_2O_3, SnO_2, but their quantities are relatively lower, and the X-Ray diffraction pattern at Figure 1 shows Ilmenite ($FeTiO_3$) as the main mineral exist at the raw material for experiment.

Table 1. Chemical composition of Ilmenite.

COMPOUND	Fe_2O_3	TiO_2	SiO_2	Al_2O_3	MnO	MgO	CaO	K_2O	P_2O_5	Cr_2O_3	SnO_2
%	49,44	38,30	1,76	1,78	2,00	1,44	0,08	0,03	0,17	2,66	1,16

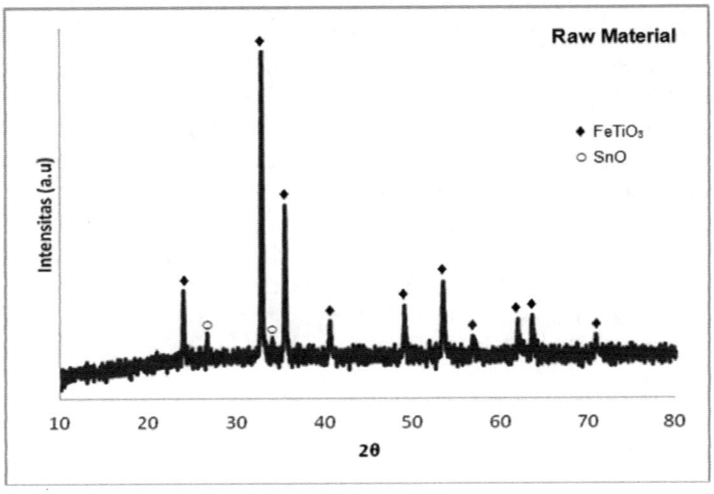

Figure 1. X-Ray Diffraction pattern of Ilmenite.

2.2. The Equipment for experiment

There were two main equipments were used in experiment. The 1st equipment was Programmable Muffle Furnace for decomposition of Ilmenite with NaOH at elevated temperature to produce decomposed Ilmenite and 2nd equipment was 1-liter capacities of glass reactor for the titanium dissolution from the decomposed Ilmenite into the aqueous sulphuric acid solutions. The glass reactor was mantled with controllable heater to control the reaction temperatures during dissolution process.

2.3. Procedure of experiment

The procedure of experiment was covering the procedure for decomposition of Ilmenite with NaOH at elevated temperature and the procedure for titanium dissolution from the decomposed Ilmenite.

2.3.1. The procedure for Ilmenite decomposition process

The Ilmenite decomposition process was carried through the following steps: Firstly, Ilmenite was mixed with NaOH in a mass ratio of 2:1 in Stainless steel crucible. The mixed Ilmenite-NaOH were then roasted in the programmable Muffle Furnace, in air atmosphere at temperature 800°C for 5 hours. That roasting temperature was choose based on our previous experiment for the decomposition of Ilmenite with NaOH which show that the best temperature for the decomposition of Ilmenite to sodium titanate and sodium ferrite was at temperature 800°C [11]. At elevated temperature, Ilmenite will be decomposed by sodium hydroxide to produce sodium ferrite ($NaFeO_2$) and sodium titanate (Na_2TiO_3) according to the reaction (1). The calculated Gibbs free energy reaction using HSC 5.1 software for reaction 1 gives a negative value at temperature lower than 1000°C [12], therefore the formation of sodium titanate and sodium ferrite through the reaction 1 will be thermodynamically favorable at temperature 800°C:

$$2\ NaOH + 2/3\ FeTiO_3 + 1/6\ O_2\ (g) = 2/3\ NaFeO_2 + 2/3\ Na_2TiO_3 + H_2O \qquad (1)$$

The generated water at reaction 1 creates trans granular cracks throughout Ilmenite structure which promote the diffusion oh Na^+ ions throughout the structure, leading to the formation of sodium titanate [12].

The decomposed Ilmenite from reaction 1 was used further as the raw material for titanium dissolution experiment.

2.3.2. The procedure for dissolution process

The procedure for titanium dissolution from decomposed Ilmenite was carried out through the following steps: the washing process of decomposed Ilmenite with water and the titanium dissolution process from the washed decomposed Ilmenite in to the aqueous sulphuric acid solutions.

In the washing process, the decomposed Ilmenite samples from decomposition experiment were washed by demineralized water until the washed solution pH reached 7. After washing process, the washed decomposed Ilmenite samples were added into the aqueous sulphuric acid solutions in 1-liter glass reactor. The dissolution process was carried out at certainties period of time and temperature, under agitation speed 300 rpm. At this dissolution process, the sulphuric acid concentration were varied from 5% up to 15%, dissolution temperature was varied from 30°C up to 70°C, and dissolution time were from 30 minutes to 300 minutes. After the dissolution experiment finished, the mixed solution and solid were discharged out from the glass reactor and it was filtered to separate solid from solution. The solid cake was then washed by water and dried in oven at 100°C for 24 hours prior to X-Ray Diffraction (XRD) analysis, and the titanium content in solution was analyzed by using atomic absorption spectrophotometer.

3. Results and Discussion

3.1. X Ray diffraction pattern of the decomposed Ilmenite

The decomposed Ilmenite produced by roasting process of the mixed Ilmenite-NaOH was analyzed using X-Ray diffractometer. Sample was analyzed over an angle 2θ with a maximum range of 10 to 80°. The result of analysis by X Ray diffractometer at figure 2 shows that the formed phases during the roasting process of the mixed Ilmenite-NaOH at 800°C were sodium titanate (Na_2TiO_3) and sodium Ferrite ($NaFeO_2$). This result of experiment agrees with findings by Stephen Parirenyatwa et al which conducted roasting experiments on the mixed Ilmenite and NaOH [12]. The formation of sodium titanate and sodium ferrite are taking place through the reaction 1 which is thermodynamically favorable at temperature 800°C.

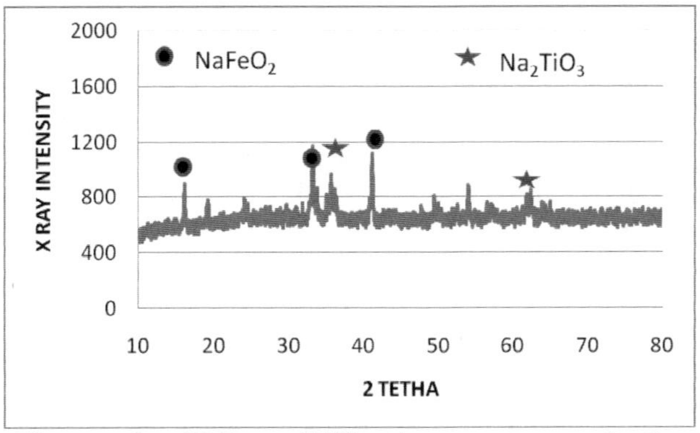

Figure 2. X-Ray Diffraction pattern of the decomposed Ilmenite

3.2. Effect of dissolution time on Ti dissolution

During the titanium dissolution process from the decomposed Ilmenite into the sulphuric acid solutions, the sodium titanate containing in the decomposed Ilmenite reacts with sulphuric acid to produce titanium sulphate and sodium sulphate according to reactions 2 [13]:

$$Na_2TiO_3 + 3\ H_2SO_4 \rightarrow Ti(SO_4)_2 + Na_2SO_4 + 3\ H_2O \quad (2)$$

Reaction 2 is influenced by dissolution time, temperature and sulphuric acid concentrations. The effect of dissolution time on titanium dissolution from the decomposed Ilmenite in 10% aqueous sulphuric acid solutions, at dissolution temperature 60°C is presented at figure 3. The figure shows that at the beginning of the dissolution process, fast titanium dissolution process happened, however when the dissolution time were extended from 0.5 hours to 3 hours, the dissolved titanium in 10% aqueous sulphuric acid solutions increased from 58% at 0.5 hours dissolution time to 82% at 3 hours dissolution time and slowly titanium dissolution process happened, that probably due to the formation of undissolved iron material at the decomposed Ilmenite. Referring to Stephen Parirenyatwa finding the presence of iron in the sample roasted with NaOH may be due to the presence of unreacted Ilmenite [12].

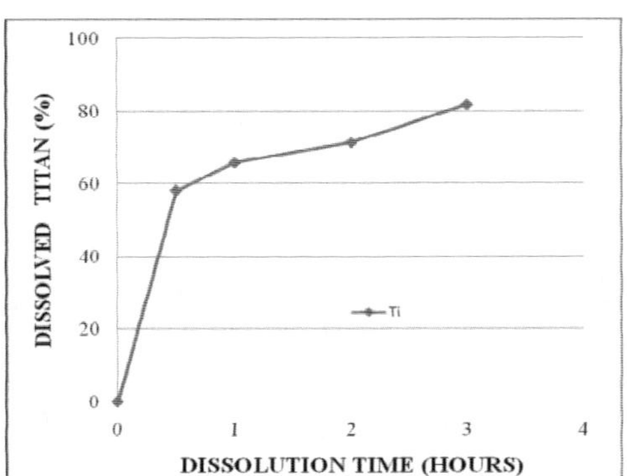

Figure 3. The effect of dissolution time on Ti dissolution at 10% H_2SO_4

3.3. The effect of temperature on Ti dissolution

The influence of temperatures on titanium dissolution from the decomposed Ilmenite were investigated at the temperature range 30°C – 70°C, in 10% aqueous sulphuric acid solutions. The result of experiment at figure 4 shows that temperature gives a significant effect on the titanium dissolution from the decomposed Ilmenite. Almost at all dissolution times, the dissolved titanium increased with the increased of the dissolution temperatures from 30°C up to 60°C. However, when the dissolution temperature was increased from 60°C to 70°C, the dissolved titanium from the decomposed Ilmenite decreased that probably due to the precipitation of some of the dissolved titanium from the solutions at higher temperature than 60°C.

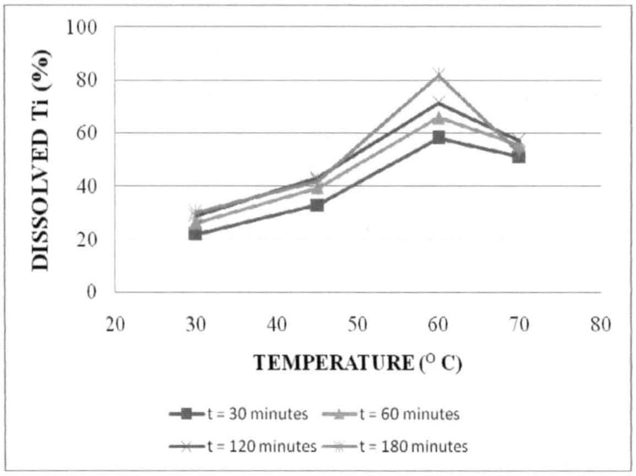

Figure 4. The effect of temperature on titanium dissolution.

3.4. The effect of H_2SO_4 on Ti dissolution

The influence of sulphuric acid concentration on titanium dissolution from the decomposed Ilmenite in aqueous sulphuric acid solutions were investigated at temperature 60°C, the sulphuric acid concentration were varied from 5% to 15%, and the dissolution time were from 30 minutes to 180 minutes. The result of experiment at figure 5 shows that at temperature 60°C and 3 hours dissolution time, the dissolved titanium from decomposed Ilmenite increased from 30% in 5% sulphuric acid concentration to 88% in 15% sulphuric acid concentration. This result of experiment shows that the sulphuric acid concentration has a great effect on titanium dissolution.

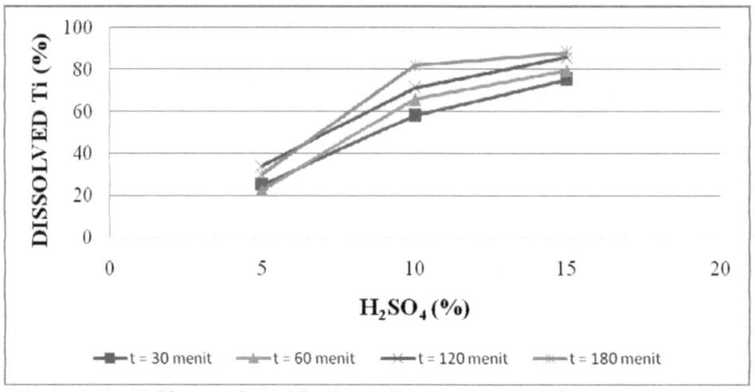

Figure 5. Effect of Sulfuric acid concentration on titanium dissolution.

4. Conclusion

The dissolution experiment to dissolve titanium from Indonesian Ilmenite has been investigated through the decomposition process of Ilmenite with NaOH at temperature 800°C for 5 hours, followed with titanium dissolution from the decomposed Ilmenite in the aqueous sulfuric acid solutions.

The result of experiment shows the titanium dissolution from the decomposed Ilmenite in aqueous sulfuric acid solution were influenced by dissolution time, dissolution temperature and sulphuric acid concentration. The titanium dissolution increased when dissolution time, dissolution temperature or sulphuric acid concentration were increased.

When the dissolution time were extended from 0,5 hours to 3 hours, the dissolved titanium into 10% aqueous sulphuric acid solutions increased from 58% at 0,5 hours dissolution time to 82% at 3 hours dissolution time.

In 10% aqueous sulphuric acid solutions at 3 hours dissolution times, when the dissolution temperatures were increased from 30°C up to 60°C, the dissolved titanium increased from 30% at dissolution temperature 30°C to 82% at dissolution temperature 60°C.

At temperature 60°C and 3 hours dissolution time, the dissolved titanium from decomposed Ilmenite increased from 30% in 5% sulphuric acid concentration to 88% in 15% sulphuric acid concentration.

References

[1] Dimitrios F and Guillaume H 2009 *Journal of Metal* **61** 36
[2] Rosebaum J B 1982 *Journal of Metal* 76-79
[3] Zhang W, Zhu Z, Cheng C Y 2011 *Hydrometallurgy* **108**, pp 177–188
[4] Fujishima A, Honda K 1972 *Nature* **238** 37
[5] Fujishima A, Kohayakawa K and Honda K 1975 *J. Electrochem. Soc.* **122**, 1487
[6] Fujishima A, Rao T N and Tryk D A 2000 *J. Photochem. Photobiol. C: Photochem* (1), **1**
[7] Adachi M, Murata Y, Harada M, Yoshikawa S 2000 *Chemistry Letters*, **29** (8) 942-943
[8] Royani A, Lalasari L H ,Firdiyono Subagja R and Solihin 2011 *Prosiding Seminar Material Metalurgi* 101-106
[9] Subagja R, Andriyah L, and Lalasari L H 2013 *Asian Transaction on Basic and applied Sciences*, **2** 59 -64
[10] Subagja R, Andriyah L, and Lalasari L H 2013 *International Journal of Basic & Applied Sciences-IJENS* **13** 97- 103
[11] Subagja R, Royani A, and Prasetyo P 2012 *Metalurgi* **27** 241-249
[12] Parirenyatwa S, Castejon L E, Segado S S, Hara Y, Jha A 2016 *Hydrometallurgy* **165** 213-226
[13] Sayed E and A. Manaa 2016 *Journal of Saudi Chemical Society* **20** 673-679

Determination of compressive strength of 3D polymeric lattice structure as template in powder metallurgy

M S Utomo[1], Y Whulanza[2,3], F P Lestari[1], A Erryani[1], I Kartika[1] and N A Alief[3]
[1]Research Center for Metallurgy and Materials – Indonesian Institute of Sciences, Indonesia
[2]Research Center for Biomedical Engineering – Universitas Indonesia, Indonesia
[3]Department of Mechanical Engineering – Universitas Indonesia, Indonesia

E-mail: muha209@lipi.go.id

Abstract. Powder metallurgy has been developed to fabricate metal foams from various materials including magnesium. One particular challenge on powder metallurgy is fabrication of highly-ordered and interconnected porous products. Pores interconnectivity is important in application of porous magnesium alloy for biodegradable orthopedic implants. Meanwhile, 3D polymeric printing technology offers capability to build precise and rapid lattice wireframe products in a simple and low-cost manner. Here, we design and validate the capability of 3D polymeric lattice as template in powder metallurgy for fabrication of magnesium-based biodegradable implants. Lattice structures were made of ABS, PLA, and PVA filaments. Lattice structures were cubical-shaped with uniform dimension and variations in pore size are included in the study. Both computational and experimental tests are performed to determine the compressive strength of the lattice structures. Uniaxial stress with uniform magnitude is applied to test the lattice structures. The resulting stress, strain, and deformation of the 3D polymeric lattice are observed. Variations in materials and pore size affect the stress, strain and deformation of the 3D polymeric lattice. Parameters can be further optimized to meet the requirement of the design and fabrication process in consideration of the tolerable stress, strain and deformation.

1. Introduction

The need for orthopedic implants is increasing over time. Degenerative diseases such as osteoarthritis and osteoporosis among growing population have driven the growth of the orthopedic implant market in global. In addition, accidental cases such as work accidents, traffic accidents, and sports injuries have been the cause of the increasing demand for orthopedic implants [1].

To fulfill its function to replace bone tissue, implants have certain criteria that must be met through the selection of materials and appropriate fabrication techniques. Based on their use, implants are divided into two categories: hard tissue / bone replacement implants such as fracture plates, screws, pins, and dental implants; and soft tissue replacement implants such as vascular implants [2].

The advantage of metal-based orthopedic implants compared to non-metals is that mechanical properties are better and more suitable for their function [3]. However, metal-based implants have a disadvantage when the implant must be planted for a long time in the patient's body. This can increase the risk of complications due to implant toxicity caused by the accumulation of metal ions formed from the corrosion process. Another disadvantage is different mechanical parameters with bone.

The difference in Young's modulus between implants and bone can make loss of bone tissue due to the shielding stress phenomenon. This phenomenon occurs because of the accumulation of load on the implant that reduces the burden on the bone in accordance with Wolff's law which states that the formation of human bones will adjust to the supported load, if the load received by the bone decreases then the bone density will decrease and weaken because there is no stimulus to maintain bone mass [4].

Biodegradable implants are an alternative to permanent implants as implants that are temporary and able to decay within a period of time after bone tissue grows so that implant removal surgery is not required anymore from the patient's body. Metallic implants can be developed from titanium or magnesium materials which have mechanical properties, biocompatibility, and overall compatibility for the human body [5].

In designing a biodegradable implant, there is one important factor that must be considered, namely its ability to become scaffold or to grow and develop new bone tissue. Hopefully, the implant is capable of having a balanced overall rate of growth of new bone tissue so that when the implant decays perfectly, the bone tissue is mature enough and able to function like normal adult bone tissue. To be able to meet this criterion, the implant should have pores with certain dimensions that are connected to each other as a place to grow new bone tissue. Studies show that osteointegration processes occur best at pore sizes of 100-700 microns and porosity level at 60-80%. Interconnected pores facilitate fluid flow that gives cells to proliferate inside the implant and fully integrate with metal implants [6].

The porosity level is defined as the ratio between the volume of the pore cavity to the volume of implants which affects the mass and mechanical properties of implants such as Young's modulus, tensile strength, and compressive strength. Increasing the porosity level affects the mechanical strength of implants and implants with the same degree of porosity and different pore sizes have different mechanical properties. By modifying the porosity level, metal-based implants can be designed to have mechanical properties similar to human bones.

The process of making implants with complex internal architecture such as interconnected pores can be done, among others, through three-dimensional printing techniques. Three-dimensional printing techniques for metal materials are carried out through a pattern sintering process such as SLS (selective laser sintering), SLM (selective laser melting), or EBM (electron beam melting) [7]. SLS and SLM use lasers while EBM uses electron shots [8]. However, these processes require relatively expensive tools so that the cost of producing a complete implant cannot be reduced. Moreover, these processes also do not allow the manufacture of complete implants with magnesium-based materials because of their reactive properties so that the commonly used fabrication techniques are powder metallurgy with compacting and sintering.

Meanwhile, 3D polymeric printing technology offers capability to build precise and rapid lattice wireframe products in a simple and low-cost manner. Here, we numerically study the capability of 3D polymeric lattice as template in powder metallurgy for fabrication of magnesium-based biodegradable implants.

2. Materials and Methods

3D models of polymeric lattices are built on SolidWorks 2013. The models are then converted and transferred into Comsol Multiphysics 5.3 for numerical simulation of compressive stress test. Cubical shape is chosen to represent basic shapes of lattice structures. The lattices dimension are 15 mm x 15 mm x 15 mm. Although, to reduce simulation load, the dimension of the tested lattices are reduced into 5 mm x 5 mm x 15 mm. To compenstate this reduction, the applied compressive load is proportionally adjusted. During compression process of powder metallurgy, the polymeric lattice structure will be subjected to uniaxial 300 psi compressive load. Two lattice pore sizes are used which are 1000 and 1500 microns. They refer to the requirement for the bone tissue growth and possibility of shrinkage during the fabrication process. The shape of the lattice is square. The wire of the lattice is approximately 200 microns in diameter. The wire diameter of the lattice is obtained based on the default nozzle size of the 3D polymer printer.

Three materials are used as the polymeric lattice materials which are acetyl butyl styrene (ABS), poly lactic acid (PLA), and poly vinyl alcohol (PVA). These materials are selected since all of them are common materials to be used as filaments in polymeric 3D printer. The mechanical properties of each materials are presented in Table 1. Another aspect on the material selection beside mechanical properties and practicality is the biocompatibility and biodegradability properties. PLA is known for its biocompatibility and biodegradibility. PLA itself has been developed as polymer-based implant. All selected materials are soluble in water-based solvent such as ethanol, chloroform, and DMO, each with their own dissolution rates.

Table 1. Mechanical properties of ABS, PLA, and PVA

Parameters	ABS	PLA	PVA
Density (kg/m^3)	1010 - 1200	1250	1120
Ultimate tensile strength (MPa)	33 - 110	35	32 - 66.2
Young's Modulus (GPa)	1.1 – 2.9	3.5	1.0 - 2.2
Poisson's ratio	0.35	0.36	0.45

3. Results and Discussion

Finite element analysis was performed using Comsol. Beforehand, 3D model of polymeric lattices was built using Solidworks. The 3D model was converted into STL format and imported as 3D component domain. Materials for the model were ABS, PLA, and PVA. Structural mechanics was applied as the governing physics for the environment. Two surface boundary conditions were defined: load and fixed constraints. The actual uniaxial compression was represented as loads located on top of the model and directed downward along y-axis. The applied load is 300 psi which represent the actual compression load. Fixed constraints were located at the bottom of the model. Discretization was then performed on the 3D model. The discretization generated 35000 cells and 40000 nodes. The calculation steps were formulated and the computational analysis was executed. Mesh convergence is performed to validate the numerical calculation result and discussed separately. Distribution of von Mises stress, strain, and displacement were selected as the output parameters visualized by color scales and minimum-maximum values. Figure 1 shows the 3D visualization of the proposed polymeric lattice designs. There are two designs, two cubical-based geometries with 1000X1000X1000 and 1500X1500X1500 cu.µm open pores. Designs with basic geometrical shapes are proposed as references to more complex designs in the future. Figure 2 shows the discretization result of the 3D model of the polymeric lattice. Figure 3 shows the distribution of von Mises stress on the 3D model of the polymeric lattice made from ABS under uniaxial compressive load of 300 psi. Figure 4 shows the distribution of strain on the 3D model of the polymeric lattice made from ABS under uniaxial compressive load of 300 psi. Figure 5 shows the distribution of displacement on the 3D model of the polymeric lattice made from ABS under uniaxial compressive load of 300 psi. Figure 6 shows graphs comparing the maximum von Mises stress for all variants. Moreover Figure 7 shows graphs comparing the displacement for all variants.

Computational simulation results show that the polymeric lattice structure is deformed mainly in the middle layer. It means that the middle layer is the weakest part of current lattice structure. It suggests that when the lattice structure is compressed during fabrication of porous magnesium alloy, the middle layer will have highest potential to be deformed. This becomes important for the fabrication of porous magnesium alloy due to potential of arising non-uniformity. Non-uniformity at the porous magnesium alloy might arises from two mechanisms. First mechanism is that the compressed and deformed middle layer compacts the powder around the area. This issue might produce a solid non-porous middle layer in the porous magnesium alloy. The other mechanism is that the compressed and deformed middle layer inhibits powder to infiltrate into the gaps. This might produce a cavity in the middle layer of the porous magnesium alloy. The prediction of stress is not as accurate as prediction of displacement due to stress assumption over the object. In addition, factors such as element/mesh size, might affect the accuracy of the output value [15].

Effect of lattice shape to compressive properties on 3D-printed ABS material has been carried out. Honeycomb-like cellular lattice structure has the highest mechanical properties among other shapes such as rectangular, triangular, and circular. Although, the difference was not significant. Meanwhile, increase in porosity also decrease mechanical strength [14]. PLA has also been observed to exhibit shape-memory effect which can be triggered by thermal stimuli [17]. The 3D printing fabrication technique also introduce internal stress/strain to the polymer material due to the heating/cooling process. There are several ways to improve the mechanical properties of 3D-printed product, either by modifying the process or the material. Material modification can be performed through addition of fibers or particles made from glass, carbon, or polymers. Such treatment can improve the tensile strength of 3D-printed product up to 50% [16].

Foams material and lattice structures have fundamental difference. Foams material has a solid structure filled with cavities with varying size and shape. It can form closed cell or open cell foam material. Meanwhile, lattice structures are assembled by identical skeletal structure forming three dimensional geometrical shapes such as cubes or cylinders. Pores are shaped due to the spatial between skeletal structures. There are several types of lattice structures such as Kagome, hexagonal diamond, and cross pyramidal. Each lattice structures has its own mechanical behavior to withstand compressive stress which can be observed through the stress-strain curve [9].

Fabrication of metallic lattices using additive manufacturing by selective laser melting analyzed by numerical and experimental compressive tests has been conducted [10,11,12]. Theoretical approach to determine mechanical behavior of lattice structures under compressive loading has been proposed [13]. The lattice are subjected to compression or tension load and bending without twisting moment when subjected to uniaxial compressive loading, which according to Euler-Bernoulli beam theory is linearly distributed over the cross section.

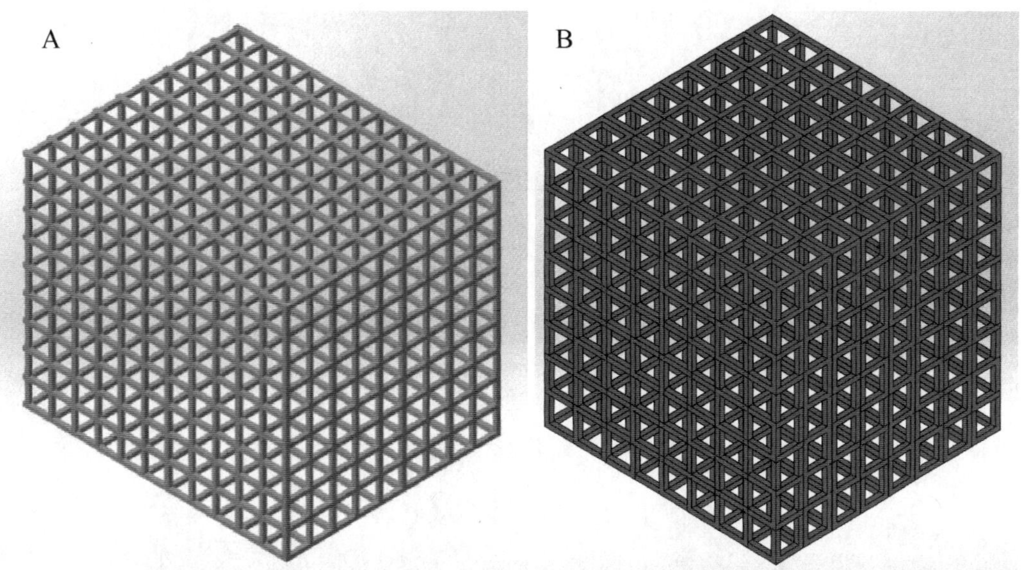

Figure 1. 3D visualization of cubical lattice structure with 1000X1000X1000 (A) and 1500X1500X1500 (B) cu.µm open pore volume.

A

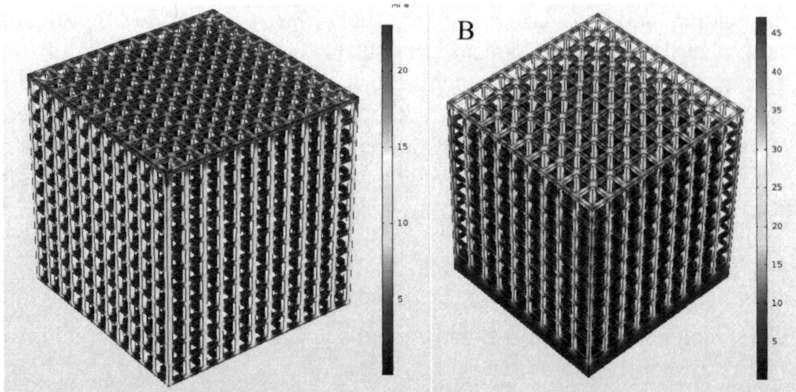

Figure 2. Visualization of von Mises stress distribution on ABS material, 1000X1000X1000 (A) and 1500X1500X1500 (B) cu.μm open pore volume.

A

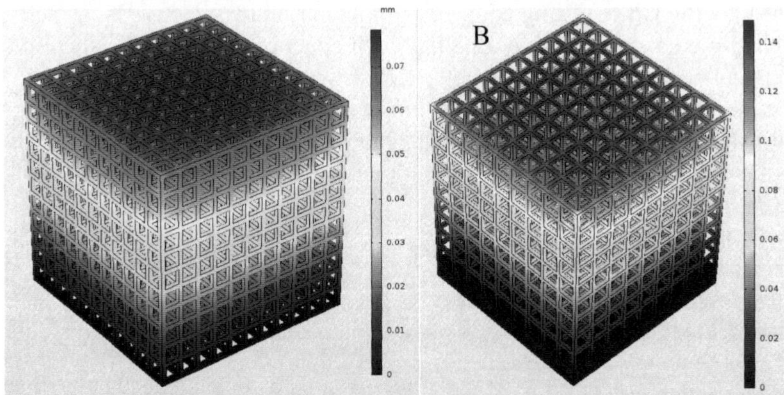

Figure 3. Visualization of displacement distribution on ABS material, 1000X1000X1000 (A) and 1500X1500X1500 (B) cu.μm open pore volume.

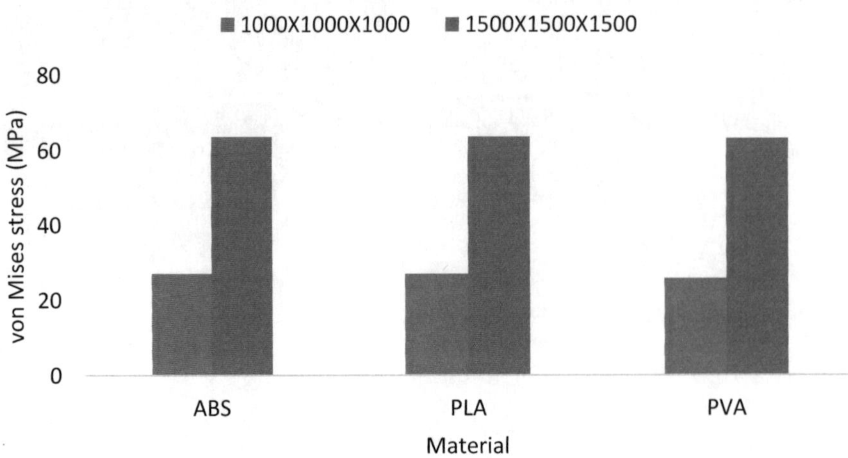

Figure 4. Graph comparing maximum von Mises stress distribution shows material variation does not affect the amount of stress acting on the structure while reduction of pore size from 1500X1500X1500 to 1000X1000X1000 cu.μm reduces the maximum value of von Mises stress to half.

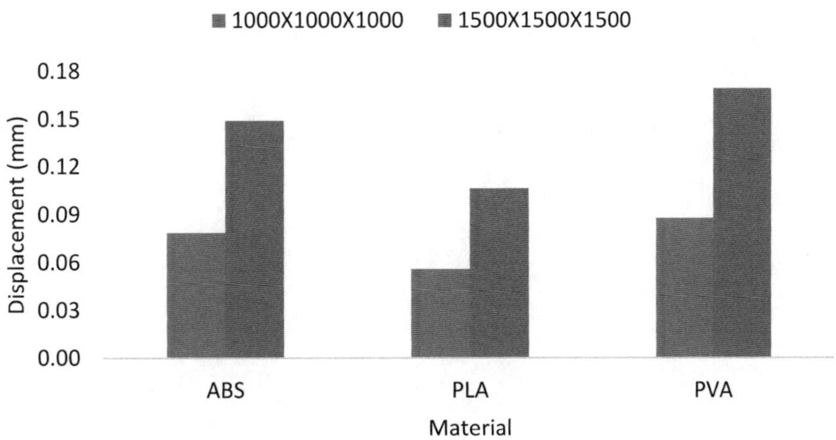

Figure 5. Graph comparing maximum displacement shows both material and pore size variation affect the maximum displacement value. PLA shows lowest value of deformation on both pore sizes. while PVA shows highest value of deformation on both pore sizes.

4. Conclusion

Design and numerical validation of 3D polymeric lattice has been presented. Three commercially-available filament materials (ABS, PLA, and PVA) are used to fabricate lattice structures with variation in shape and pore size. Each structure is subjected to uniform uniaxial compressive load. Difference in material and geometry made each lattice structures distribute compressive stress load in different ways leading to different deformation pattern. Under same outer dimension and uniform pore distribution, different pore size alters the porosity/pore density.

References

[1] R Magetsari, Suyitno, R Dharmastiti, UA Salim, L Hidayat, T Yudiman, Z A Lanodiyu, dan P Dewo, *International Journal of Morphology* **33(4)**, 2015, pp. 1255-1260.
[2] C J Pendegrass, *Doctoral thesis, University of London*, 2005.
[3] S Toghyani dan M Khodaei, *Materials Research Express* **5**, 2018.
[4] S W Kim, H D Jung, M H Kang, H E Kim, Y H Koh, Y Estrin, *Materials Science and Engineering C* **33**, 2013, pp. 2808-2815.
[5] M K Datta, D T Chou, D Hong, P Saha, S J Chung, B Lee, A Sirinterlicki, M Ramanathan, A Roy, dan P N Kumta, *Materials Science and Engineering B* **176**, 2011, pp. 1637-1643.
[6] S Dutta, K B Devi, dan M Roy, *Advanced Powder Technology*, 2017.
[7] G Jiang, Q Li, C Wang, J Dong, dan G He, *Materials and Design*, 2014.
[8] I Maskery, N T Aboulkhair, A O Aremu, C J Tuck, I A Ashcroft, R D Wildman, dan R J M Hague, *Materials Science and Engineering: A*, volume **670**, 2016, pp. 264-274.
[9] S K Moon, Y E Tan, J Hwang, and Y J Yoon, *International Journal of Precision Engineering and Manufacturing-Green Technology* vol. **1** No. 3, pp. 223-228 (2014).
[10] R Mahshid, H N Hansen, and K Loft-Hojbjerre, *Materials & Design* **104**, pp. 276-283 (2016).
[11] D Noviello, *Autodesk University*, pp. 1-22.
[12] O Rehme and C Emmelmann, *Proc. of SPIE* Vol. **6107**, pp. 1-12 (2006).
[13] K Ushijima, W J Cantwell, and D H Chen, *Journal of Computational Science and Technology* vol. **4** No. 3, pp. 159-171 (2010).
[14] O Iyibilgin, C Yigit, and M C Leu 2013 *Proceedings of Solid Freeform Fabrication Symposium*, 895-907
[15] Dutt A 2015 *Int. Jour. of Eng. Sci. and Innovative Technology* **4(3)** 181-185

[16] Wang J, Xie H, Weng Z, Senthil T, and Wu L 2016 *Materials and Design* **105**, 152-159
[17] Zhang Q, Yan D, Zhang K, and Hu G 2015 *Scientific Reports* **5** 1-6

Acknowledgement
This research is funded by Insinas grant scheme from Ministry of Research and Higher Education.

Characteristic of Ti-6Al-6Nb alloys following solution treatment with cryogenic cooling for implant applications

F Rokhmanto[1], H Arief[2], Alfirano[2], and C Sutowo[1]
[1)] Research Centre for Metallurgy and Material - Indonesian Institute of Sciences Bd. 470 Kawasan Puspiptek Serpong, Setu – Tangerang Selatan.
[2)]Teknik Metalurgi, Fakultas Teknik, Universitas Sultan Ageng Tirtayasa, Cilegon, Indonesia

E-mail: fendy.rokhmanto@gmail.com

Abstract. Titanium alloys are widely used as implant materials, due to the mechanical properties, one of which is the low elastic modulus. The differences in the elastic modulus of Titanium alloys and Bones are wide enough, therefore it's necessary to investigate how to obtain a lower Titanium elastic modulus. In this paper used Ti-6Al-6Nb alloys as an implant material. The Ti-6Al-6Nb alloys Solution Treated at 1100°C, following cryogenic cooling with liquid Nitrogen. Before that The Ti-6Al-6Nb alloys Homogenized at 1100°C for 12 hours, then Hot rolled with preheated at 1100°C for 30 minutes. As comparison Solution Treatment also done in the ice water cooling and air cooling. The Ti-6Al-6Nb alloys characterized with a hardness test, metallographic and ultrasonic test to obtain the elastic modulus. The analysis shows that the lowest elastic modulus was 100,19 GPa and highest hardness was 52,54 HRc is in the cryogenic cooling process due to the beta phase formed.

1. Introduction

Titanium alloys especially alpha-beta Titanium widely used as implant materials because of an excellent corrosion resistance, an excellent biocompatibility and good mechanical properties, such as the low elastic modulus. Ti-6Al-4V known as commercially alpha-beta alloys implant material, but Vanadium is also known as a toxic element and caused allergic in the human body [1,2]. Vanadium in Ti-6Al-4V is a beta phase stabilizer, while Aluminum is an alpha phase stabilizer [1]. Therefore Vanadium as a beta phase stabilizer should be substituted, for example with Molybdenum or Niobium [3].

The mechanical properties, such as hardness, strength, and elastic modulus is depend on the alloying materials and the microstructure. Where to get the suitable mechanical properties, such as hardness and elastic modulus, material can be heat treated above the temperature alpha-beta region (beta transfuse temperature) [4]. The beta transfuse temperature of Ti-6Al-4V is 995 °C [5].

In previous studies of Ti-6Al-6Nb, the influence of solution treatment temperature on mechanical properties was studied [6], but the influence of cooling medium after solution treatment on mechanical properties has not been studied. In previous studies of Ti-6Al-6Mo, the elastic modulus affected by cooling medium [7], but the cryogenic cooling has not been studied. In this paper will be studied the influence of cryogenic cooling after solution treatment on the characteristic of Ti-6Al-6Nb for implant applications.

IOP Conf. Series: Materials Science and Engineering **541** (2019) 012043 doi:10.1088/1757-899X/541/1/012043

2. Experiment

In this paper, the composition of titanium alloys are 6% Al, 6% Nb (wt.%), the material purity, that is Titanium, Aluminum and Niobium is 99.9%. The materials, re-melted used vacuum arc furnace in the argon atmosphere. The alloys ingot dimension is ø 7.5 x 80 mm. The ingot homogenized at vacuum tube furnace at 1100°C for 12 hours and furnace cooling in the argon atmosphere. After that, thermomechanical treatment held with hot rolled up to 80% reduction, with preheating at 1100°C for 30 minutes. The solution treatment held at 1100°C for 1 hours, followed by quenching with various media, liquid Nitrogen (cryogenic cooling), water, and air cooling. The Ingots, characterized with a hardness tester, an optical microscope and SEM-EDX to investigate microstructure, and the elastic modulus obtained with an ultrasonic approach by measuring ultrasonic velocity (ASTM E494-95).

3. Result and Discussion

Figure 1. shows the microstructure of Ti-6Al-6Nb alloy in various condition, it has two phases, an alpha phase (light) and a beta phase (dark). Figure 1a. is the microstructure of the as cast, the structure is a lamellar or needle-like structure. The structure called needle-like because of the shape are like a needle (lujering). Figure 1b, 1c and 1d show the microstructure in condition after solution treatment with a various cooling medium. In the air cooling medium the structure is equiaxed or globular, but in the faster cooling, water quenching (Figure 1c) and cryogenic cooling (Figure 1d), the globular structure looks flatter and along shape.

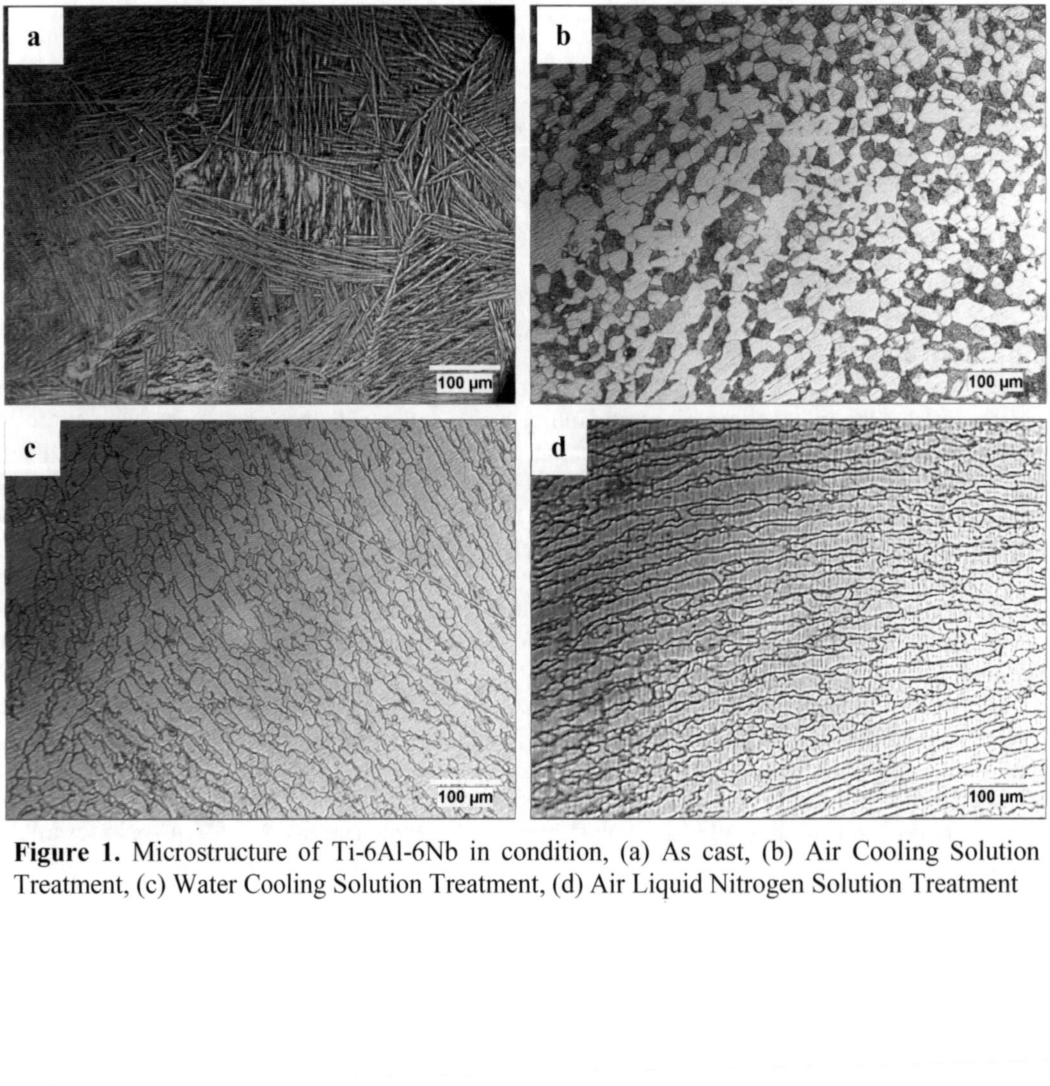

Figure 1. Microstructure of Ti-6Al-6Nb in condition, (a) As cast, (b) Air Cooling Solution Treatment, (c) Water Cooling Solution Treatment, (d) Air Liquid Nitrogen Solution Treatment

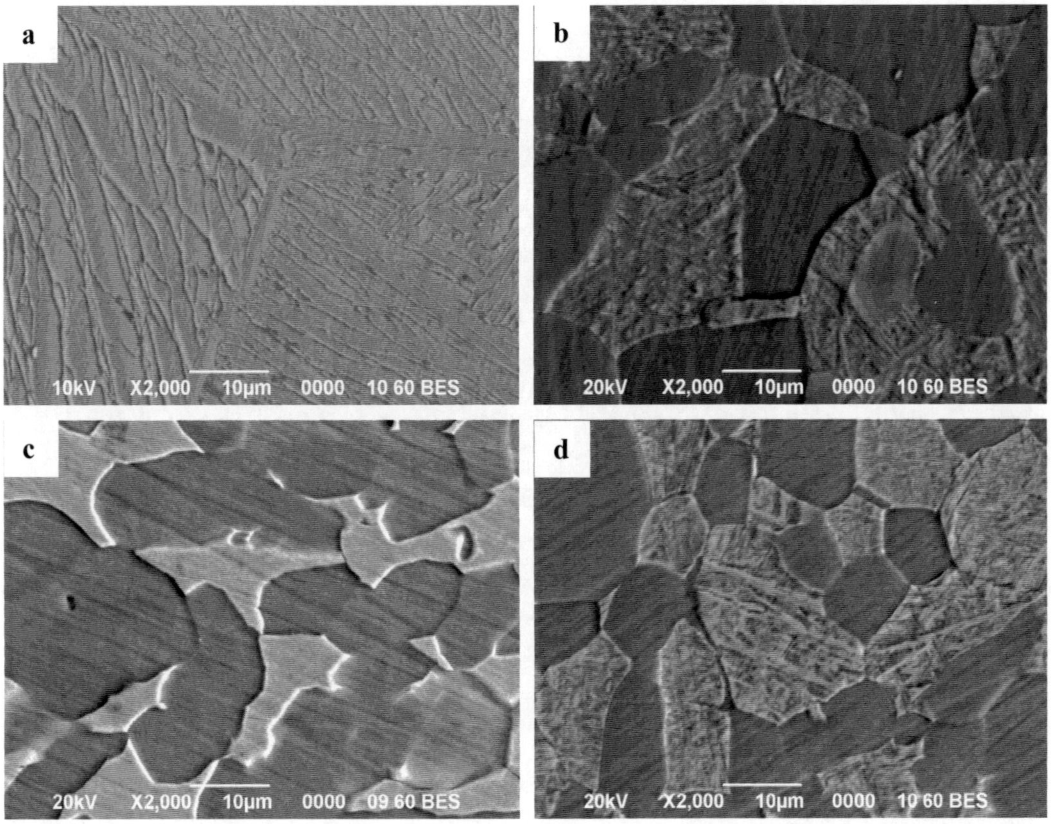

Figure 2. Microstructure of Ti-6Al-6Nb in condition, (a) As cast, (b) Air Cooling Solution Treatment, (c) Water Cooling Solution Treatment, (d) Air Liquid Nitrogen Solution Treatment

Figure 2. is an SEM picture of Ti-6Al-6Nb alloy in various condition, it has two phases, an alpha phase (dark) and a beta phase (light). The lamellar structure observed in as cast condition as shown in Figure 2a. However the equiaxed structure observed in the SEM picture after solution treatment with air cooling is Figure 2b, water quenching is 2c and cryogenic cooling is 2d. The SEM shows higher magnification than optical microscope. In rapid cooling condition (Figure 1c and 1d) the equiaxed or globular structure looks flatter and along shape, however when observed with higher magnification under SEM looks the equiaxed structure in a smaller shape (Figure 2c and 2d) than air cooling (Figure 2b). Figure 2b, 2c and 2d show the beta phase dominates the microstructure and the grain size structure became smaller when the cooling rate is faster. The cooling rate was prohibits the beta phase transform to alpha phase, because of that the alloys dominated by beta phase. The alfa phase is also formed, it is promoted by the Aluminum as known as alfa stabilizer.

Figure 3. shows the hardness of Ti-6Al-6Nb alloy in various condition. The as cast ingots have the highest hardness number 54.2 HRC, after solution treatment process the hardness is decreased. The cooling rate after the solution treatment process enhanced the hardness from 49.9 HRC in air cooling became 50 HRC in water quenching and 52.5 HRC in cryogenic cooling (liquid Nitrogen cooling). The highest number in as cast ingot promoted by the acicular structure (needle-like) that similar with martensitic structure in steel, besides that in this condition the microstructure dominated by an alpha phase (Figure 1a and 2a) that had HCP crystal structure [8]. The enhanced hardness number after the solution treatment process was promoted by the grain size structure. The smaller grain size enhanced the hardness although the Ti-6Al-6Nb dominated by a beta phase that had BCC crystal structure [8].

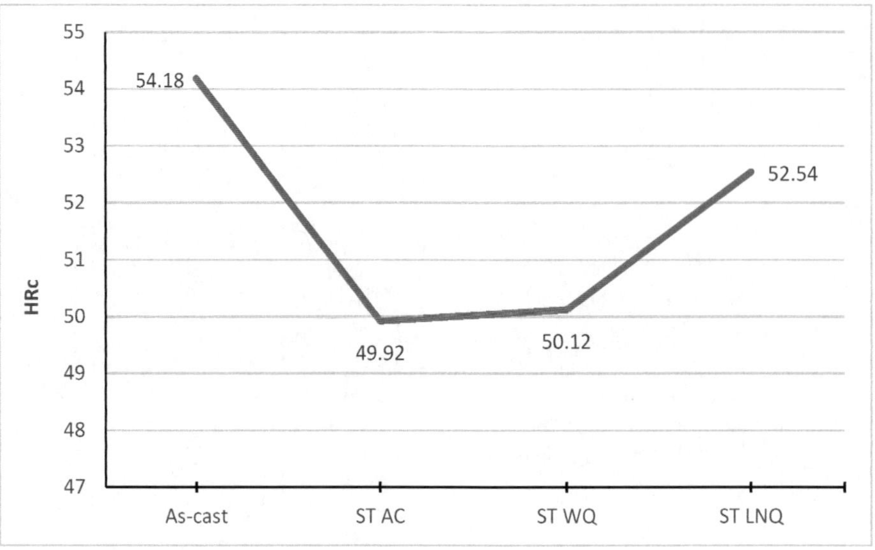

Figure 3. Hardness of Ti-6Al-6Nb

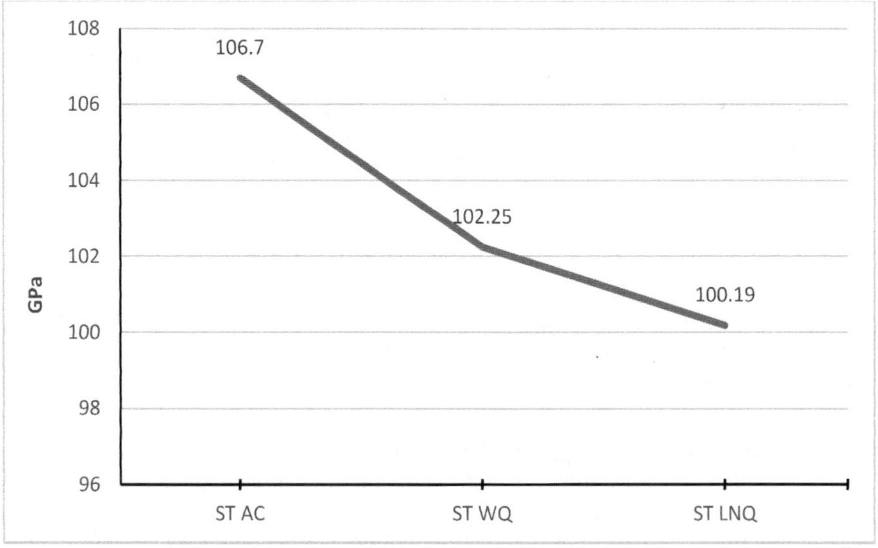

Figure 4. Elastic modulus of Ti-6Al-6Nb

Figure 4. shows the elastic modulus that obtained with an ultrasonic approach by measuring ultrasonic velocity that calculated with formula below: (ASTM E494-95)

$$E = \frac{\left[\left(\rho V_T^2\left(3V_T^2 - 4V_T^2\right)\right)\right]}{\left(V_L^2 - V_T^2\right)} \qquad 1)$$

$$\frac{V_T}{V_L} = \sqrt{\frac{1-2\sigma}{2(1-\sigma)}} \qquad 2)$$

Where:
σ = 0.36
ρ = density (4.25 g/cm3)
VL = longitudinal velocity
VT = transversal velocity
E = elastic modulus

Figure 4. shows the elastic modulus had decreased when the cooling rate is faster. In the air cooling the elastic modulus is 106.7 Gpa, than decreased to 102.25 Gpa in water quench and became 100.2 Gpa in cryogenic cooling. The decreasing elastic modulus is inversely proportional to the hardness as shown in Figure 3. Thus it can be understood that the smaller grain size can decrease the elastic modulus, as well as the phase formed, the more beta phases are formed, the smaller the elastic modulus.

4. Conclusion

The microstructures, hardness and elastic modulus of Ti-6Al-6Nb in solution treatment process with cryogenic cooling medium have been investigated. From the investigated can be concluded that the alloys had equiaxed microstructures with alpha-beta phases that dominated by beta phase. As compared with air cooling and water quenching process, the cryogenic cooling had the smallest grain size structure, lowest elastic modulus 100 GPa and highest hardness 52.5 HRc

References

[1] Keizo H, et.al *Nippon Steel Technical Report No 62*, 1994.
[2] Mitsuo N, , *Elsevier Material Science and engineering A* **243** 231-236, 1998.
[3] Marsumi Y and Pramono A W., *Research Centre for Metallurgy – Indonesian Institute of Sciences* 2014.
[4] Sutowo C, et al. 2017 *Prosiding Semnastek* 1-6
[5] Obasi G C, et al. 2012 *Acta Materialia* **60** (17), 6013–6024
[6] Senopati, G, et al. 2018 *AIP Conference Proceedings.* **1964**
[7] Rokhamanto, F. et al. 2018 *AIP Conference Proceedings.* **1964**
[8] Matthew J. and Donachie, Jr, 2000 *The Material Information Society, Second Edition*

Acknowledgement

This paper financially supported by 2018 thematic program of Indonesian Institute of Sciences.

IOP Conf. Series: Materials Science and Engineering **541** (2019) 012044 doi:10.1088/1757-899X/541/1/012044

Decomposition of spodumene mineral in granitic rocks from South Kalimantan - Indonesia by potassium sulphate

A Suharyanto [1,2], L H Lalasari[1], M Z Mubarok[2]
[1]Research Center for Metallurgy and Materials – Indonesian Institute of Sciences (LIPI), Kawasan PUSPIPTEK, Gdg 470, Jl. Raya Puspiptek, Tangerang Selatan, 15314, Banten, Indonesia
[2]Department of Metallurgical Engineering, Institut Teknologi Bandung, Indonesia, Jl. Ganesha, 10, Bandung 40132, Jawa Barat Indonesia

E-mail: ariy002@lipi.go.id

Abstract. The decomposition behaviour of spodumene mineral in granitic rock from Kalimantan Indonesia by using potassium sulphate has been studied in this research. The granitic rock from Kalimantan is indicated to contain spodumene mineral. The sspodumene is one of the main minerals which contains lithium. The purpose of this study is to decompose the α-spodumene mineral in the granitic rock of Kalimantan to β-spodumene which is subsequently easier to be dissolved in water. In this study, the decomposition of α-spodumene was performed under variations of roasting temperatures (i.e. 350°C, 450°C, 550°C, 600°C, 700°C, 800°C, 900°C and 1000°C) and roasting times (20 min, 40 min and 60 min). The results of the decomposition products were characterized by X-ray diffraction (XRD), X-ray fluorescence (XRF), inductively coupled plasma optical emission spectroscopy (ICP-OES), scanning electron microscopy (SEM) and energy dispersive X-ray spectroscopy (EDS). It was identified that the granitic rock from Kalimantan consists of α-spodumene ($LiAlSi_2O_6$) with lithium content of 3.09 ppm and the other major minerals are quartz (SiO_2) and albite ($NaAlSiO_3O_8$). The α–spodumene was completely converted to β-spodumene at 700°C for 60 minutes.

1. Introduction

Indonesia is a country that has been blessed with abundant mineral resources, part of them are igneous rocks such as granitic rock [1]. Granite is a bright colour igneous rock which forms from slow crystallization of magma under the surface of the earth. Granite mainly consists of quartz (SiO_2) and feldspar ($KAlSi_3O_8$ – $NaAlSi_3O_8$ – $CaAl_2Si_2O_8$) as well as small amount of mica ($KAl_2(AlSi_3O_{10})(F,OH)_2$), amphiboles ($Ca_2(Mg,Fe,Al)_5(Al,Si)_8O_{22}(OH)_2$) and spodumene ($LiAlSi_2O_6$). The content of spodumene mineral in granitic rocks leads to the interest of granitic rock utilization since the spodumen contains lithium.

The most important lithium mineral is spodumene, which is generally present together with quartz, feldspar, and mica. Along with brines, spodumene mineral has been considered as one of the major resources used for lithium extraction [2]. In terms of production cost, lithium extraction from brines is still more competitive in comparison to lithium extraction from spodumene, lepidolite and other solid minerals[3]. Various processes for lithium extraction from spodumene ore have been proposed.

Lithium is a light metal that has been utilized in the manufacture of glass, ceramics, and lubricating greases since its discovery in 1817. Lithium currently plays an important role in the high-tech industry and has been used in several fields which includes pharmaceuticals, rechargeable batteries, and electronics. More than 30% of total lithium consumption in 2015 is for batteries and since the year of 2000, lithium consumption increased by around 20% every year. The production of electric vehicle has been drastically increasing in recent years and will continue to grow in the next couple decades; therefore the demands of lithium metals as one of the main components of batteries for electric vehicle will be also increasing [4]. The global production of lithium was estimated to be 35,000 tonnes in 2013 which represents an increase of 6% compared to 2012 [5]. Many studies of lithium extraction have been carried out by several researchers. Barbosa et.al. (2014) have studied the extraction of lithium by means of the chlorination roasting of β -spodumene with temperature range of 1000 to 1100 °C for periods of time of 0 to 180 min [6]. Natasha et.al. (2018) have investigated the decomposition of Schistmica rock from Kebumen, Center Java, Indonesia by using sodium sulfate as decomposition agent [7].

Spodumene occurs naturally in α-phase with a monoclinic structure of the pyroxene type [8]. This structure is resistant to the attack of chemical agents, either gaseous or liquid. Spodumene can be transformed into its β-phase by calcination at 1100 °C [8]. This phase is more reactive and less resistant to ordinary chemical agents [9]. A suitable amount of K_2SO_4 and CaO has been added during the roasting process to avoid the fusion of the roasted product and enhancingthe efficiency of lithium extraction [10]. Yan et.al., 2012, have conducted the experimental work by adding appropriate amount of CaO and K_2SO_4 during the roasting process of lepidolite to optimize the extraction of lithium [11].

In this study, the decomposition of α-spodumene in the granitic rock from South Kalimantan was performed under variations of roasting temperatures and roasting duration. Potassium sulfate (K_2SO_4) was used as a decomposition agent. The decomposition of α-spodumene was aimed at transforming alpha (α) phase in the spodumen mineral into beta (β) spodumene phase. The beta-spodumene is expected to be readily soluble in water leaching step which is done after the roasting. During water leaching, lithium is released from spodumene mineral and dissolved in water.

2. Experimental and Method

2.1 Materials and Characterization

The granitic rock sample used in this investigation was obtained from South Kalimantan. Potassium sulfate as decomposition reagent with a specification of pro analysis (P.A) grade was purchased from Merck. The characterization of samples was performed by X-ray fluorescence (XRF), inductively coupled plasma optical emission spectroscopy (ICP-OES) and X-ray diffraction (XRD). The morphology of raw material and the decomposition products were analysed by Scanning Electron Microscope (SEM).

2.2 Methods

In this study, granitic rock samples from South Kalimantan were crushed with a jaw crusher, milled with a ball mill and sieved to obtain a particle size distribution of -200 mesh. The sample was then analyzed by X-ray diffraction (XRD), X-ray fluorescence (XRF), inductively coupled plasma optical emission spectroscopy (ICP-OES) and scanning electron microscopy (SEM). The decomposition of α-spodumene in the ore sample was carried out by roasting of the mixture of the ore sample and potassium sulfate (K_2SO_4) in the "Carbolite" muffle furnace at various temperatures and roasting times. Prior to the roasting, the ore samples were mixed homogenously with potassium sulfate with a mass ratio of 1:1. The sample was then analyzed by Simultaneous Thermal Analysis (STA) to determine the temperature range which will be investigated in the decomposition experiments. The STA pattern of the mixture of granitic ore sample and K_2SO_4 is presented in Figure 1. HSC Chemistry software was also used to predict the reaction occurs in the decomposition process, its Gibbs free energy and equilibrium constant at various temperatures. As can be seen in Figure 1, a mass loss occurs in the mixture of granitic ore and K_2SO_4 as it was heated up to 700°C. Mass removal takes place sharply as the heating was carried out at temperature range of 320°C to 600 °C indicated by the appearance of the peak that represent positive enthalpy changes (ΔH). This peak indicates that the reaction between the granitic ore sample and K_2SO_4 is an endothermic reaction. Based on the results of STA analysis and simulation by HSC chemistry software, the authors varied roasting temperatures at 350 °C, 450 °C, 550 °C, 600 °C, 700 °C, 800 °C, 900 °C and 1000 °C. Meanwhile, the roasting times was varied at 20 min, 40 min and 60 min.The products of decomposition were characterized by XRD, XRF, SEM and Energy Dispersive X-ray Spectroscopy (EDS). The flowsheet of the experimental procedure is summarized in Figure 2.

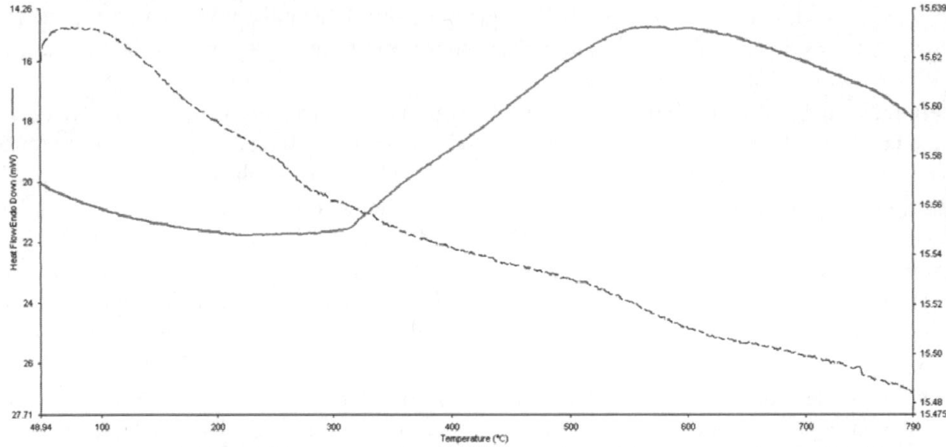

Figure 1. The STA pattern of the mixture of granitic ore sample and K_2SO_4.

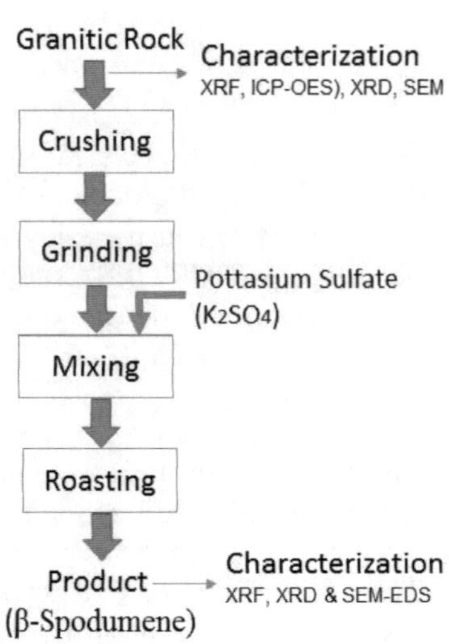

Figure 2. Flowsheet of the decomposition experiment of α-spodumene in granitic ore sample from South Kalimantan.

3. Result and discussion

3.1. Raw Material Characterization

After ore preparation stage, the sample was analyzed by XRF and the analysis result is presented in Table 1. The XRF analysis did not detect the presence of lithium in the granitic ore sample due to its low concentration, therefore, the lithium content was measured by using ICP-OES. The result of ICP-OES analysis showed that the content of lithium in the granitic rock sample from South Kalimantan was 3.09 ppm.

Table 1. The results of elemental analysis of granitic rock sample from South Kalimantan by XRF.

Element	Mg	Al	Si	P	Cl	K	Ca	Ti	Mn	Fe
Content (%)	6.19	9.20	48.62	0.72	0.96	2.03	14.27	1.19	0.43	13.87

The result of XRF analysis shows that Si, Ca, Fe, Al, and Mg are the major elements of the granitic rock sample from South Kalimantan. The presence of Si, Al, and Li elements in granitic rock indicates that South Kalimantan's granitic rocks contain lithium mineral as spodumene ($LiAlSi_2O_6$)) at a fairly small content. Until recently, the data of major source of the lihium-bearing minerals in Indonesia has not yet been clearly reported. In addition to the granitic rock samples from Kalimantan used in this research, the other source of lithium metal in Indonesia is sekismika rock such as that discovered in Kebumen Distric in Central Java Province [7].

The result of XRD analysis of the ore sample was shown in Figure 3. The XRD pattern shows that the major minerals in granitic rock sample from South Kalimantan are quartz (SiO_2), albite ($NaAlSiO_3O_8$) and α-spodumene ($LiAlSi_2O_6$). According to ICSD no. 98-009-0145, quartz can be found at 2Θ position of around 23°, 39°, 47° and 55° with a hexagonal crystal system. According to ICSD no. 98-005-2343, albite minerals can be found at 2Θ of around 21°, 27° and 28° with anorthic crystal systems. Meanwhile, according to ICSD no. 98-001-0238, spodumene minerals can be found at 2Θ of around 14°, 24°, 32°, 36°, 64° and 76° with a monoclinic crystal system. The spodumene-associated peaks appearing at XRD pattern in Figure 3 indicated that lithium mineral was present in the South Kalimantan's granitic rock. This XRD result is in aggrement with the result of ICP-OES analysis that measured the content of lithium of 3.09 ppm in the digested ore sample. Results of SEM-EDS analyses of the ore sample are shown in Figure 4 and Table 2.

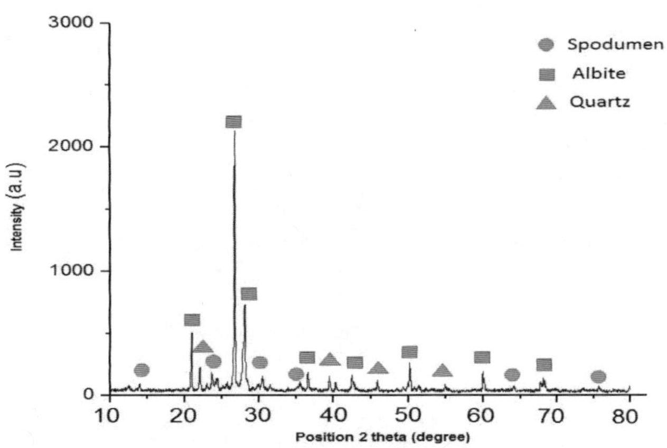

Figure 3. Diffractogram of the granitic ore sample

Figure 4. SEM micrograph of the granitic ore sample before roasting.

Table 2. Result of elemental analysis by ZAF Method Standardless Quantitative Analysis of EDS-SEM.

Element	Mass (%)
O	43.54
Mg	0.46
Al	1.67
Si	54.48
Ca	0.18
Fe	0.12

The SEM micrograph indicates that the morphology of granitic ore has irregular grains with non-uniform grain size. The grain size of the original granitic ore sample is at the range of 7 - 19 μm. The result of EDS-SEM in Table 2 reveals that the major elements within the sample are Si, O and Al. The result of the EDS analysis is in aggrement with the result of XRD analysis that detected quartz, albine and spodumenen (which contains Si, O and Al elements) as the major minerals in the granitic ore sample from South Kalimantan.

3.2 Effect of roasting temperature

To investigate the effect of roasting temperatures on the decomposition behaviour of the granitic ore sample, the products of decomposition at various temperature with ore: K_2SO_4 mass ratio of 1:1 were analyzed by XRD and the results are shown in Figure 5.

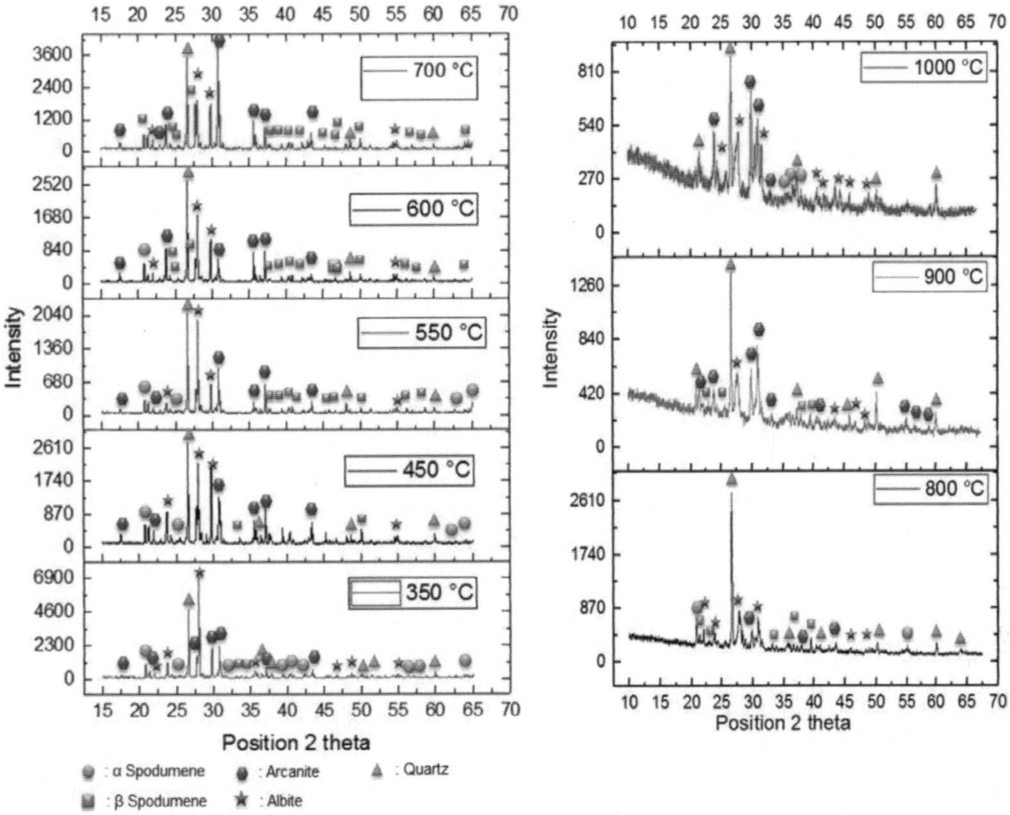

Figure 5. Diffractogram of the decomposition products with various temperatures (ore: K_2SO_4 mass ratio of 1:1, roasting time 60 min)

The XRD pattern in Figure 5 indicates that the roasting process induced the change of the structure and phase of minerals in the granitic ore sample. The spodumene minerals which is initially has alpha phase with monoclinic crystal system was transformed into beta phase with tetragonal crystal system. The change in the structure and phase starts to be detected at roasting temperature of 350 °C with 60 minutes roasting time. This β-spodumene was detected at 2 thetas of around 31° and 33.5° according to ICSD 98-001-4235. However, this phase disappeared at roasting temperature of 1000°C and roasting time of 60 minute. At this temperature, arcanite (K_2SO_4) was detected with the peak position at 2 thetas of around 31° and 33.5°. The arcanite has orthorhombic crystal according to ICSD 98-000-2827. The optimum temperature to generate β-spodumene is 700°C (roasting time 60 minute). At this condition, α-spodumene has been well converted to β-spodumene as indicated by the XRD result of decomposition product (Figure 5). Natasha et.al. also reported an optimum temperature of 700°C with 40 minute roasting time to form β-spodumene from Kebumen sekismika rock[7].

The possible reaction mechanisms between spodumene in granitic rocks with K_2SO_4 as decomposition agent are as follows: 1) alpha spodumene is converted to beta spodumene at temperatures of above 350°C 2) Beta spodumene phase reacts with K_2SO_4 to form Li_2SO_4 phase. At this stage, the potassium in potassium sulphate compound is replaced by lithium atom from spodumen to form lithium sulphate.

The chemical reactions proposed for the conversion of α-spodumene to β-spodumene is as follow:

IOP Conf. Series: Materials Science and Engineering **541** (2019) 012044 doi:10.1088/1757-899X/541/1/012044

$$\alpha\ LiAlSi_2O_{6(s)} + K_2SO_{4(s)} \rightarrow \beta LiAlSi_2O_{6(s)} + K_2SO_{4(a)} \quad (1)$$

$$2\beta LiAlSiO_{6(s)} + K_2SO_{4(s)} \rightarrow 2KAlSi_2O_{6(s)} + Li_2SO_{4(s)} \quad (2)$$

The mechanism and the reaction steps during conversion of α-spodumene to β-spodumene is illustrated in Figure 6. Profile of Gibbs free energy and logarithmic of equillibrium constant of Reaction (2) as a function of temperature according to the data from HSC Chemistry Software is illustrated in Figure 7.

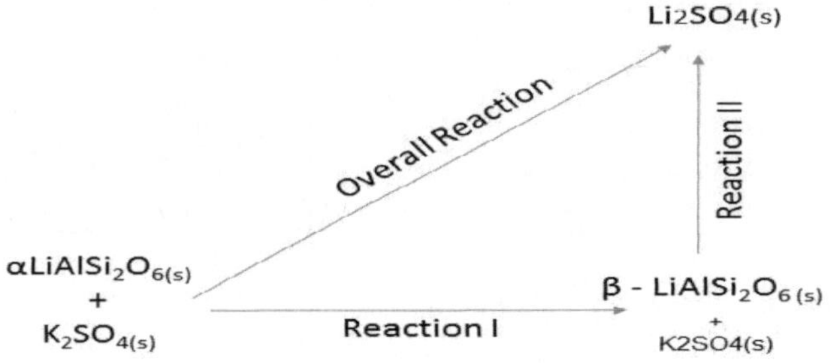

Figure 6. The mechanism reaction between α sprodumene and K_2SO_4 additives.

The SEM micrographs of decomposition products resulted from roasting experiments at temperatures of 350°C and 700°C for 60 minutes are illustrated in Figure 8.a and 8.b, respectively. It was found roasting tends to reduce the grain size of the granitic ore sample. At higher roasting temperatures, smaller grain size of the decomposition product was obtained. The results of SEM analysis shows that the grain size of the decomposition product resulted from the decomposition at temperature of 350 °C is at the range of 4 – 10 μm which is notably smaller than the initial grain size of un-roasted ore which has grain size at the range of 7 - 19 μm. Further increased of roasting temperature to 700°C resulted in smaller grain size at the range of 2 – 6 μm.

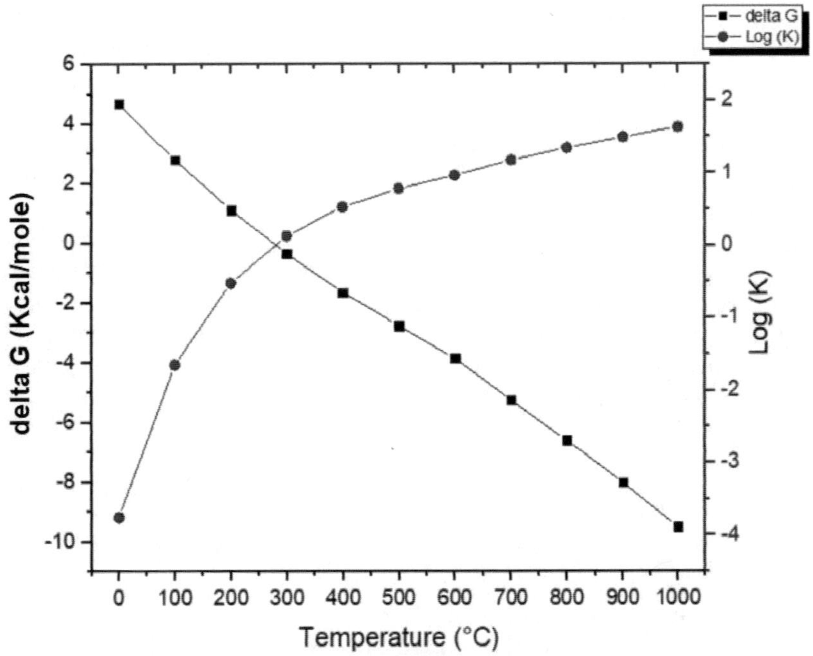

Figure 7. Profile of Gibbs free energy and logarithmic of equillibrium constant of Reaction (2) as a function of temperature according to the data from HSC Chemistry Software.

(a) (b)

Figure 8. SEM micrographs of the decomposition products resulted from roasting experiments at temperatures of (a) 350°C and (b) 700°C for 60 minutes.

3.3 Effect of roasting time

Roasting experiments at temperature of 700°C for 20, 40 and 60 minutes were conducted to evaluate the effect of roasting times on the formation of β-spodumen-phase. XRD Diffractogram of the decomposition products resulted by roasting at temperature of 700°C 20 min, 40 min and 60 min is presented in Figure 9. The XRD pattern indicates the effect of roasting times on the formation of β-spodumen-phase. At roasting temperature of 700°C, duration of roasting for 60 minute was found to be an optimal condition in which conversion of α-spodumene to β-spodumene took place completely. The α-spodumene began to be transformed to β-spodumene after 20 minutes of roasting at 700°C, at which β spodumene was found at peaks with 2 thetas of 22°, 23°, 26° and 33.5° according to ICSD 98-001-4235. For 40 minute roasting duration at 700°C, β spodumene was detected at peaks with 2 thetas of 22°, 23°, 26° and 33.5° according to ICSD 98-001-4235. Meanwhile for the roasting duration of 40 minute at 700°C, β spodumene was detected at peaks with 2 thetas of 22°, 23°, 26°, 33.5°, 36°, 37°, 40°, 42°, 45°, 56°, 59° and 62° according to ICSD 98-001-4235. For the roasting duration of 60 minute at 700°C, α-spodumene was completely converted to β-spodumene in which β-spodumene was detected at peaks with 2 thetas of of 21° and 47°. On the other hand, roasting duration of 20 minute and 40 minute resulted in the decomposition product which is still contains α-spodumene phase which was detected at peaks with 2 thetas of of 21° and 47° according to ICSD no. 98-001-0238.

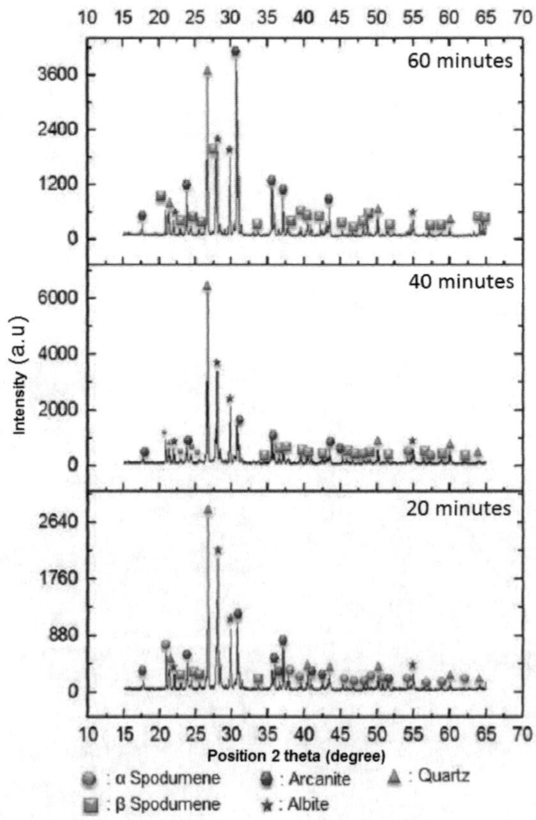

Figure 9. XRD Diffractogram of the decomposition products of the granitic ore sample at 700°C for 20 min, 40 min and 60 min

4. Conclusions

The granitic rock from South Kalimantan has major mineral components of albite ($NaAlSiO_3O_8$), quartz (SiO_2) and spodumene ($LiAlSi_2O_6$). Decomposition process of the granitic ore with K_2SO_4 at temperatures of above 350°C for 60 minutes changes the crystal structure of the spodumene from monoclinic to tetragonal and its phase from α-spodumene to β-spodumene. The α-spodumene was completely converted to β-spodumene (which is more soluble in water) at roasting temperature of 700°C for 60 minute. SEM analysis of the decomposition products indicated that

IOP Conf. Series: Materials Science and Engineering **541** (2019) 012044 doi:10.1088/1757-899X/541/1/012044

roasting of the granitic ore sample reduces grain size and at higher roasting temperatures, smaller grain size of the decomposition product was obtained.

5. References

[1]. Darman H, and Sidi F H, 2000, *An Outline of The Geology of Indonesia*, Ikatan Ahli Geologi Indonesia
[2]. Habashi F, 1980. *Principles of Extractive Metallurgy*, **IV**. Gordon and Breach, USA.
[3]. Helvaci C, 2003. *Int. Geol. Rev.* **45**, 1–14
[4]. P K Choubey, M Kim, R R Srivastava, J Lee, dan J Lee, 2016 *Minerals Engineering*. **89**, 119-137
[5]. Virginia, pp. 94 – 95 http://www.meridian-int-res.com/Projects/Lithium_Problem_2.pdf (accessed on Juli 10, 2018)
[6]. Barbosa L I, Valente G, Orosco R P, & Gonzalez J. A. (2014). *Minerals Engineering*, **56**, 29-34
[7]. Natashaa N C, Lalasari L H, Rohmah M, and Sudarsono J W 2018. Majalah *Metalurgi* (2018) **2**: 69 – 78.
[8]. Peltosaari O, Tanskanen P, Heikkinen E, and Fabritius T, 2015. *Minerals Engineering*. **82**, 54-60
[9]. N. K. Salakjani, P. Singh, dan A. N. Nikoloski, *Minerals Engineering*, **98**, 71-79, 2016
[10]. Wang, Jianling, Wang, Jimin, Zhu, Jianchun, 1994. *Nonferrous Met.* **4** (2), 107–112
[11]. Yan, Q., Li, X., Wang, Z., Wu, X., Wang, J., Guo, H., Hu, Q., Peng, W., 2012. *Int. J. Miner. Process*

Acknowledgements
The authors of this work would like to acknowledge Research Center for Metalurgy and Materials, Indonesian Institute of Sciences (P2MM - LIPI) and Ministry of Research, Technology and Higher Education of the Republic of Indonesia for the financial support. The authors are also grateful for the main financial supports from INSINAS budget 2018.

Effect of shot peening and anodizing on fatigue crack growth of aircraft material Al 7050-T7651

H Ardianto[1], R Yudhono[1] and A A Erissonia[2]
[1]Program Studi Teknik Dirgantara, [2]Program Studi Aeronautika
Sekolah Tinggi Teknologi Kedirgantaraan
Jl. Parangtritis Km. 4,5 Yogyakarta, Indonesia

E-mail: haris.ardianto@gmail.com

Abstract. The rate of fatigue crack propagation of the 7050-T7651 aircraft material subjected to automatic shot peening machine and chromic acid anodizing (CAA) has been studied and compared with previous research results. Al 7050-T7651 is made into specimens according to ASTM E647 standard treated with automatic shot peening machine (with steel shot material, 0.017inch diameter, shot flow 30%, 3 bar pressure, Almen intensity 0.0082 A with 100 mm shooting range and 90° angle of shot) and CAA (operating standard) in surface treatment area of Indonesian Aerospace. The fatigue cracking test with a load of about 200 kg, with a stress ratio of R = 0.1. The results of the fatigue crack growth test resulted in the constant of Paris A = 3.654×10^{-9} and n = 1.67, indicating that the automatic shot peening machine and chromic acid anodizing treatment decreased the fatigue life and increased the fatigue crack growth rate compared to the automatic shot peening machine, manual machine with or without CAA, and without treatment.

1. Introduction

Al 7050-T7651 belongs to the type of aluminum alloy with the composition of Al 87.3-90.3%; Zn 5.7-6.7%; 1.9-2.6% MG; Cu 2-2,6%; Zr 0.08-0.15% and other alloy elements which are given heat treatment with a 477°C solution temperature and aging temperature 121-177°C based on ASM (Aerospace Specification Metal) Material Data Sheet [1].

Shot peening is the surface treatment by firing shot material so that the dimple is formed and residual stress arises on the surface of the working specimen to increase resistance to fatigue failure. Whereas CAA Chromic Acid Anodizing is the treatment of aluminum surface oxidation or its electrolysis alloy, chromic acid is used as an electrolyte solution to form a protective aluminum oxide / hydroxide layer and is applied to improve corrosion resistance and paint adhesion, but hard and brittle. Several specimen tests with shot peening and CAA combined treatments have been shown to increase fatigue life and reduce the rate of fatigue crack propagation [2-3].

Types of materials similar to Al 7050-T7651 treated with manual shot peening and CAA have also been tested for fatigue, and showed that the combination of treatments increased fatigue life and reduced the fatigue crack growth rate [4]. Fatigue crack propagation test is one standard that can be used in most aircraft constructions, such as the fuselage component, upper and lower spar, upper and lower wing, horizontal stabilizer, etc. [5].

The purpose of this paper is as a follow-up study to deepen the analysis of previous author's research, in this case the specimen was treated with automatic shot peening [6] and added data analysis of the combination treatment of automatic shot peening and CAA applied to the surface treatment at Indonesian Aerospace.

2. Experimental Procedures

Al 7050-T7651 specimen material was formed according to ASTM E647 standard type L-T [7] as shown below

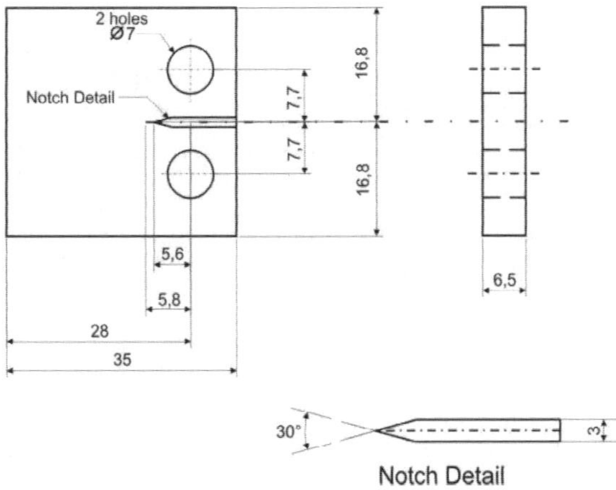

Figure 1. Specimen according to ASTM E647 [8].

Automatic machine shot peening process and CAA are carried out at the Directorate of Aerostructure, Sub Directorate of Surface Treatment Indonesian Aerospace with the following shot peening treatment parameters

Shot material : Steel
Shot ball diameter : 0,017 inch
Shot flow : 30%
Pressure : 3 bar
Almen Intensity : 0,0082 A
Shooting distance : 100 mm
Shooting angle : 90°

Anodizing Process with specifications

Tank Volume : 37000 Lt
Level : 20 cm
Free CrO_3 : 30 gr/L min
Total CrO_3 : 30-100 gr/L
Cl-as NaCl : 0,2 gr/L max
SO_4-as Na_2SO_4 : 0,5 gr/L max
Temperatur : 40±2 ^{0}C
Time : 40 minutes
Voltage : 40±1 Volt
Catode Plate : Titanium

While the fatigue testing process is carried out at the Material Engineering Lab. Mechanical and Industrial Engineering Department, Engineering Faculty, UGM with a servopulser machine

Merk : Shimadzu
Machine Type : Mesin Servopulser
Model : 4825

Crack Visual : *microscope travelling optic*
Pull-press loading parameters, 200 kg and R (ratio) = 0.1.

Fatigue crack growth rate can be calculated using the Paris equation with the ASTM E647 specimen [9],

$$\frac{da}{dN} = A(\Delta K)^n$$

$$\left(\frac{da}{dN}\right) = \frac{b_1}{A_2} + 2b_2\left(\frac{N_i - A_1}{A_2}\right)$$

$$\Delta K = \frac{\Delta P}{B\sqrt{W}} \frac{\left(2+\frac{a}{W}\right)}{\left(1-\frac{a}{W}\right)^{\frac{3}{2}}}\left[0{,}866 + 4{,}64\left(\frac{a}{W}\right) - 13{,}32\left(\frac{a}{W}\right)^2 + 14{,}72\left(\frac{a}{W}\right)^3 - 5{,}6\left(\frac{a}{W}\right)^4\right]$$

$$\Delta P = P_{max} - P_{min}$$

with

a	= crack length
N	= cycles amount
A, n	= Paris constant
b_1,b_2	= regression constant
ΔK	= different stress intensity factors
P	= load
W	= width of specimen
B	= thick of specimen

Fatigue crack propagation test is carried out by observing each increase in crack length that occurs on the front side of the specimen (ai), where these data are used to see the relationship between the addition of crack length (a) and the number of cycles (N), processed using incremental polynomial methods for get relationship with / dN-ΔK. Trendline relationships with / dN-ΔK are obtained by making a graph with log scale on d / dN and ΔK, so that a line equation can be used to determine the characteristics of fatigue crack propagation (Paris A and n constants).

3. Results and Discussion

Figures 2 and 3 and Table 1 show a comparison of data distribution, Paris trendline and constants from previous studies [4] and [6] and additional recent research data. In naming the type of specimen some new names are used to clarify the different treatments used.

Previous analysis stated that the automatic shot peening machine treatment was effective in reducing the rate of fatigue crack propagation compared to manual shot peening machines, although the data distribution and BMSPAuto trendline position were above and BMSP with a higher Paris A constant, but the gradient line for the BMSPAuto trendline was lower or higher slope than BMSP seen from the lower Paris constant n. This shows that there is a decrease in the rate of fatigue crack propagation and increasing fatigue life. This will be obvious if applied to predict fatigue life using the Paris constant.

The effect of Almen intensity increase (one of the shot peening parameters) from 0.006 A [4] to 0.0082 A [6] turned out to be able to reduce the rate of fatigue crack creeping, but to some extent if the intensity of Almen increases, it will increase surface roughness and reduce resistance. fatigue (increasing crack propagation rate) [9].

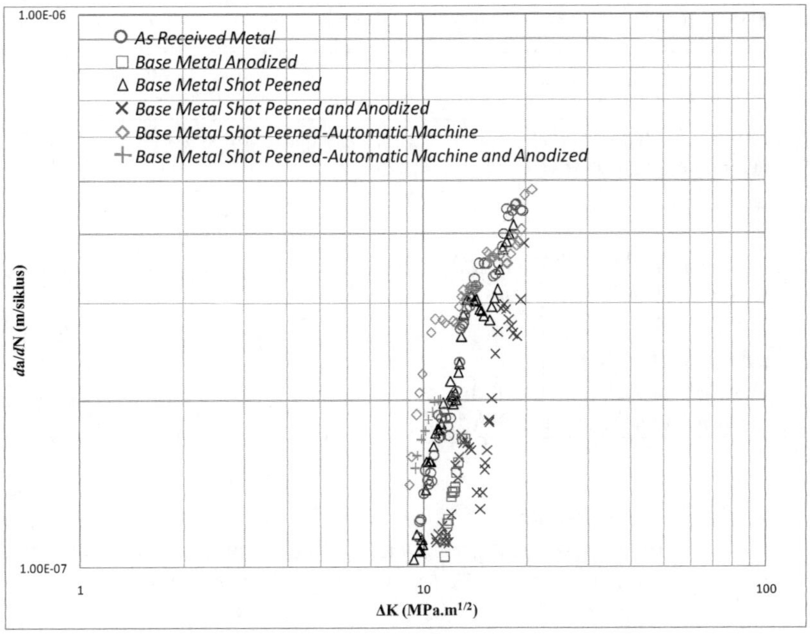

Figure 2. Relationship between da/dN and ΔK.

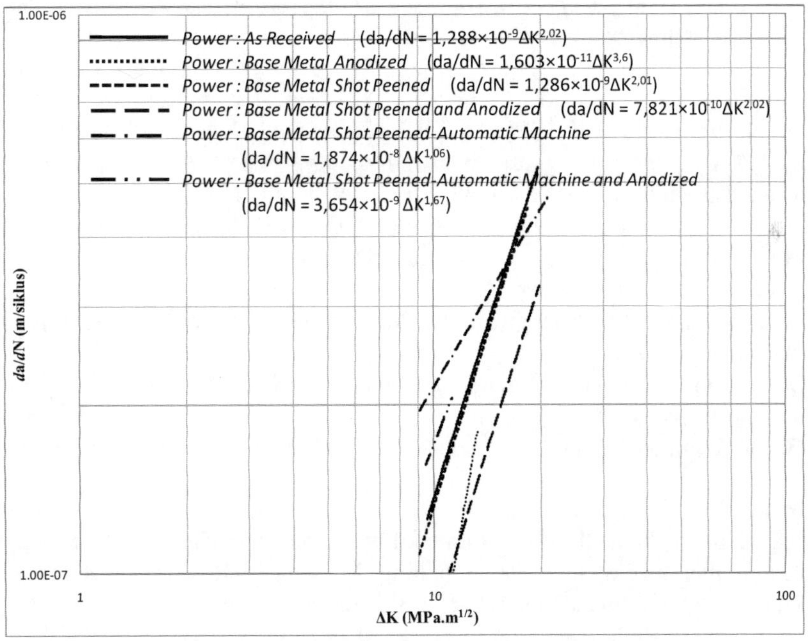

Figure 3. Relationship between da/dN and ΔK in trendline

However, when testing the BMSPAutoA fatigue crack rate compared to BMSPA, it appears that the data distribution and BMPAuto trendline position are quite far above BMSPA with the Paris A constant higher, but the line gradient is not too significant difference, then BMSPAutoA treatment actually decreases fatigue life and Fatigue crack rate is increasing. As a result of the larger dimple and increasing surface hardness, due to the effect of increasing Almen intensity plus anodizing treatment it actually reduces fatigue resistance.

Table 1. Test results of fatigue crack propagation Al 7050-T7651.

Kategori	Paris Constants	
	A	n
As Received Material (AR)	1.288×10^{-9}	2.02
Base Metal Anodized (BMA)	1.603×10^{-11}	3.6
Base Metal Shot Peened-Manual Machine (BMSP)	1.286×10^{-9}	2.01
Base Metal Shot Peened-Manual Machine and Anodized (BMSPA)	$7{,}821 \times 10^{-10}$	2.02
Base Metal Shot Peened-Automatic Machine (BMSPAuto)	1.874×10^{-8}	1.06
Base Metal Shot Peened-Automatic Machine and Anodized (BMSPAutoA)	3.654×10^{-9}	1.67

4. Conclusions

Basically, the lower the relationship trendline position da/dN dan ΔK and the more gradient sloping line, the lower the rate of propagation of fatigue cracks, both of them are interconnected and influence each other. Shot peening treatments and CAA reduce fatigue life and increase the rate of creeping fatigue cracks with Paris constant $A = 3.654 \times 10^{-9}$ and $n = 1.67$.

References

[1] Metals Handbook 1990. *Properties and Selection: Nonferrous Alloys and Special-Purpose Materials*. ASM International 10th Ed. Vol. **2**

[2] Giummarra C and Zonker H R 2005 *ICAF 2005 Preceedings* (Hamburg Germany : *ICAF 2005 Preceedings*)

[3] Costa M Y P Voorwald H J C Pigatin W L Guimaraes V A Cioffi M O H 2006 *Material Research* **9** No 1 107-109

[4] Ardianto H and Iswanto P T 2015 *Jurnal Teknika STTKD* **2** No. 1 1-11

[5] Starke E A Jr and Staley J T 1996 Application of Modern Aluminum Alloys to Aircraft, *Elsevier Science Ltd. Prog. Aerospace Sci.* **32** 131-172

[6] Ardianto H Shot 2017 *Proceeding SENATIK* **3**

[7] ASTM International 1997 Standard Test Method for Plane-Strain Fracture Toughness of Metallic Materials (USA)

[8] ASTM Designation : E647-00 Standard Test Method for Measurement of Fatigue Crack Growth Rates. (United States : ASTM)

[9] Diep H Bae H Ramulu M 2011 *Conf Proc 2011: ICSP-11* (South Bend, IN USA)

Acknowledgement

The author would like to thank the Directorate of Research and Community Service; Directorate General of Research and Development Strengthening; Ministry of Research, Technology and Higher Education which has provided financial support for this research.

IOP Conf. Series: Materials Science and Engineering **541** (2019) 012046 doi:10.1088/1757-899X/541/1/012046

Study of leaf spring fracture behavior used in the suspension systems in the diesel truck vehicles

Husaini[1], N Ali[1], R Riantoni[1], T E Putra[1], and H Husin[2]
[1]Laboratory of Computational Mechanics
Department of Mechanical Engineering – Syiah Kuala University
Darussalam – Banda Aceh 23111, Indonesia
[2]Department of Chemical Engineering, Syiah Kuala University, Darussalam - Banda Aceh 23111, Indonesia

E-mail: husainiftm@unsyiah.ac.id

Abstract. The leaf spring is frequently applied on a vehicle such as used on the truck and it is one of the machine element suspension systems. It needs to have an excellent resistance to the fracture propagation to avoid failure that could lead to severe accidents. Fracture failure on the leaf spring is frequently experienced especially on diesel trucks, known as premature failure. Therefore, this case is needed to be studied and investigated. The purpose of this study is to investigate the cause of failure on the broken leaf spring by counting analytically the stress intensity factor (K_I) occurred near the crack tip. Moreover fracture surface observation conducted by Scanning Electron Microscope (SEM) to find out whether there is or not an initial crack and crack propagation direction. This present study describes the experimental work and analytical analysis of theleaf spring used on Diesel Truck Vehicle. Hardness testing was performed using the Rockwell method. The chemical composition testing on the leaf spring material was found that the material in accordance with AISI 5150 standard. The hardness value of leaf springs tested of about 106.9 HRB. The result of analytical analysis shows that the maximum value of stress intensity factor (K_I) occurred near the crack tip at the near the hole of leaf spring. In this case the value of K_I is higher than thevalue of the fracture toughness (K_{IC}) of the leaf spring material. Therefore, this result could be inferred that crack propagation happened starting from initial crack tip which is located near the hole on the leaf spring. The failure of leaf springs also caused by fatigue crack propagation due to dynamic loading during the operation marked by appearance the beach mark on the fracture surface.

1. Introduction

One of the components for the suspension system that is often used in a vehicle such as Truck is a leaf spring [1]. The use of leaf springs on truck vehicles in addition to holding the load also serves as a vibration damper. Furthermore, it is known that loading on the truck is at the midpoint on the back of the vehicle cart [2].

In general the using of the leaf spring on the truck often experience failure due to fatigue load. In this case, fracture occurs because of an initial fine crack in the part of high the stress concentration on the spring surface or in other parts. The crack propagation behavior occurred usually starts from crack initiation and then it is followed by crack propagation and caused a final fracture [3-5].

Since the failure by fatigue fracture is the most common failure in automotive components such as crankshaft [6] as well as those involving spring on trucks. In addition to, leaf spring especially as automotive components often experience over load and vibration that occur due to the unevenness of the road. Therefore, the load occurs suddenly so that the wheels up and down with irregular frequencies. Many cases indicated that Leaf springs are affected by the dynamic loading [1].

The material used for leaf springs is usually a plain carbon steel having 0.90 to 1.0% carbon. The leaves are heat treated after the forming process. Heat treatment of steel spring products will produce greater strength and therefore greater load capacity, greater deflection range and better fatigue properties [7]. The leaf spring is produced using hot rolling depending on the expected specification [2].

This study will examine the fracture of leaf springs used in a Colt diesel truck. The earliest known data is a small capacity truck with a container tank behind its part. Several times, trucks carry more load capacity and operate on uneven roads. Furthermore, the truck was operated for two years with a distance of 572 km, until the leaf spring number 8 experienced a broken failure. Therefore, the failure of leaf springs needs to be investigated experimentally and analyzed analytically.

The purpose of this study is to investigate the causes of spring failure used on the diesel trucks experimentally and analytically. When the causes of failure found, accordingly the same failure can be anticipated in the future and can also prevent vehicle damage.

2. Experimental Procedure

2.1. Broken leaf spring and visual inspection

The object research used in this study is the leaf fracture spring used in colt diesel trucks. Leaf springs experience fracture after the vehicle runs up to a distance of about 572 km. Broken leaf spring as shown in Figure 1. This material was obtained from a mechanical workshop, and then identified the type of vehicle and the allowable load capacity. The fracture that occurs in the leaf spring is near the bolt hole as indicated by the arrow.

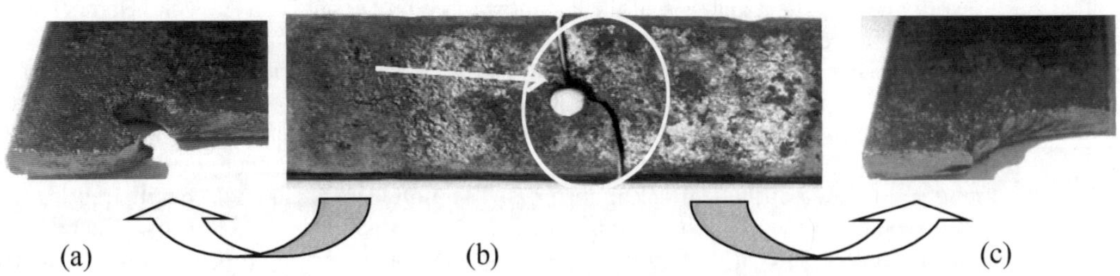

Figure 1. (a) Broken leaf springs on the left side of Fig. 1(b). (b) Broken leaf spring for research object. (c) Broken leaf springs on the right side of Fig. 1(b).

Figure 2 shows a schematic drawing that explains the arrangement of springs from number 1 to number 8. In this case the broken one is spring number 8 as shown in Figure 2.

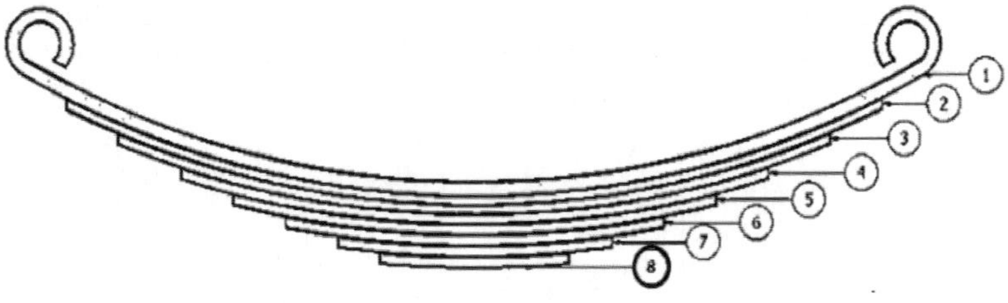

Figure 2. Schematic drawing of the arrangement of the leaf springs from No. 1 to No. 8 (broken spring).

Visual observations on the fracture surface of leaf spring evaluated were around the bolt hole area. From visual observation, it was found that there was an initial crack on the side of the bolt hole as shown in Figure 3. Furthermore, the beach mark can be seen on the fracture surface. This finding indicates that the spring has experienced fatigue fracture [8].

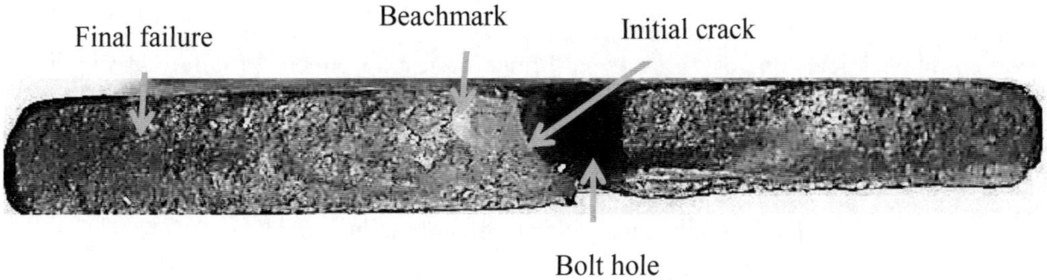

Figure 3. Picture of the fracture surface of the leaf spring of number 8.

Figure 4 shows that the image taken from initial crack tip area as shown in Figure 3. It can be seen that the crack initiationoccurred here which is characterized by relatively smooth fracture surfaces which indicates that fatigue cracking has occurred.

Figure 4. Fracture surface image taken from initial crack tip area of Figure 3.

Fracture surfaces of leaf springs (Figure 3 and 4) show that fatigue crack propagation has occurred which starts from the initial crack, then crack propagation occurs in line with increasing dynamic load so that the final fracture is split in two parts [8][9].

In this study also conducted a hardness test using Rockwell Type Zwick / Roell ZHR to measure hardness on the surface and core of leaf spring specimens [10]. The indentor used is a steel ball with a diameter of 1/16 in and 100 kg of loading. The hardness test specimen is shown in Figure 5.

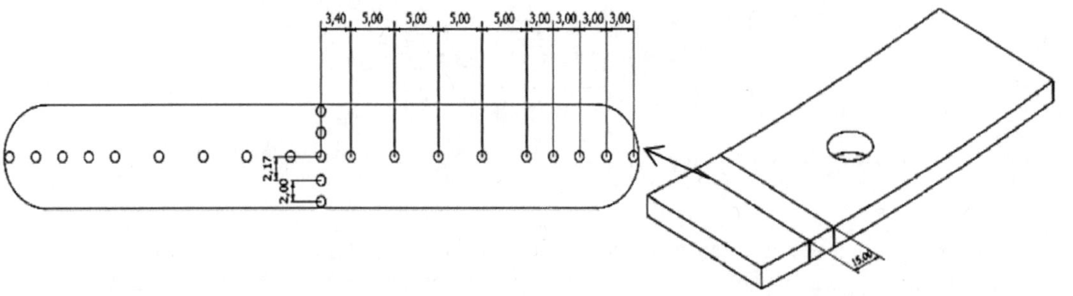

Figure 5. Specimen configuration with hardness testing points marked with small circles (o).

2.2. Material Properties Chemical Composition

Assessment of chemical composition was performed on test specimens of failure leaf spring. A small part of spring sample was taken from the broken spring and then tested. The elements of spring material are shown in Table 1. Furthermore, the results found comparing to the chemical composition of the ASTM standard AISI 5150. The mechanical properties of AISI 5150 as shown in Table 2.

Table 1. Chemical composition of the material (%wt).

Element	Test Result	AISI 5150 [9]
C	0,529	0,480 - 0,530
Si	0,293	0,150 - 0,300
Mn	0,762	0,700 - 0,900
P	0,0021	$\leq$ 0,0350
S	0,02	$\leq$ 0,04

Table 2. Mechanical properties of AISI 5150 steel.

Mechanical Properties of Materials	Value	Unit
Modulus Young (E)	140	GPa
Poison's ration ($\boldsymbol{\mu}$)	0,28	-
Density (ρ)	7850	kg/m^3
Yield Strength (σ_y)	360	MPa
Tensile Strength (σ_B)	675	MPa
Shear Modulus (S)	80	GPa
Fracture Toughness, K_{IC} [11]	23	MPa.m$^{1/2}$

2.3. Analytical Analysis

Analytical analysis is conducted to get the load distribution on each wheel, and then calculates to find the stress (σ) that occurs in the spring. By knowing the amount of stress that occurs in the spring, the value of stress intensity factor that occurs around the crack tip can be determined. As it is known that if there is a fine crack in a material such as a spring, then around the crack tip will appear the stress intensity factor (K_I). If the K_I value that occurs near the crack tip is greater than the value of fracture toughness (K_{IC}) of the spring material, it will cause crack propagation.

Based on this understanding, the stress intensity factor is calculated using the data available from the broken spring [12][13]. Furthermore, analytical solutions will be explained in detail in sub sections 3.3.

3. Results and Discussion

3.1. Hardness on the Leaf Spring

The hardness test results in the horizontal direction from point A to B and the vertical direction from point C to D as shown in Figure 7. The distance between test points is approximately 2 to 5 mm (see Figure 6.a). Hardness in the area close to the bolt hole or at the core is lower than the hardness value in the part close to the side of the spring surface. It shows that the toughness of the material in the center of the leaf spring is higher than the part on the outer side of the spring. Therefore this spring is very susceptible to fracture. This result is in accordance with the results reported by previous researchers [2][7].

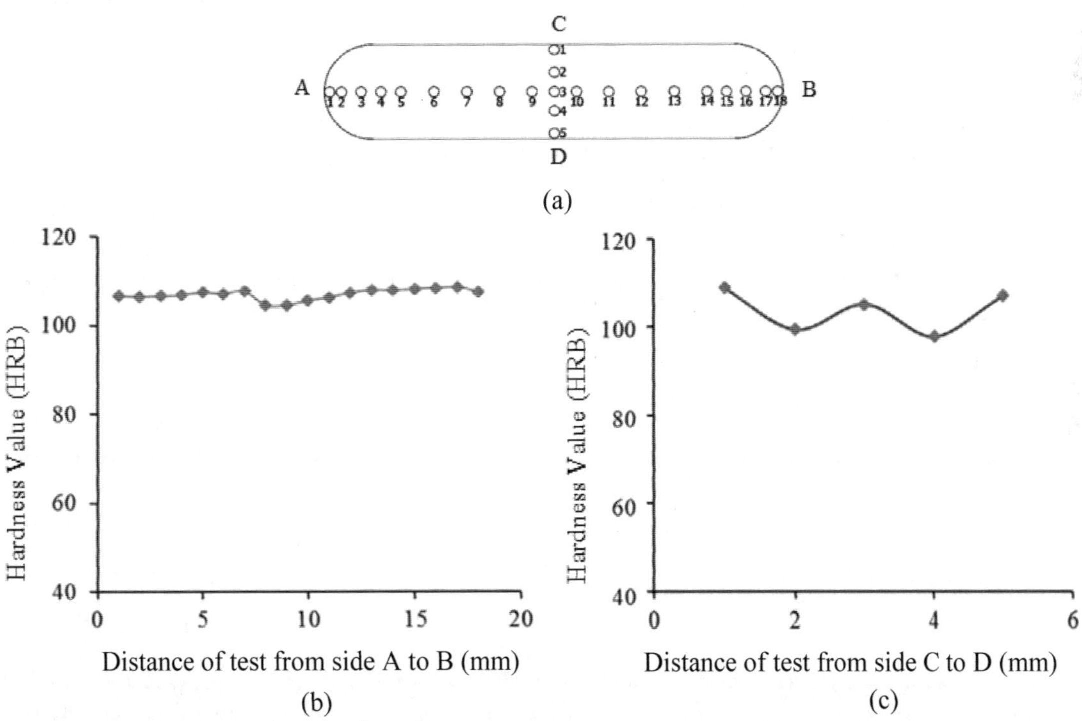

Figure 6. Curve of hardness on the Leaf Spring. (a) Hardness test point at the spring. (b) Curve of hardness in the inner section in the direction from A to B (see Figure 6 (a)). (c) Curve of hardness in the inner section in the direction from C to D (see Figure 6 (a)).

The leaf spring cross section hardness yields different values between the center and the edge side. Hardness test results show that, in the horizontal directions from A to B are 104.6 HRB for minimum value and 108.5 HRB for the maximum value. The difference in hardness value is insignificant only 3.6%. The same results are encountered in directions from C to D are 97.8 HRB minimum and 107.1 HRB maximum where the results remain under 10% namely 8.7%. The mean hardness was 106.5

HRB in the A to B direction and 102.5 HRB for the C side to D direction. The mean of the overall hardness value is 104.5 HRB. It can be concluded that there is no hardening that cause the leaf springs damaged. This matter as seen from hardness value is not significantly different between the points around the fracture and at the point on the edge.

The high hardness difference between the value of the standard and the test results can decrease the ductility. This may cause the spring elasticity function to tend to decrease and when the maximum loading on the spring will easily happened crack extension [7][13][14].

3.2. SEM Observation on the Fracture Surface

In reality, leaf springs as a suspension system in vehicles always receives dynamic loads and sometimes until they experience failure of fatigue fractures. Therefore, the fracture surface was necessary to investigated using scanning electron microscope (SEM) [14]. Microscopic observations with magnitudes of 6000 on the fracture surface of the spring identified that the crack occur around the bolt hole. The crack starts on the bolt portion then propagate to the cross section of the spring. Furthermoreit can be explained as in Figure 7. Where, Figure 7 shows a series of SEM images taken at the crack surface (Figure 7 (d)). The initial crack and the direction of crack propagation as shown in Figure 7 (a), then crack propagation occurred (Figure 7 (b)), until the end final fracture occurred as shown in Figure 7 (c).

3.3. Analytical Analysis

Analytical analysis aims to obtain the stress and the stress intensity factor value that appear due to the applied load working on the spring. The amount of force received by each suspension both front and rear is the total force divided by four wheels. For this purpose, it is needed the weight of unloaded truck of 3000 kg, the vehicle weight during operation is 11000 kg and maximum allowable vehicle load standard is 7500 kg. The load distributed on each wheels (*W*) is calculated using equation (1) [2][15].

$$W = \frac{m \times g}{4} \quad (N) \tag{1}$$

Where *m* is a vehicle weight of 11000 (kg), *g* is the gravity of 9.8 (m/s^2). The maximum load leaded by each wheelsis 26950N.

Then the stress (σ) that occurs in the spring can be calculated using equation (2) [15].

$$\sigma = \frac{6\,x\,W\,x\,L}{n\,x\,b\,x\,t^2} \quad (MPa) \tag{2}$$

Where *W* is the load worked on each wheels of 26950N, *L* is half span of the spring is 430 mm, *n* is the total number of leaves 8 pieces, *b* is width of the leaves 70 mm, ***t*** is thickness of master leaf 14.5 mm. From the calculation results obtained that the stress that occurs in the spring is 590.55 MPa. It can be seen here that the stress that occurs exceeds the yield stress of the material (σ_y) as shown in Table 1. Therefore, it can be stated that the spring will fail due to deformed.

Furthermore, it will also be examined from the crack analysis by finding out how much the stress intensity factor (K_I) occurs in the spring due to the load received. Using a Single-edge-crack-at-hole specimen as shown in Figure 8, the value of K_I is can be calculated by means of equation (3) [16].

$$K_I = \sigma\sqrt{(\pi a)}F_{hs} \tag{3}$$

IOP Conf. Series: Materials Science and Engineering **541** (2019) 012046 doi:10.1088/1757-899X/541/1/012046

Where a is the crack length. The crack length is obtained from the visual observation with the value is about 1.5 mm, σ is the stress that occurs in the spring as it found from equation (2). F_{hs} is the boundary-correction factor calculated referred to previous research methods [16]. Then, the results of boundary correction factor (F_{hs}) obtained of about 2.49. It was found that the K_I value occurred near the crack tip on the spring is about 63.95 MPa$\sqrt{m}$. This result shows that K_I is higher than the fracture toughness K_{IC} of AISI 5150 material equal to 23 MPa$\sqrt{m}$ (Table 2). Therefore, this can cause crack propagation occurred [11][13][15] as well as this fact indicates that this is one of the causes of leaf spring failure.

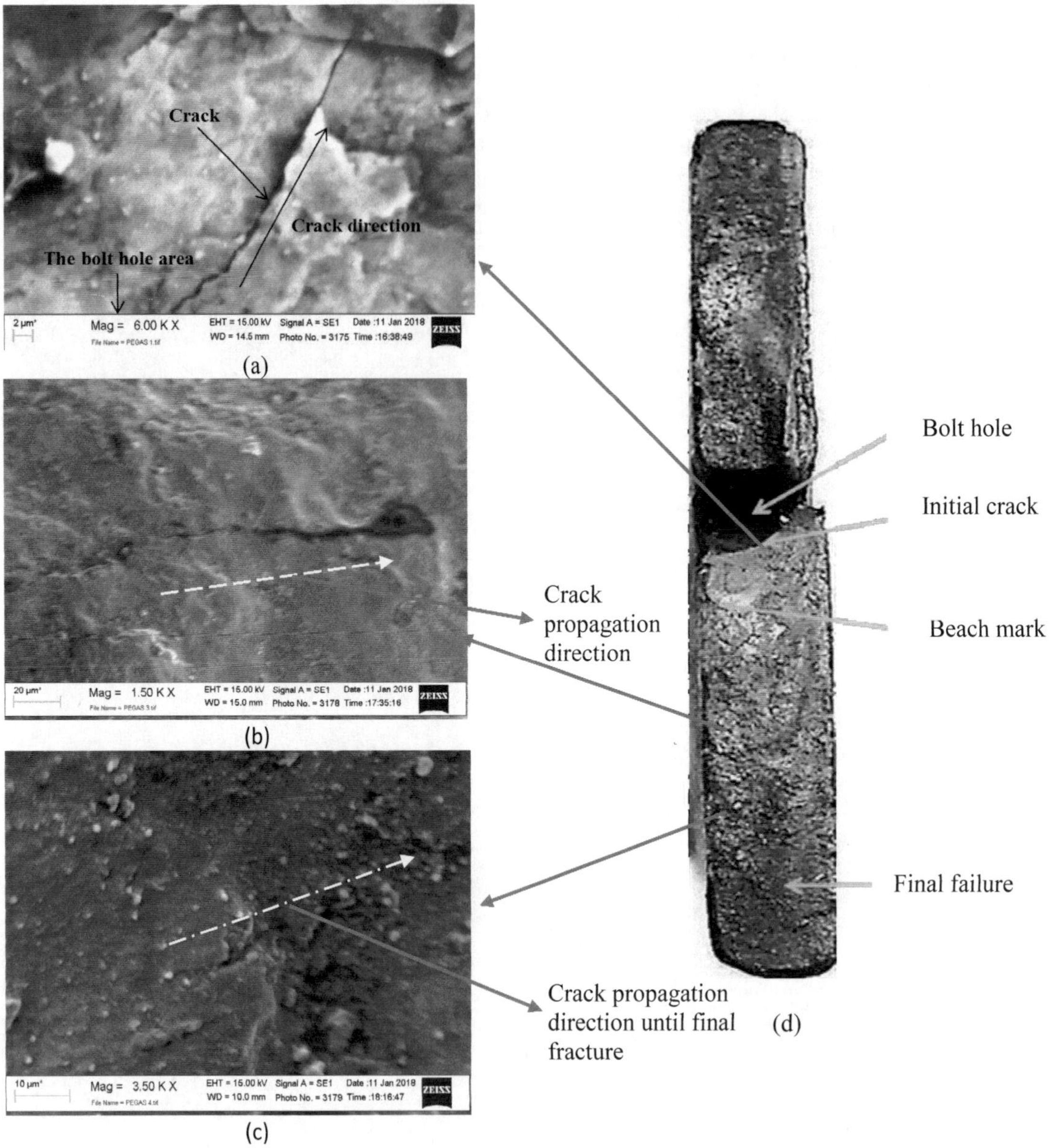

Figure 7. A series of SEM images taken at the crack surface. (a) The initial crack and the direction of crack propagation. (b) Crack propagation occurred. (c) Final fracture occurred. (d) Leaf spring fracture surface.

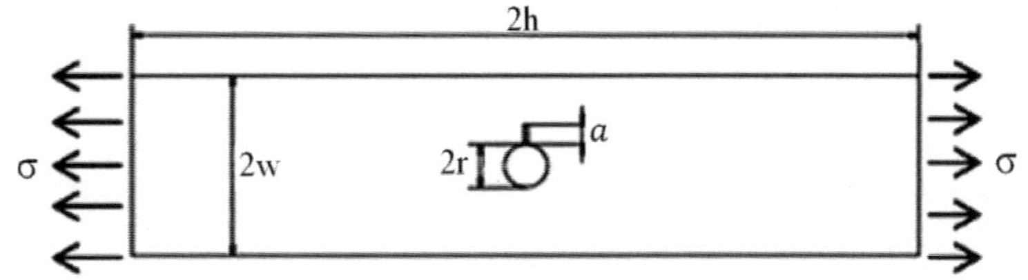

Figure 8. Configurations of Single-edge-crack-at-hole specimen (radius, $r = 5$ mm, width of the leaves, $w = 25$ mm).

4. Conclusions

Study on the leaf spring fracture behavior used in the suspension systems in the Diesel Truck vehicles has been studied and can be drawn several conclusions as follows:

1. From SEM observation, it was found that on the fracture surface of the spring identified the initial crack occur around the bolt hole. The crack extension starts on the bolt part then propagate to the cross section of the spring until final failure.
2. The value of stress intensity factor K_I is higher than the value of fracture toughness K_{IC} of AISI 5150 material. This result indicated that crack propagation occurred and this fact also indicated that this is one of the causes of leaf spring failure.
3. Fracture of the spring was also induced by fatigue fracture that characterized by existing the beach mark on the fracture surface.

References

[1] Aher V K and Sonawane P M 2012 *Int. J. of Eng. Res. and App. (IJERA)* **2** 1786
[2] Sepfitrahand R Y 2013 *JURNAL APTEK* **5** 2
[3] Todinov M T 1999 *Int. J. Mech. Sci.* **41** 357
[4] Husaini Nurdin A Agustian B 2016 *Proc. SNTR* III 111
[5] Husaini and Zuhaimi 2016 *Int. J. of Tech.* **3** 456
[6] Fonte M Anes V Duarte P Reis L Freitas M 2015 *Eng. Fail. Anal.* **56** 109
[7] Pradhan A Rathore Y 2016 *Int. J. of Late. Tren. In Eng. and Tech.* **7** 221
[8] Richard C R 2002 *2nd. Edt. Fatigue design handbook* Society of Automotive Engineers Inc
[9] Qian C Shi W Chen Z Yang Song 2017 *Eng. Fail. Anal.* **168** 40
[10] E Giannakis 2016 *IOP Conf. Ser.: Mater. Sci. Eng.* **161** 012065
[11] Samoila A 2011 *Liftning for lifting of the locomotive with bogies,* Karlstads University Press.
[12] Husaini Kishimoto K Hanji M Notomi M 2016 *ARPN J. of Eng. and Appl. Sci.* **11** 885
[13] Akhyar H and Husaini 2016, *Int. J. of Met. Cast.* **10** 452
[14] Husaini Kishimoto K 2001 *Proc. of SPIE - The Int. Soci. for Opt. Eng.* **4317** 111
[15] Khurmi RS Gupta JK 2005 *Machine Design* (India: Eurasia Publishing House Ltd New Delhi) 110055
[16] Newman Jr J C and Daniewicz S R 2014 *Eng. Fract. Mech.* **127** 252

Acknowledgements
The authors would like to express their gratitude to Syiah Kuala University for financial support for this research through the grant No. 03/UN11.2/PP/PNBP/SP3/2018.

IOP Conf. Series: Materials Science and Engineering **541** (2019) 012047 doi:10.1088/1757-899X/541/1/012047

Efficiency and characterization study of low voltage current sacrificial anode on Al-Zn-Cu and Al-Zn-Si alloy

D Ferdian[1], D Yudha R[2], M Ario W[1], T Yunita[2], and Sulistijono[2]

[1] Department of Metallurgy and Materials Engineering, Universitas Indonesia, Kampus Baru UI Depok, 16424, Indonesia

[2] Materials and Metallurgical Engineering Department, Kalimantan Institute of Technology, Kampus ITK Karang Joang, Balikpapan, 76127, Indonesia

E-mail: 06141008@itk.ac.id

Abstract. The sacrifice of the anode is the cathodic protection used to prevent corrosion. Al-Zn-In alloys are often used it causes excessive protection that can increase the possibility of hydrogen embrittlement in high-strength steel. Therefore, the solution is to use aluminum alloys with low voltage current. The purpose of this research is to know the efficiency of aluminum sacrificial anode with alloy elements Zn, Si and Cu as low voltage current. As well as studying the potential of Al alloys with elements Zn, Cu and Si can be used as sacrificial anodes at sea water conditions. Testing methods performed are Optical Emission Spectroscopy testing, weight loss testing, efficiency testing, Electrochemical Impedance Spectroscopy testing, and metallographic testing. From this research, we found the result of Al-Zn-Cu alloy which has electrochemical capacity and best efficiency is Al-5Zn-0,1Cu Alloy for 2795 Ah / kg with 98% efficiency and Al-Zn-Si alloy with capacity value electrochemistry and the best efficiency is Al-5Zn-0.1Si alloy for 2808 Ah / kg with 98% efficiency. The lowest value of transfer resistance is found in Al-5Zn-0.1Cu alloy and Al-5Zn-0.1Si alloy with the value of 1012.7 Ω and 1092.9 Ω and the double capacitance value in the smallest Al-Zn-Cu alloy is alloy Al-5Zn-0.5Cu with the value of 13.2 μMho while in Al-Zn-Si alloys are present in Al-5Zn-0.5Si with the value of 19 μMho. With the structure obtained is the greater the addition of Cu and Si, the more rough and corroded surface sacrificial anodes.

1. Introduction

Corrosion is a very important problem throughout the world that greatly affects the natural and industrial environment. Corrosion and pollution are interconnected where more pollution accelerates corrosion and results as rust where both of these will affect the efficiency of industrial development and the resilience of infrastructure asset [1]. Corrosion prevention can be done in various ways such as inhibition, cathodic protection of the material, material selection, coating and so forth. Cathodic protection is the reduction or elimination of corrosion by making metals into cathodes by impressed currents [2]. The sacrifice of the anode means sacrificing the anode so the anode is corroded and protect the cathode metal. The addition of alloying elements to aluminum will change sacrificial anodes properties. Addition of indium to aluminum alloy will make the electrode potential more negative and current efficiency will decrease more. The indium contained in the aluminum anode will also increase its activity, so that its electrochemical performance will improve [3]. However, the addition of Ti and Zr as a very high quantity purifying counter is contra-productive. Bifilms are precipitated prematurely from molten alloys with excess Ti or Zr prior to pouring, exhaust alloying and corrosion distribution in corrosion areas [4]. The addition of Ga or / and Bi into Al-Zn-Sn alloys can improve the uniformity distribution of the compacted structure. In addition, the addition of Ga or / and Bi can improve the efficiency of current in Al-Zn-Sn [5]. Furthermore, the addition of copper in Al-5Zn alloy showed that copper will increase the corrosion rate and the potential of the alloy, and also will reduce the charge transfer of the Al-5Zn alloy [6]. In this research we will use weight loss testing method and efficiency test to find the efficiency that can be produced by low voltage current anode at Al-Zn-Cu and Al-Zn-Si alloy, material characterization by optical emission spectrometer to know chemical composition of sacrificial anode as well the electrochemical impedance spectroscopy test and metallography to determine the value of electrical elements and the morphological structure of low voltage current anodes on Al-Zn-Cu and Al-Zn-Si alloys.

2. Experimental

2.1. Materials
The material used for this research was cast from high purity Al, Zn and Cu ingot with 99% grade and also Si and In in graphite crucible under atmospheric condition. The material that will be prepared for are Al-5Zn-0.5Cu, Al-5Zn-0.3Cu, Al-5Zn-0.1Cu Al-5Zn-0.5 Si, Al-5Zn-0.3 Si, Al-5Zn-0.1 Si and Al-5Zn-0.02 In. The aluminum is fed into the graphite crucible and fused in a furnace with a temperature of 750°C for 90 min. After the aluminum melting, furnace temperature is decreased to 710°C follow by alloying process by weighted amount of Zn, Cu, and Si added to the molten aluminum. During the blending process, the melted metal is stirred manually for 20 seconds after 15 minutes it is poured into mold.

2.2. Materials Characterization
The alloying will be characterized for the chemical composition, metallographic test and electrochemical impedance spectroscopy (EIS). Chemical composition was performed by using optical emission spectrometer (OES). After that, the sample preparation will be performed for metallographic testing according to the standard. The corrosion behavior of the alloys was also tested by using electrochemical test technique in seawater at room temperature. EIS were performed at range of frequency of 100 kHz-10mHz.

2.3. Efficiency Testing
This efficiency test was performed using DNV RP-B401 standard. The circuit is composed of Al alloy anodes, steel cathodes, aerators, and seawater. The sacrificial anode shall be positioned at the center of the steel cathode as well as the temperature for this test shall be maintained, at 20°C with a tolerance of 3°C. The principle of weight loss was used for testing this efficiency, so that the initial and final weights are calculated and can be determined. Measurements of potential values were performed according to DNV RP - B401 standard, where the current is given from the power supply and the measurement is carried out for 24 hours for 4 days. After the efficiency testing is completed, the specimen will be cleaned again with running water and soaked in a solution of 20 grams of chromium trioxide and 30 ml of phosphoric acid per liter of water for 10 minutes at a temperature of 80°C. The specimen is then washed with water and alcohol and then dried again. The final weight calculation is performed after the cleaning of the specimen is performed. The weight calculation was performed on each specimen of Al-sacrificial anode Al, Cu rod, Cu plate on coulometer to know the final weight and can be calculated to lose weight.

3. Result and Discussion

3.1. Electrochemical Impendence Spectroscopy

EIS test was performed to analyze the surfaced electrochemical behavior. In Figure 1 show the nyquist plot result from this research.

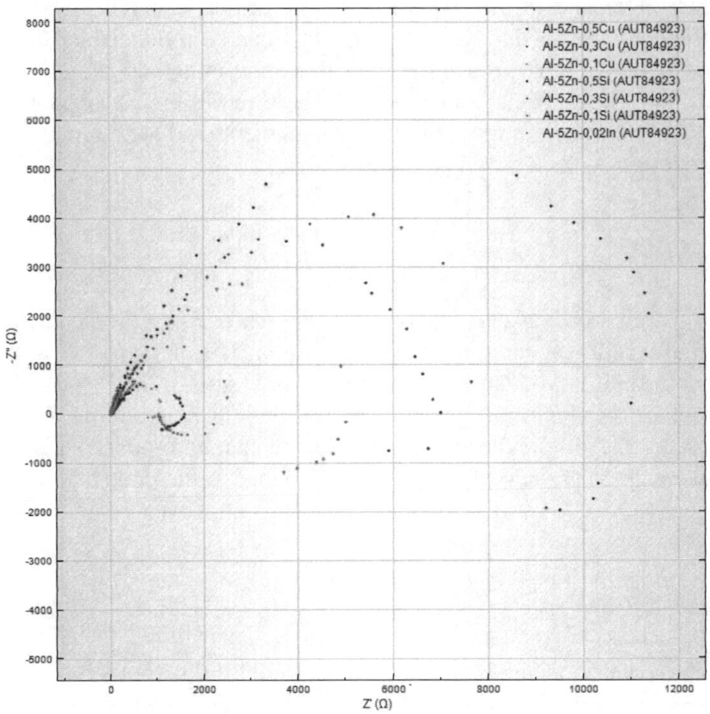

Figure 1. EIS result from Al-Zn-Cu, Al-Zn-Si and Al-Zn-In Alloy.

Based on the graph of nyquist plot Figure 1 we can see that the more elements of Cu and Si, the greater the impedance value obtained in each sacrificial anode. This proves that the more addition of Cu and Si elements in the aluminum sacrificial anode make corrosion resistant of the material higher. We observed a semicircle curve in the plot nyquist graph where the larger the diameter of the semicircle curve the greater the value of charge transfer resistance (Rct) in the sacrificial anode. The value of the electrical elements obtained can be seen in Table 1.

Table 1. EIS Fitting Circuit Result.

Alloy	R_{ct} (Ω)	N	C_{dl}/CPE (μMho)
Al-5 Zn- 0.5 Cu	6623.4	0.90722	13.2
Al-5 Zn- 0.3 Cu	8770.5	0.83448	22.5
Al-5 Zn- 0.1 Cu	1012.7	1.1	40.4
Al-5 Zn- 0.5 Si	10887	0.81177	19
Al-5 Zn- 0.3 Si	1189.9	0.9768	33.8
Al-5 Zn- 0.1 Si	1092.9	0.051251	108
Al-5Zn-0.02In	692.43	0.8651	35.7

3.2. Metallographic Examination

The aluminum sacrificial anode microstructure was observed using an optical microscope in as-cast condition. Figure 2 shows the microstructure formed on the aluminum sacrificial anode with the variation of Cu and Si concentrations and the sacrificial anode of aluminum with the addition of the In element. From the microstructure we can see that bright color is the grain boundary of the sacrificial anode alloy, while the aluminum has a darker color than the grain boundary. The precipitates formed at the grain boundary have white color. Because of the addition of elements Cu and Si, the formation of precipitate appear in the grain boundary, where the more precipitate appear in the material, the material will be more easily corroded.

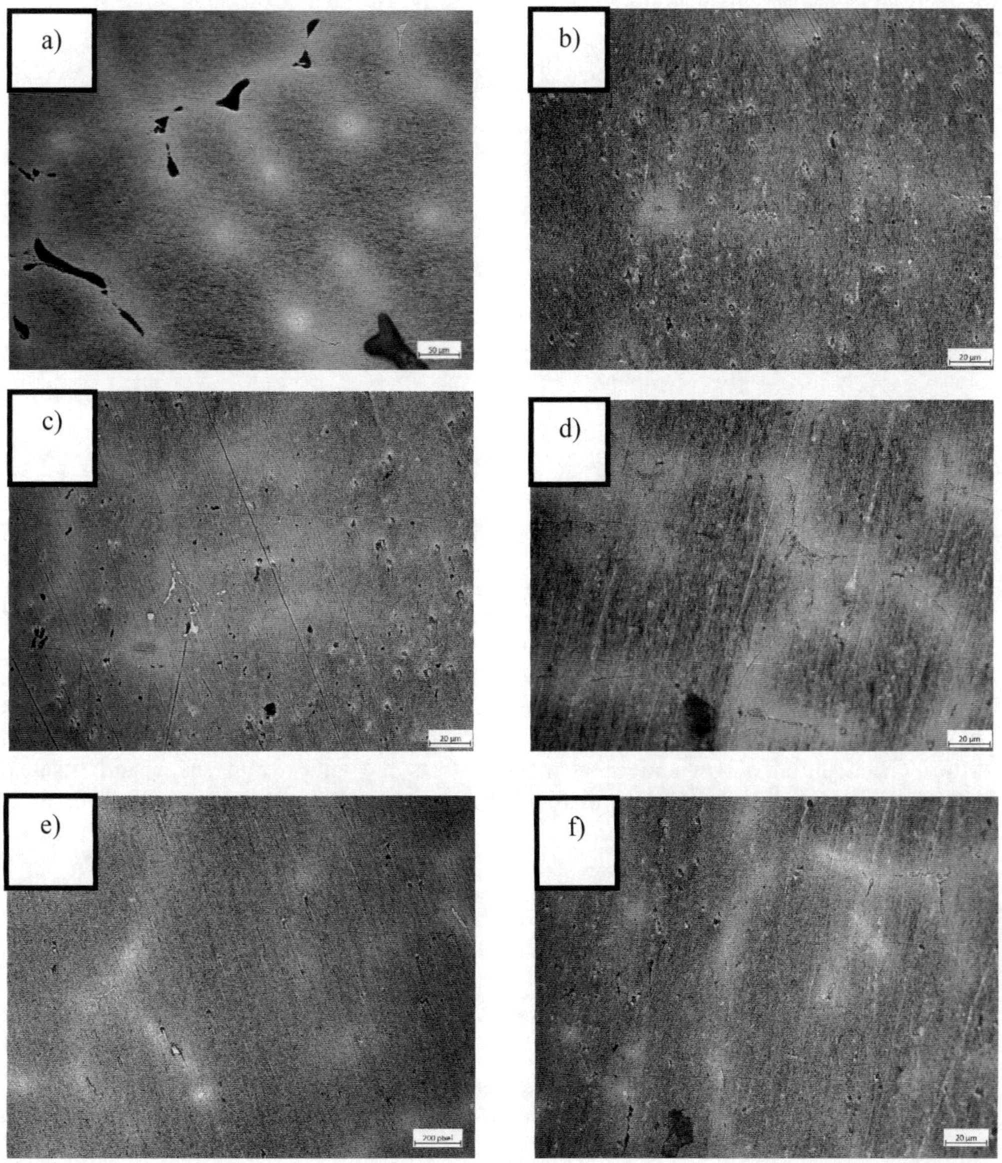

Figure 2. Metallographic result of a)Al-5Zn-0.1Cu b) Al-5Zn-0.3Cu c) Al-5Zn-0.5Cu d) Al-5Zn-0.1Si e) Al-5Zn-0.3Si f) Al-5Zn-0.5Si that showed precipitation (200x magnification).

3.3. Efficiency Testing

The sacrificial anode efficiency testing was performed using the sacrificial anode samples after casting with diameter of 10 mm and length of 50 mm. The external electric current applied to the sacrificial anode is determined using the surface area of the victim anode sample. From Table 2 it can be seen that the increasing levels of Cu and Si in the alloy composition decrease the electrochemical capacity. The value of electrochemical capacity is the sacrificial anode's performance capability in protecting or protecting a protected structure. The higher the electrochemical capacity, the better the sacrificial anode protects the structure. In Al-Zn-Cu alloys, the highest and best value electrochemical capacity is the Al-5Zn-0.1Cu alloy with the value of electrochemical capacity of the test of 2795 Ah/kg, while in Al-Zn-Si alloy, the electrochemical capacity value highest and best is Al-5Zn-0.1Si alloy with electrochemical capacity test value of 2808 Ah / kg.

Table 2. Value of Electrochemical Capacity Testing, Theoretical Capacity and Efficiency of Sacrificial Anodes.

Alloy	Theoretical Capacity (A.h/kg)	Electrochemical Capacity Testing (A.h/kg)	Anode Efficiency (%)
Al-5Zn-0.5Cu	2847	1853	65
Al-5zn-0.3Cu	2860	2024	71
Al-5Zn-0.1Cu	2867	2795	98
Al-5Zn-0.5Si	2877	2035	71
Al-5Zn-0.3Si	2870	2676	93
Al-5Zn-0.1Si	2867	2808	98
Al-5Zn-0.02In	2872	2399	84

4. Conclusion

Al-Zn-Cu alloy which has the best value of electrochemical capacity and efficiency is Al-5Zn-0.1Cu alloy for 2795 Ah / kg with 98% efficiency and Al-Zn-Si alloy which has the best electrochemical capacity and efficiency value. The lowest charge transfer resistance is found in Al-5Zn-0.1Cu alloys and Al-5Zn-0.1Si alloys of 1012.7 Ω and 1092.9 Ω. The double capacitance value of the smallest Al-Zn-Cu alloy is Al -5Zn-0.5Cu of 13.2 μMho while in Al-Zn-Si alloys are present in Al-5Zn-0.5Si of 19 μMho. The addition of Cu and Si elements affects the alloy grain shape and gives rise to precipitates. Higher additions of Cu and Si elements make the grain shape smoother and make precipitates appear more at grain boundaries.

References

[1] Valdez B Schorr M Zlatev R Carrillo M S Alvarez L Eliezer A and Rosas N 2012 *Corrosion Control in Industry* InTech
[2] Popov B N 2015 *Corrosion Engineering Principles and Solved Problems* Columbia Elsevier B V
[3] Hejian S and Shizhong H 1990 *Chin. J. Oceanol. Limnol* **8** 354
[4] Sina H, Emamy M, Saremi M, Keyvani A, Mahta M and Campbell J 2006 *Mater. Sci. and Eng A* **431** 263
[5] Jun-guang H, Jiu-ba W, Xu-dong L, Guo-wei W and Chun-hua X 2011 *Trans. Nonferrous Met. Soc. China* **21** 1580
[6] Ferdian D, Pratesa Y, Togina I, and Adelia I 2017 *Procedia Eng.* **184** 418

Acknowledgements
We would like to give our thanks and appreciation to Hibah PITTA 2018 Nomor: 2413/UN2.R3.1/HKP.05.00/2018 for supporting the funding in this research.

IOP Conf. Series: Materials Science and Engineering **541** (2019) 012048 doi:10.1088/1757-899X/541/1/012048

Synthesis and characterization of nano rosette TiO_2 and $CH_3NH_3PbCl_2I$ for potential use in perovskite solar cell

N Sofyan[1], A Ridhova[1], Salman[1], A H Yuwono[1], A Udhiarto[2]
[1]Department of Metallurgical and Materials Engineering, Faculty of Engineering, Universitas Indonesia, Depok 16424 Indonesia
[2]Department of Electrical Engineering, Faculty of Engineering, Universitas Indonesia, Depok 16424, Indonesia

E-mail : nofrijon.sofyan@ui.ac.id

Abstract. The characteristics of nano rosette TiO_2 hydrothermally grown on a glass substrate with potential application in perovskite solar cell has been examined. Nano rosette TiO_2 was synthesized through deposition on top of a fluorine-doped tin oxide glass substrate via hydrothermal reaction at 170°C for 6 hours. Formation and growth mechanism were characterized using X-ray diffraction (XRD), whereas the morphology was examined using a field emission scanning electron microscope (FE-SEM). The results showed that the formation of nano rosette TiO_2 has been completed after 6 hours. Structural study from X-ray diffraction also showed that the crystal structure formation has been completed after 6 hours. This nano rosette TiO_2 was tested for its performance as an electron transport layer (ETL) in the perovskite solar cell device. For this purpose, the perovskite $CH_3NH_3PbCl_2I$ 0.8 M was deposited on top of the ETL layer and the performance was tested using a semiconductor parameter analyser (SPA). The result showed an efficiency of about 3.4%, which is promising for the next development.

1. Introduction

High energy requirements continue to drive intensive research in the field of solar cells. One of the latest generations of solar cells is perovskite solar cell (PSC), some of which still focus on the utilization of perovskite structured compounds, especially of lead-inorganic organic hybrids [1]. Perovskite compound was discovered in 1839 by Gustav Rose on the chemical compound $CaTiO_3$ with the general formula of ABX3, where A and B are cations of normally different radius sizes, whereas X is an anion binding to both cations [2]. The compound with this perovskite structure could be used as an active layer of light catcher in a solar cell device, which is then converted into electrons to generate electrical current.

The efficiency of the PSC has grown tremendously since it was first recorded in 2009 of 3.8% until 2017 at around 22.7% and it is still expected to continue to increase as new progress in the area emerges [3]. Many investigators have put their efforts in the improvement of the PSC. For examples, Xiao et al. [4] have reported that the use of a combination of TiO_2 compact layers with the growth of TiO_2 nanorods as electron transporting layer (ETL) and partial substitution of CH_3NH_3I (MA-I) with MA-Br was able to increase the efficiency of PSC. Govindasamy et al. [5] reported that the addition of acetic acid to TiO_2 during chemical bath deposition process decreased the activation energy of the material. Liu and Aydil [6] have also reported that the addition of NaCl and two-step hydrothermal route could restraint the growth of nanorods during the synthesis and improved the electron transfer efficiency of the PSC device. In another work, Jin et al. [7] have also found that by partial substitution of PbI_2 with 3% $ZnCl_2$ in the perovskite layer was capable of increasing the electron transport properties and increasing the efficiency of the PSC device.

Based on the aforementioned description, it can be understood that the structure and orientation of the semiconductor material in the PSC can improve the electron transfer process and thus the efficiency of the PSC device. Meanwhile, efficiency can be increased even more by substituting partial substitution of perovskite coating compounds. Therefore, a method that can combine the existing methods to improve PSC efficiency is needed. In this work, we combined the two methods, namely the use of combination of compact layers of TiO_2 nano rosette as an ETL and partial substitution of perovskite coating compounds by mixing the precursors of lead (II) chloride ($PbCl_2$) and methylammonium iodide (CH_3NH_3I) to form $CH_3NH_3PbCl_2I$.

2. Experimental Setup

2.1. Materials

The precursors were titanium tetra isopropoxide (TTIP, ≥97.0%, Sigma-Aldrich) and hydrochloric acid (HCl 37%, Merck) all with analytical grade reagents and used without further purification. For the solvents, acetone, ethanol and distilled water were used throughout the experiment except stated otherwise.

2.2. Preparation of Nano Rosette TiO_2

The synthesis of nano rosette TiO_2 was performed through deposition on top of a fluorine-doped tin oxide (FTO, 30 Ω/cm^2) glass substrate via modified hydrothermal reaction similar to the previously reported procedure by others [4]. The process was begun by mixing distilled water with hydrochloric acid in a beaker glass at a fix volume ratio of 1:1 to reach a total volume of 40 mL at room temperature condition. The solution was mixed under a magnetic stirrer for 5 minutes before being added with a volume of 450 μL TTIP and the agitation was continued for another 5 minutes to get a clear solution.

Meanwhile, the glass substrate was prepared by firstly cleaning it ultrasonically in a mixed solution of distilled water, acetone, and ethanol with volume ratios of 1:1:1 for 10 minutes. The cleaned substrate was placed horizontally at the bottom of 100 mL Teflon-lined stainless-steel autoclave before being poured with the previous TTIP prepared solution. The synthesis was conducted at 170°C for 6 hours in a vacuum oven (DZF-6050, Berkeley Scientific). After the synthesis, the autoclave was cooled to room temperature under flowing water for approximately 5 minutes. The glass substrate was then taken out and rinsed with distilled water and allowed to dry in air for 15 minutes before being calcined using hot-plate (Cimarec SP 131320-33, Thermo Scientific) at 450°C for 90 minutes and ready for the characterization.

2.3. Device Fabrication

The perovskite layer acting as an absorber was deposited on top of nano rosette TiO_2 layer through a solution drop and spin-coating process (K-359SD-1 Spinner, Kyowa Riken). Perovskite $CH_3NH_3PbCl_{3-x}I_x$ 0.8 M where $x = 1$ was obtained by mixing precursors of lead (II) chloride ($PbCl_2$, Merck) and methylammonium iodide (CH_3NH_3I, Greatcell Solar) with a mass ratio of 1 : 3, i.e. 278 mg of $PbCl_2$ and 470 mg CH_3NH_3I in 5 mL dimethylformamide (DMF, Merck) at 80°C under a magnetic stirrer for 2 hours. A 50 μL of this perovskite solution was dropped onto synthesized nano rosette TiO_2 layer and allowed to stand for 1 minute before being spin-coated at 3000 rpm for 30 seconds to remove the residual solvent. The perovskite layer from the spin-coating deposition was the calcined in a muffle furnace (LEF-1xxS type, Labtech) at 150°C for 60 minutes.

The hole transport layer was further deposited on top of the perovskite layer. This layer consisted of 80 mg of Spiro-OMeTAD (Greatcell Solar) dissolved in 1 mL of chlorobenzene. This solution was dropped on top of the perovskite layer and spin-coating was performed at 3000 rpm for 30 seconds.

The last layer, the gold metal electrode, was deposited using sputter coater (SC7620 Sputter Coater, Quorum) under a pressure of 10^{-1} mbar and a current of 10 mA for 120 seconds following the pattern on the FTO glass. The device was then ready for performance characterization.

2.4. Characterization

The morphology of the reaction product was examined using a field emission scanning electron microscope (FE-SEM, FEI Inspect F50) whereas the structural characteristics of the sample was observed using an X-ray diffractometer (XRD, PANalytical X'Pert PRO) operating at 40 kV 30 mA with Cu Kα radiation (λ = 1.5406 Å). Device performance was measured through an I-V curve characteristic using a semiconductor parameter analyser. All characterizations were performed under an ambient condition with no control on environmental humidity and or pressure.

3. Results and Discussion

The morphology of TiO_2 hydrothermally synthesized at 170°C for 6 hours was studied through secondary electron images. Figure. 1 shows the morphology of the product in the forms of nano rosette TiO_2 under different conditions. As can be seen in Figure. 1, complete formation of nano rosette has been obtained at 6 hours of hydrothermal reaction time. Figure. 1(a) and 1(b) are the as-synthesized nano rosette TiO_2 at two different magnifications, whereas Figure. 1(c) and 1(d) are morphology of the nano rosette TiO_2 after being calcined at 450°C for 90 minutes. Not much differences that can be observed from these two morphologies except that at higher magnification, the one that has been calcined has sharp and smooth surface morphology. This could be because of the calcination process has resulted in the crystalline form of the nano rosette. After the completion of reaction time, the petal of the rosette has reached an average size of about 200 nm. These results are in agreement with our previous results [8].

(a) (b)

IOP Conf. Series: Materials Science and Engineering **541** (2019) 012048 doi:10.1088/1757-899X/541/1/012048

(c) (d)

Figure 1. Secondary electron image of nano rosette TiO_2 hydrothermally synthesized at 170 °C for 6 hours: (a) as-synthesized at low magnification, (b) as-synthesized at high magnification, (c) calcined at 450 °C for 90 minutes at low magnification, and (c) calcined at 450 °C for 90 minutes at high magnification. Low magnification has the bar scale of 10 μm and high magnification has the bar scale of 10 μm.

In order to examine that the nano rosette has formed a more crystalline phase after the calcination process, X-ray diffraction examination has also been performed and the results are given in Figure. 2. As can be seen in Figure. 2(a), the pattern is X-ray diffractogram of rutile TiO_2 reference (JPDS 01-082-0154), Figure. 2(b) is diffractogram of the as-synthesized of hydrothermally grown TiO_2, and Figure. 2(c) is the calcined diffractogram of hydrothermally grown TiO_2.

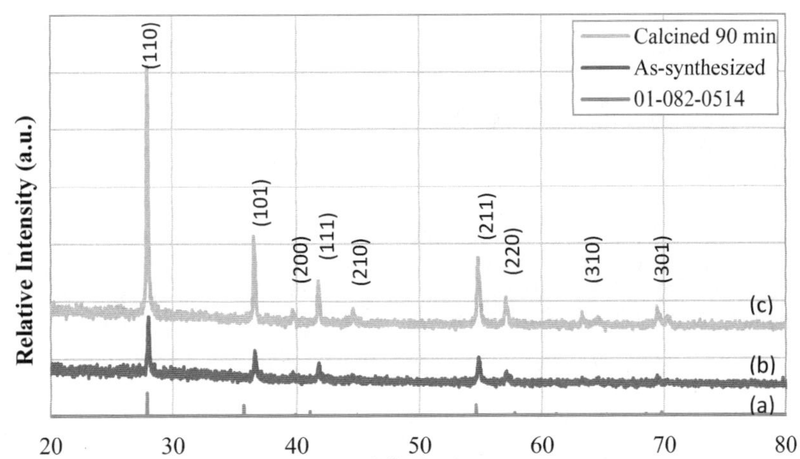

Figure 2. X-ray diffractograms of (a) rutile TiO_2 reference (JPDS 01-082-0154), (b) as-synthesized, and (c) calcined at 450°C for 90 minutes of the hydrothermally grown nano rosette TiO_2 for 6 hours.

The diffractogram of the as-synthesized nano rosette TiO_2, as can be seen in Figure. 2(b), the phase form has shown a complete structure of crystalline phase but it is still with low intensity. All peaks have been identified and show the trend and confirmed the pure rutile crystal structure (JCPDS file No. 01-082-0514) indexed to tetragonal $P4_2/mnm$. Some peaks, however, shift to high angle due to the differences in the lattice parameters. There are four distinct diffraction peaks observed at 2□ 27.659°, 36.579°, 41.742°, and 54.776° corresponding to (110), (101), (111), and (211) of rutile crystal planes, respectively. No other characteristic peaks of other phases and or impurities are detected in the diffractograms. This tendency would be understandable since the formation of nano rosette TiO_2 has been completed. After calcination, the intensity has increased and has shown most of the corresponding peaks. Calculation using Scherrer formula [9] results in an average crystallite size of about 77 nm.

For the peak shifting, refinement of the crystal structure has been carried out using an initial input from the known lattice parameters of the rutile reference with a = 4.508 Å and c = 3.027 Å. The results of calculation showed that the lattice parameters of the obtained nano rosette TiO_2 are a = 4.557(6) Å and c = 2.940(5) Å. This finding agrees with the results found by others [10].

Morphology of the perovskite layer deposited on top of the nano rosette TiO_2 was observed before being coated using gold metal electrode. Figure. 3a shows low magnification backscattered electron images (bar scale 100 μm) of the perovskite layer coated on top of the nano rosette TiO_2 hydrothermally synthesized at 170°C for 6 hours. Inset in Figure. 3a is high magnification backscattered electron image with bar scale 2 μm. The morphology reveals the profile of the perovskite layer on top of the nano rosette TiO_2. As can be seen from the Figureure, the formation of perovskite layer has not been completely homogeneous indicated by uncovered area of nano rosette TiO_2. Moreover, as can be seen in Figure. 3(b), the spectrum of energy dispersive X-ray spectroscopy from the backscattered image is still showing the presence of titanium element indicated that some areas on the surface of the electron transport layer have not been fully covered by the perovskite layer. Hence, this deposition process still needs to be further optimized. Nonetheless, the performance of this result has been tested and the result is given in Figure. 4.

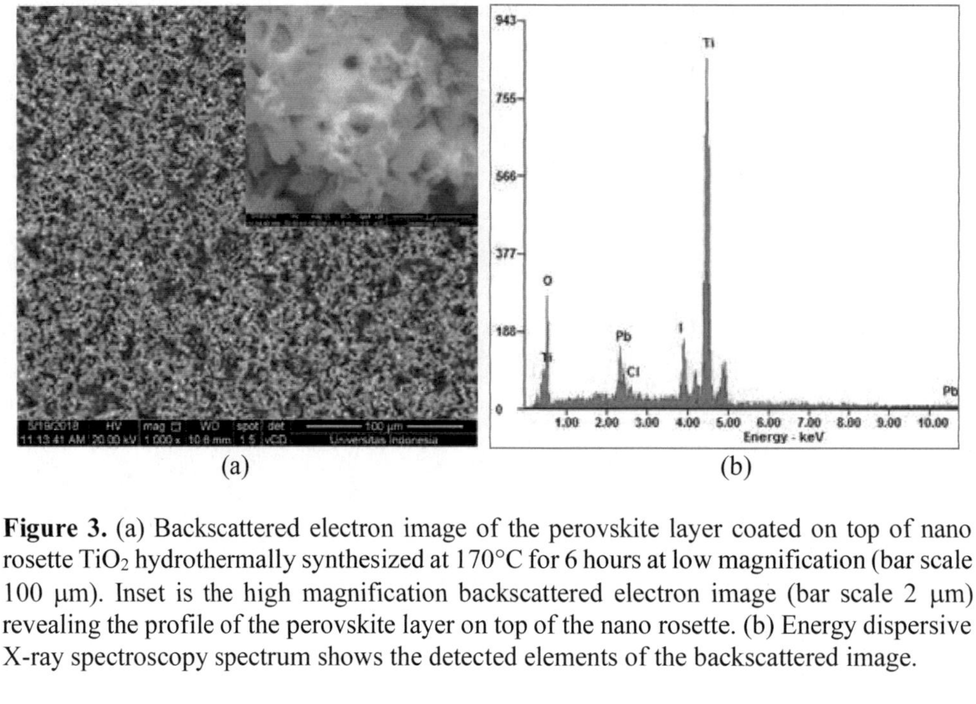

(a) (b)

Figure 3. (a) Backscattered electron image of the perovskite layer coated on top of nano rosette TiO_2 hydrothermally synthesized at 170°C for 6 hours at low magnification (bar scale 100 μm). Inset is the high magnification backscattered electron image (bar scale 2 μm) revealing the profile of the perovskite layer on top of the nano rosette. (b) Energy dispersive X-ray spectroscopy spectrum shows the detected elements of the backscattered image.

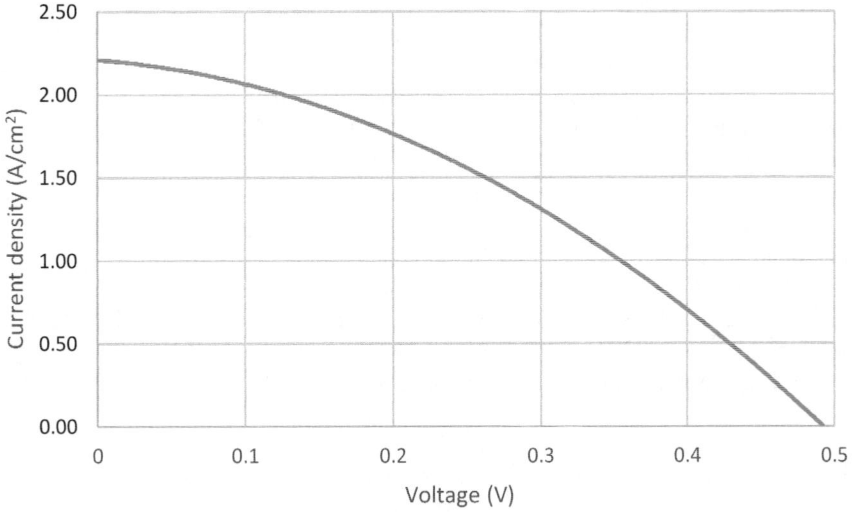

Figure 4. I-V curve characteristics of the PSC device fabricated from nano rosette TiO_2 hydrothermally synthesized at 170 °C for 6 hours.

Power conversion efficiency was calculated based on this I-V curve characteristic, in which [11]:

$$\eta = \frac{FF \times I_{SC} \times V_{OC}}{I_{in}} x100 \quad (1)$$

where I_{sc} is the short-circuit photocurrent density (A cm^{-2}), V_{oc} is the open-circuit voltage (volts), I_{in} is the intensity of the incident light (W cm^{-2}) and FF is the fill factor defined as

$$FF = \frac{I_{max} V_{max}}{I_{SC} V_{OC}} \quad (2)$$

where I_{max} and V_{max} are maximum photocurrent and voltage, respectively. Their values can be extracted from the maximum power of the I-V characteristics. Based on the data obtained from the photocurrent-voltage examination of the PSC device, the maximum power conversion efficiency (PCE) is found to be 3.4%. Compared to the existing efficiency [3], this value is still relatively low. This low value is expected because of inhomogeneous distribution of the perovskite layer that is yet to be improved, however, this current result is convincing and promising for the next development.

4. Conclusion

The characteristics of nano rosette TiO_2 grown on glass substrate via hydrothermal at different reaction time for potential use as electron transport layer in perovskite solar cell have been examined. Secondary electron images showed that the formation of crystal structure has been completed after 6 hours. X-ray diffractograms supported the morphological study in which after 6 hours of hydrothermal reaction time, the formation of nano rosette has been completed indicated by high intensity of the crystal structure indexed to rutile $P4_2/mnm$ with lattice parameters of $a = 4.557(6)$ Å and $c = 2.940(5)$ Å. The performance of the materials used as electron transport layer in PSC tested using a semiconductor parameter analyser showed an efficiency of about 3.4%.

References

[1] Niu G, Guo X and Wang L 2015 *J. Mater. Chem.* A **3** 8970

[2] Giorgi G and Yamashita K 2015 *J. Mater. Chem.* A **3** 8981
Shaikh J S, Shaikh N S, Sheikh A D, Mali S S, Kale A J, Kanjanaboos P, Hong C K, Kime J H, Patil P S, *Mater. Design* 2017 **136** 54
[3] Xiao G, Shi C, Li L, Zhang Z, Ma C, and Lv K 2017 *Ceram. Int.* **43** 12534
[4] Govindasamy G, Murugasen P, Sagadevan S 2016 *Mater. Res.* **19**(2) 413
[5] Liu B and Aydil E S 2009 *J. Am. Chem. Soc.* **131** 3985
[6] Jin J, Li H, Chen C, Zhang B, Xu L, Dong B, Song H, and Dai Q 2017 *ACS Appl. Mater. Interfaces* **9** 42875
[7] Sofyan N, Ridhova A, Yuwono A H, Wu J 2018 *Int. J. Technol.* **9**(6) 1196
[8] Patterson A L 1939 *Phys. Rev.* **56** 978
[9] Facchin G, Carturan G, Campostrini R, Gialanella S, Lutterotti L, Armelao L, Marcì G, Palmisano L and Sclafani A 2000 *J. Sol-Gel Sci. Technol.* **18** 29
[10] Fernando J M R C and Senadeera G K R 2008 *Curr. Sci.* **95** 663

Acknowledgments

This work is funded by the Directorate of Research and Community Services (DRPM) Universitas Indonesia through the grant of Hibah PITTA No. 2504/UN2.R3.1/HKP.05.00/2018.

IOP Conf. Series: Materials Science and Engineering **541** (2019) 012049 doi:10.1088/1757-899X/541/1/012049

The microstructure and mechanical hardness of cast Ti-30Nb-5Sn after solution treatment

D R Adiningsih[1], **E P Utomo**[1], **Anawati**[1]
[1]Department of Physics, Faculty of Mathematics and Natural Sciences, Universitas Indonesia, Depok 16424, Indonesia.

E-mail: anawati@sci.ui.ac.id

Abstract. This work reports the investigation on the development of microstructure and the resulting mechanical hardness of a metastable β Ti-30Nb-5Sn alloy as a result of solution treatment. The solution treatment was conducted at 1000°C for 5 h followed by water quenching. The microstructure was studied by a field emission-scanning electron microscope (FE-SEM) which equipped with EDX analysis for identifying the elemental composition. The phase change occurred in the alloy was analyzed by XRD while the mechanical hardness was measured by Vickers's hardness technique. The results showed that the solution treatment induced the nucleation of intragranular α-phase precipitation in the alloy which formerly composed of only single β-phase with dendritic structure in the as-cast condition. The α-phase exhibited a micro size plate-like morphology distributed randomly in the metallic grain and along grain boundaries of the solution-treated alloy. The α phase precipitate contained a significantly lower Nb concentration (25.07 wt%) than the surrounding matrix (33.78 wt.%). The presence of α-phase precipitate enhanced the mechanical hardness from 274.3 to 335.8 HV. The results implied that the microstructure and hardness of the Ti-30Nb-5Sn were significantly altered by the solution treatment.

1. Introduction

The current commercial metallic biomaterials used in the clinic are made of stainless steel, Co-Cr alloy, pure-Ti, and Ti-6Al-4V. Two major problems which are found to be the reason for implant failure in the application are toxicity and stress shielding effect [1-4]. Ti and its alloys exhibited high specific density, good mechanical properties, and excellent corrosion resistance in a harsh human body environment [5]. The alloying elements of Al and V in commercial Ti-6Al-4V was reported to cause neurological disorders and nerve's disease [6-7]. There some research investigating the development of Ti composite by mixing Ti and ceramic or glass material which revealed a good biocompatibility but counteracts by the significantly increased hardness [8]. Recent investigations of Ti alloy was focused on the development of non-allergic alloy with low elastic modulus such as the metastable β Ti which includes Ti-Mo, Ti-Nb, and Ti-Zr [9-11]. A large difference in elastic modulus between implant material and the adjacent bone may lead to stress shielding effect.

The β Ti-Nb-Sn alloys are extensively investigated [9-14] in recent years considering their non-toxic property and excellent biocompatibility. Niobium serves as a β stabilizer while the Sn prevents the precipitation of ω phase resulting in the lower elastic modulus [9]. Both Nb and Sn had been proved to have a good cytocompatibility [12]. An optimization of the concentration of Nb and Sn in the alloy is necessary to avoid the phase transformation of the β-phase. Previous work [13] had shown that the stable β-phase in Ti-30Nb was achieved by alloying with 5 wt% Sn. Besides, the concentration of the alloying element, the microstructure of Ti-Nb-Sn alloy was sensitive to a heat treatment which further affected the mechanical properties. The ω precipitate was reported [14] to form in Ti-35Nb as a result of aging for short time at 573-773 K but returned to its original properties at room temperature. The formation of ω precipitate resulted in a superelasticity phenomenon where the modulus elasticity dropped significantly. The ability to exhibit superelasticity shows potential for application as a shape memory alloy.

This work aimed to investigate the effect of heat treatment at 1000°C for 5 h on the microstructure and mechanical hardness of Ti-30Nb-5Sn. The treatment is typically performed to obtain homogeneous distribution of alloying element in the alloy, which is also called as solution treatment. The phase transformation which may occur as a result of solution treatment was analyzed by X-ray diffractometer.

2. Experimental Procedure

The Ti-30Nb-5Sn alloy was prepared from raw materials of Ti scrap (99.5%), Nb rod (99.8%), and Sn scrap (99.99%). The raw material was weighed 100 g in total to make an alloy. The arc vacuum furnace was used to melt the material and cast into an ingot. The ingot was remelted 5 times to ensure the homogenization of the components. The fabrication method was similar to an earlier report [13]. Some of the as-cast alloys was further solution treated at 1000°C for 5 h followed by water quenched. The solution treatment was conducted in a furnace under air exposure.

Prior further use, the specimen was ground with #800 to #2000 grit paper and then polished with alumina paste until mirror finished. The specimen was then washed in DI water. Before using for characterization, the specimen was degreased in acetone and ethanol in an ultrasonic bath for 3 min.

For microstructure observation, the specimen was etched in 5 M HF solution and desmutted in 30% HNO_3 solution. The microstructure was studied by FE-SEM (FEI Inspect F-50). The alloy composition was confirmed by EDX. The crystalline phases existed in the alloy was analyzed by using X-ray diffractometer (XRD, Panalytical X'Pert Pro MPD). Meanwhile, the mechanical hardness was measured by means of Vickers microhardness indenter (Struers DuraScan). Five-points measurements were conducted on each specimen to obtain the average value.

3. Results and Discussion

The composition of the as-cast alloy was analyzed by EDX on a large area on the specimen to make sure the ratio of each component. The alloy composition is listed in Table 1. The concentrations of Nb and Sn were slightly higher at 36.8 wt% and 6.9 wt%, respectively, but still in the tolerable range. The EDX analysis was done at three points on the specimen and the results were not so different indicating that the distribution of the element was homogeneous. The alloying element, in particular, Sn was volatile at a high temperature under vacuum condition inside the arc furnace chamber, which potentially caused significant mass loss during the process. However, the EDX result confirmed that the Ti-30Nb-5Sn alloy was successfully fabricated by using an arc furnace.

Table 1. Elemental composition of the Ti-30Nb-5Sn alloy.

Element	Wt%	At%
Nb	36.8	24.3
Sn	6.9	3.6
Ti	56.3	72.1

Figure 1 shows the plane-view FE-SEM images of the Ti-30Nb-5Sn alloy before and after solution treatment. The as-cast alloy exhibited a lamellar with branching structure forming dendrites (Fig. 1a). The space between lamellar was about 30 μm. A higher magnification image in the inset Fig. 1a shows more clear the dendrites. The structure indicated that alloy had only a single phase. Diffusion of the molten alloy during casting gave lamellar and dendrites path in the microstructure [5]. Meanwhile, after solution treatment, the dendritic structure did not exist any longer. The microstructure of the alloy after solution treatment showed a clear metallic grain structure decorated by plate-like precipitates of about 10 μm size. The discontinuous grain boundaries exhibited similar morphology as the plate-like precipitates. The plate-like precipitates and the grain boundaries corresponded to the α-phase. Both grain boundaries and precipitates appeared darker indicating lower average atomic weight than the surrounding matrix. A higher magnification image in the inset Fig. 1b revealed some grain boundary lines existed inside the precipitates. Similar plate-like morphology of the α-phase precipitates was reported [11] to form as a result of aging at temperature 500° and 600°C.

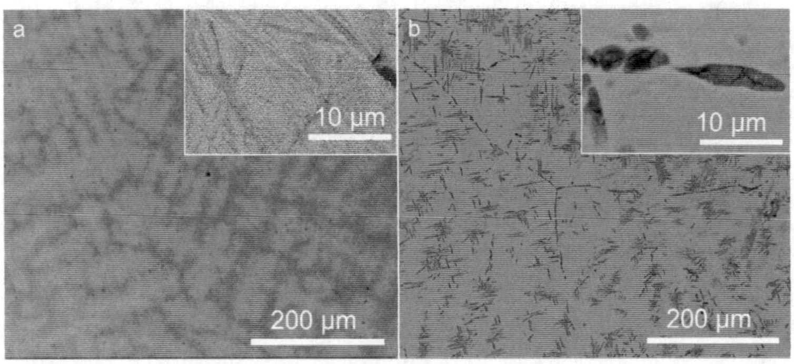

Figure 1. FE-SEM images of the a) as-cast and b) solution treated Ti-30Nb-5Sn alloy showing the transformation in the microstructure.

To confirm the composition of the plate-like precipitate, EDX point analysis was conducted on the precipitate as well as the matrix for comparison purpose. The results are presented in Fig. 2. The Nb concentration in the precipitate was only 25.07 wt% which was 8.71 wt% lower than that of the matrix. Oppositely, the Ti concentration was higher at 67.43 wt% at the precipitate than at the matrix (59.77 wt%). The low Nb and higher Ti concentration proved that the precipitate consisted of α-phase. It was reported that the martensitic α-phase typically has low Nb and higher Ti content [7, 11]. The low Nb concentration at the precipitate reduced the average atomic weight and hence gave darker appearance than the surrounding matrix.

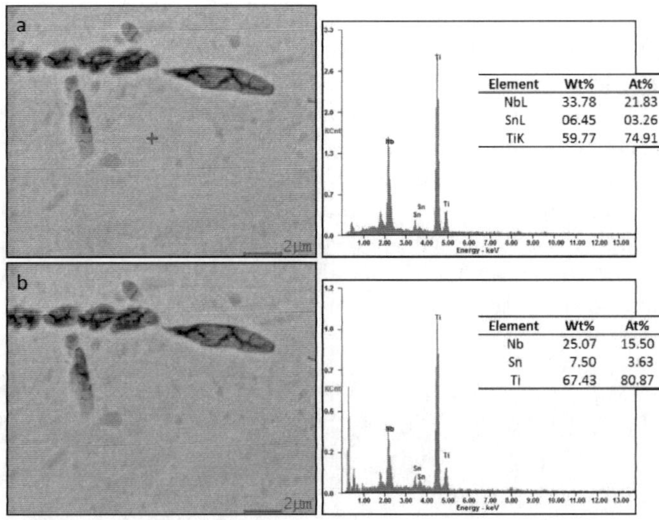

Figure 2. FE-SEM images and the corresponding EDX point analysis at the a) matrix and b) plate-like precipitate.

Figure 3 shows the XRD pattern of the Ti-30Nb-5Sn alloy as-cast and after solution treatment. The as-cast specimen composed of mainly β-phase. The distinct peaks corresponded to the β-phase aroused at the angles of 38.9°, 55.9°, 69.5°, and 82.1° in the as-cast specimen. The as-cast alloy entirely composed of β-phase. Meanwhile, the XRD pattern of the solution-treated specimen exhibited numerous peaks of α-phase in addition to the β-phase. The α-phase was identified by the serial peaks appeared at the angles of 32.5°, 35.9°, 41.1°, 54.1°, 55.4°, 62.7°, 68.9°, and 75.8°. The XRD results were in agreement with the FE-SEM images in Fig.1 that the solution-treated specimen consisted of β-phase and numerous α-phase precipitates while the as-cast specimen composed only β-phase. The solution treatment conducted on the Ti-30Nb-5Sn induced the formation of α-phase precipitates in the metastable β-phase matrix.

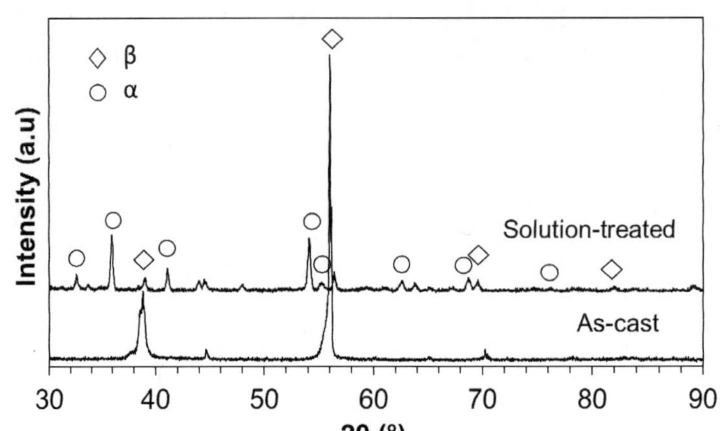

Figure 3. X-ray diffraction pattern of as-cast and solution-treated Ti-30Nb-5Sn alloys.

Table 2 shows the effect of solution treatment on the mechanical hardness of Ti-30Nb-5Sn alloy. The presence of α-phase precipitates in the metallic grain of β-phase of Ti-30Nb-5Sn alloy enhanced the mechanical hardness of the alloy. The as-cast alloy which exhibited a hardness of 274.3 HV became significantly harder with hardness value 335.8 HV after solution treatment. The presence of α-phase as a result of solution treatment enhanced the mechanical hardness of about 61.5 HV. The martensitic α phase is brittle and hence tends to raise the hardness of Ti alloy [10, 11].

Table 2. The average mechanical hardness of as-cast and solution-treated Ti-30Nb-5Sn alloys.

Specimen	Microhardness (HV)
As-cast	274.3
Solution-treated	335.8

4. Conclusion

The effect of solution treatment on the microstructure and hardness of Ti-30Nb-5Sn alloy have been investigated. Surface characterization and elemental analysis unambiguously showed that the solution treatment induced the formation of α-phase precipitates as confirmed by the XRD analysis. The precipitate existed in the form of micro-size plate-like morphology segregated in the metallic grain of β-phase. The α-phase was also enriched along the grain boundaries. Prior to solution treatment, the as-cast alloy exhibited a lamellar and dendritic structure of single β-phase. The presence of α-phase precipitate in the solution-treated specimen resulted in an increase of mechanical hardness from 274.3 to 335.8 HV. The microstructure of metastable β-phase is very sensitive to heat treatment. The α-phase played an important role in the precipitation hardening mechanism of the β Ti alloy which opened the possibility to control the mechanical properties of β Ti alloy.

5. References

[1] Miura K, Yamada N, Hanada S, Jung T, and Itoi E 2011 *Acta Biomater.* **7** 2320-2326
[2] Niinomi M, Nakai M, and Hieda J 2012 *Acta Biomater.* **8** (11), 3888-3903
[3] Rosalbino F, Maccio D, Scavino G 2012 *J. Mater. Sci: Mater. Med.* **23**, 865-871
[4] Hsu S, Ho W, Wu S, Chen Y, Hsu H 2016 *Thin Solid Films* **620** 139-144
[5] Leyen C and Peters M, 2003 Titanium and Titanium alloy: Fundamentals and Applications (Wiley-VCH GmbH&Co., Weinheim, Germany)
[6] Hao Y L, Li S J, Sun S Y, Yang R 2006 *Mater. Sci. Eng. A* **441** 112-118
[7] Moraes P E L, Contieri R J, Lopes E S N, Robin A, and Caram R 2014 *Materials Characterization*, **96**, 273-281
[8] Anawati A, Tanigawa H, Asoh H, Ohno T, Kubota M, Ono S 2013 *Corros. Sci.* **70**, 212-220
[9] Wang B L, Zheng Y F, Zhao L C 2008 *Mater. Sci. Eng. A* **486** 146-151
[10] Griza S, Sa D H G, Batista W, Blas J, Pereira L C 2014 *Mater. Des.* 200-208
[11] Bahl S, Krishnamurthy A S, Suwas S, Chatterjee K 2017 *Mater. Des.* **126** 226-237
[12] Wang X, Chen Y, Xu L, Liu Z, Woo K 2013 *Mater. Des.* **49** 511-519
[13] Utomo E P, Kartika I, Anawati A 2018 *AIP Conference Proc.* **1964** 020046
[14] Kim H Y, Ikehara Y, Kim J I, Hosoda H, Miyazaki S 2006 *Acta Mater.* **54** 2419-2429

Acknowledgement
This work was fully supported by Hibah Publikasi Internasional Terindeks untuk Tugas Akhir Mahasiswa (PITTA) Universitas Indonesia with contract no. 2226/UN2.R3.1/HKP.05.00/2018.

IOP Conf. Series: Materials Science and Engineering **541** (2019) 012050 doi:10.1088/1757-899X/541/1/012050

Effect of nickel additions on microstructure evolution and mechanical properties of low-alloy Cr-Mo cast steel

S B Pratomo[1,2], H Oktadinata[3] and T W Widodo[4]
[1] Metal Industries Development Center, Jalan Sangkuriang 12, Bandung, Indonesia
[2] Department of Metallurgical Engineering, Universitas Jenderal Achmad Yani, Bandung, Indonesia
[3] Department of Mechanical Engineering, Universitas Jenderal Achmad Yani, Bandung, Indonesia
[4] Indonesian Center for Agricultural Engineering Research and Development, Indonesia

E-mail: bimo_bblm@yahoo.com

Abstract. This paper presents investigation of nickel additions on the low-alloy Cr-Mo cast steel which was normalized at temperature of 920 °C. In this experiment, approximately 0.3, 0.5, and 1.0 wt% Ni was added into the Fe-1.0Mn-0.8Cr-0.4Mo cast steel. The microstructure of low-alloy Cr-Mo cast steel was observed by optical microscope and scanning electron microscope, and the phase compositions were identified by EDX analysis. Tensile, hardness, and Charpy impact tests were conducted to investigate correlation between nickel additions to microstructure characteristics and mechanical properties. The results show that increasing nickel from 0.3 to 1.0 wt% on the alloys has improved the strength without sacrificing the impact toughness. Ni addition into low alloy steel increased the austenite stability due to the grain refinement. The strength was found increase linearly with Ni addition which may caused mainly by solid solution strengthening due to Ni dissolved into the ferrite matrix.

1. Introduction

Low- alloy Cr-Mo steels are used in various application such as the components of power station boilers with application in extreme temperature up to about 560°C [1]. These steels also widely used in application as advanced high strength steel (AHSS) which have microstructure of carbide free bainitic steel in rail and protective armor steel [2,3]. Alloying is one of techniques to increase the materials properties such as ductility, strength, toughness, corrosion and oxidation resistance. Small amount of Cr, Mo, Ni and C were added into steel to make better strength, toughness, and corrosion resistance [4]. One method of the steel strengthening mechanism which is known as the solid solution is performed by alloying of small amounts of dissolved atoms [5]. The mechanical performance of metals which resulted solute atoms dissolved in solvent depending on the amount and type of the solutes [4].

Addition of Mn and Ni increased the hardness which was attributed mainly to solid solution strengthening due to part to formation of hard phases [6]. Mn and Ni have similar effect on phase transformation and grain size, both were also responsible for hardening but effect of Mn is more noticeable [6,7]. Ni is known as an element which contribute to improve the hardenability and heat resistance of steel [8]. K. Kim et al. suggested increasing Ni content on Q & P treated steel improved the tensile strength, elongation and hardness due to solid solution mechanism and the grain refinement effect [9].

Mn and Ni influence impact toughness in different ways. B.Y. Kang et al. suggested for low Mn content (0.5%), ductile-to-brittle transition temperature was not any increase even with the higher hardness by Ni addition. But, in the case of Ni was co-present with a high Mn content (1.6%), with a large addition of Ni the microstructure became predominantly martensite which susceptible to intergranular brittle crack propagation [6]. When Mn and Ni increased then the proportion of acicular ferrite was increased, bainite and M-A constituent began to arise. Furthermore, the higher Mn and Ni tend to promote micro segregation of Mn, Ni, and Si elements in a network or parallel to grain boundaries [7].

Compared to the other carbide-forming elements (Cr, Mo, W), Ni tends dissolved into the ferrite matrix instead of forming carbides. By addition of Ni into the material, carbide-forming elements were suppressed to be the solid solution in the matrix [8]. The effect of Ni addition on the low alloy steel provides some benefits. Ni does not form secondary phases, it remains in the solid solution strengthening mechanism in the ferrite matrix [10]. Ni distributed in the ferrite matrix uniformly. Ni which dissolved in the ferrite matrix increases the stacking fault energy, so that at low temperature promotes dislocation crossing slip, and thus boosts power consumption of crack propagation, finally increase of the low-temperature impact toughness [11]. The retained austenite is part of M-A constituent in steel, it is mainly found in the lower Ni content that indicated that the transformation process could be more completely by increasing Ni content. The large volume fraction of M-A was mainly found in the lower Ni content effect to reduce the toughness. The more Ni content, the less retained austenite and the more the transformation completed [12].

Ni additions increased austenite grains stability by expands the austenite regions due to the grain refinement and lowered the ferrite transformation temperature so that refined the low alloy steel grains size [9,10,12]. A consequence of the lower ferrite transformation temperature is that Ni promotes the formation of acicular ferrite at the expense of other kinds of ferrite such as side plate ferrite and grain boundary ferrite [7,10,12]. Ni improves the impact toughness by the grain size refinement in the columnar zone as well as by Ni by itself [6]. Y. Sun et al. reported increasing 0-1.8% Ni on high ductility ductile iron increased low temperature impact toughness significantly, the highest toughness reached with 0.71% Ni addition, then descents [11].

This paper investigated effect of nickel additions to improve the hardness of low alloy Fe-1.0Mn-0.8Cr-0.4Mo steel without deteriorate impact toughness. This study including observations of microstructure and phase compositions, tensile and hardness measurements, and performance of impact toughness at room temperature.

2. Experimental method

2.1 Melting and casting

The medium carbon steel scraps were melted in a medium frequency induction furnace of 200 kg capacity. Additional alloying which consist of Fe-Mn, Fe-Cr, Fe-Mo, and pure Ni were supplied into the molten metal in induction furnace until low alloy Fe-1.0Mn-0.8Cr-0.4Mo cast steel produced. Pure Ni was added to the melt to produce three different content of nickel (0.3%, 0.5%, and 1.0%). Furthermore, the melt was heated to the temperature of 1700°C.

The small amount of melt was poured as test specimens for spectrometric analysis. The chemical composition was measured by optical emission spectrometer during controlling the levels of nickel content in the low-alloy Cr-Mo cast steel. When the specified nickel content achieved, the melt was poured at temperature range of 1620-1650°C for 6 to 7 seconds into Y-block sand molds which made from sand cast and alpha resin binder, and finally the melt was allowed cooling to room temperature. After completely solidification, the Y-block samples were heat treated of normalizing by electric furnace

IOP Conf. Series: Materials Science and Engineering **541** (2019) 012050 doi:10.1088/1757-899X/541/1/012050

to temperature of 920°C for 10 minutes and allowed cooling to room temperature in the atmosphere. The chemical analysis result of the casting is presented in Table 1.

(a) (b)

Figure 1. (a) Pouring into Y-block sand molds and (b) Y-block sample

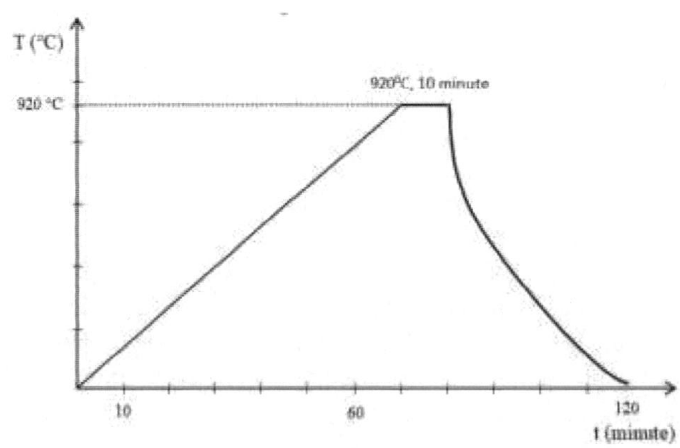

Figure 2. Heat treatment schematic

Table 1. Chemical analysis of the low-alloy Cr-Mo cast steel samples (%wt.)

Sample No.	C	Si	Mn	P	S	Cr	Mo	Ni	Fe
1	0.40	0.20	1.00	0.01	0.01	0.80	0.40	0.30	Bal.
2	0.40	0.20	1.00	0.01	0.01	0.80	0.40	0.50	Bal.
3	0.40	0.20	1.00	0.01	0.01	0.80	0.40	1.00	Bal.

2.2. Microstructure
The test specimens for metallography examination were prepared in the standard procedure, roughed by abrasive paper, polish using diamond paste, and etched with 3% Nital (3 ml HNO_3 + 97 ml ethanol). Microstructure of the as cast low-alloy Cr-Mo samples were observed using an optical microscope and continued with a scanning electron microscope (SEM) which equipped with an EDX analyzer.

2.3. Mechanical tests
The mechanical tests i.e. tensile, hardness, and impact toughness were performed on each test specimens of 0.3%, 0.5%, and 1% Ni addition. The tensile tests were conducted at room temperature using a tensile testing machine with reference of JIS Z 2241 standard specification. To evaluate the impact toughness, the Charpy V-notch samples of size 10 mm x 10 mm x 55 mm were prepared and tested using impact testing machine at room temperature according to JIS Z 2242 standard specification. Hardness measurements were carried out at room temperature using Rockwell C hardness tester with reference of JIS Z 2245 standard specification.

3. Results and discussion

3.1 Microstructure investigation
The typical microstructures of Ni additions into the low alloy Fe-1.0Mn-0.8Cr-0.4Mo steel after normalizing at 920°C for 10 minutes were observed by optical microscope and back scatter electron-scanning electron microscope (BSE-SEM). The typical microstructures of Ni-added low alloy Cr-Mo cast steel are shown in Figure 3 (observation by optical microscope) and Figure 4 (observation by scanning electron microscope). It seems that the microstructure contains carbide-free bainitic ferrite phase. Some elements such as manganese, silicon, chromium and molybdenum are important elements in carbide-free bainitic steel of high strength low alloy steels [2,3]. Base on optical microscope images in Figure 3, it was observed that increasing Ni affect to enlarge light areas. The BSE-SEM results as shown in Figure 4 also indicated that increasing Ni resulting light bainitic ferrite phase increased.

Analysis by image analyzer was conducted to measure light and dark area percentage. The result show that the light areas increased when nickel content increased. The light areas percentage of each Ni content were 47.5% (0.3% Ni), 60.8% (0.5% Ni), and 73.5% (1% Ni). It seems that increasing Ni affect to increasing light bainitic ferrite region.

Analysis of Fe, C, Ni, Cr, Mo, and Mn elements distribution was performed by using color mapping by x-ray diffraction-scanning electron microscope (EDX-SEM) for the sample of 1% Ni addition. Color mapping analysis in Figure 5 shows Ni and other elements (Fe, C, Cr, Mo, and Mn) did not segregated in the certain area, but they are uniformly distributed on both light and dark areas. It could be these elements dissolved in the bainitic ferrite phases, and substitute Fe in the crystal lattice with atom substitutional solid solution mechanism. More Ni added makes more Fe substituted by Ni in crystal structure. Similar result was reported by Y. Sun et al. in their study with ductile iron containing 0-1.8 wt% Ni. According to them Ni element distributed uniformly in the ferrite matrix, furthermore the tensile strength and the low temperature impact toughness increased with Ni addition up to 0.71 wt% [11].

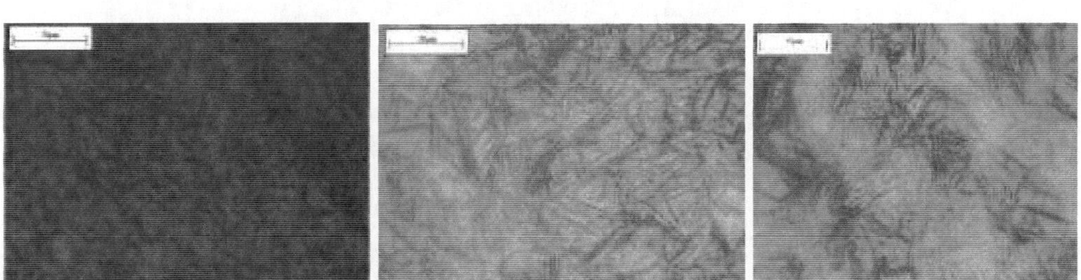

IOP Conf. Series: Materials Science and Engineering **541** (2019) 012050 doi:10.1088/1757-899X/541/1/012050

(a) (b) (c)

Figure 3. Optical microscope images of sample with: (a) 0.3 % Ni, (b) 0.5 % Ni, (c) 1.0 % Ni

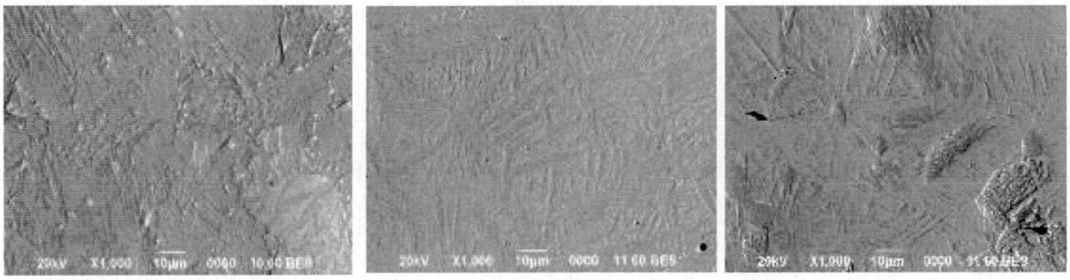

(a) (b) (c)

Figure 4. BSE-SEM images of sample with: (a) 0.3 % Ni, (b) 0.5 % Ni, (c) a) 1.0 % Ni

Although Cr, Mn, and Mo elements are carbide-forming elements, they are also dissolved in the ferrite matrix with substitutional mechanism. It could be presence of Ni suppressed these carbide-forming elements to be solid solution in the ferrite matrix and avoid carbide segregation. In an earlier study reported by Y. Zhang et al., Ni tends dissolved into the matrix instead of forming carbides compared to the other carbide-forming elements such as Cr, Mo, W). By addition of Ni into the material, carbide-forming elements were suppressed to be the solid solution in the ferrite matrix [8].

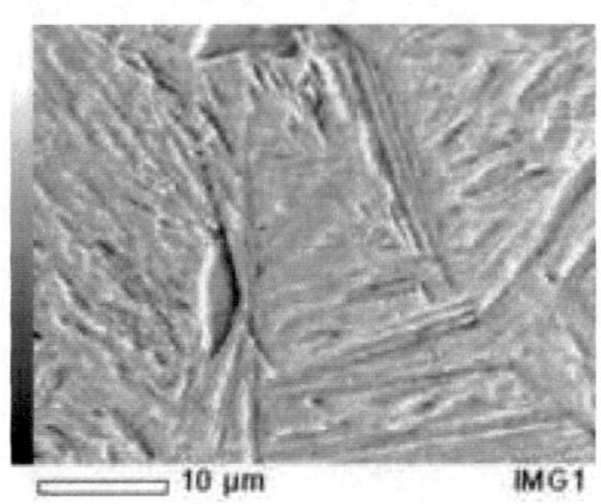

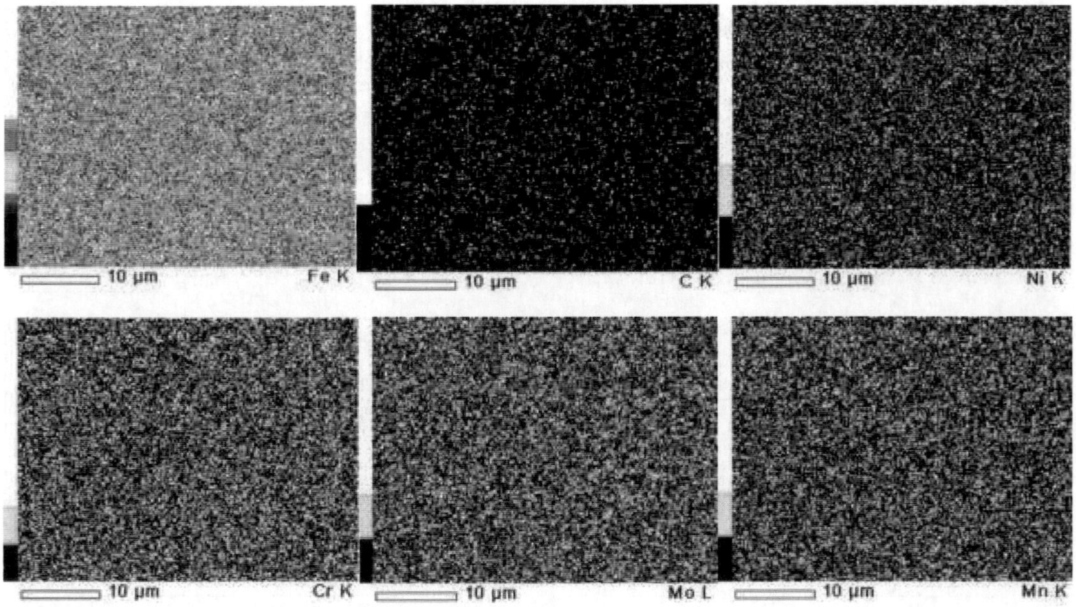

Figure 5. Color mapping (EDX-SEM analysis) to observe distribution of elements in microstructure

Further analysis by energy dispersive X-ray spectroscopy (EDX) was carried out to identify certain elements which was visually observed in the BSE image. The observation result show that the dark areas containing Cr and Mn, while light areas only containing Ni without Cr and Mn. Since the atomic weight of Ni higher than Cr and Mn, the phase that containing Ni shown lighter image. It seems that Ni dissolved only in light areas. Figure 6 shows Ni-dissolved bainitic ferrite (light phase) and Ni-free bainitic ferrite (dark phase). It seems that presence of Ni formed Ni-dissolved bainitic ferrite and suppressed Cr and Mn elements to be dissolved into the Ni-free bainitic ferrite.

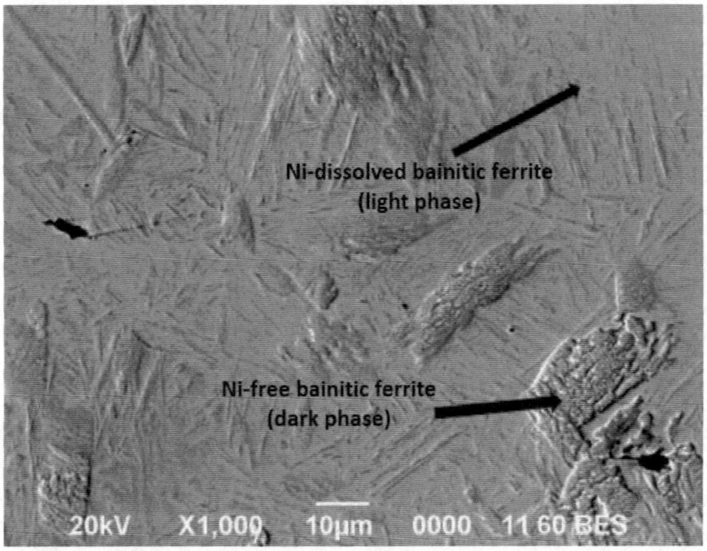

Figure 6. Ni-dissolved bainitic ferrite phase and Ni-free bainitic ferrite phase in 1.0% Ni added Cr-Mo steel.

3.2 Mechanical properties

Mechanical test results of the samples with various Ni content are presented in Table 2. The mechanical test results indicated that the higher Ni content increased the yield strength, tensile strength and hardness, while the impact toughness relatively unaffected. Figure 7 shows effect of Ni additions on the tensile strength, hardness, and impact energy of the samples material. Tensile strength increased significantly by increasing of Ni, which shows 1040 N/mm^2 (0.3% Ni), 1200 N/mm^2 (0.5% Ni), and 1350 N/mm^2 (1% Ni). The hardness also tends to increased linearly by increasing Ni. The hardness improved from 30 HRC at 0.3% Ni to 44 HRC at 1% Ni. Ni additions in low-alloy Cr-Mo cast steel from 0.3% to 1% have improved strength and hardness without sacrificing impact toughness.

Table 2. Yield strength, tensile strength and impact energy of the low-alloy Cr-Mo cast steel samples

Sample No.	Yield Strength (MPa)	Tensile Strength (MPa)	Total Elongation (%)	Hardness (HRC)	25 °C Impact Energy (J)
1	791	1037	9	30	11
2	1052	1202	2	40	10
3	1177	1351	1.7	44	9

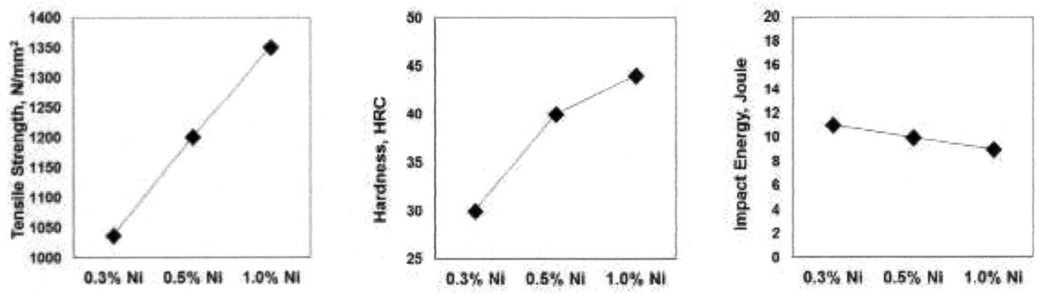

Figure 7. Mechanical test results of samples with 0.3%, 0.5%, and 1% Ni : (a) tensile strength, (b) hardness, (c) impact energy

To verify the hardness of Ni-dissolved bainitic ferrite phase and Ni-free bainitic ferrite phase, micro hardness Vickers test were performed on samples. The result shown that the average hardness of Ni-dissolved bainitic ferrite phases were 610 VHN (56 HRC) compared to only 350 VHN (36 HRC) on Ni-free bainitic ferrite phases. This is indicated that the solid solution strengthening with Ni resulting higher strength compared to Cr and Mn. Stress field within lattice caused by substitutional of larger atoms than Fe (Ni) resulting the compressive strains, whereas caused by substitution of smaller atoms (Cr, Mn) resulting the tensile strains. Stress field increases the lattice strain by dislocation, resulting strengthening of nickel dissolved bainitic ferrite phase. The strength was improved due to the interaction of local stress fields and dislocation that impede dislocation movement. B.Y Kang et al. suggested that addition of Ni increased the hardness which was attributed mainly to solid solution strengthening due to part to formation of hard phases [6].

When conducted the impact test, compressive strain within the lattice resulting by substitution Ni in crystal lattice would absorb the impact energy, causing the toughness of Cr-Mo steel was not significantly change. Other reason, strength and hardness increased without sacrifice impact toughness due to presence of Ni that promoted the grain size refinement. Finer grains size causing more grain boundaries present. Furthermore, more energy required for the crack propagation path so that improves the strength and toughness.

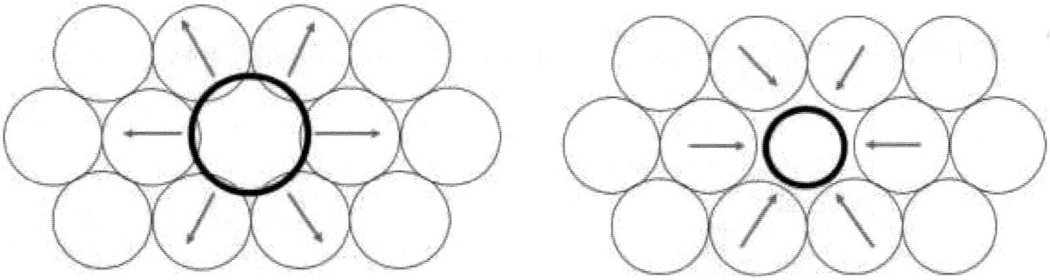

(a) Ni (b) Cr/Mn

Figure 8. Tensile strain field within lattice caused by substitutional of Ni and Cr/Mn

4. Conclusions

Effect of Ni additions into low-alloy Cr-Mo cast steel has been investigated. The results show that increasing Ni from 0.3% to 1% has improved the tensile strength and hardness while the impact toughness did not change significantly. Ni addition contributed to improve the strength by solid solution strengthening mechanism. Fe substituted by Ni in ferrite matrix by substitution solid solution mechanism resulting compressive strain field within crystal lattice, and causing strengthening of nickel dissolved bainitic ferrite phase. Compressive stress would absorb the impact energy resulting the improvement of toughness resistance. Ni additions into low-alloy Cr-Mo cast steel promotes finer grains size and more grain boundaries present. Consequently, there will be more energy required for the crack propagation path, it improves the strength and toughness.

References

[1] Dobrzanski J 2004 *Journal of Materials Processing Technology* **157–158** 297–303
[2] Bhadeshia H.K.D.H 2005 *Materials Science Forum* **500-502**: 63-74
[3] Zhou M, Xu G, Tian J, Hu H and Yuan Q 2017 *Metals* **7** 263-275
[4] Wu Z, Troparevsky M C, Gao Y F, Morris J R, Stocks G M and Bei H 2017 *Current Opinion in Solid State and Materials Science* **21** 267-284
[5] Schmauder S and Kohler Ch 2011 *Comput. Mater. Sci.* **50** 1238–1243
[6] Kang B Y, Kim H J and Hwang S K 2000 *ISIJ International* **40**, 1237–1245
[7] Jaberi F S and Kokabi A H 2012 *J. Mater. Eng. Perform* **2** 1447–1454
[8] Zhang Y, Shimizu K, Kusumoto K, Hara H, and Higuchi C 2017 *Wear* **376-377** 452–457
[9] Kim K and Lee S J 2017 *Materials Science & Engineering* A **698** 183-190
[10] Husby H, Iannuzzi M, Johnsen R, Kappes M, and Barnoush A 2018 *Int. J. Hydrogen Energ.* **43** 3845-3861
[11] Sun Y, Hu S, Xiao Z, You S, Zhao J and Lv Y 2012 *Materials and Design* **41** 37-42
[12] Wang Z Q, Wang X L, Nan Y R, Shang C J, Wang X M, Liu K and Chen B 2018 *Materials Characterization* **138** 67-77

The application of zinc oxide layer as ethylene sensor

A Sholehah[1,2], D A Pusparasmi[1,2], B Yuliarto[3]
[1]Advanced Material and Tomography Laboratory, Faculty of Engineering, Universitas Sultan Ageng Tirtayasa, Jl. Jend. Sudirman Km 3, Cilegon, Banten – Indonesia 42435
[2]Departement of Metallurgical Engineering, Faculty of Engineering, Universitas Sultan Ageng Tirtayasa, Jl. Jend. Sudirman Km 3, Cilegon, Banten – Indonesia 42435
[3]Departement of Engineering Physics, Faculty of Industrial Technology, Institut Teknologi Bandung, Jl. Ganesha No. 10, Bandung, Jawa Barat – Indonesia 41032

E-mail: amalia.sholehah@untirta.ac.id

Abstract. As an emerging *n*-type semiconductor, ZnO has been widely applied in sensor area. In this study, we prepared an ethylene sensor using ZnO layer on FTO glass substrate. The seed layer was deposited using simple electrodeposition method using voltage of -1.1 V for 2 hours in cold bath (6ºC). The ZnO layer was further grown using CBD technique at 90˚C for 2 hours. In order to observe the effect of structural manipulation on the sensor performance, some of the ZnO layer samples were undergoing hydrothermal treatment at 100ºC for 1 hour under 1 bar N_2 gas. After annealing process, all ZnO layers were exposed with 50 ppm ethylene gas in closed chamber. Based on the results, it is shown that the optimum ZnO layer has been successfully identified the ethylene gas in concentration of 50 ppm, with response value of 2.40% at 200˚C.

1. Introduction

Sensor is an electronic circuit component, which can sense physical and chemical signals coming from a stimulant existing around precisely and can convert them into measurable quantities and generally, electrical signals. There are various sensors according to where they are used like biosensors, gas sensor, humidity sensor, pressure sensor, vibration sensor, temperature sensor, speed sensor, and magnetic field sensor [1]. Because it has been considered that advanced semiconductor technology allows the desired superior features for the sensors to be ensured, the use of especially metal oxide semiconductors as a sensing material has increased recently and this has positively contributed to development of sensors. Among the sensing materials, metal oxide semiconductors like ZnO, CdO, SnO_2, TiO_2, CuO, and W_3O have become preferable than other sensor types due to their higher sensitivity and their ability in working under severe conditions and their longer lifecycle superior [1].

Of all available types of semiconductor sensors for different types of chemicals and biological toxins, ZnO-based sensors have gained extensive attention around the world. The presence of a good response rate towards the chemical toxins with outstanding selectivity and sensitivity makes it one of the most significant materials for preparing low cost sensors [2]. Being an *n*-type semiconductor with a wide band gap of 3.37 eV and a large exciton binding energy of 60 meV, the electron mobility in ZnO nanomaterials is enhanced [2]. In addition, ZnO nanomaterials have the advantages of a low cost of production, a harmless nature and a simple mode of large-scale production. Furthermore, ZnO nanomaterials are chemo-resistive in nature, and their sensing aptitude is principally restricted by the change in the chemical signal when the respective analyte molecules encounter its exterior surface [2].

In recent years, much attention has been given to ZnO-based nanostructures for sensing. The presence of the numerous properties of ZnO has been utilized for the development of effective and

IOP Conf. Series: Materials Science and Engineering **541** (2019) 012051 doi:10.1088/1757-899X/541/1/012051

highly selective sensors [2]. Undoped and doped ZnO materials with different metals have been studies for sensor applications. To this date, the use of ZnO as a sensing material has been investigated for many sensors, e.g. humidity [3], formaldehyde [4], and SO_2 [5] sensor.

Ethylene is a small hormone in plant which controls the ripening process. Previously, Esser et al [6] has been investigating the role of carbon nanotube as an ethylene sensor in fruit ripening process. Inspired by Esser's results, in this study, we prepared a ZnO layer which applied as a sensing material for ethylene gas. The ZnO layer was grown on conductive glass surface using simple electrodeposition process. The morphologies, optical, electrical, and structural properties of the ZnO layers were further investigated. Using the data, the correlation among the properties and sensing behavior were concluded afterward.

2. Experimental Methods

The synthesis procedure for fabricating ZnO nanorods was started firstly with preparation of substrates. The fluorine-doped tin oxide (FTO) glasses (Dyesol, Ω 7 ohm/m^2) were cleaned using aquades, aceton and ethanol in ultrasonic cleaner for 480 seconds in each liquid. The cleaned glasses were dried and stored in a clean container until further use. The seed solution was prepared using 0.1 M zinc nitrate tetrahydrate ($Zn(NO_3)_2.4H_2O$/Zn-nitrate, Merck), hexamethylenetetraamine ($C_6H_{12}N_4$/HMTA, Merck) and dissolved in cold water (0°C) in a beaker glass, then aged for 1 hour. The electrodeposition system was constructed using FTO glass as the cathode and Cu metal as the anode. The experiment was conducted using voltage of -1.1 V for 2 hours in cold bath (6°C).

The ZnO layer on FTO glass was grown using chemical bath deposition (CBD) method afterwards. The seeded FTO glass was placed vertically in equimolar Zn-nitrate solution and HMTA 0.05 M at 90°C for 2 hours in the oven. After that, the FTO glass was washed using distilled water and dried in air. In order to observe the effect of structural manipulation on the sensor performance, some of the as-grown ZnO layer samples were undergoing hydrothermal (HT) treatment by placing the layer on top of boiling water in a reactor. The ZnO layer was placed upside down, facing the bottom of the reactor. This process was carried out used a closed reactor at 100°C for 1 hour under 1 bar nitrogen gas (N_2) pressure. The as-grown and HT ZnO layer samples were annealed at 200°C for 10 minutes after each processes of synthesis subsequently. The layers were further exposed to 50 ppm ethylene gas in closed chamber under 2 different temperatures, i.e. 100 and 200°C.

The morphology of the obtained ZnO nanorods was examined using field emission-scanning electron microscopy (FE- SEM, FEI Quantana 650), while the crystal structural information was gained using X-ray diffraction (XRD, Pan Analytical X-Pert Pro), and the optical properties were analyzed using diffuse reflectance (DRS) UV-Vis spectroscopy (Shimadzu 2450). The respective E_g of the samples was determined by using Tauc equation.

3. Results and Discussion

SEM results for ZnO thin films on FTO glass are shown in figure 1. Most of the ZnO were grown in rod structure, with hexagonal cross-section. Hydrothermal (HT) method produces more variation of diameter nanorods than CBD (Figure 1). The average diameter of nanorods resulted from CBD process is 0.704 μm, while hydrothermal samples had an average diameter of 1.157 μm. From the cross-sectional view (Figure 2), it is revealed after CBD process, the ZnO layer experiences growth in different directions. Despite structures that grow on the *z*-axis (perpendicular), ZnO rods that grow in diagonal or horizontal direction is still visible in the structures. From the measurement, the thickness of ZnO after CBD was 1.98 μm. However, the thickness of ZnO layer after HT treatment was not convenient to determine. Figure 2.b. has revealed that after HT treatment, the vertical rod structure was collapsed, hence resulting a stack of ZnO rods. Thus, we can assume that using hydrothermal treatment may damage the rod structure, due to the pressure applied during the process.

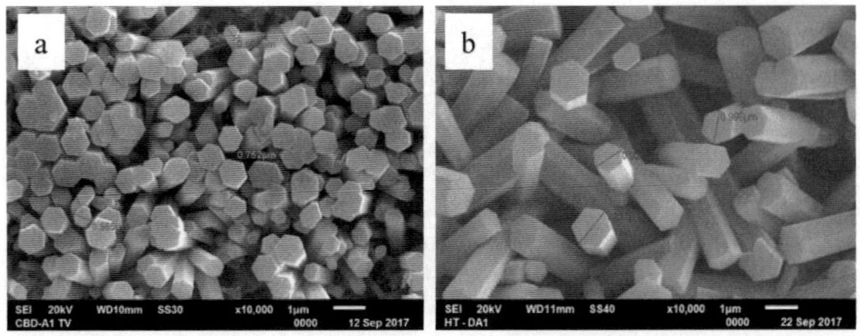

Figure 1. Top view of ZnO layer on FTO substrate a) as-grown using CBD method, b) after hydrothermal treatment, with magnification of 10,000x.

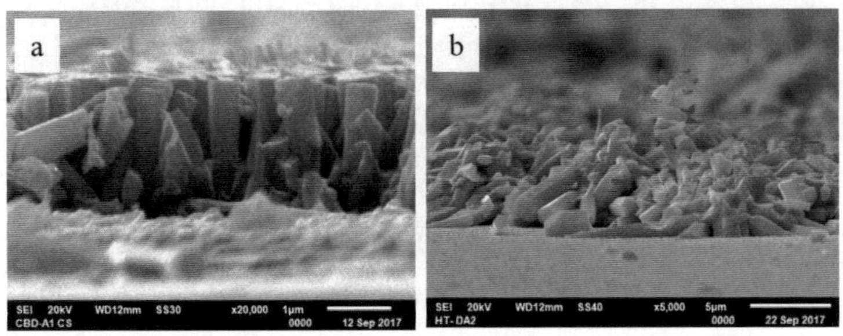

Figure 2. Cross-sectional view of of ZnO layer on FTO substrate a) as-grown using CBD method, b) after hydrothermal treatment, with magnification of 20,000x.

Diffractogram patterns of ZnO layer is presented in Figure 3. From the figure, it is shown that the diffractograms not only presenting ZnO peaks, but also the peaks for FTO substrate. The ZnO layer has a polycrystalline structure, which confirmed with ICSD number 98-018-0052. As seen in the patterns, there are some peaks having different orientations and, existence of the planes corresponding to the three different main peaks, which are (100), (002) and (101). In both samples, the most dominant peak for ZnO was measured at 2θ 31.67°, which correspond to (100) crystal plane. The weak peaks (102), (110), and (103) were also observed, in addition to, these three main peaks.

The crystalline size of the samples is determined using Debye–Scherrer relation [7]:

$$D = \frac{0.9\lambda}{\beta_{hkl}\cos\theta} \tag{1}$$

where D is the crystal size, β_{hkl} is the peak width of half maximum intensity, θ is the diffraction angle and λ represents wavelength of the X-ray. From the calculation, it is obtained that the crystallite size for ZnO sample after CBD and HT treatment were 127.03 nm and 138.36 nm, respectively.

The optical behavior of ZnO layers are presented in Figure 4. The absorbance of the layers was in visible area, with ZnO layer after CBD and HT treatment band-gap energy (E_g) value of 3,47 and 3,41 eV, respectively. Our previous research has revealed that E_g values are very influential on the size of crystallite obtained. The larger the size of the crystallite, the smaller the value of Eg. This corresponds to a shift in the linear region of the curve towards a larger wavelength, this phenomenon is known as red shift [8].

The presence of ethylene around the sensor will reduce its resistance. This phenomenon is influenced by chemical composition, surface modification, surface sensing microstructure and environmental conditions of the sensor (temperature and humidity). The change in the response on exposure to the target gas was calculated using equation [9]

$$\Delta R\ (\%) = \frac{R_a - R_g}{R_a} x100\% \quad (2)$$

where R_a is the resistance before exposure and R_g is the resistance with the target gas flowing through the chamber. This response change is usually called the response sensitivity; hence the term sensitivity is used here. Besides the sensitivity, we also calculated the response and recovery times, i.e., the time needed for the sensor to reach 90% of its steady-state value [9].

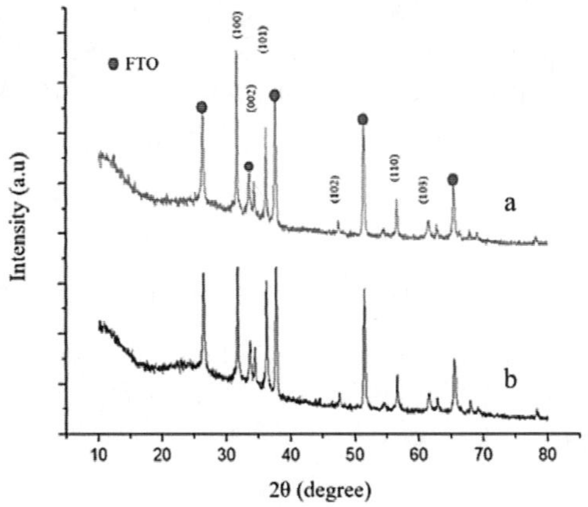

Figure 3. Diffractogram of ZnO layer on FTO substrate a) as-grown using CBD method, b) after hydrothermal treatment.

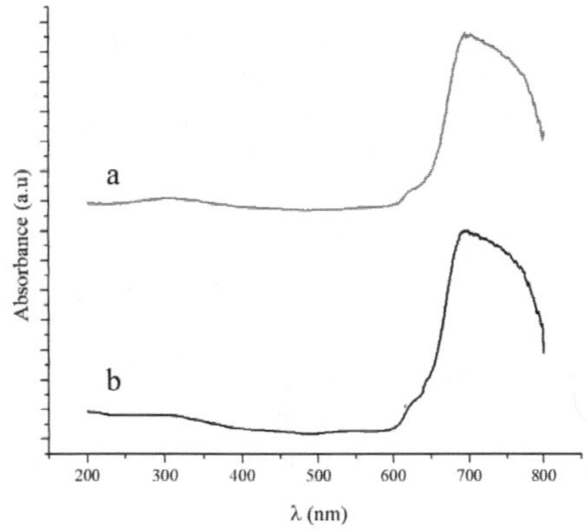

Figure 4. Absorption spectra of ZnO layer on FTO substrate a) as-grown using CBD method, b) after hydrothermal treatment.

In this study, the ethylene concentration used was 50 ppm, and the sensing performance test was conducted in two different temperature, i.e. 100 and 200°C. From the calculation, the ΔR value can be obtained, and listed in Table 1. The response during ethylene exposure can be seen in Figure 5.

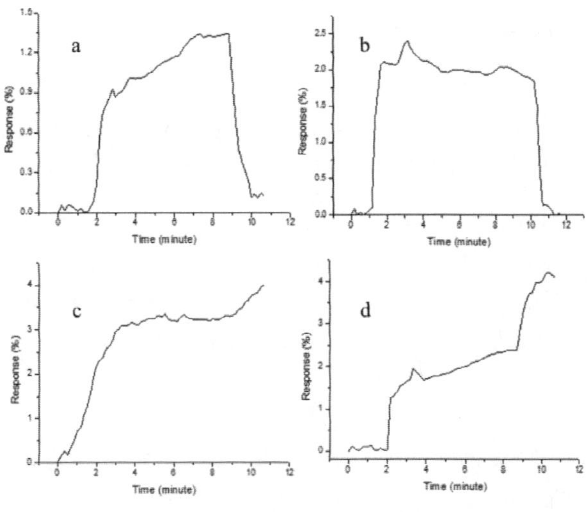

Figure 5. Sensor response during ethylene gas exposure for ZnO layers a) as-grown with CBD treatment, at 100°C, b) as-grown with CBD treatment, at 200°C, c) after HT treatment, at 100°C, and d) after HT treatment, at 200°C.

Table 1. The sensors' response on ethylene gas.

Sample	Ethylene Concentration (ppm)	Responses (%)	
		100°C	200°C
CBD	50	1.34	2.40
HT	50	undefined	undefined

From Table 1, it is noted that the resistance of the ZnO layer in the CBD sample has returned to its original value after the exposure of ethylene gas was stopped. However, this did not occur in the ZnO sample undergoing the HT process. Considering the collapsed structure displayed previously in Figure 2.b., we can assume that damage in the structure may contribute to the sensing performance. Hence, in order to obtain higher response, it is necessary to conduct improvement and sensitizing of the ZnO layer's surface.

4. Conclusion

In summary, we have developed a ZnO-based sensor for ethylene gas. The as-grown ZnO layer is simple to prepare and has been successfully identified the ethylene gas in concentration of 50 ppm. The sensor response is assumed to be related to the morphology and structure of the ZnO layer on the substrate. From the study, it is revealed that the as-grown ZnO layer has been successfully identified the ethylene gas in concentration of 50 ppm, with response value of 2.40% at 200°C.

References

[1] Ates T, Ates C, and Yakuphanoglu F 2013 *Sensors Actuators A. Phys.* **190** 153–160
[2] Chaudhary S, Umar A, Bhasin K K, and Baskoutas S 2018 *Materials (Basel).* **11** 1–38
[3] Ates T, Tatar C, and Yakuphanoglu F 2013 *Sensors Actuators, A Phys.* **190** 153–160
[4] Hu P *et al.* 2011 *J. Mater. Chem.* **21**, 14277
[5] Septiani N. L. W. *et al.* 2018 *Sensors Actuators, B Chem.*, **261** 41–251
[6] Esser B., J. M. Schnorr, and T. M. Swager,2010 *Angew. Chemie - Int. Ed.* **51** 5752–5756
[7] Venkateswarlu K., Chandra Bose A., and N. Rameshbabu 2010 *Phys. B Condens. Matter*, vol. **405** 4256–4261
[8] Narayanan G. N., Sankar Ganesh R., and Karthigeyan A. 2015 *Thin Solid Films*, **598**, 39–45
[9] Muchtar A. R., Septiani N. L. W., Iqbal M., Nuruddin A., and Yuliarto B., 2018 *J. Electron. Mater.* **47** 3647–3656

Ackowledgments

The research was funded by Post-doctoral Research Grant of Ministry of Research, Technology, and Higher Education of Indonesia with contract number 240/UN43.9/PL/KT/2018.

IOP Conf. Series: Materials Science and Engineering **541** (2019) 012052 doi:10.1088/1757-899X/541/1/012052

Corrosion resistance of epoxy primer, polyurethane, and silyl acrylate anti-fouling on carbon steel

S Devanny[1] **and R Riastuti**[1]
[1]Department of Metallurgy and Materials Engineering, Faculty of Engineering, Universitas Indonesia, Depok, West Java, Indonesia

E-mail: riastuti@metal.ui.ac.id

Abstract. Anti-fouling paint is commonly applied as a protective coating on steel structures in the marine environment; one of the anti-fouling paint materials is silyl acrylate copolymer which can hydrolyse with seawater and form leaching layers to release biocide pigments. This study evaluates a three-layer paint comprised of epoxy primer, polyurethane, and silyl acrylate anti-fouling on carbon steel plates through polarization measurements and a salt spray test. Linear polarization measurements are carried out on an unscribed specimen with a potential range of -1.0 to 1.0 Volts. The salt spray test is conducted on scribed specimens, lasts 168 hours, and involves periodic visual observation. The polarization analysis shows pseudo-passive behavior in a system that represents, first, the less protective passive layer of primer and, second, the progressive erosion of a leaching layer of the anti-fouling paint which exposes a new layer beneath. The salt spray test shows that the three-layer paint has poor corrosion resistance in the atmospheric environment indicated by peeled layers and the occurrence of underfilm corrosion since the 48-hour test period.

1. Introduction

Seawater generally contains about 3.4% salt and is somewhat alkaline with a pH of 8. Seawater is a good electrolyte and can cause galvanic and crevice corrosion that is also affected by oxygen contents, speed, temperature, and biological organisms [1]. The largest marine organisms that lead to corrosion, such as barnacles, mollusks, and related species, are mostly attached to structures around the sea such as docks, bridge posts, and ship bodies. This fouling inhibits the speed of the ship, reduces the flow of seawater in pipes and tubing, and even damages the wood structure. Products produced by the growth of organisms or metabolic processes are often acidic and thus accelerate corrosion on metal substrates. These organisms also envelop the metal surface from dissolved oxygen and forms differential aeration cells, which contribute to the acceleration of corrosion. The oxygen-free or anaerobic environment under these macroorganism is a place for the growth of sulfate-reducing bacteria [2].

One of the ways to prevent the attachment of marine organisms bringing about corrosion is to apply an anti-fouling paint to parts that are submerged in water. The anti-fouling paint is formulated with pigments that release biocides to kill micro and macroorganisms on the surface of the structure. Based on the biocide release mechanism, the anti-fouling paint is divided into a paint with a water-insoluble matrix and that with a water-soluble matrix [3]. Tin, copper and zinc are the most widely added metal biocides in anti-fouling paint [4]. Until now, a self-polishing copolymer paint—a development of a soluble matrix paint and a substitute tin-based paint that is harmful to the environment—is still widely used since the formulation allows the release of biocide compounds at controlled rates [5].

Most of the anti-fouling paints are organic coatings constructed of a primer and a final coat—both also have an anticorrosive function—, but the anti-fouling paint as a final coat is often porous [6]. The use of anti-fouling paint as a final coat in a multilayer paint is usually combined with other anticorrosive paints which have (1) a barrier function, such as epoxy, polyurethane, and acrylic, (2) a passivation function with phosphate and chromate pigments, or (3) a sacrificial function such as zinc-rich. These anticorrosive paints provide direct protection to metal substrates by forming thick and tight paint films or by galvanic protection which hinders water penetration and other aggressive species that trigger corrosion.

Prior explorations investigated a self-polishing anti-fouling paint with several matrixes such as silicon [7], epoxy [5], polyurethane and acrylic [8], and zinc-based acrylate [9]. Different approaches were also carried out to determine the rate of paint matrix degradation and biocide pigment release in water, the rust morphology formed in metals, and the development of more environmentally-friendly paint technology. However, there have not been many researches on anti-fouling paints in a multilayer-paint system.

Therefore, the investigation on the corrosion resistance of epoxy primer, polyurethane, and silyl acrylate anti-fouling on structural carbon steel is conducted. The research method includes polarization measurements of the unscribed specimen and a salt spray test on the scribed specimens. Through these two tests, the corrosion resistance of the multilayer paint, a combination of anticorrosive and anti-fouling paints, in the atmospheric environment will be revealed.

2. Research Procedure

A JIS SS400 carbon steel plate with dimensions of 15 cm x 10 cm x 0.2 cm was used as a metal substrate. Before being painted, the plates were sandblasted in accordance with the SSPC SP-10 standard to achieve a cleanliness level of 2½ Sa (an ISO 8501-1 standard). The multilayer paint was comprised of three types of commercial paint applied through manual spraying with dry film thickness (DFT) as shown on Figure 1. The drying process of each layer took approximately 4 hours followed by a final curing taking 7 days in an open space.

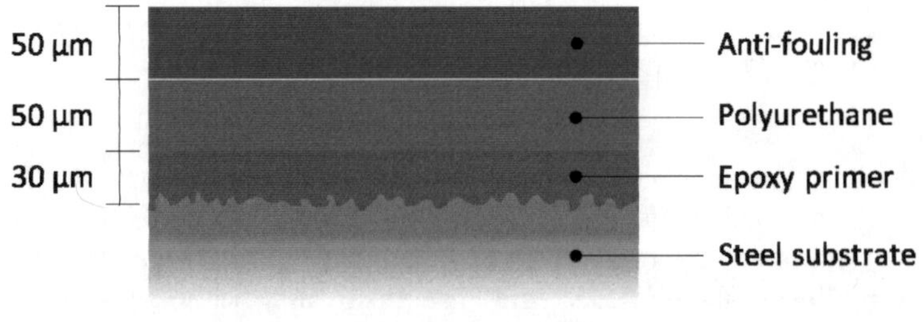

Figure 1. The three-layer paint.

In the process of polarization measurements, the painted specimens were cut to 1 x 1 cm, soldered with a copper wire, and mounted with resin. These specimens functioned as the working electrode, the platinum wire as the counter electrode, and Ag/AgCl as the reference electrode. All the electrodes were immersed in 3.5% NaCl solution and connected to a PGSTAT-302N Autolab potentiostat controlled by Nova 1.11 software. Electrochemical cells were put in a Faraday cage to minimize the capture of external noise that may interferes with the measurement process. The polarization measurements were conducted at a potential of -1.0 to 1.0 Volts and repeated for 3 times to ensure the accuracy.

Meanwhile, other specimens were scribed across until the scribe penetrated the entire layer and revealed a steel substrate. The specimens were then exposed to 5% NaCl solution in a 168-hour salt spray cabinet based on an ASTM B 177/ISO 9227 standard. Visual observations were done periodically at 48, 96 and 168 hours to assess rust, blisters, and other kinds of damage inside and outside the scribes.

3. Result and Discussion

3.1. Polarization measurements

The polarization measurements, carried out 3 times on the same specimen, produce a potential curve *E* (V) and a current density *i* (A/cm^2) as shown on Figure 2. This curve informs anodic and cathodic reactions, corrosion potential (E_{corr}), and passive conditions that may occur in the system. On the other hand, the Tafel extrapolation of the curve shows the corrosion current density (i_{corr}) to calculate the corrosion rate. The result of these polarization measurements shows that the layer is actively corroded after -0.9 and -1.05 V vs Ag/AgCl (E_{corr}) potential while the current density is relatively constant or only increases slightly as the passivation potential rises. In the 1st measurement, passivation looks quite clear at a potential of about -0.92 V vs. Ag/AgCl. Whereas in the 2nd and 3rd measurements, the passivation curve is formed twice at the potential around -0.95 V and -0.60 V vs. Ag/AgCl.

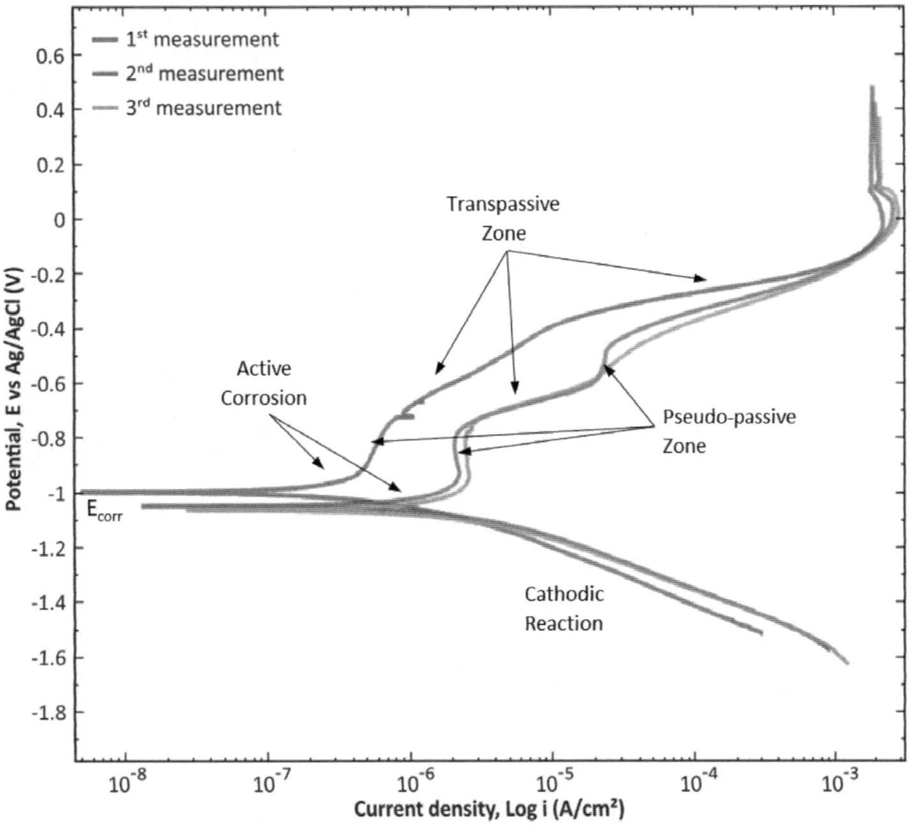

Figure 2. The polarization curve of specimens in 3.5% NaCl.

Table 1. The results of Tafel extrapolation.

Measurement	E_{corr} (V)	i_{corr} (µA/cm^2)	β_a (mV/dec)	β_c (mV/dec)	R_p (kΩ)	CR (mm/year)
1st	-0.998	1.107	165.25	-624.17	88.1260	0.01287
2nd	-1.040	1.167	124.30	299.99	32.7070	0.01356
3rd	-1.057	1.021	193.57	112.29	30.2070	0.01187

IOP Conf. Series: Materials Science and Engineering **541** (2019) 012052 doi:10.1088/1757-899X/541/1/012052

This passivation is linked to zinc phosphate contents, $Zn_3(PO_4)_2$, as an anti-corrosion pigment in the epoxy primer paint. According to a study by Hao et al. [10], zinc phosphate pigment has an inhibitory mechanism in the form of a shielding effect combination of zinc phosphate and the formation of a passive phosphate layer consisting of $FePO_4$, Fe_2O_3, and FeO on the metal surface that covers corrosion products and prevents further corrosion. However, the pseudo-passive behaviour of the system shows that the passive layer of phosphate formed at the steel paint-substrate interface is weak, unstable, and not tight.

Another factor that influences the occurrence of pseudo-passive condition is the characteristic of the anti-fouling paint materials that can start chemical hydrolysis with seawater, forming a thin leaching layer—which then eroded by the flow of seawater—and releasing the biocides. Silyl acrylate polymers are relatively difficult to be permeated so that the cycle of both formation and erosion of the leaching layer takes place at a relatively constant rate. In other words, passivation occurs when the leaching layer forms and erodes, leading to a decrease in the thickness of the paint. This also explains a bit more about the difference of the curve shape from the first run of polarization measurement with the next two, where during the initial immersion the paint is still intact or the process of forming and erosion of the leaching layer has not yet occurred. As the measurement took place and the specimen has been immersed for some time, the leaching layer is formed and eroded thus affects the passivation curves on subsequent measurements.

3.2. Salt spray test

The rating of specimen condition after the salt spray test is shown in Table 2. The average scribe width of the specimens before the test (0 hour) is around 0.5 mm while after the 168-hour test, the rate becomes 3.8 mm or rises by approximately 3.3 mm. This indicates that the paint film loses adhesion in the scribes. Even so, in areas that are not scribed there are almost no visible signs of paint damage such as rust, flaking, and blisters.

Table 2. The results of salt spray test.

Test Periods (hours)	Specimen Code	Rating of Failure at Scribe		Rating of Unscribed Areas	
		Mean Creepage from Scribe (mm)	Rating Numbers(*)	Area Failed (%)	Blister Rating(**)
0	A	-	10/10	-	10/10
	B	-	10/10	-	10/10
	C	-	10/10	-	10/10
48	A	0.2	9/10	-	10/10
	B	0.3	9/10	-	10/10
	C	0.1	9/10	-	10/10
96	A	1.0	8/10	-	10/10
	B	0.8	8/10	-	10/10
	C	1.0	8/10	-	10/10
168	A	2.5	6/10	0.1	8F/10
	B	2.5	6/10	0.1	8F/10
	C	5.5	4/10	0.1	8F/10

(*) Rating based on an ASTM D 1654 standard on a scale of 10 – 0.
(**) Rating based on an ASTM D 714 standard on a scale of 10 – 0.

Figure 3 shows the visual condition of the specimens before and after salt spray test. All the specimens suffer a severe damage to the scribes and the formation of rust has clearly been visible from the 48-hour test period. Most of the corrosion products are brown-orange-reddish and black which are the characteristics of iron oxide rust. The iron rust seems to decay until it comes out of the scribed areas, leaving rust-brown stain on the specimen surface. In addition, there are also white patches around the scribes which are salt deposits from NaCl solution that dissolve in the anti-fouling paint and then dry.

0 hour **48 hours** **96 hours** **168 hours**

Figure 3. The condition of specimens before and after salt spray test.

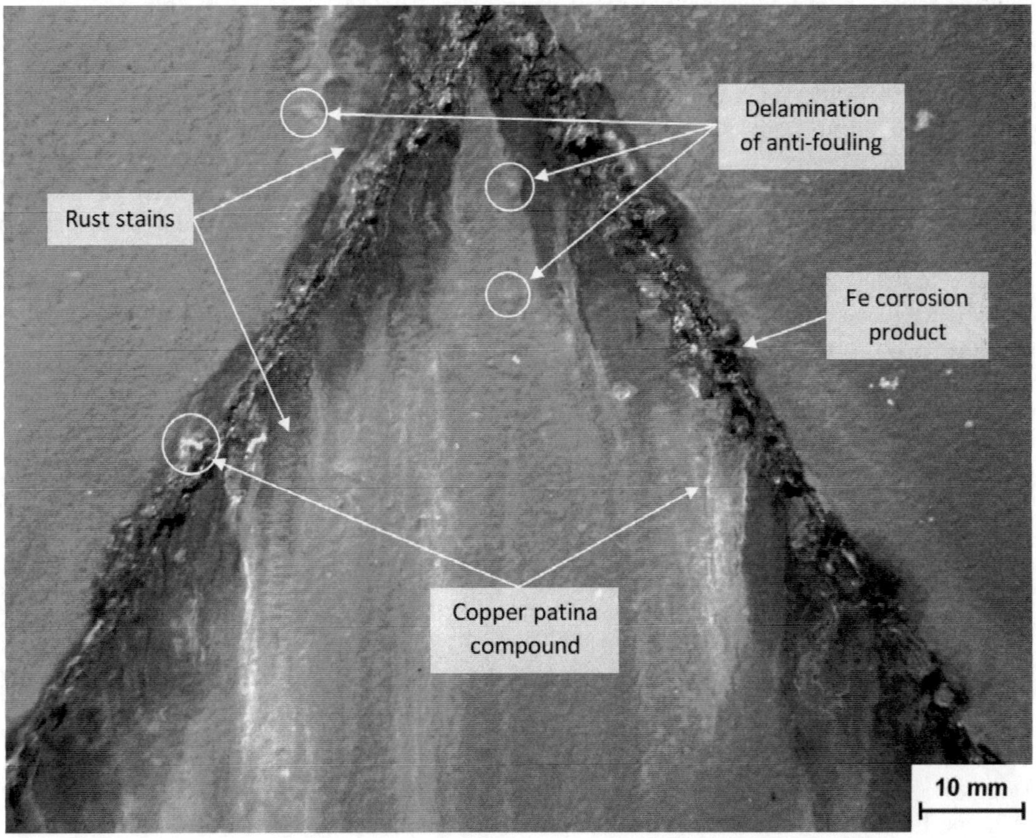

Figure 4. The detailed rust in the middle of scribe.

A number of greenish products appear in several locations around the scribes, which are suspected as copper hydroxy chloride compounds. If the paint is in its actual service condition, which is seawater, the pigments released by the anti-fouling paint should be carried away by the flow of seawater. However, in this salt spray test, the drops of electrolyte solution actually remain attached to the surface of the paint and seep into the paint film instead of flowing down. The reason is that the self-polishing anti-fouling paint formulated to hydrolyze with water forms a leaching layer. Therefore, the copper (I) oxide (Cu_2O) pigment released remains attached to the specimen surface and reacts with NaCl and oxygen solution to form the following copper hydroxy chloride compounds ($Cu_2(OH)_3Cl$) [11]:

$$Cu_2O + Cl^- + H_2O + \frac{1}{2}O_2 \rightarrow Cu_2(OH)_3Cl + OH^- \quad (1)$$

However, Cu_2O is said to be a semiconductor with relatively weak electrochemical activity [12]. Another study also reveals that the existence of Cu_2O in the anti-fouling paint does not have any role in the corrosion process of metal underneath, as corrosion only occurred at areas where crevices or paint damage had formed [13]. While in this study, no corrosion was seen in which anti-fouling paint was intact, proving that corrosion attacks occurred due to differential aeration cells in the paint crevice or scribe. Polyurethane and epoxy primer paint also act as insulation layers between anti-fouling paint layer and steel substrate, so that neither copper nor seawater can rapidly penetrate through coating pores. Moreover the Cu_2O anti-fouling paint contains virtually no metallic copper. Hence, the presence of Cu_2O as a paint pigment has no significant effect on the corrosion rate of carbon steel.

In short, the system of the three-layer paint has poor corrosion resistance especially when the paint has defects such as scribes and other types of mechanical damage. The blisters and the peeled parts around the scribes indicate that low adhesion reduces the paint protection. Electrolyte solutions that are absorbed horizontally into the paint film through the edge of the scribe also get the steel substrate to corrode. Corrosion products formed at the paint-steel interface push the layer until it blisters and is peeled. The lifting of the layer from the steel substrate allows corrosion products to be eroded by the flow of water and thus the steel surface is exposed to electrolyte attacks and forms new corrosion products.

In addition, anticorrosive zinc phosphate pigments contained in the epoxy primer also work less effectively due to the amount of rust formed in the scribes. As a result, the specimens experience a significant increase in the scribe width. This strengthens the analysis of pseudo-passive behavior on the polarization curve, that a protective passive layer that should be formed from phosphate and iron reaction with oxygen cannot protect the steel substrate, especially, in the defective or non-coated parts.

Several studies suggest that salinity has an effect on the rise in the diffusion of biocide pigments such as Cu_2O and the solubility of rosin in water, in addition to the thickness of the leaching layer which is the result of the pigment dissolution at the water-paint interface and of the polishing of the paint binder [14,15]. Therefore, the biocide release in seawater is higher than in brackish and fresh water, especially in the rosin-based anti-fouling paint or the self-polishing paint [16]. Based on this theory, high NaCl concentration in the test medium (5% NaCl) makes the anti-fouling paint degrade so fast that the thickness of the layer decreases. The thinning of the paint layer—the paint barrier function reduces—causes the water to enter the paint layer more quickly and attack the steel substrate.

4. Conclusion

Polarization measurements and a salt spray test on epoxy primer, polyurethane, and silyl acrylate anti-fouling coated on carbon steel plates show that, first, the pseudo-passive condition on the polarization curve indicates a passive layer of the phosphate pigment reactions on the epoxy primer which is both less protective and affected by (a) the formation cycle and (b) the erosion of the leaching layer of the anti-fouling paint system. Second, in atmospheric condition, the paint corrosion resistance on the scribed specimens is low since they are not protected by a passive phosphate layer. The corrosion process occurs when the electrolyte solution enters the paint layer horizontally and weakens its adhesion. The loss of adhesion then has the layer easily lifted and triggers underfilm corrosion and blisters.

References

[1] Fontana M G, 1987 *Corrosion Engineering*, 3rd ed. Singapore: McGraw-Hill, Inc.
[2] Jones D A, 1996 *Principles and Prevention of Corrosion*, New Jersey: Prentice Hall, Inc.,
[3] Nurioglu A G, Esteves A C C, and de With G 2015 *J. Mater. Chem. B* **3** 6547–6570
[4] Karlsson J, Ytreberg E, and Eklund B 2010 *Environ. Pollut.* **158** 681–687
[5] Kiosidou E, Karantonis A, and Pantelis D 2014 *Chem. Eng. Trans.* **41** 301–306
[6] Chambers L D, Stokes K R, Walsh F C, and Wood R J K 2016 *Surf. Coat. Technol.* **201**, 3642–3652
[7] Kiosidou E D, Karantonis A, Pantelis D I, Silva E R, and Bordado J C M 2017 *J. Coat. Technol. Res.* **14**, 333–345
[8] Kiosidou E D, Karantonis A, Pantelis D I, Silva E R, and Bordado J C M 2017 *J. Coat. Technol. Res.* **14**, 1381–1395
[9] Chen R *et al.* 2017 *RSC Adv.* **7** 40020–40027
[10] Hao Y, Liu F, Han E H, Anjum S, and Xu G 2013 *Corros. Sci.* **69** 77–86
[11] Strandberg H 1998 *Atmos. Environ.* **32** 3521–3526
[12] Ding R, Li X, Wang J, Li W, Wang X, and Gui T 2017 *J. Mater. Eng. Perform.* **27** 5947–5963
[13] Bagley F, Atlar M, Charles A, and Anderson C 2015 *Ocean Eng.* **109** 595–602
[14] Singh N. and Turner A. 2009 *Environ. Pollut.* **157** 371–376
[15] Lagerström M, Lindgren J F, Holmqvist A, Dahlström M, and Ytreberg E 2018 *Mar. Pollut. Bull.* **127** 289–296
[16] Ytreberg E *et al.* 2017 *Environ. Pollut.* **225** 490–496

Acknowledgements

The authors would like to thank the financial support provided by Universitas Indonesia through the PITTA 2018 funding scheme No. 2520/UN2.R3.1/HKP.05.00/2018 managed by the Directorate for Research and Community Engagement (DRPM UI) and also to PT. Gunanusa Utama Fabricators for all facilities that offered to perform this work.

The effect of saccharin addition to nickel electroplating on the formation of nanocrystalline nickel deposits

R Riastuti[1], **S T Siallagan**[1], **A Rifki**[1], **F Herdino**[1], **C Ramadini**[1]
Department of Metallurgy and Materials Engineering, Faculty of Engineering, Universitas Indonesia, Depok, West Java, Indonesia

E-mail: riastuti@metal.ui.ac.id

Abstract. This study investigates the effect of saccharin addition on nickel deposits. Coating substrates with nickel is carried out through electroplating. This method involves Watts bath, 0, 1, 3, and 5 g/L saccharin, and a current density of 6 A/dm^2. The electroplating process lasts for 20 minutes at 50°C. The plated substrate is a Steel Plate Cold Rolled Coil (SPCC). The deposits attached to the SPCC surface is characterized by a Scanning Electron Microscope (SEM) to observe the structural changes of the deposits and an X-Ray Diffraction (XRD) to calculate their grain size. The results show that, first, nickel deposits without saccharin addition has a pyramidal and crystal polyhedral structure while those with saccharin reflect a globular structure and colonial patterns. Secondly, the addition of 5g/L saccharin produces the smallest grain, 17.39 nm.

1. Introduction

Coatings are often and easily applied to prevent metal corrosion. This technique has several variants such as electroplating, Physical Vapor Deposition (PVD), Chemical Vapor Deposition (CVD), laser beam deposition, and ion implantation. Of all the types of coating, electroplating is the most economical and its production rate is so high that this method fit the use in industry [1]. On the other hand, the material for coating or plating the object is nickel and thus, the method is completely named 'nickel electroplating'.

Nickel electroplating is the process of depositing nickel onto the surface of a conductive material electrochemically through direct electric current. This technique gets the metal surface more shining and more resistant to corrosion in the atmosphere [2]. Nanocrystalline nickel deposits can be produced through electroplating which results in low porosity residuals [1]. In addition to enhancing corrosion resistance, the nano-sized nickel grains also raise the metal strength and tribological properties [5].

The scale of nanometer on the deposits can be obtained by adding saccharin to nickel electroplating. Besides electroplating, several parameters capable of producing nano-sized particles include electrodeposition mode, temperature, current density, and stirring [5, 6, 7]. Here are some previous investigations that involve saccharin and the aforementioned parameters. Rashidi and Amadeh (2008) found that the addition of 5 g/L saccharin in Watts bath reduced the nickel grain size, from 182 nm to 24 nm, by increasing the current density from 10 to 75 mA/cm^2 [1]. Next, El-Sherik (1995) concluded that the decline in nickel electrodeposition grain size between 40-10 nm occurs when saccharin from 0.5 to 10 g/L is put to Watts bath, using pulsed current [5].

This research explores the effect of saccharin addition to nickel electroplating on nickel deposits. The characterization of nickel particles uses a Scanning Electron Microscope (SEM) to observe their structural changes and an X-Ray Diffraction (XRD) to calculate their grain size. The object to plate is a Steel Roll Cold Rolled Coil (SPCC).

2. Methodology

2.1. Materials

The plated material in this study is a 50x50x2 mm Steel Plate Cold Rolled Coil (SPCC) which plays as a cathode. SPCC is a carbon steel produced through cold rolling and extensively exploited in automotive and electrical equipment. Carbon rods perform the function as an anode. Figure 1 below is the picture of a steel plate cold rolled coil surface.

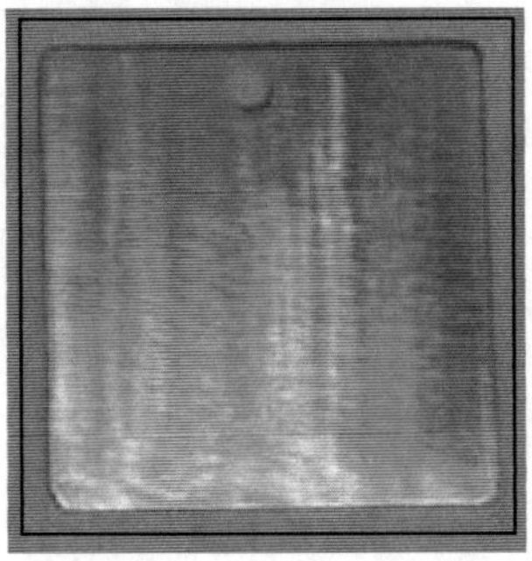

Figure 1. SPCC Substrate.

Table 1 lists the results of SPCC substrates composition analysis through an Optical Emission Microscope (OES). The results have satisfied the JIS G3141 standard: commercial quality, where the contents of carbon are less than 0.12%.

Table 1. SPCC Chemical Composition.

Material	Composition			
	C (max)	Mn (max)	P (max)	S (Max)
SPCC	0.0482	0.182	0.011	0.0079

2.2. Research Procedure

Surface preparation is the initial stage of nickel electroplating. This preparation aims to remove rust, oil, and other inhibiting substances from the surface of the substrate. The stages in the steel surface preparation are as follows. First, the surface was rubbed with a sandpaper, grit of 120-1000, to get rid of the oxide layer and dirt. Then, the object underwent the 10% NaOH-degreasing process for 11 minutes at 65°C. Next, the surface was 32% HCl-pickled for 10 minutes at a room temperature. Lastly, the substrate surface was etched with 10% H_2SO_4 for 4 minutes at a room temperature.

After the surface preparation, the process came to nickel electroplating. This process engaged (**a**) Watts bath solution containing 300 g/L nickel sulphate ($NiSO_4$), 30 g/L nickel chloride ($NiCl_2.6H_2O$), and 30 g/L boric acid (H_3BO_3), (**b**) 0.2 cc/L wetting agent, as well as (**c**) 0, 1, 3, and 5g/L saccharin. The pH of the solution ranges from 4 - 4.2. This plating process lasted for 20 minutes at 50°C with a current density of 6A/dm^2. The process then left nickel deposits on the substrate surface. Nickel deposits were then observed with an optical microscope, a Scanning Electron Microscope (SEM), and an X-Ray Diffraction (XRD) to find the form and pattern of the nickel deposits and their grain size.

Only in XRD, the observation collaborates with Schererr equation. The formula for this equation is provided bellow:

$$D = \frac{k\gamma}{\beta \sin \theta} \quad (1)$$

D is the grain size at the nanometer scale, *k* the Scherrer constant, γ the X-Ray wavelength, and β the Full Width at Half of Peaks (FWHP). The analysis of grain size through Scherrer equation was conducted with the help of an X'Pert High Score Plus.

3. Results and Discussion

This chapter is comprised of such sub-chapters as optical microscopic observation, Scanning Electron Microscope (SEM) observation, and X-Ray Diffraction (XRD) observation.

3.1. Optical Microscopic Observation

Figure 2a shows the results of optical microscopic observation on the macro photography of nickel-electroplated steel without saccharin addition. On the contrary, those with saccharin addition is on figure 2b (1 g/L), c (3 g/L), and d (5 g/L).

The no-saccharin addition nickel surface has a darker surface than the saccharin-added nickel layer, as shown on figure b, c, and d. After being investigated, the saccharin-added nickel layers are smoother than that which is not saccharin-added. Inside the finer nickel layers exist very small grains which have the light bounce at the same angles. Therefore, the surface appears brighter. Meanwhile, the rough surfaces get the light to scatter to different angles and thus, look more gloomy [8]. In other words, the addition of saccharin to nickle electroplating leads to smaller and smoother nickel particles and even layers.

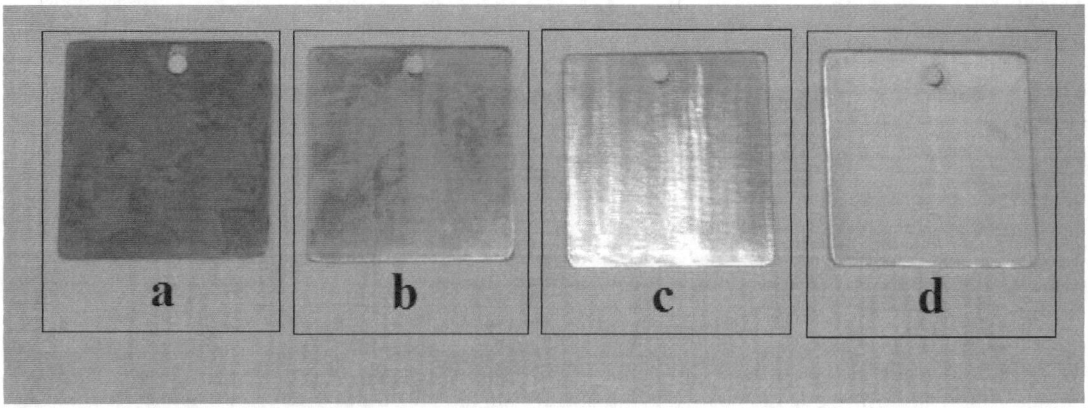

Figure 2. The optical microscopic observation on the macro photography of nickel-electroplated steel with saccharin addition; a) 0 g/L, b) 1 g/L, c) 3 g/L, and d) 5 g/L.

Figure 3 presents the results of optical microscopic observation on the 'cross-section' macro photography. The formed nickel layers were examined using an optical microscope at 200x magnification. Figure 3a, b, c and d show that the substrates have been completely plated with nickel either with the addition of saccharin or not. Nickel coatings look yellowish-white while SPCC is shown in ferrite (bright) and perlite (dark)-phased macro structures.

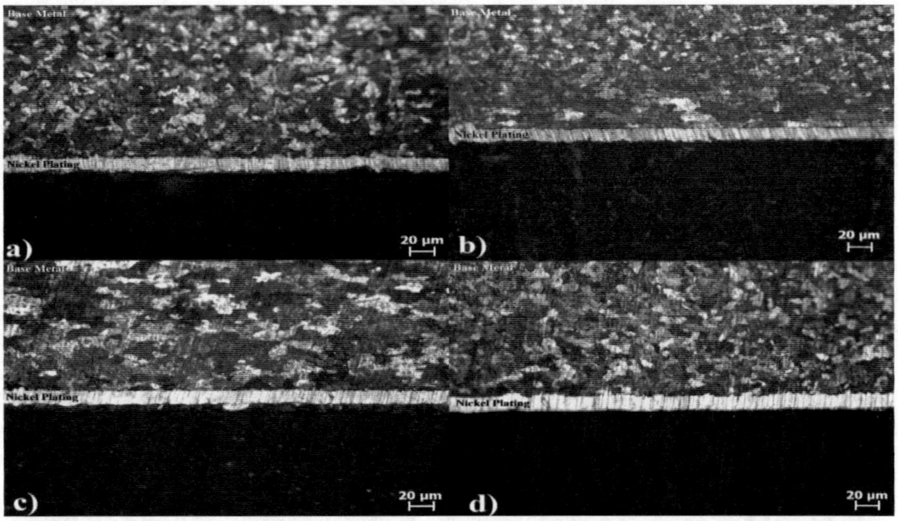

Figure 3. The cross-section of nickel-electroplated steel with saccharin addition; a) 0 g/L, b) 1 g/L, c) 3 g/L, and d) 5 g/L at 200x magnification.

3.2. Scanning Electron Microscope (SEM) Observation

The observation of the form of nickel deposits uses a scanning electron microscope at 5000x magnification. The results are shown on figure 4. On the one hand, figure 4(a) presents that no-saccharin addition nickel coatings have pyramidal and polyhedral crystals and an uneven surface.

On the other hand, the addition of saccharin to nickel electroplating as shown on figure 4(b), (c), and (d) produces globular and colonial particles. The formation of colonies on these deposits improves the density between particles. This phenomenon is an inhibitory effect of organic additives on the reduction of deposited nickel ions. In addition, additives in the electroplating process also prevent the formation and growth of pyramidal structures and reduce the roughness of the coatings, having the surface be glossy [6, 9]. Similar changes in the deposit structures occur in a research by Wang, Gao, Xu, and Xue (2005). In this exploration utilizing electrodepositions, pyramidal particles and polyhedral crystals get the surface of polycrystalline nickel rougher than that of nanocrystalline nickel [10].

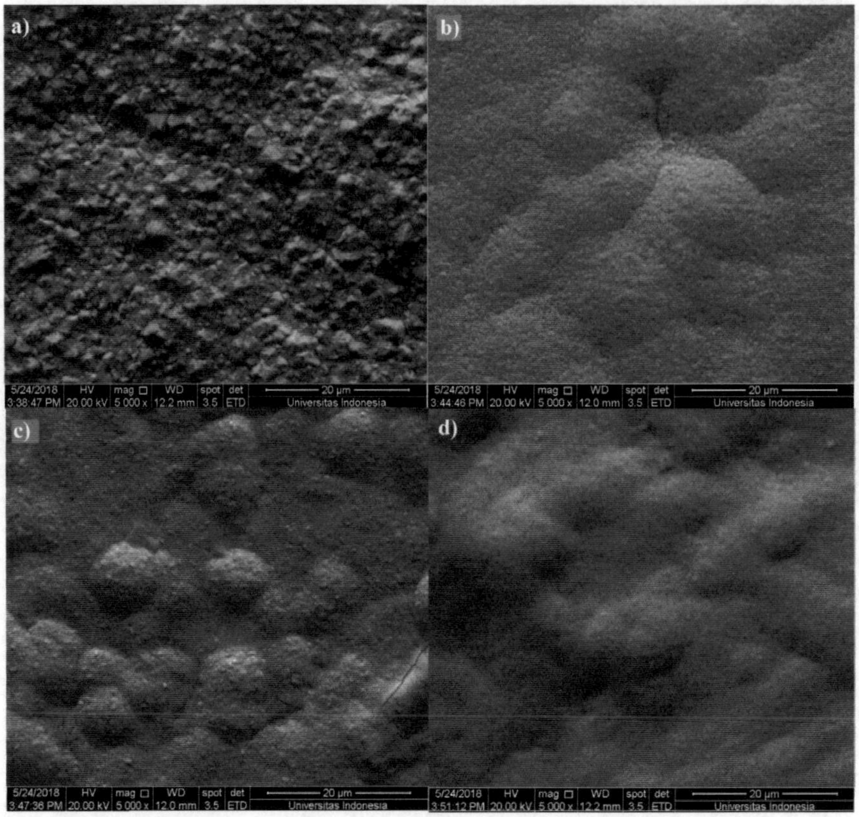

Figure 4. The micro photography of nickel-electroplated steel with saccharin addition; a) 0 g/L, b) 1 g/L, c) 3 g/L, and d) 5 g/L.

3.3. X-Ray Diffraction (XRD) Observation

Figure 5 shows the trends of deposits particles size through the XRD analysis based on saccharin concentrations. The analysis reveals that the nickel deposits have a FCC crystal structure: fields (111) and (200). The addition of saccharin during electroplating has proven to widen the peak, indicating that the grain size of the deposits becomes smaller.

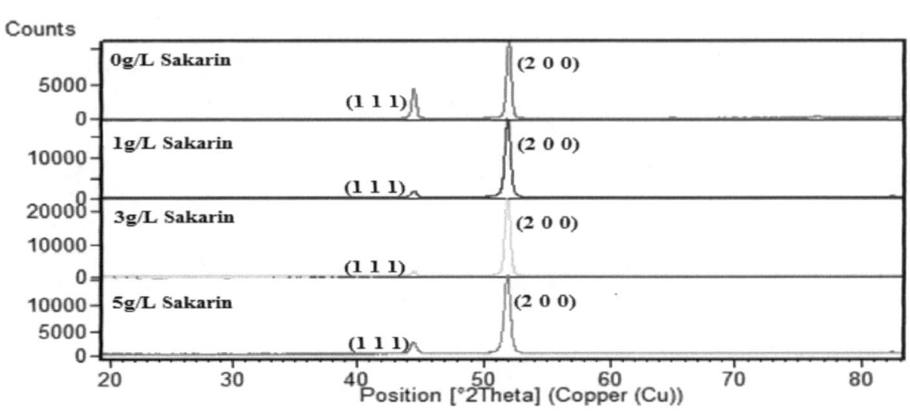

Figure 5. The trends of nickel particle size based on saccharin concentrations

IOP Conf. Series: Materials Science and Engineering **541** (2019) 012053 doi:10.1088/1757-899X/541/1/012053

Measuring nickel particle size through Scherrer equation was carried out on fields (111) and (200) because both have the highest intensity. Table 2 and figure 6 provide the details of the nickel grain size calculation based on Scherrer equation and the details of the changes in the grain size.

Table 2. The nickel grain size.

Saccharin Concentrations (g/L)	Grain Size (nm)		
	Peak/2θ		Rates
	(111)/44.5°	(200)/51.8°	
0	24.1	25.5	24.8
1	18.3	24.7	21.48
3	18.3	20.03	19.43
5	17.55	17.22	17.39

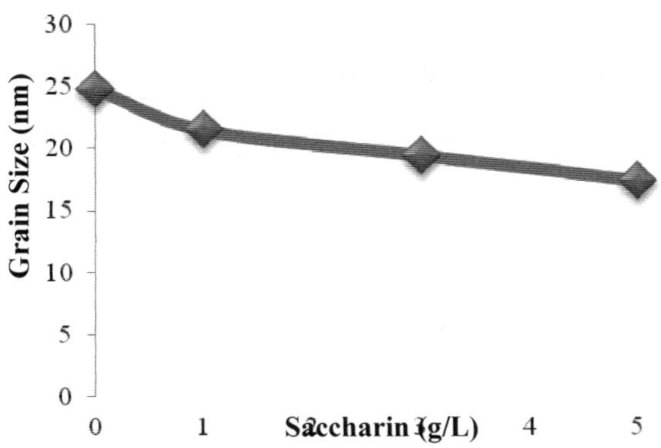

Figure 6. The effect of saccharin concentrations on the grain size.

The addition of saccharin to nickel coatings through electroplating has proven to reduce the nickel grain size. The small grain size leads to smoother layers. These smooth surfaces indicate the evenness of nickel layers. This further raises the brightness of the layers. Among all saccharin concentration variations, 5 g/L results in the smallest grain size, 17.39 nm, and the brightest light. In connection with brightness, Yli-pentti (2014) suggested that saccharin is one of the traditional brighteners widely used.

The workings of saccharin in decreasing the grain size of the nickel deposits are as follows. Saccharin blocks the surface of the substrates through the formation of complex compounds that effectively increases the frequency of nucleation, but decreases the diffusion of ions absorbed on the cathode surface. This phenomenon not only hinders the growth of nickel particle size but also affects the hydrogen evolution and distorts the cathodic overpotential [6].

4. Conclusion

The conclusion of all off the above investigations is, first, the addition of saccharin to nickel deposits through electroplating alters the deposits structure, from pyramidal to globular and colonial. The layers with the globular structure have a more even appearance than those with the pyramidal one. Secondly, the addition of 5g/L saccharin in nickel electroplating most effectively reduces the nickel grain size, from 24.8nm to 17.39nm, among all concentrations.

References

[1] Rashidi A M and Amadeh A, 2008 *Surf. Coatings Technol.* **202** 3772–3776
[2] Di Bari G. A., 2011 *Mod. Electroplat. Fifth Ed.* 79–114
[3] Kumar, H K, Swygenhoven S V, and Suresh S 2003 **51** 5743–5774
[4] Bhardwaj M, Balani K, Balasubramaniam R, Pandey S, and Agarwal A 2011 *Surf. Eng.* **27** 642–648
[5] Wasekar N P, Haridoss P, Seshadri S K, and Sundararajan G, 2016 *Surf. Coatings Technol.* **291** 130–140
[6] Rashidi A M and Amadeh A, 2009 *Surf. Coatings Technol.* **204** 353–358
[7] Rashidi A M and Amadeh A, 2010 *J. Mater. Sci. Technol.* **26** 82–86
[8] Yli-Pentti A, 2014 *Electroplating and Electroless Plating*, **4**. Elsevier
[9] Meng G. *et al.* 2015 *J. Mater. Sci. Technol.* **31** 1186–1192
[10] Gao L Y, Xu T, and Xue Q, 2005 *Materials Chemistry and Physics 99* 96–103

Acknowledgments

This research was financed by PITTA UI 2017 while research facilities were provided by PT. Main Radian Motor. We thank the two institutions for fully supporting this research.

IOP Conf. Series: Materials Science and Engineering **541** (2019) 012054 doi:10.1088/1757-899X/541/1/012054

The effect of alcohol media in the synthesis of calcium carbonate nanoparticles by ultrasonic waves

E Sulistiyono[1], M Handayani[1], M R Arwanda[2], T Kusumaningsih[2], F. Firdiyono[1]
[1]Research Center of Metallurgy and Materials, Indonesian Institute of Sciences, Building 470, Puspiptek Area, Tangerang Selatan
[2]Department of Chemistry, Faculty of Mathematics and Natural Sciences, Sebelas Maret University, Surakarta

E-mail: eko221068@gmail.com

Abstract. The process of destroying kidney stones using ultrasonic waves at certain frequencies has long been known in the medical field. The process of granular destruction using ultrasonic waves is very possible to produce a very fine grain size due to the large number of micro bubbles from the cavitation effect on the ultrasonic probe. The purpose of this research is to determine the effect of alcohol media in the destruction of calcium carbonate granules by ultrasonic waves to obtain carbonate calcium with nano-sized particles. The alcohol used in this study was 2-butanol, Ethylene glycol and n-propanol. The raw material of calcium carbonate was calcium carbonate precipitates originating from the Bayat region, Klaten Regency, Central Java, which was processed first by the carbonation method into calcium carbonate precipitates. The best result of experiments on the destruction of precipitated calcium carbonate by ultrasonic waves was the use of ethylene glycol media with ultrasonic time of 10 minutes and 30% amplitude. The result of the particle size analysis (PSA) at the best condition obtained with the average size of the granules of calcium carbonate precipitate was 72 nm.

1. Introduction

The concept of nanotechnology is a technology that presents phenomena or material properties on a nanometer scale that have special properties. Because it has a very small particle size in the nano meter scale, the experts agree that so-called nanoscale materials are materials that have dimensions of 1 nm up to 100 nm in size [1]. One way to produce nano-size material from a suspension of a solution is to break down particles in suspension with the help of ultrasonic waves [2]. Generally ultrasonic waves used are greater than 20,000 Hz and up to 1 MHz units, depending on the material used [3].

Several studies have been developed to explain sonic radiation of 20 KHz can break chemical bonds. The essence of sonochemical events is the formation, growth and breakdown of microbubble produced during the process of radiation in liquids. The growth level of microbubble occurs by diffusion from evaporation of the solution in the microbubble volume. The last stage is the breakdown of microbubble, which occurs when the microbubble is in the maximum size [2].

Calcite and Aragonite are the two main constituents of $CaCO_3$, which are known to be easily formed in large quantities in nature [4]. Natural calcite is a mineral formed through the crystallization process in shallow sea areas. Calcites are included in the carbonate group with the same chemical composition, carbonate ($CaCO_3$), but sometimes these minerals also compose rocks that have different compositions [5].

Limestone ($CaCO_3$) is one of the many industrial minerals that are needed and used both directly and indirectly. The direct use of limestone is in the cement, sugar, ceramic, and additive materials in the process of smelter and building materials, while indirect use still needs to be processed in advance which can be in the form of light calcium carbonate or quicklime (CaO). Light calcium carbonate and calcium lime are as much needed for Industries of the paint, toothpaste, paper bleaching, cosmetics, soil neutralization, building materials and others [6].

The Research center of metallurgy and materials (P2MM) has succeeded in making light calcium carbonate precipitate derived from limestone in various regions in Indonesia [7-8]. If this light calcium carbonate material can be made in nano size, this will increase the added value because it has superior properties. In this study, calcium carbonate precipitate nanoparticles were synthesized by Ultrasonic method using alcohol-based media such as propanol, butanol and ethylene glycol.

2. Method

2.1. Materials

The raw material used in this research is light calcium carbonate which comes from the processing of limestone originating from Bayat, Klaten, Central Java. The raw material for limestone from the Bayat area, Klaten Region, Central Java has superior quality and can be made of high quality calcium carbonate precipitate with little impurity [9]. As a medium for ultrasonic processes in this activity using materials 2-Propanol, 2-Butanol and Ethylene Glycol with analytical grade (Merck) were obtained from commercially available. The use of ultrasonic media 2-Propanol, 2-Butanol and Ethylene Glycol with analytical grade (Merck) was carried out on magnesium carbonate and was able to produce nano particle hydromagnesite with a size below 100 nm [10].

2.2. Experiment

Limestone is processed in the stages of calcination, slaking, carbonation and drying processes. The detail synthesis of light calcium carbonate nanoparticles was performed using the previous method [7-8]. The process of mixing between Light Calcium Carbonate that has been produced with various types of alcohols: 2-Propanol, 2-Butanol and Ethylene Glycol. In this study, the experimental variables were the type of solvents and amplitude of the ultrasonic process. The solvents used in this experiment were 2-Propanol, 2-Butanol and Ethylene Glycol, with the amplitude were 20%, 25%, 30%, 35% and 40%. The fixed variable used in this experiment was the temperature of ultrasonic process which was maintained at a maximum of $90^{O}C$, the media volume was 200 ml, the addition of 10 g material and the ultrasonic time was about 10 minutes. The result was then measured of sediment rate using the Stokes method, measurement of the particle size with PSA Nano and characterization of the surface morphology of particles using SEM.

3. Result and Discussion

3.1. Analysis of limestone as raw materials by X-Ray Fluorescence and X-Ray Diffraction

X-Ray Fluoresence (XRF) analysis results show that limestone from the Bayat region has a large amount of impurities such as magnesium, silicon and iron. Calcium contents in the limestone oxide are only 90%, then magnesium impurities are 6.54% and iron impurities are 0.71%. After a series of processes obtained light calcium carbonate with a high purity level of 99.16%, and impurities are only very small amounts of phosphorus and sulfur.

Table 1.XRF analysis on limestone from Bayat, Klaten Regency and after processing into light calcium carbonate precipitate and compared with Merck.

Component	Before Process (% wt)	After Process(% wt)	Comparison with$CaCO_3$ from Merck (% wt)
CaO	90,93	99.16	99.40
MgO	6,54	Nd	Nd
SiO_2	0.74	Nd	Nd
Al_2O_3	0.47	Nd	Nd
P_2O_5	0.27	0.14	Nd
SO_3	0.05	0.14	0.16
Fe_2O_3	0.71	Nd	Nd

[a]laboratorium of Mathematics and Natural Sciences (MIPA)
-UI. Nd = No Detector

The X-Ray Diffraction analysis conducted at NanoTech Laboratory shows that limestone from the Bayat region shows peak of calcite and impurities were not detected. This can be seen in the XRD analysis image as shown in the figure 1.

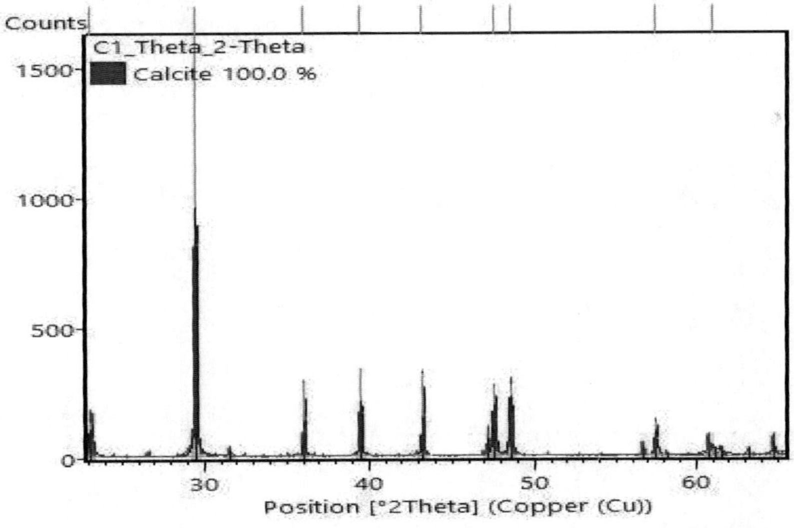

Figure 1. The X-Ray Diffraction analysis of the limestone as raw materials

After a series of work processes on raw calcium carbonate are carried out, a light calcium carbonate precipitate product is obtained. The results of the XRD spectra at 2 theta shows that there are no significant shifts in the sample resulted from the experiments compared with the natural limestone, but there are differences in the intensity of $CaCO_3$. The results of the matching-lab analysis, it is seen that only the peak of $CaCO_3$presented in the spectra while the peak of other calcium compounds such as

CaO, $Ca(OH)_2$, and $Ca(HCO_3)_2$ is undetectable. The reaction take place perfectly by forming $CaCO_3$again, and the side reaction of bicarbonate formation does not occur. The below spectra showed that CaO has been formed with a percentage of 86.6% and the by-product is a portlandite which is an impurity in the form of CaO_2H_2.

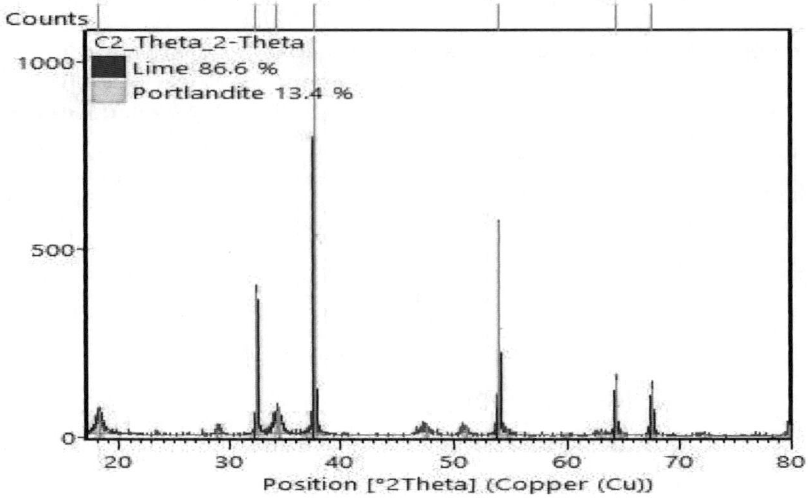

Figure 2. The X-Ray Diffraction analysis of light calcium carbonate precipitate

3.2. Measurement of precipitation rate by Stokes Method

The stokes method is a method of measuring the precipitation speed of solids in a liquid medium. According to Stokes method, if the particle size is smaller, the deposition process is longer so that the the grain become smaller and the precipitation speed is also smaller. In liquid media, in addition to grain size, the deposition process is also influenced by solid density, media density and binding capacity between media molecules and solid molecules.

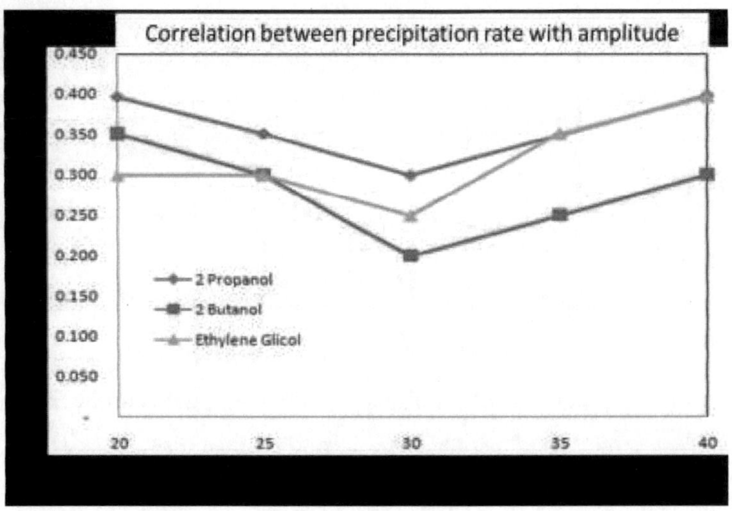

Figure 3.The results of precipitation rate measurements on the amplitude of ultrasonic waves in alcoholic media

In general, the experimental results on the three-alcohols media shows a similar tendency, the higher of the amplitude, the precipitation rate become smaller so that the size of the particle become smaller. From the results of precipitation rate measurements, it is seen that the use of 30% ultrasonic waves shows the lowest deposition speed. But the tendency changes at amplitude 30%, where above 30% of amplitude the precipitation speed rises again. So it can be concluded that the use of 30% amplitudes is the best point that produces the smallest average size of particles.

3.3. Analisys of particle size by particle size analyzer (PSA)
In this study, the particle size of calcium carbonates before ultrasonic process and after ultrasonic process are measured by PSA in alcohol media, The results of the PSA measurement can be seen in the table below:

Tabel 2. Analysis result of particle size by PSA (Source: Laboratorium of Nanotech LIPI Serpong)

No	Alcohol Media	Particle size by PSA (Nanometer)		
		Before ultrasonic	After Ultrasonic	Size ratio
1	2-Propanol	524	195	2,687
2	2-Butanol	234	133	1,759
3	Ethylene glycol	207	72	2,875

The measurement of PSA shows that using ethylene glycol media gives the best results, where the average grain size is 72 nm, compared to 2-propanol which gives a grain size average of 195 nm. Based on the size ratio between before being subjected to ultrasonic and after being subjected to ultrasonic, it is also seen that the use of ethylene glycol as the media is the most effective.In terms of solution density, the highest density is ethylene glycol 1.11 g / cm^2 while butanol 0.806 g / cm^3 and 2 propanol 0.786 g / cm^3. Seeing the density of each solvent, it shows that ethylene glycol should have been the longest. By looking at the density of each solvent, it shows that indeed using ethylene glycol will result in the longest deposition due to the highest density, thus giving the smallest particle size.

3.4. Analysis of surface morphology and particle crsytal by scanning electron microscopy (SEM)
The result of SEM analysis for calcium carbonate precipitate without ultrasonic process can be seen in figure 4.

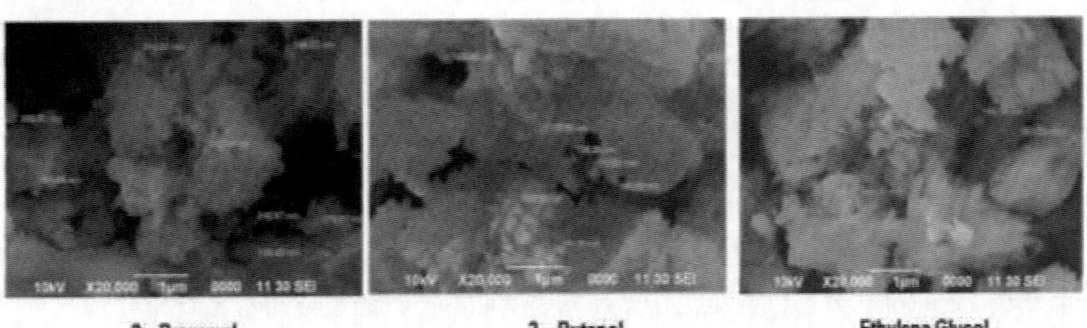

Figure 4. Measurement of SEM analysis of calcium carbonate precipitate in alcohol media without ultrasonic process

It can be seen that the granules appeared to be large enough indicating that the granules undergo a coagulation process in drying process. This is because the granules cluster together to form a fairly large lump and not in a single grain with a compact shape. The results of SEM analysis of calcium carbonate precipitate show a grain size range of 100 nm to 300 nm.

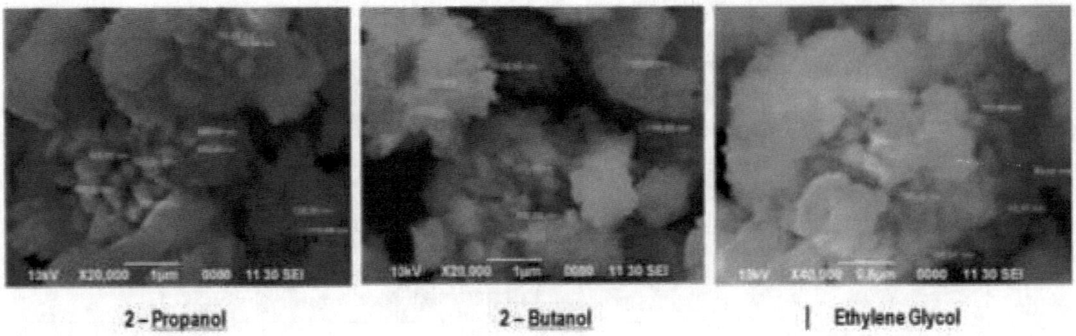

Figure 5. Measurement of SEM analysis of calcium carbonate precipitate in alcohol media with ultrasonic process for 10 minutes

The results of SEM analysis for samples in alcohol media subjected to ultrasonic waves within 10 minutes can be seen in figure 5. It shows that there is a tendency for reduction of particle size of calcium carbonate precipitate, although the coagulation process still occurs. the drying process carried out at 150°C, which is above the boiling point of alcohol, was still unable to break down the coagulant. If the temperature is raised above 600°C, it will be able to break the coagulant because all organic compounds undergo the pyrolysis process. But if the drying process is raised above 600°C, the calcination process will occur so that calcium carbonate changes to calcium oxide which is alkaline and hygroscopic. In SEM observations, it can be seen that the particle size of calcium carbonate precipitate in ethylene glycol media shows the best decrease after being subjected to ultrasonic waves.

4. Conclusion

The ultrasonic process has succeeded in reducing the size of calcium carbonate precipitates using alcohol media. From the results of precipitation rate measurements, it can be seen that there has been a decrease in precipitation speed with the best results on the use of 30% amplitude. Based on the analysis of the particle size analyser, it is seen that the reduction in particle size is best in the use of ethylene glycol media which produces particle size up to 72 nm and a particle reduction ratio of 2.875. The use of alcohol media has a disadvantage, among others, the emergence of the coagulation process in drying process of calcium carbonate precipitates. This coagulation process can be proven based on SEM analysis of the granules which is seen as large clumps with a crystal structure that is not in a single grain.

References

[1] Taniguchi N 1974 *Proc. Intl. Conf. Prod. Eng.* Tokyo. Part II. — Japan Society of Precision Engineering. p18-23
[2] Gedanken A 2004 *Ultrasonic Sonochem.* **11** 47-55.
[3] Crocker M J 1997 *Enclycopedia of Accoustic* (John Wiley and Sons, Inc)
[4] Wang C, Liu Y, Bala H, Pan Y, Zhao J, Zhao X, Wang Z 2007 *Colloids and Surfaces A: Physicochem. Eng. Aspects*. **297** 179 – 182.
[5] Nishida I 2004 *Ultrasonics Sonochem.* **11** 423–428

[6] Lailiyah Q, Baqiya M, Darminto 2012 *Jurnal Sains dan Seni ITS,* 1, No. 1 ISSN: 2301-928X. Surabaya : ITS.
[7] Handayani M, Sulistiyono E, Firdiyono F, and Fajariani E N 2018 *IOP. Conf. Ser.: Mater. Sci.* Eng **333** 01243
[8] Sulistiyono E, Handayani M, Firdiyono F, and Fajariani E N 2018 *IOP. Conf. Ser.: Mater. Sci.* Eng **333** 012042
[9] Handayani M and Eko S 2010 *Makalah Pendamping*, ISBN: 979-498-547-3: 271-275
[10] Sulistiyono, E, Azwar, M, Firdiyono, F 2012 *Majalah Metalurgi*, ISSN 0216-3188: 135-140

IOP Conf. Series: Materials Science and Engineering **541** (2019) 012055 doi:10.1088/1757-899X/541/1/012055

Synthesis and characterization of SiC and CNT doped MgB_2 superconducting wire

N Sofyan[1], M T Setiawan[1], S Herbirowo[2], A Imaduddin[2]
[1]Department of Metallurgical and Materials Engineering, Faculty of Engineering, Universitas Indonesia, Depok 16424 Indonesia
[2]Research Center for Metallurgy and Materials–Indonesian Institute of Sciences, Kawasan PUSPIPTEK, Tangerang Selatan 15314, Indonesia

E-mail : nofrijon.sofyan@ui.ac.id

Abstract. Silica nanoparticles and carbon nanotubes-doped magnesium diboride (MgB_2) superconducting wires have been successfully prepared via powder-in-tube (PIT) method. In this work, pure MgB_2 was doped using various dopants concentrations at 1 and 2 wt.% of each silica nanoparticles (SiC) and carbon nanotubes (CNT). The mix was sintered at 800°C for 3 hours before putting it in a 6 mm diameter stainless steel tube wire and drawn to get 3 mm diameter of stainless-steel wire. The materials were characterized using X-ray diffraction (XRD) for phase and structural analysis and scanning electron microscope (SEM) for surface morphology. Further, superconductivity characteristic of the wires was examined through a transition temperature using a cryogenic magnet. Structural characterization examined using X-ray diffraction showed that no other impurities of other phases were detected. Microstructural investigation using electron microscope showed even distribution of the particles with inherent porosities. Measurement at cryogenic temperature showed that pure MgB_2 and 2 wt.% CNT doped MgB_2 showed superconductivity characteristic after being sintered whereas others showed unique resistivity behaviors. After deformation by drawing process to form wires, all of the samples showed a superconducting behavior, however, the presence of SiC and CNT decreased the critical temperature, Tc, of MgB_2. Although the samples doped with CNT decrease the Tc, CNT doped samples showed higher Tc than that of MgB_2/SiC wires.

1. Introduction

Superconductor is defined as a material that has zero resistivity below a certain temperature. In this condition, the superconductor flows current without any obstacle. The temperature when the resistivity vanished is called as critical temperature (Tc) [1]. Magnesium diboride (MgB_2) is a superconductor with relatively high critical temperature (39 K). It has a simple structure by which it consists of magnesium and boron, and superconductivity in this MgB_2 was discovered by Nagamatsu and his coworkers in 2001 [2].

Superconductivity can be enhanced by determining the optimum microstructure of MgB_2 which includes the grain size, grain orientation, voids, cell structure and impurities [2]. Later on, researchers found that carbon addition in MgB_2 superconductor increases the current density significantly due to the substitution of carbon to boron lattices [4]. Therefore, SiC and Carbon Nano Tube (CNT) is given to synthesize MgB_2 superconductor with addition of carbon that make possible for MgB_2 as an HTS application [4]. Unfortunately, carbon addition has a side effect to the critical temperature. The higher the carbon content lowers the critical temperature of MgB_2. As a result, addition below 20% are done by researcher [5]. Although below addition below 2% has not be proceeded due to the insignificant amount of addition. A journal of Vajpayee et al uses the lowest addition of SiC of 2% in powder in tube (PIT) method [6]. The current density enhances significantly and the critical temperature decreases insignificantly compared to their other sample conducted with higher amount of SiC. As well as the result of Vajpayee et al that uses 3% CNT addition to MgB_2 bulk, resulting highest critical temperature among other samples and higher current density than samples added by 15% and 20% CNT [7].

MgB_2 is an intermetallic compound, which means it has simpler crystal structure compared to the typical copper oxide superconductors that are often used in electronics, energy, transportation and medical industries [4]. Compared to Nb-based superconductor material, the MgB_2 superconductor has the advantages in terms of ease of manufacturing process, lightweight, cheap, but with good mechanical properties

Doping were found helpful in some aspect and lowers the initial properties of pure MgB_2 superconductor [4].

The common and useful doping is SiC and CNT. These substances effects on homogeneity of the compound [4]. In any case impurity will achieved during the process but it would not agglomerate in one particular area by the presence of C and too much SiC/CNT will decrease the critical temperature since it will agglomerate with itself [5]. The journals of Varghese et al confirm that the highest effective percentage of SiC is on 2% and from Arpita et al is on 3% addition of CNT. Therefore, in this research we will implementing low-doped SiC/CNT to achieve deep knowledge how SiC and CNT affect this superconductor material.

2. Experimental Setup

2.1. Stainless Steel Tube Preparation

Superconductor wire was prepared by firstly preparing the stainless tube with outer diameter of 6 mm and inner diameter of 3 mm as cover tubes in order to enhance its formability during fabrication. The stainless-steel tube was cut with a length of 100 mm for each sample. The tube was annealed in a muffle furnace at 850 °C with a heating rate of 10 °C/min and was held for 3 hours. After the annealing, the tube was slowly cooled in the furnace to reduce the residual stress and to achieve good formability during the drawing process. Finally, the tube was cleaned using an abrasive paper, alcohol and compressed air to eliminate any oxide that might form during the annealing. To achieve a closed condition of the tube as a requirement for PIST (powder-in-sealed-tubes) fabrication method, one end of the tube was sealed through an electrical resistance welding (ERW) using E308-16 electrode. After the welding, the tubes were checked for any leaking and ready for the next treatment.

2.2. Superconducting Wire Preparation

In this work, five MgB_2 samples were prepared, i.e. pure MgB_2 sample, MgB_2 with the addition of 1 wt.% and 2 wt.% SiC, and MgB_2 with the addition of 1 wt.% and 2 wt.% CNT. Firstly, MgB_2 powder (Aldrich, ~100 mesh, > 99%) and the dopants (SiC, Iolitech Nanomaterials, 50 – 60 nm, 99% and carbon nanotubes, CNT, Cheap Tubes Inc., ~ 20-40 nm) were prepared by weighing stoichiometric amounts of MgB_2 at 1 wt. % and 2 wt. % dopants. After the weighing, the materials were mixed using an agate mortar for 30 minutes before being loaded at an ambient condition into the stainless-steel tube and sintered in a muffle furnace at 800 °C for 3 hours. After the sintering, the stainless-steel tube was rolled and drawn at room temperature to reduce the diameter of the tube to wire dimension. The final diameter of the wire after the drawing was 2 mm. This diameter reduction increased the length of the wire whereas the structure of the MgB_2 will be more compact.

3. Results and Discussion

The magnesium diboride was shown to be a dominant phase in the XRD result. Based on the source of JCPDS PDF-4 No. 04-018-7387 2014 edition system, peaks of MgB_2 are at 34°, 42°, 51°, 60°, 63°, 76° and 83°. As we can see at figure 4.3, that there are 7 peaks indicating the MgB_2 phase. The same case stated by Varghese et al. SiC phase is found 0.9% in the result on addition of 2% SiC. The low amount of SiC is resulting low coverage to the MgB_2 matrix. In this case, inhomogeneous mixing of SiC and amorphous crystallinity behavior also is resulting less intensity of the phase in the XRD result.

Although SiC phase is not detected, the addition of SiC is affecting the peak of MgB_2. The higher amount of SiC addition decreases the intensity of the MgB_2 particularly at 42°and acts as an impurity. Besides the MgB_2 phase, we can also see Fe-C-Cr-Ni phase at the 43° and 74° peaks. The sample were tested in wire condition with the width of MgB_2 matrix less than 1 mm that cause the phase of stainless steel 304 detected during the XRD characterization

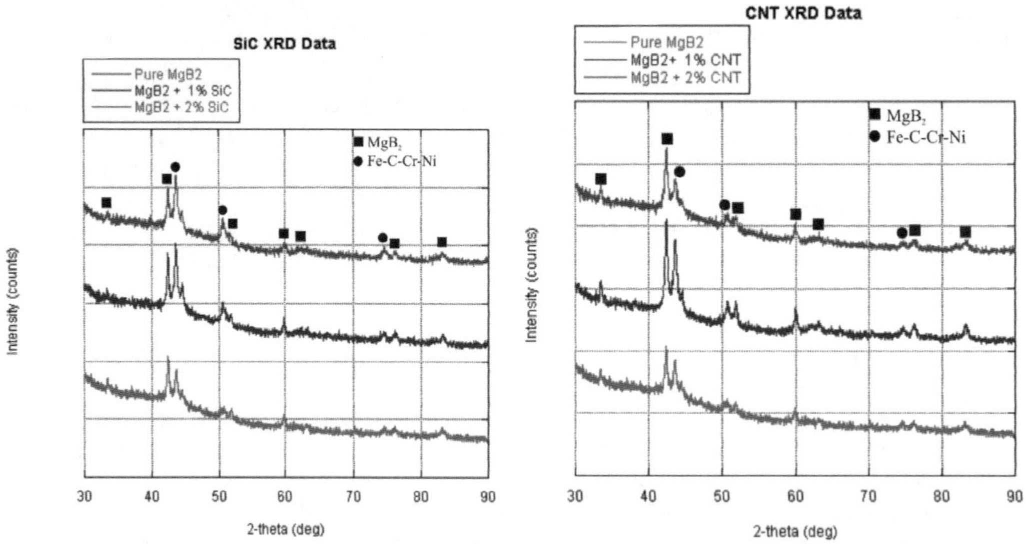

Figure 1. X-ray diffractograms of MgB_2 + Sic (left) and MgB_2 + CNT (right).

Scanning Electron Microscopy (SEM) were done to 8 samples. Which were 3 samples before rolling, and 5 after the drawing process. By having SEM result before and after sintering we saw the difference of microstructure based on the metal deformation. Various magnification was taken to achieve better image since the surface was rough. Mapping was also ensued to see substance distribution in the sample's surface.

The SEM result of pure MgB2 can be seen in fig. 4.2(left). As can be seen, there is inhomogeneity of diffusion during sintering, resulting a deposition of rich magnesium area. This phenomenon may occur because of the presence of impurities or a corrosion product that decrease the diffusion ability of magnesium towards boron particles. Small voids and porosity can also be observed on the specimen. These voids are a common phenomenon for PIT product particularly that uses small inner diameter tubes as its media [10]. By using smaller tube, it would be harder to determine the compaction force as the pressing machine has bigger surface area. So, this voids and porosity are caused by poor compaction process resulting from air entrapment and high free volume.

The pure MgB2 sample after been rolled, shows better morphology for superconductor as seen in figure 2 (right). Crystal structure are more dominated at the MgB2 matrix. Void's size is decreased but we can still find it with 10000x magnification. The rolling process found to be helpful to compacting the MgB2 structure due to dislocation movement the matrix is denser as the rolling process occur. As we can see that agglomeration of small particles are occurred at the magnification of 10000x and

IOP Conf. Series: Materials Science and Engineering **541** (2019) 012055 doi:10.1088/1757-899X/541/1/012055

5000x. The journal written by S. K Chen et al expressed that precipitation and agglomeration are the advantages for superconducting properties since it has higher density. Although it also has lower mechanical properties for wire application [6].

Figure 2. Secondary electron images of pure MgB_2 after sintering only (left) and after rolling (right).

As we can see in figure 3(top left), that irregularity of grain size is huge and the crystals are attached with small particles. This is caused by different initial particle size of magnesium and boron with Nano-SiC. Hand milling process resulting irregularity of particle size after process affects the crystal morphology of the MgB2. The voids are also detected on 10000x magnification reveals that compaction process is not optimum.

Figure 3. Secondary electron images of MgB2 + 1% SiC (top left), MgB_2+2%SiC (top right), MgB2 + 1% CNT (bottom left) and MgB_2 + CNT 2% (bottom right) after rolling.

IOP Conf. Series: Materials Science and Engineering **541** (2019) 012055 doi:10.1088/1757-899X/541/1/012055

In the Figure 4(a), the MgB_2 added with SiC formed a superconductor material after been rolled. The effect of SiC addition to MgB_2 superconductor is decreasment of critical temperature. Where the onset critical temperature of MgB_2 added with 1% SiC is at 36,1 K and 36,2 K for addition of 2% SiC. As we know that applied pressure on MgB_2 decreases the Tc, this result shows that this theory does not apply on the MgB_2 sample added with SiC [7]. Before the sample with addition of SiC has been rolled, the cryogenic result shows non-superconductor behavior. And after rolling, the sample shows a transition from non-superconductor to a superconductor material. This shows that rolling process will enhance the density of MgB_2 sample with SiC addition and able to transform the material to a superconductor wire.

As we can see at table 1 that the onset critical temperature is at 42,3 and it has a zero-critical temperature of 37,3. These values indicates that after the sample have been rolled the critical slope pattern is shifting to the lower temperature causing the rolled sample have a huge temperature value gap on the T_c onset and T_c zero. Proven a research by V. G. Tissen et al. that applied pressure on MgB_2 with a density of 2.23 g/cm^{-3} decreases its critical temperature [7]. A Related research "Superconductivity in Laser Ablated $LaBa_2Cu_3O_z$" conducted by M. Murgesan et al. proves that there is a minimum thickness for an optimal superconductor. When the thickness is less than the minimum value it will decrease the critical temperature and the superconductor properties [8]. Decreasing of electrical properties occurs when a powder metallurgy product is rolled after exceeding its true density of the material [9]. Internal crack occurs after rolling is proceeded after exceeding the true density of a powder metallurgy product [9]. Internal crack is initiated on the porosity or void

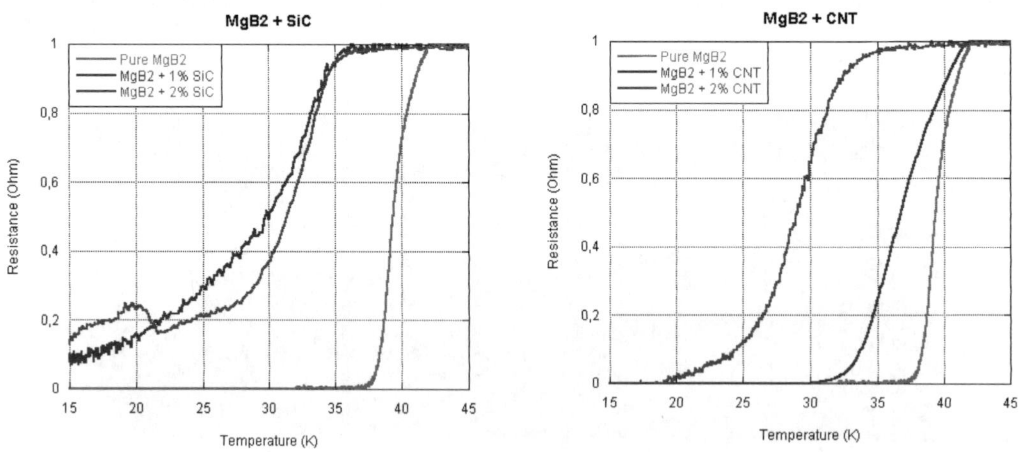

(a) (b)

Figure 4. R-T measurement result of MgB_2 + SiC (a) and MgB_2 + CNT (b).

Table 1. Critical temperature of MgB_2 at various doping concentration.

Sample	Tc onset (K)	Tc zero (K)
Pure MgB_2	42,3	37,3
MgB_2 + 1% CNT	36,1	N/A
MgB_2 + 2% CNT	36,2	N/A
MgB_2 + 1% CNT	42,1	28,2
MgB_2 + 2% CNT	34,6	19

Figure 4 shows us the normalized cryogenic result of MgB_2 with addition of CNT sample. As we can see that the presence of CNT in the MgB_2 decreases the critical temperature. The higher addition amount of CNT shifts the Tc to lower temperature. The MgB_2 with the addition of 1% CNT has a onset critical temperature of 42,1 and a zero critical temperature of 28,2 K. Although the onset critical temperature is almost similar to the pure MgB_2 sample, the zero critical temperature is much lower than the onset. This is not favorable for a superconductor since the resistivity does not decreases rapidly.

The MgB_2 sample added with 2% CNT critical temperature is much lower than the sample added with 1% CNT. The added 2% CNT sample has an onset critical temperature of 34,6K and a zero critical temperature of 19K. This shows that the higher amount of CNT addition will decreases the critical temperature significantly. Decreasement of critical temperature by addition of CNT is mainly due to the interaction of both MgB2 and CNT in which both of the materials acting as different conductors[29].

Where I_{max} and V_{max} are maximum photocurrent and voltage, respectively. Their values can be extracted from the maximum power of the I-V characteristics. Based on the data obtained from the photocurrent-voltage examination of the PSC device, the maximum power conversion efficiency (PCE) is found to be 3.4%. Compared to the existing efficiency [9], this value is still relatively low. This low value is expected because of inhomogeneous distribution of the perovskite layer that is yet to be improved, however, this current result is convincing and promising for the next development.

4. Conclusion

Finally, after finishing the data procurement and analysis of the acquired data of the research of synthesis and characterization of MgB2/SiC and MgB2/CNT superconductor wire, there are some points that could be concluded that synthesis of MgB_2 using PIT method as a superconductor was achieved, rolling process increases brittleness and density of the samples. The addition of SiC/CNT in small amount may not be detected in XRD, rolled sample has less porosity compared to before rolled specimen, the rolled sample has higher critical temperature compared to unrolled sample, the addition of CNT samples has higher critical temperature than the addition of SiC samples, the presence of SiC or CNT in the MgB_2 superconductor decreases the critical temperature.

References

[1] Hummel R. E., Electronic Properties of Material 3th Edition, Berlin: Springer-Verlag, 2011.
[2] Adriana S and German S 2009 *Nova Science* **2** 21163
[3] Nagamatsu J, Nakagawa N, Muranaka T, Zenitani Y and Akimatsu 2001 *Nature* **410** 1-5
[4] Wei D 2009 *Mgb2 Superconductor Research* (China: Yantai University)
[5] Varghese N, Vinod K, Syamaprasad U and Roy SB 2009 *Journal of Alloy and Comp* **484** 734
[6] Vajpayee A, Awana V P S, Yu S, Bhalla G L and Kishan H 2009 *Phys C* **470** S653
[7] Murgesan M, Gnanasekar K I and Rao M S R 1997 *Adv in Supercond Proc Sym* **391** 5-7
[8] Hausner H H, Roll K H and Johnson P K 1968 Perspective in Powder Metallurgy "Fundamentals, Methods, Applications, B.V: Springer-Science Business
[9] Zhang, Y Ma, Z Gao, D Wang, S Awaji 2007 *IEEE Trans. Appl. Supercond,* **17** 2915

Acknowledgments

The authors would like to thanks Indonesian Institute of Sciences for supporting this research and Universitas Indonesia for financial support through Hibah PITTA No. 2104/UN2.R12/HKP. 05.00/2016